COSMIC PLASMA PHYSICS

ASTROPHYSICS AND SPACE SCIENCE LIBRARY

VOLUME 251

COSMIC PLASMA PHYSICS

by

BORIS V. SOMOV

Astronomical Institute and Faculty of Physics,
Moscow State University

KLUWER ACADEMIC PUBLISHERS

DORDRECHT / BOSTON / LONDON

A C.I.P. Catalogue record for this book is available from the Library of Congress.

ISBN 0-7923-6512-7

Published by Kluwer Academic Publishers,
P.O. Box 17, 3300 AA Dordrecht, The Netherlands.

Sold and distributed in North, Central and South America
by Kluwer Academic Publishers,
101 Philip Drive, Norwell, MA 02061, U.S.A.

In all other countries, sold and distributed
by Kluwer Academic Publishers,
P.O. Box 322, 3300 AH Dordrecht, The Netherlands.

Coverphotograph reproduced with kind permission
from the European Southern Observatory

Printed on acid-free paper

Printed in the Netherlands.

Contents

Physics of Cosmic Plasma: Classics, Practice, Perspectives

Preface

This book is addressed to young people without a background in plasma physics; it grew from the lectures given many times in the Faculty of General and Applied Physics at the Moscow Institute of Physics and Technics (the well known 'fiz-tekh') since 1977. A similar full-year course was also offered to the students of the Astronomical Division of the Faculty of Physics at the Moscow State University over the years after 1990. A considerable amount of new material, related to modern astrophysics, has been added to the lectures. So the contents of the book can hardly be presented during a one-year lecture course, without additional seminars.

In fact, just the seminars with the topics '**how to make a cake**' were especially pleasant for the author and useful for students. In part, the text of the book retains the imprint of the seminar form, implying a more lively dialogue with the reader and more visual representation of individual notions and statements. At the same time, the author's desire is that these digressions from the academic language of the monograph will not harm the rigour of presentation of this textbook's subject – the physical and mathematical introduction to cosmic plasma physics.

The idea of the book is not typical for the majority of textbooks on cosmic plasma physics. Its idea is

> the consecutive consideration of physical principles, starting from the most general ones, and of simplifying assumptions which give us a simpler description of plasma under cosmic conditions.

Thus I would recommend the students to read the book straight through each chapter to see the central line of the cosmic plasma physics, its **classic fundamentals**. In so doing, the boundaries of the domain of applicability of the approximation at hand will be outlined from the viewpoint of physics rather than of many possible astronomical applications. After that, as an aid to detailed understanding, please return with pencil and paper to work out the missing steps (if any) in the formal mathematics.

On the basis of such an approach the student interested in modern astrophysics, its **current practice**, will find the answers to two key questions: (1) what approximation is the best one (the simplest but sufficient) for description of a phenomenon in cosmic plasma; (2) how to build an adequate model for the phenomenon, for example, a solar flare. Practice is really important for understanding the theory of cosmic plasma. Related exercises (problems and answers supplemented to each chapter) to improve skill do not thwart the theory but serve to better understanding of cosmic plasma physics.

As for the applications, preference evidently is given to physical processes in the solar plasma. Why? – Much attention to solar plasma physics is conditioned by the possibility of the all-round observational test of theoretical models. This statement primarily relates to the processes in the solar atmosphere. For instance, flares on the Sun, in contrast to those on other stars as well as a lot of other analogous phenomena in the Universe, *can be seen* in their development, i.e. we can obtain a sequence of images during the flare's evolution, not only in the optical and radio ranges but also in the ultraviolet, soft and hard X-ray ranges.

This book is mainly intended for students who have mastered a course of general physics and have some initial knowledge of theoretical physics. For beginning students, who may not know in which subfields of space physics they wish to specialize, I believe

> it is better to cover a lot of fundamental theories thoroughly than to dig deeply into any particular astrophysical subject or object,

even a very interesting one, for example black holes. Astronomers, or astrophysicists, of the future will need tools that allow them to explore in many different directions. Moreover, astronomy of the future will be, more than hitherto, *precise science* similar to mathematics, physics or plasma physics.

The beginning graduate students are usually confronted with a confusing amount of work on cosmic plasma physics published in a widely dispersed literature. Knowing this difficulty, the author has tried as far as possible to represent the material in a self-contained form which does not require the reading of additional literature. However, there is an extensive bibliography in the end of the book, allowing one to find the original works. In many cases, particularly where a paper in Russian is involved, the author has aimed to give the full bibliographic description of the work, including its title, etc.

Furthermore, the book contains recommendations as to introductory (unavoidable) reading needed to refresh the memory about a particular fact, as well as to additional (futher) reading to refine one's understanding of the subject. Separate **remarks of an historical character** are included in many places. It is sometimes simpler to explain the interrelation of discoveries by representing the subject in its development. It is the author's opinion that the outstanding discoveries in cosmic plasma physics are by no means governed by chance. With the same thought in mind, the author gives preference to original papers on a topic under consideration; it happens in science, as in art, that an original is better than nice-looking modernizations.

The majority of the book's chapters begin from an 'elementary account' and illustrative simple examples but finish with the most modern results of scientific importance. New problems determine the most interesting perspectives of cosmic plasma physics as a new developing science. The author hopes, in this context, that professionals in the field of cosmic plasma physics and adjacent sciences will enjoy reading this book too. Open issues are the focus of our attention in many places where they are. In this way, **perspectives of the cosmic plasma physics** with its many applications will be also of interest for readers.

The author is grateful to his young colleagues Sergei A. Bogachev, Sergei V. Diakonov, Yuri E. Litvinenko, Sergei A. Markovskii, Anna V. Oreshina, Inna V. Oreshina, Alexandr I. Podgornii, Yuri I. Skrynnikov, Andrei R. Spektor, Vyacheslav S. Titov, and Alexandr I. Verneta for generous help and valuable remarks. He is also happy to acknowledge helpful discussions with many of his colleagues in the world.

Moscow, 2000 Boris V. Somov

Introduction
History and Neighbours

Cosmic plasma physics studies electromagnetic processes and phenomena in space, mainly the role of forces of an electromagnetic nature in the dynamics of cosmic matter. Two factors are specific to the latter: its gaseous state and high conductivity. As you know, such a combination is unlikely to be found under natural conditions on earth; the matter is either a non-conducting gas (the case of gas dynamics or hydrodynamics) or a liquid or a solid conductor. By contrast, **plasma is the main state of cosmic matter**. It is precisely the poor knowledge of cosmic phenomena and cosmic plasma properties that explains the retarded development of cosmic electrodynamics. It has been distinguished as an independent branch of physics in the pioneering works of Alfvén (see Alfvén, 1950).

Soon after that, the problem of thermonuclear reactions initiated a great advance in plasma research (e.g., Simon, 1959; Glasstone and Loveberg, 1960; Leontovich, 1960). This branch has been developing rather independently, although being partly 'fed' by astrophysical ideas. They contributed to the growth of plasma physics, for example, the idea of stelarators. Presently, the reverse influence of laboratory plasma physics on cosmic electrodynamics is also important.

From the physical viewpoint,

> cosmic plasma physics is a part of plasma theory related in the first place to the dynamics of a high-conductivity plasma in space.

However it is this part that is the most poorly studied one under laboratory conditions. During the 1930s, scientists began to realize that the Sun and other stars are powered by nuclear fusion and they began to think of re-creating the process in the laboratory. The ideas of astro- and geophysics dominate here, as before. At present time, they mainly come from many space experiments and fine astronomical observations. From this viewpoint, physics of cosmic plasma belongs to experimental science.

Magnetic fields are easily generated in the cosmic plasma owing to its high conductivity. The strongest magnets in the Galaxy are presumably the so-called magnetars, the highly magnetized (with the strength of the field of about 10^{15} G) neutron stars formed in the supernova explosions. The energy of magnetic fields is accumulated in cosmic plasma, and the sudden release of this energy – an original electrodynamical 'burst' or 'explosion' – takes place under definite but quite general conditions (Peratt, 1992; Kivelson and Russell, 1995; Rose, 1998). It is accompanied by fast directed plasma ejections, powerful flows of heat and radiation and impulsive acceleration of particles to high energies.

This phenomenon is quite a widespread one. It can be observed in flares on the Sun and other stars (e.g., Haisch, Strong, and Rodonò, 1991), in the Earth's magnetosphere as **magnetic storms** and substorms (Nishida and Nagayama, 1973; Tsurutani *et al.*, 1997; Kokubun and Kamide, 1998; Nagai *et al.*, 1998; Nishida, Baker, and Cowley, 1998), in nuclei of active galaxies and quasars (e.g., Ozernoy and Somov, 1971; Begelman, Blandford and Rees, 1984). However, this process, while being typical of cosmic plasma, can be directly and fully studied on the Sun.

We observe how magnetic fields are generated (strictly speaking, how they come to the surface of the Sun, called the photosphere). We observe the development of **solar flares** and other non-stationary large-scale phenomena, such as coronal transients, coronal mass ejections into the interplanetary medium (e.g., Crooker, Joselyn, and Feynman, 1997), by means of ground observatories (in radio and optical wavelength ranges) and spaceships (practically in the whole electromagnetic spectrum). For example, the *Yohkoh* two telescopes working in soft and hard X-ray bands, respectively, allow us to study the creation and development of non-steady processes in the solar atmosphere (Ichimoto *et al.*, 1992; Tsuneta *et al.*, 1992; Tsuneta, 1993).

The LASCO experiment on board the SOHO satellite makes observations of such events in the solar corona out to 30 solar radii. Moreover, SOHO is equipped with an instrument, the MDI magnetograph, for observing the surface magnetic fields of the Sun. Following SOHO, the satellite TRACE was launched to obtain high spatial resolution X-ray images (see Golub *et al.*, 1999). With the solar maximum approaching, we have an unprecedented opportunity to use the three satellites for coordinated observations.

The link between the solar flares observed and **topology** of the magnetic field in *active* regions, in which these flares occured, was investigated by

Gorbachev and Somov (1989, 1990). They developed the first model of an actual flare, the flare on 1980, November 5, and have shown that the all characteristic features of this flare can be explained by the presence of a current sheet formed on the so-called *separator* which is the intersection of the separatrix surfaces. In particular, the flare ribbons in the chromosphere as well as the 'intersecting' soft X-ray loops in the corona are the consequences of a topological structure of a magnetic field near the separator.

An increasing number of investigations clearly relates the location of a chromospheric flare – the flare's manifestation in the solar chromosphere – with the topological magnetic features of active regions (Mandrini *et al.*, 1991 and 1993; Démoulin *et al.*, 1993; Bagalá *et al.*, 1995; Longcope and Silva, 1998). In all these works it is confirmed that the flares can be considered as a result of the interaction of large-scale magnetic structures; the authors derived the location of the separatrices – surfaces that separate cells of different field line connectivities – and of the *separator*.

These studies strongly support **the concept of magnetic reconnection** in solar flares (Giovanelli, 1946; Dungey, 1958; Sweet, 1958). Solar observations with the Hard X-ray Telescope (HXT) and the Soft X-ray Telescope (SXT) on board the *Yohkoh* satellite clearly showed that

> the reconnection process is common to impulsive (compact) and gradual (large scale) solar flares

(e.g., Masuda *et al.*, 1994, 1995). However, in the interpretation of the *Yohkoh* data, the basic physics of magnetic reconnection in the solar atmosphere remained uncertain (see Kosugi and Somov, 1998). Significant parts of the book are devoted to the physics of the reconnection process, a fundamental feature of cosmic and laboratory plasmas.

Solar flares and coronal mass ejections strongly influence interplanetary and terrestrial space by virtue of shock waves, hard electromagnetic radiation and accelerated particles (e.g., Kivelson and Russell, 1995). That is why the problem of 'weather and climate' prediction in near space becomes more and more important. The term 'near space' refers to the space that is within the reach of orbiting stations, both manned and automated. The number of satellites (meteorological, geophysical, navigational ones) with electronic systems sensitive to the ionizing radiation of solar flares is steadily growing.

It is no mere chance that solar flares and coronal mass ejections are of interest to physicians, biologists and climatologists. Flares influence not only *geospace* – the terrestrial magnetosphere, ionosphere and upper atmosphere

(Hargreaves, 1992; Horwitz, Gallagher, and Peterson, 1998) but also the biosphere and the atmosphere of the Earth. They are therefore not only of pure scientific importance; they also have an **applied or practical relevance**. However, the latter aspect is certainly beyond the scope of this book.

The subject of the present book is the systematic description of the most important topics of cosmic plasma physics. However, the aim of the book is not the strict substantiation of the main principals and basic equations of modern plasma physics; this can be found in many wonderful monographs (Balescu, 1963; Liboff, 1969; Lifshits and Pitaevskii, 1981; Klimontovich, 1986, Schram, 1991). There are also many nice textbooks (Boyd and Sanderson, 1969; Goldston and Rutherford, 1995; Parks, 1991; Sturrock, 1994; Choudhuri, 1998) to learn general plasma physics without or with some astrophysical applications.

The primary aim of the book in your hands is rather the solution of a much more modest but still important problem, namely to help the students of astrophysics to understand the interrelation and limits of applicability of different approximations which are used in cosmic electrodynamics. If, on his/her way, the reader will continously try, following the author, to reproduce all mathematical transformation, he/she finally will find the pleasant feeling of real knowledge of the subject and the real desire for constructive work in the physics of cosmic plasmas.

The author believes that the book will help the young reader to master the modern methods of cosmic plasma physics and will teach the application of these methods while solving concrete problems in the physics of the Sun and many other astronomical objects. A good working knowledge of cosmic plasma physics is essential for the modern astrophysicist.

Chapter 1

Particles and Fields: Exact Self-Consistent Description

There exist two different ways to describe *exactly* the behaviour of a charged particle system in electromagnetic and gravitational fields. We discuss both of them in this chapter.

1.1 Charged Particles in the Electromagnetic Field

1.1.1 A general formulation of the problem

Maxwell's equations for the electric field **E** and magnetic field **B** are well known to have the form (e.g., Landau and Lifshitz, *Classical Theory of Field*, 1971, Ch. 4):

$$\operatorname{curl} \mathbf{B} = \frac{4\pi}{c} \mathbf{j} + \frac{1}{c} \frac{\partial \mathbf{E}}{\partial t} , \tag{1.1}$$

$$\operatorname{curl} \mathbf{E} = -\frac{1}{c} \frac{\partial \mathbf{B}}{\partial t} , \tag{1.2}$$

$$\operatorname{div} \mathbf{B} = 0 , \tag{1.3}$$

$$\operatorname{div} \mathbf{E} = 4\pi \rho^{\mathrm{q}} . \tag{1.4}$$

The fields are completely determined by electric charges and electric currents. Note that, in general, Maxwell's equations imply the continuity equation for electric charge (see Problem 1.1) as well as the conservation law for electromagnetic field energy (Problem 1.2).

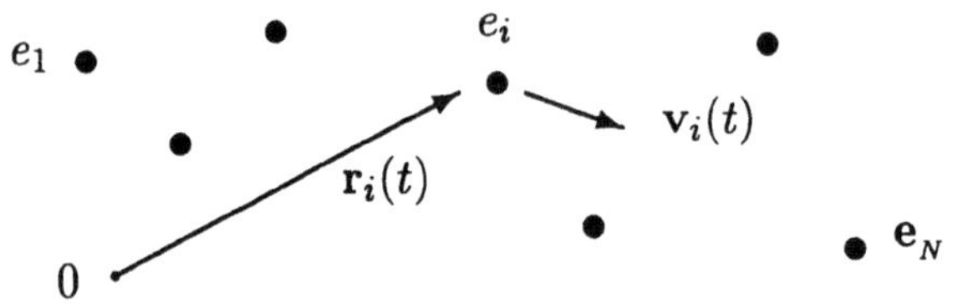

Figure 1.1: A system of N charged particles.

Let there be N particles with charges $e_1, e_2, \dots e_i, \dots e_N$, coordinates $\mathbf{r}_i(t)$ and velocities $\mathbf{v}_i(t)$, see Figure 1.1. By definition, the electric charge density

$$\rho^{\mathrm{q}}(\mathbf{r}, t) = \sum_{i=1}^{N} e_i \, \delta(\mathbf{r} - \mathbf{r}_i(t)) \tag{1.5}$$

and the density of electric current

$$\mathbf{j}(\mathbf{r}, t) = \sum_{i=1}^{N} e_i \, \mathbf{v}_i(t) \, \delta(\mathbf{r} - \mathbf{r}_i(t)) \,. \tag{1.6}$$

The delta function of the vector-argument is defined as usually (see definition and properties in Vladimirov, 1967, Ch. 2):

$$\delta(\mathbf{r} - \mathbf{r}_i(t)) = \prod_{\alpha=1}^{3} \delta_\alpha = \delta\left(r_x - r_x^i(t)\right) \, \delta\left(r_y - r_y^i(t)\right) \, \delta\left(r_z - r_z^i(t)\right). \tag{1.7}$$

The coordinates and velocities of particles can be found by integrating the equations of motion – the Newton equations:

$$\dot{\mathbf{r}}_i \equiv \frac{d\mathbf{r}_i}{dt} = \mathbf{v}_i(t) \,, \tag{1.8}$$

$$\dot{\mathbf{v}}_i \equiv \frac{d\mathbf{v}_i}{dt} = \frac{1}{m_i} \, e_i \left[\mathbf{E}(\mathbf{r}_i(t)) + \frac{1}{c} \, \mathbf{v}_i \times \mathbf{B}(\mathbf{r}_i(t)) \right]. \tag{1.9}$$

Let us count the number of unknown quantities: the vectors $\mathbf{B}$, $\mathbf{E}$, $\mathbf{r}_i$, and $\mathbf{v}_i$. We obtain: $3 + 3 + 3N + 3N = 6(N+1)$. The number of equations is equal to $8 + 6N = 6(N+1) + 2$. Therefore **two equations seem to be unnecessary**. Why is this so?

1.1.2 The continuity equation for electric charge

Let us make sure that the definitions (1.5) and (1.6) conform to the conservation law for electric charge. Differentiating (1.5) with respect to time gives

(see Problem 1.3)

$$\frac{\partial \rho^{\mathrm{q}}}{\partial t} = -\sum_i e_i\, \delta'_\alpha\, \dot{r}^{\,i}_\alpha\,. \tag{1.10}$$

Here the index $\alpha = 1,2,3$ or (x,y,z). The prime denotes the derivative with respect to the argument of a function, the overdot denotes differentiation with respect to time t. Summation over the repeated index α (contraction) is implied:

$$\delta'_\alpha\, \dot{r}^{\,i}_\alpha = \delta'_x\, \dot{r}^{\,i}_x + \delta'_y\, \dot{r}^{\,i}_y + \delta'_z\, \dot{r}^{\,i}_z\,.$$

For the electric current density (1.6) we have the divergence

$$\operatorname{div}\mathbf{j} = \frac{\partial}{\partial r_\alpha}\, j_\alpha = \sum_i e_i\, v^i_\alpha\, \delta'_\alpha\,. \tag{1.11}$$

Comparing formula (1.10) with (1.11) we see that

$$\boxed{\frac{\partial \rho^{\mathrm{q}}}{\partial t} + \operatorname{div}\mathbf{j} = 0\,.} \tag{1.12}$$

Therefore the definitions for ρ^{q} and $\mathbf{j}$ conform to the continuity Equation (1.12).

As we shall see it in Problem 1.1, conservation of electric charge follows also directly from the Maxwell Equations (1.1) and (1.4). The difference is that above we have not used Equation (1.4).

1.1.3 Initial equations and initial conditions

Operating with the divergence on Equation (1.1) and using the continuity Equation (1.12) we obtain

$$0 = \frac{4\pi}{c}\left(-\frac{\partial \rho^{\mathrm{q}}}{\partial t}\right) + \frac{1}{c}\frac{\partial}{\partial t}\operatorname{div}\mathbf{E}\,.$$

Thus, by postulating the definitions (1.5) and (1.6), by virtue of the continuity Equation (1.12) and without using the Maxwell Equation (1.4), we find that

$$\frac{\partial}{\partial t}\left(\operatorname{div}\mathbf{E} - 4\pi\rho^{\mathrm{q}}\right) = 0\,. \tag{1.13}$$

Hence Equation (1.4) will be valid at any moment of time, provided it is true at the initial moment.

Operate with the divergence on Equation (1.2):

$$\frac{\partial}{\partial t} \operatorname{div} \mathbf{B} = 0 \,. \tag{1.14}$$

We come to the conclusion that the Equations (1.3) and (1.4) play the role of *initial conditions* for the time-dependent equations

$$\frac{\partial}{\partial t} \mathbf{B} = -\, c \operatorname{curl} \mathbf{E} \tag{1.15}$$

and

$$\frac{\partial}{\partial t} \mathbf{E} = +\, c \operatorname{curl} \mathbf{B} - 4\pi \, \mathbf{j} \,. \tag{1.16}$$

Equation (1.3) implies the absence of magnetic charges or, which is the same, the *solenoidal* character of the magnetic field.

Thus, in order to describe the gas consisting of N charged particles, we consider the time-dependent problem of N bodies with a given interaction law.

> The electromagnetic part of the interaction is described by Maxwell's equations, the time-independent scalar equations playing the role of initial conditions for the time-dependent problem.

Therefore the set consisting of eight Maxwell's equations and $6N$ Newton's equations is neither over- nor underdetermined. It is *closed* with respect to the time-dependent problem, i.e. it consists of $6\,(N+1)$ equations for $6\,(N+1)$ variables, once the initial and boundary conditions are given.

1.1.4 Cosmic plasma applications

The set of equations described above can be solved analytically in just three cases:

1. $N = 1$, the motion of a charged particle in a given electromagnetic field, for example, drift motions and the so-called adiabatic invariants, wave-particle interaction and the problem of particle acceleration in space plasma; see Chapters 4, 5, 13, 16, and 18.

2. $N = 2$, Coulomb collisions of two charged particles. This is important for the kinetic description of physical processes, for example, the kinetic effects under propagation of accelerated particles in cosmic plasma, collisional heating of plasma by a beam of accelerated electrons or/and ions, see Chapters 3 and 6.

3. $N \to \infty$, a very large number of particles. This case is the frequently considered one in cosmic electrodynamics, because it allows us to introduce statistical and macroscopic descriptions of space plasma, the widely-used magnetohydrodynamic (MHD) approximation; see Chapters 2, 7, and 8.

Numerical integration of Equations (1.1)–(1.9) in the case of large but finite N, like $N \approx 3 \times 10^6$, is possible by using powerful modern computers. Such computations called 'particle simulations' have proved to be increasingly useful for understanding physics of cosmic plasma. One important example of a simulation is magnetic reconnection in a collisionless plasma (Horiuchi and Sato, 1994, 1997; Cai and Lee, 1997). This process often leads to fast energy conversion from field energy to particle energy as well as a topological change of magnetic field in cosmic plasma; see Chapters 16 to 18.

Note also that the set of equations described above can be generalized to include consideration of neutral particles. This is necessary, for instance, in the study of the generalized Ohm's law (Chapter 7) which can be applied in the investigation of physical processes in the solar photosphere and prominences (Sections 21.2 and 21.3). Dusty and *self-gravitational* plasmas in space are interesting in view of the diverse and often surprising facts about planetary rings and comet environments (e.g., Bliokh *et al.*, 1995).

1.1.5 Gravitational systems

Gravity plays a central role in the dynamics of many astrophysical systems – from stars to the Universe as a whole (e.g., Lahav *et al.*, 1996; Rose, 1998; Bertin, 1999). It is important for astrophysical applications that a *gravitational* force as well as an electromagnetic force acts on the particles:

$$m_i \, \dot{\mathbf{v}}_i = \mathbf{F}_{\text{em}} - m_i \, \nabla \phi \, . \tag{1.17}$$

Here the gravitational potential

$$\phi(t, \mathbf{r}) = - \sum_{n=1}^{N} \frac{G \, m_n}{|\, \mathbf{r}_n(t) - \mathbf{r} \,|} \, , \qquad n \neq i \, , \tag{1.18}$$

G is the gravitational constant. We shall return to this subject many times, for example, while studding the virial theorem in MHD (Chapter 14). This theorem is widely used in astrophysics. In particular, it is applied in the physics of the Sun for the analysis of equilibrium states of large-scale electric currents observed via their magnetic fields in the solar photosphere

(Gopasyuk, 1990; Lin *et al.*, 1993; Zhang, 1995; Wang *et al.*, 1996) that are thought to be the source of flare energy (see also Section 14.2).

Though the potential (1.18) of the gravitational interaction looks similar to the Coulomb potential (see (6.1)),

> physical properties of gravitational systems differ so much from properties of plasma.

We shall see this fundamental difference, for example, in Section 2.5.3.

1.2 Liouville's theorem

1.2.1 Continuity in phase space

Let us introduce the distribution function

$$f = f(\mathbf{r}, \mathbf{v}, t) \tag{1.19}$$

for particles as follows. Consider the six-dimensional space called *phase space* $X = \{\mathbf{r}, \mathbf{v}\}$.

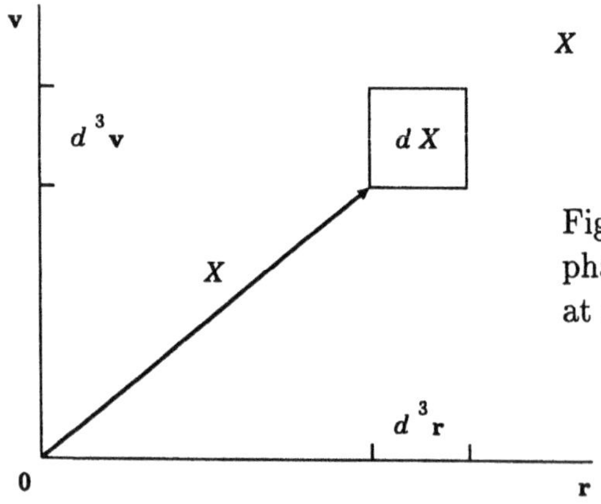

Figure 1.2: The six-dimensional phase space X. A small volume dX at a point X.

The number of particles present in a small volume $dX = d^3\mathbf{r}\, d^3\mathbf{v}$ at a point X (see Figure 1.2) at a moment of time t is defined to be

$$dN(X,t) = f(X,t)\, dX. \tag{1.20}$$

Accordingly, the total number of the particles at this moment is

$$N(t) = \int f(X,t)\, dX \equiv \iint f(\mathbf{r}, \mathbf{v}, t)\, d^3\mathbf{r}\, d^3\mathbf{v}. \tag{1.21}$$

If, for definiteness, we use Cartesian coordinates, then

$$X = \{ x, y, z, v_x, v_y, v_z \}$$

is a point of the phase space (Figure 1.3) and

$$\dot{X} = \{ v_x, v_y, v_z, \dot{v}_x, \dot{v}_y, \dot{v}_z \} \tag{1.22}$$

is the velocity of this point in the phase space.

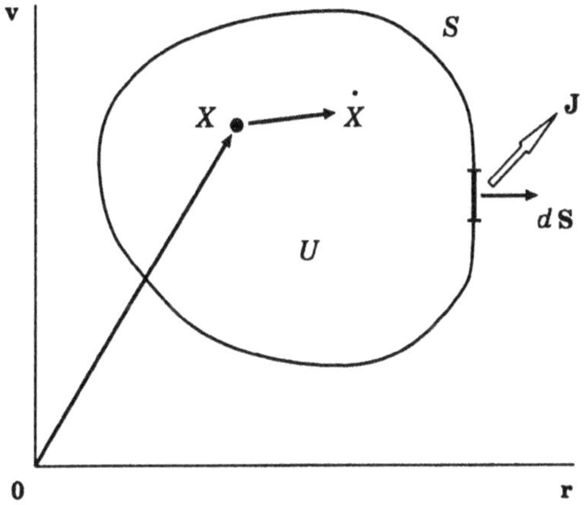

Figure 1.3: The six-dimensional phase space X. The volume U is enclosed by the surface S.

Suppose the coordinates and velocities of the particles are changing *continuously* – 'from point to point'. This corresponds to a continuous motion of the particles in phase space and can be expressed by the *continuity equation*:

$$\boxed{\frac{\partial f}{\partial t} + \mathrm{div}_X \, (\dot{X} f) = 0} \tag{1.23}$$

or

$$\frac{\partial f}{\partial t} + \mathrm{div}_{\mathbf{r}} \, (\mathbf{v} f) + \mathrm{div}_{\mathbf{v}} \, (\dot{\mathbf{v}} f) = 0 \, .$$

Equation (1.23) expresses the *conservation law* for the particles, since the integration of (1.23) over a volume U enclosed by the surface S gives

$$\frac{\partial}{\partial t} \int\limits_U f \, dX + \int\limits_U \mathrm{div}_X \, (\dot{X} f) \, dX =$$

by virtue of definition (1.20) and the Ostrogradskii-Gauss theorem

$$= \frac{\partial}{\partial t} N(t) \Big|_U + \int_S (\dot{X} f)\, dS = \frac{\partial}{\partial t} N(t) \Big|_U + \int_S \mathbf{J} \cdot d\mathbf{S} = 0 \,. \tag{1.24}$$

Here $\mathbf{J} = \dot{X} f$ is the *particle flux density* in phase space. Thus

> a change of the particle number in a given phase space volume U is defined by the particle flux through the boundary surface S only.

The reason is clear. There are no sources or sinks for the particles inside the volume.

1.2.2 The character of particle interactions

Let us rewrite Equation (1.23) in another form in order to understand the meaning of divergent terms. The first of them is

$$\operatorname{div}_{\mathbf{r}} (\mathbf{v} f) = f \operatorname{div}_{\mathbf{r}} \mathbf{v} + (\mathbf{v} \cdot \nabla_{\mathbf{r}})\, f = 0 + (\mathbf{v} \cdot \nabla_{\mathbf{r}})\, f \,,$$

since $\mathbf{r}$ and $\mathbf{v}$ are independent variables in phase space X. The second divergent term is

$$\operatorname{div}_{\mathbf{v}} (\dot{\mathbf{v}} f) = f \operatorname{div}_{\mathbf{v}} \dot{\mathbf{v}} + \dot{\mathbf{v}} \cdot \nabla_{\mathbf{v}} f \,.$$

So far no assumption has been made as to the character of particle interactions. It is worth doing here. Restrict our consideration to the interactions with

$$\boxed{\operatorname{div}_{\mathbf{v}} \dot{\mathbf{v}} = 0 \,,} \tag{1.25}$$

then Equation (1.23) can be rewritten in the equivalent form:

$$\frac{\partial f}{\partial t} + \mathbf{v} \cdot \nabla_{\mathbf{r}} f + \frac{\mathbf{F}}{m} \cdot \nabla_{\mathbf{v}} f = 0$$

or

$$\frac{\partial f}{\partial t} + \dot{X}\, \nabla_X f = 0 \,, \tag{1.26}$$

where

$$\dot{X} = \left\{ v_x, \, v_y, \, v_z, \, \frac{F_x}{m}, \, \frac{F_y}{m}, \, \frac{F_z}{m} \right\}.$$

Having written that, we 'trace' the particle phase trajectories. Thus Liouville's theorem is found to have the following formulation:

$$\frac{\partial f}{\partial t} + \mathbf{v} \cdot \nabla_{\mathbf{r}} f + \frac{\mathbf{F}}{m} \cdot \nabla_{\mathbf{v}} f = 0 \,. \tag{1.27}$$

Liouville's theorem: *The distribution function remains constant on particle phase trajectories* if condition (1.25) is satisfied.

We shall call Equation (1.27) the Liouville equation. Let us define also the Liouville operator

$$\frac{D}{Dt} \equiv \frac{\partial}{\partial t} + \dot{X}\,\frac{\partial}{\partial X} \equiv \frac{\partial}{\partial t} + \mathbf{v}\cdot\nabla_{\mathbf{r}} + \frac{\mathbf{F}}{m}\cdot\nabla_{\mathbf{v}}\,. \tag{1.28}$$

This operator is just the total time derivative following a particle motion in the phase space X. By using definition (1.28), we rewrite Liouville's theorem as follows:

$$\boxed{\frac{Df}{Dt} = 0.} \tag{1.29}$$

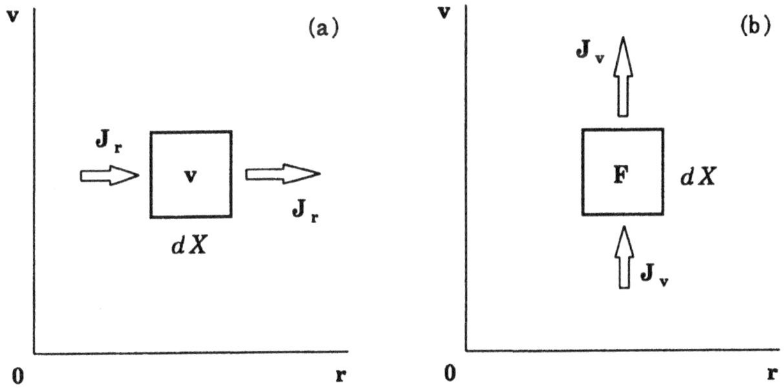

Figure 1.4: Action of the two different terms of the Liouville operator in the six-dimensional phase space X.

What factors lead to the changes in the distribution function? Let dX be a volume element in the phase space X. The second term in Equation (1.27), $\mathbf{v}\cdot\nabla_{\mathbf{r}} f$, means that the particles go into and out of the phase volume element considered, because their velocities are not zero (Figure 1.4a). So, this term describes a simple kinematic effect. The third term, $(\mathbf{F}/m)\cdot\nabla_{\mathbf{v}} f$, means that the particles escape from the phase volume element dX or come into this element due to their acceleration or deceleration under the influence of forces (Figure 1.4b).

1.2.3 The Lorentz force, gravity

Recall that the forces have to satisfy condition (1.25). Rewrite it as follows:

$$\frac{\partial \dot{v}_\alpha}{\partial v_\alpha} = \frac{1}{m} \frac{\partial F_\alpha}{\partial v_\alpha} = 0$$

or

$$\frac{\partial F_\alpha}{\partial v_\alpha} = 0 . \tag{1.30}$$

In other words, the component of the force F_α ($\alpha = 1, 2, 3$) should not depend upon the velocity component v_α. This is a sufficient condition.

The classical Lorentz force

$$F_\alpha = e \left[E_\alpha + \frac{1}{c} \left(\mathbf{v} \times \mathbf{B} \right)_\alpha \right] \tag{1.31}$$

obviously has that property. The gravitational force in the classical approximation is entirely independent of velocity.

1.2.4 Collisional friction in plasma

As a contrary example consider the friction force (cf. formula (6.61) for the collisional drag force in plasma):

$$\mathbf{F} = -k\mathbf{v} , \tag{1.32}$$

where the constant $k > 0$. In this case the right-hand side of Liouville's equation is not zero:

$$-f \operatorname{div}_{\mathbf{v}} \dot{\mathbf{v}} = -f \operatorname{div}_{\mathbf{v}} \frac{\mathbf{F}}{m} = \frac{3k}{m} f ,$$

because

$$\frac{\partial v_\alpha}{\partial v_\alpha} = \delta_{\alpha\alpha} = 3 .$$

Instead of Liouville's equation we have

$$\frac{Df}{Dt} = \frac{3k}{m} f > 0 . \tag{1.33}$$

The distribution function (that is the particle density) does not remain constant on particle trajectories but increases as the time elapses. In the operational sense, along the phase trajectories, it increases exponentially:

$$f(t, \mathbf{r}, \mathbf{v}) \sim f(0, \mathbf{r}, \mathbf{v}) \exp\left(\frac{3k}{m} t\right) . \tag{1.34}$$

The physical sense of this phenomenon is obvious. As the particles are decelerated by the friction force, they move down in Figure 1.5. By so doing, they are concentrated in the constantly diminishing region of phase space situated in the vicinity of the axis $\mathbf{v} = \mathbf{0}$.

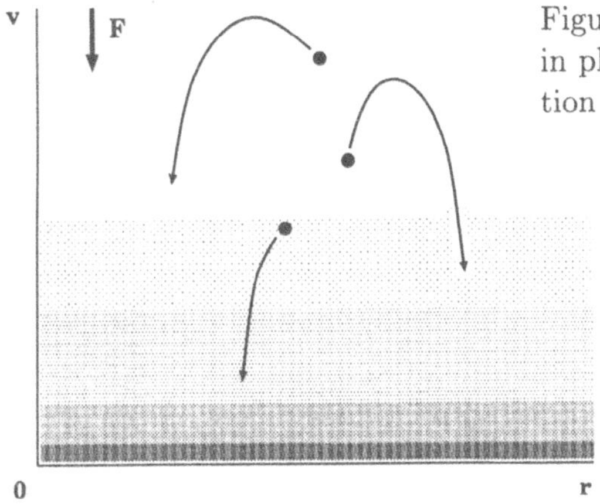

Figure 1.5: Particle density increase in phase space as a result of the action of the friction force $\mathbf{F}$.

There is a viewpoint that the Liouville theorem is valid for the forces that *do not disperse* particle velocities (Shkarofsky, Johnston, and Bachynski, 1966, Ch. 2). Why? It is usually implied that particle *collisions* enlarge such a dispersion: $\mathrm{div}_{\mathbf{v}}\, \dot{\mathbf{v}} > 0$. So

$$\frac{Df}{Dt} = \left(\frac{\partial f}{\partial t} \right)_{\mathrm{c}} = -f \, \mathrm{div}_{\mathbf{v}}\, \dot{\mathbf{v}} \, < 0 \,. \tag{1.35}$$

In this case the right-hand side of Equation (1.35) is called the *collisional* integral (see Sections 2.1 and 2.2). In contrast to the right-hand side of (1.33), that of Equation (1.35) is usually negative.

The above example of the friction force is instructive in that it shows how the forces that are diminishing the velocity dispersion ($\mathrm{div}_{\mathbf{v}}\, \dot{\mathbf{v}} < 0$) lead to the violation of Liouville's theorem; in other words, how they lead to a change of the distribution function along the particle trajectories. For the validity of Liouville's theorem only the condition (1.25) is important; **the divergence of the forces acting in the velocity space has to equal zero**. The sign of this divergence is unimportant. So, in general, Liouville's theorem is valid for conservative forces and magnetic fields (e.g., Shkarofsky, Johnston, and Bachynski, 1966).

1.3 The exact distribution function

Let us consider another property of the Liouville theorem. Introduce the N-particle distribution function of the form

$$\hat{f}(t, \mathbf{r}, \mathbf{v}) = \sum_{i=1}^{N} \delta\left(\mathbf{r} - \mathbf{r}_i(t)\right) \delta\left(\mathbf{v} - \mathbf{v}_i(t)\right). \tag{1.36}$$

We shall call such a distribution function the *exact* one. Substitute this expression for the distribution function in Equation (1.27). The resulting three terms are

$$\frac{\partial \hat{f}}{\partial t} = \sum_i (-1)\, \delta'_\alpha\left(\mathbf{r} - \mathbf{r}_i(t)\right) \dot{r}^i_\alpha\, \delta\left(\mathbf{v} - \mathbf{v}_i(t)\right) +$$

$$+ \sum_i (-1)\, \delta\left(\mathbf{r} - \mathbf{r}_i(t)\right) \delta'_\alpha\left(\mathbf{v} - \mathbf{v}_i(t)\right) \dot{v}^i_\alpha ,$$

$$\mathbf{v} \cdot \nabla_{\mathbf{r}} \hat{f} \equiv v_\alpha \frac{\partial \hat{f}}{\partial r_\alpha} = \sum_i v_\alpha\, \delta'_\alpha\left(\mathbf{r} - \mathbf{r}_i(t)\right) \delta\left(\mathbf{v} - \mathbf{v}_i(t)\right),$$

$$\frac{\mathbf{F}}{m} \cdot \nabla_{\mathbf{v}} \hat{f} \equiv \frac{F_\alpha}{m} \frac{\partial \hat{f}}{\partial v_\alpha} = \sum_i \frac{F_\alpha}{m}\, \delta\left(\mathbf{r} - \mathbf{r}_i(t)\right) \delta'_\alpha\left(\mathbf{v} - \mathbf{v}_i(t)\right).$$

The sum of these terms equals zero. It can occur just then that all the coefficients of different delta function arguments equal zero as well. From this we find

$$\frac{dr^i_\alpha}{dt} = v^i_\alpha(t), \qquad \frac{dv^i_\alpha}{dt} = \frac{1}{m_i} F_\alpha\left(\mathbf{r}_i(t), \mathbf{v}_i(t)\right). \tag{1.37}$$

Thus

> the Liouville equation for an exact distribution function is *equivalent* to the Newton set of equations for a particle motion, both describing a purely dynamic behaviour of the particles.

It is natural, since this distribution function is exact. No statistical averaging has been done so far. It is for this reason that both descriptions – namely, the Newton set and the Liouville theorem for the exact distribution function – are dynamic (as well as reversible, of course) and equivalent. Statistics will appear in the next Chapter when, instead of the exact description of a system, we begin to use some mean characteristics such as temperature, density etc.

This is the statistical description that is valid for systems containing a large number of particles.

We have shown that finding a solution of the Liouville equation for an exact distribution function is the same as the integration of the motion equations. Therefore, for systems of large numbers of particles, it is more advantageous to deal with the single Liouville equation which describes the entire system.

Recommended Reading: Landau and Lifshitz, *Mechanics* (1960) Ch. 2 and 7; Landau and Lifshitz, *Statistical Physics* (1959) Ch. 1, § 1–3.

1.4 Practice: Problems and Answers

Problem 1.1. Show that Maxwell's equations imply the continuity equation for the electric charge.

Answer. Operating with the divergence on Equation (1.1), we have

$$0 = \frac{4\pi}{c}\operatorname{div}\mathbf{j} + \frac{1}{c}\frac{\partial}{\partial t}\operatorname{div}\mathbf{E}\,.$$

Substituting (1.4) in this equation gives us the continuity equation for the electric charge

$$\frac{\partial}{\partial t}\rho^{\mathrm{q}} + \operatorname{div}\mathbf{j} = 0\,. \tag{1.38}$$

Thus Maxwell's equations conform to the charge continuity equation.

Problem 1.2. Starting from Maxwell's equation, derive the energy conservation law for an electromagnetic field.

Answer. Let us multiply Equation (1.1) by the electric field vector **E** and add it to Equation (1.2) multiplied by the magnetic field vector **B**. The result is

$$\frac{1}{c}\mathbf{E}\frac{\partial\mathbf{E}}{\partial t} + \frac{1}{c}\mathbf{B}\frac{\partial\mathbf{B}}{\partial t} = -\frac{4\pi}{c}\mathbf{j}\mathbf{E} - (\mathbf{B}\operatorname{curl}\mathbf{E} - \mathbf{E}\operatorname{curl}\mathbf{B})\,.$$

By using the known formula from vector analysis

$$\operatorname{div}\,[\mathbf{a}\times\mathbf{b}] = \mathbf{b}\operatorname{curl}\mathbf{a} - \mathbf{a}\operatorname{curl}\mathbf{b}\,,$$

we rewrite the last equation as follows

$$\frac{1}{2c}\frac{\partial}{\partial t}\left(E^2 + B^2\right) = -\frac{4\pi}{c}\mathbf{j}\mathbf{E} - \operatorname{div}\,[\mathbf{E}\times\mathbf{B}]$$

or

$$\boxed{\frac{\partial}{\partial t} W = -\mathbf{j}\,\mathbf{E} - \operatorname{div}\mathbf{G}\,.} \tag{1.39}$$

Here

$$W = \frac{E^2 + B^2}{8\pi} \tag{1.40}$$

is the energy of electromagnetic field in a unit volume of space;

$$\mathbf{G} = \frac{c}{4\pi}\,[\,\mathbf{E}\times\mathbf{B}\,] \tag{1.41}$$

is the flux of electromagnetic field energy through a unit surface in space, i.e. the energy flux density for electromagnetic field. This is called the Poynting vector.

The first term on the right-hand side of Equation (1.39) is the power of work done by the electric field on all the charged particles in the unit volume of space. In the simplest approximation

$$e\,\mathbf{v}\,\mathbf{E} = \frac{d}{dt}\,\mathcal{E}\,, \tag{1.42}$$

where $\mathcal{E}$ is the particle kinetic energy (see Equation (4.2)). Hence instead of Equation (1.39) we write the following form of the energy conservation law:

$$\frac{\partial}{\partial t}\left(\frac{E^2 + B^2}{8\pi} + \frac{\rho v^2}{2}\right) + \operatorname{div}\left(\frac{c}{4\pi}\,[\,\mathbf{E}\times\mathbf{B}\,]\right) = 0\,. \tag{1.43}$$

Compare this simple approach to the energy conservation law for charged particles and an electromagnetic field with the more general situation considered in Section 7.2.3.

Problem 1.3. Clarify the meaning of the right-hand side of Equation (1.10).

Answer. Substitute definition (1.7) of the delta-function in definition (1.5) of the electric charge density and differentiate the result over time t:

$$\frac{\partial \rho^{\mathrm{q}}}{\partial t} = \sum_{i=1}^{N} e_i \sum_{\alpha=1}^{3} \left[\frac{\partial}{\partial\left(r_\alpha - r_\alpha^i(t)\right)} \prod_{\beta=1}^{3} \delta\left(r_\beta - r_\beta^i(t)\right)\right] \frac{\partial}{\partial t}\left(r_\alpha - r_\alpha^i(t)\right) =$$

$$= -\sum_{i=1}^{N} e_i \sum_{\alpha=1}^{3} \left[\frac{\partial}{\partial\left(r_\alpha - r_\alpha^i(t)\right)} \prod_{\beta=1}^{3} \delta\left(r_\beta - r_\beta^i(t)\right)\right] \frac{dr_\alpha^i(t)}{dt}\,. \tag{1.44}$$

This is the right-hand side of Equation (1.10).

Problem 1.4. Show that any distribution function that is a function of the constants of the motion – the invariants of motion – satisfies Liouville's equation (1.29).

Answer. A general solution of the equations of motion (1.37) depends on $2N$ constants C_i where $i = 1, 2, \dots 2N$. If we assume that the distribution function is a function of these constants of the motion

$$f = f\,(C_1, \dots C_i, \dots C_{2N})\,, \tag{1.45}$$

we can rewrite the left-hand side of Equation (1.29) as

$$\frac{Df}{Dt} = \sum_{i=1}^{2N} \left(\frac{DC_i}{Dt}\right)\left(\frac{\partial f}{\partial C_i}\right). \tag{1.46}$$

Because C_i are constants of the motion, $DC_i/Dt = 0$. Therefore the right-hand side of Equation (1.46) is also zero, and the distribution function (1.45) satisfies the Liouville equation. This is the so-called *Jeans theorem*. It will be used, for example, in the theory of wave-particle interaction in cosmic plasma (Section 5.1).

Problem 1.5. Rewrite the Liouville theorem by using the Hamilton equations instead of the Newton equations.

Answer. Rewrite the Newton set of the motion Equations (1.37) in the Hamilton form (e.g., Landau and Lifshitz, *Mechanics*, 1960, Ch. 7, § 40):

$$\dot{q_\alpha} = \frac{\partial H}{\partial P_\alpha}\,, \qquad \dot{P_\alpha} = -\frac{\partial H}{\partial q_\alpha} \qquad (\alpha = 1, 2, 3)\,, \tag{1.47}$$

where $H(P, q)$ is the Hamiltonian of the system under consideration, q_α and P_α are the generalized coordinates and momemta, respectively.

Let us substitute the variables $\mathbf{r}$ and $\mathbf{v}$ in the Liouville equation (1.27) by the generalized variables $\mathbf{q}$ and $\mathbf{P}$. By doing so and using Equations (1.47), we obtain the following form of the Liouville equation

$$\frac{\partial f}{\partial t} + \nabla_{\mathbf{P}} H \cdot \nabla_{\mathbf{q}} f - \nabla_{\mathbf{q}} H \cdot \nabla_{\mathbf{P}} f = 0\,. \tag{1.48}$$

Because of symmetry of the last equation, it is convenient here to use the Poisson brackets (see Landau and Lifshitz, *Mechanics*, 1960, Ch. 7, § 42).

Recall that the Poisson brackets for arbitrary quantities A and B are defined to be

$$[A, B] = \sum_{\alpha=0}^{3} \left(\frac{\partial A}{\partial q_\alpha} \frac{\partial B}{\partial P_\alpha} - \frac{\partial A}{\partial P_\alpha} \frac{\partial B}{\partial q_\alpha} \right). \tag{1.49}$$

Appling definition (1.49) to Equation (1.48), we find the final form of the Lioville theorem

$$\boxed{\frac{\partial f}{\partial t} + [f, H] = 0.} \tag{1.50}$$

Q.E.D. Note that for a system in equilibrium

$$[f, H] = 0. \tag{1.51}$$

Problem 1.6. Recall the Liouville theorem in a course of mechanics – the phase volume is independent of t, i.e. it is the invariant of motion (e.g., Landau and Lifshitz, *Mechanics*, 1960, Ch. 7, § 46). Show that this formulation is equivalent to Equation (1.29).

Chapter 2

A Statistical Description of Cosmic Plasma

In a system which consists of many interacting particles, the statistical mechanism of 'mixing' in phase space works and makes the system's behaviour *on average* more simple.

2.1 The averaging of Liouville's equation

2.1.1 Averaging over phase space

As was shown in the first chapter, the exact state of a system consisting of N charged particles can be given by the *exact* distribution function (see definition (1.36)) in six-dimensional phase space $X = \{\mathbf{r}, \mathbf{v}\}$. This is defined as the sum of δ-functions in N points of phase space:

$$\hat{f}(\mathbf{r}, \mathbf{v}, t) = \sum_{i=1}^{N} \delta\left(\mathbf{r} - \mathbf{r}_i(t)\right) \delta\left(\mathbf{v} - \mathbf{v}_i(t)\right). \tag{2.1}$$

Instead of the equations of motion, we use Liouville's equation to describe the change of the system state (Section 1.3):

$$\frac{\partial \hat{f}}{\partial t} + \mathbf{v} \cdot \nabla_{\mathbf{r}} \hat{f} + \frac{\mathbf{F}}{m} \cdot \nabla_{\mathbf{v}} \hat{f} = 0. \tag{2.2}$$

Once the exact initial state of all the particles is known, it can be represented by N points in the six-dimensional phase space X (Figure 2.1). The motion of these points is described by Liouville's equation (1.27) or by the $6N$ equations of motion (1.37).

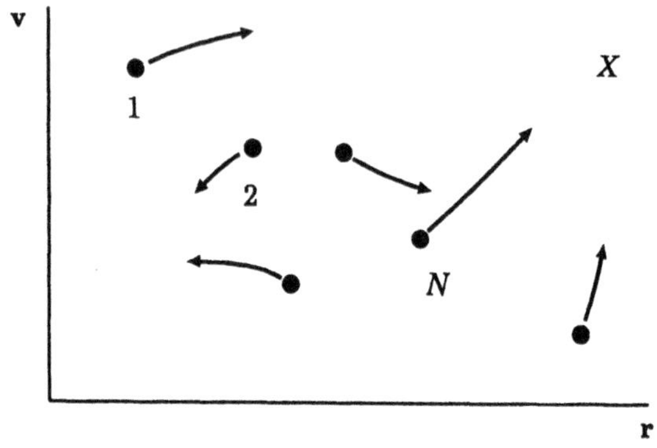

Figure 2.1: Particle trajectories in the six-dimensional phase space X.

In fact we usually know only some average characteristics of the system's state, such as the temperature, density, etc. Moreover the behaviour of each single particle is in general of no interest. For this reason, instead of the exact distribution function (2.1), let us introduce the distribution function averaged over a small volume ΔX of phase space, i.e. over a small interval of coordinates $\Delta \mathbf{r}$ and velocities $\Delta \mathbf{v}$ centered at the point $(\mathbf{r}, \mathbf{v})$, at a moment of time t:

$$\langle \hat{f}(\mathbf{r}, \mathbf{v}, t) \rangle_X = \frac{1}{\Delta X} \int\limits_{\Delta X} \hat{f}(X, t)\, dX =$$

$$= \frac{1}{\Delta \mathbf{r}\, \Delta \mathbf{v}} \int\limits_{\Delta \mathbf{r}\, \Delta \mathbf{v}} \hat{f}(\mathbf{r}, \mathbf{v}, t)\, d^3\mathbf{r}\, d^3\mathbf{v}\,. \tag{2.3}$$

Here $d^3\mathbf{r} = dx\, dy\, dz$ and $d^3\mathbf{v} = dv_x\, dv_y\, dv_z$, if use is made of Cartesian coordinates.

To put the same in another way, the mean number of particles present at a moment of time t in the element of phase volume ΔX is

$$\langle \hat{f}(\mathbf{r}, \mathbf{v}, t) \rangle_X \cdot \Delta X = \int\limits_{\Delta X} \hat{f}(\mathbf{r}, \mathbf{v}, t)\, dX\,.$$

The total number N of particles in the system is the integral over the whole phase space X.

Obviously the distribution function averaged over phase volume differs from the exact one as shown in Figure 2.2.

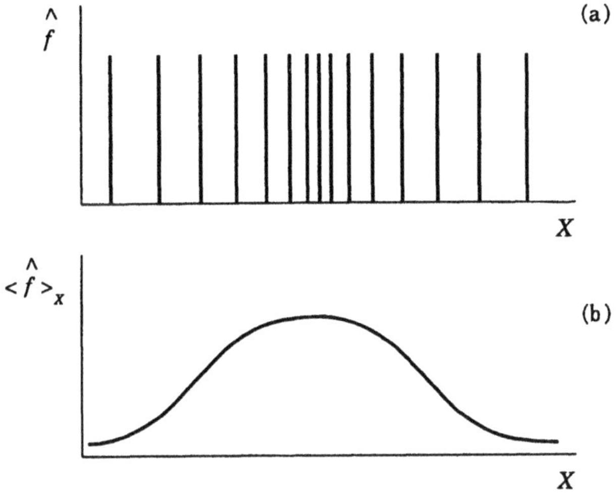

Figure 2.2: The one-dimensional analogy of the distribution function averaging over phase space X: (a) the exact distribution function (2.1), (b) the averaged function (2.3).

2.1.2 Two statistical postulates

Let us average the same exact distribution function (2.1) over a small time interval Δt centred at a moment of time t:

$$\langle \hat{f}(\mathbf{r}, \mathbf{v}, t) \rangle_t = \frac{1}{\Delta t} \int\limits_{\Delta t} \hat{f}(\mathbf{r}, \mathbf{v}, t) \, dt \, . \tag{2.4}$$

We assume that the following *two statistical postulates* concerning systems containing a large number of particles are applicable to the system considered.

The first postulate. The mean values $\langle \hat{f} \rangle_X$ and $\langle \hat{f} \rangle_t$ exist for sufficiently small ΔX and Δt and are *independent* of the averaging scales ΔX and Δt.

Clearly the first postulate implies that the number of particles should be large. For a small number of particles the mean value depends upon the averaging scale: if, for instance, $N = 1$ then the exact distribution function (2.1) is simply a δ-function, and the average over the variable X is $\langle \hat{f} \rangle_X = 1/\Delta X$. For illustration, the case $(\Delta X)_1 > \Delta X$ is shown in Figure 2.3.

The second postulate is

$$\langle \hat{f}(X, t) \rangle_X = \langle \hat{f}(X, t) \rangle_t = f(X, t) \, . \tag{2.5}$$

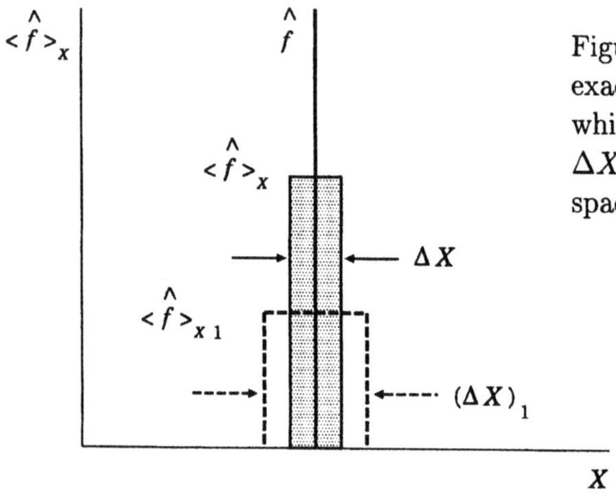

Figure 2.3: Averaging of the exact distribution function $\hat{f}$ which is equal to a δ-function. ΔX is a small volume of phase space X.

In other words, the averaging of the distribution function over phase space is *equivalent* to the averaging over time.

While speaking of the small ΔX and Δt, we assume that they are not too small: ΔX must contain a reasonably large number of particles while Δt must be large in comparison with the duration of drastic changes of the exact distribution function, such as the duration of the particle Coulomb collisions. It is in this case that the statistical mechanism of particle 'mixing' in phase space is at work and

> the averaging of the exact distribution function over the time Δt is equivalent to the averaging over the phase volume ΔX.

2.1.3 A statistical mechanism of mixing in phase space

Let us try to understand qualitatively how the mixing mechanism works in phase space. We start from the dynamical description of the N-particle system in $6N$-dimensional phase space in which

$$\Gamma = \{ \mathbf{r}_i, \mathbf{v}_i \}, \quad i = 1, 2, \ldots N,$$

a point is determined ($t = 0$ in Figure 2.4) by the initial conditions of all the particles. The motion of this point, that is the dynamical evolution of the system, can be described by Liouville's equation or equations of motion. The point moves along a complicated *dynamical trajectory* because the interactions in a many-particle system are extremely intricate and complicated.

The dynamical trajectory has a remarkable property which we shall illustrate by the following example. Imagine a glass vessel containing a gas

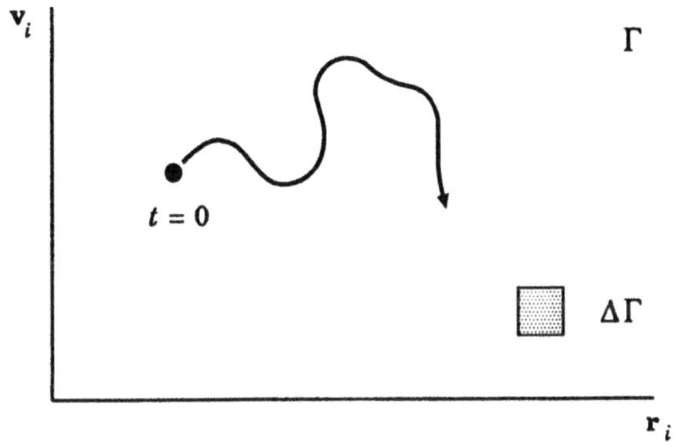

Figure 2.4: The dynamical trajectory of a system of N particles in the $6N$-dimensional phase space Γ.

consisting of a large number N of particles (molecules or charged particles). The state of this gas at any moment of time is depicted by a single point in the phase space Γ.

Imagine another vessel which is identical to the first one, with one exception, being that at any moment of time the gas state in the second vessel is different from that in the first one. These states are depicted by two different points in the space Γ. For example, at $t = 0$, they are points 1 and 2 in Figure 2.5.

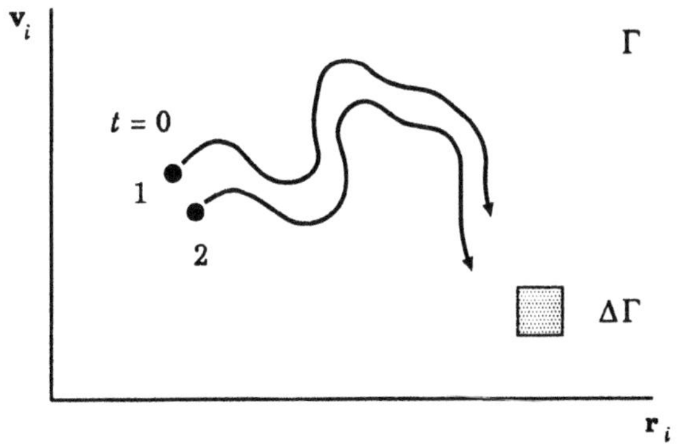

Figure 2.5: The trajectories of two systems never cross each other.

With the passage of time, the gas states in both vessels change, whereas the two points in the space Γ draw two different dynamical trajectories (Figure 2.5). These trajectories *do not* intersect. If they had intersected at just one point, then the state of the first gas, determined by $6N$ numbers $(\mathbf{r}_i, \mathbf{v}_i)$, would have coincided with the state of the second gas. These numbers could have been taken as the initial conditions which, in turn, would have uniquely determined the motion. The two trajectories would have merged into one. For the same reason the trajectory of a system cannot intersect itself. Thus we come to the conclusion that

> one and only one dynamical trajectory passes through each point of the phase space Γ.

Since the trajectories differ in initial conditions, we can introduce an infinite ensemble of systems (glass vessels) corresponding to the different initial conditions. In a finite time the ensemble of dynamical trajectories will closely fill the phase space Γ, without intersections. By averaging over the ensemble we can answer the question of what the probability is that, at a moment of time t, the system will be found in an element $\Delta\Gamma = \Delta\mathbf{r}_i\,\Delta\mathbf{v}_i$ of the phase space Γ:

$$dw = \langle\, \hat{f}(\mathbf{r}_i, \mathbf{v}_i)\,\rangle_\Gamma \; d\Gamma. \tag{2.6}$$

Here $\langle\, \hat{f}(\mathbf{r}_i, \mathbf{v}_i)\,\rangle_\Gamma$ is a function of all the coordinates and velocities. It plays the role of the *probability distribution density* in the phase space Γ and is called the statistical distribution function or simply the distribution function. It is obtained by way of statistical averaging over the ensemble and evidently corresponds to definition (2.3).

* * *

It is rather obvious that the same *probability density* can be obtained in another way – through the averaging over time. The dynamical trajectory of a system, given a sufficient time Δt, will closely cover phase space. There will be no self-intersections; but since the trajectory is very intricate it will repeatedly pass through the phase space element $\Delta\Gamma$. Let $(\Delta t)_\Gamma$ be the time during which the system locates in $\Delta\Gamma$. For a sufficiently large Δt, which is formally restricted by the characteristic time of slow evolution of the system as a whole, the ratio $(\Delta t)_\Gamma / \Delta t$ tends to the limit

$$\lim_{\Delta t \to \infty} \frac{(\Delta t)_\Gamma}{\Delta t} = \frac{dw}{d\Gamma} = \langle\, \hat{f}(\mathbf{r}_i, \mathbf{v}_i, t)\,\rangle_t \,. \tag{2.7}$$

By virtue of the role of the probability density, it is clear that

> the statistical averaging over the ensemble (2.6) is equivalent to the averaging over time (2.7) as well as to the definition (2.4).

2.1.4 The derivation of a general kinetic equation

Now we have everything what we need to average the exact Liouville Equation (2.2). Since the equation contains the derivatives with respect to time t and phase-space coordinates $(\mathbf{r}, \mathbf{v})$ the procedure of averaging over the interval $\Delta X\, \Delta t$ is defined as follows:

$$f(X,t) = \frac{1}{\Delta X\, \Delta t} \int\limits_{\Delta X} \int\limits_{\Delta t} \hat{f}(X,t)\, dX\, dt\,. \tag{2.8}$$

Averaging the first term of the Liouville equation gives

$$\frac{1}{\Delta X\, \Delta t} \int\limits_{\Delta X} \int\limits_{\Delta t} \frac{\partial \hat{f}}{\partial t}\, dX\, dt = \frac{1}{\Delta t} \int\limits_{\Delta t} \frac{\partial}{\partial t} \left[\frac{1}{\Delta X} \int\limits_{\Delta X} \hat{f}\, dX \right] dt =$$

$$= \frac{1}{\Delta t} \int\limits_{\Delta t} \frac{\partial}{\partial t} f\, dt = \frac{\partial f}{\partial t}\,. \tag{2.9}$$

In the last equality the use is made of the fact that, by virtue of the second postulate of statistics (2.5), the averaging of the smooth averaged function does not change it.

Let us average the second term in Equation (2.2):

$$\frac{1}{\Delta X\, \Delta t} \int\limits_{\Delta X} \int\limits_{\Delta t} v_\alpha \frac{\partial \hat{f}}{\partial r_\alpha}\, dX\, dt = \frac{1}{\Delta X} \int\limits_{\Delta X} v_\alpha \frac{\partial}{\partial r_\alpha} \left[\frac{1}{\Delta t} \int\limits_{\Delta t} \hat{f}\, dt \right] dX =$$

$$= \frac{1}{\Delta X} \int\limits_{\Delta X} v_\alpha \frac{\partial}{\partial r_\alpha} f\, dX = v_\alpha \frac{\partial f}{\partial r_\alpha}\,. \tag{2.10}$$

Here the index $\alpha = 1, 2, 3$.

In order to average the term containing the force $\mathbf{F}$, let us represent it as a sum of a *mean force* $\langle\, \mathbf{F}\, \rangle$ and the force due to the difference of the real force field from the mean (smooth) one:

$$\mathbf{F} = \langle\, \mathbf{F}\, \rangle + \mathbf{F}'. \tag{2.11}$$

Substituting definition (2.11) in the third term in Equation (2.2) and averaging this term, we have

$$\frac{1}{\Delta X\, \Delta t} \int\limits_{\Delta X} \int\limits_{\Delta t} \frac{F_\alpha}{m} \frac{\partial \hat{f}}{\partial v_\alpha}\, dX\, dt =$$

$$= \frac{\langle F_\alpha \rangle}{m} \frac{1}{\Delta X} \int\limits_{\Delta X} \frac{\partial}{\partial v_\alpha} \left[\frac{1}{\Delta t} \int\limits_{\Delta t} \hat{f}\, dt \right] dX + \frac{1}{\Delta X\, \Delta t} \int\limits_{\Delta X} \int\limits_{\Delta t} \frac{F'_\alpha}{m} \frac{\partial \hat{f}}{\partial v_\alpha}\, dX\, dt =$$

$$= \frac{\langle F_\alpha \rangle}{m} \frac{\partial f}{\partial v_\alpha} + \frac{1}{\Delta X\, \Delta t} \int\limits_{\Delta X} \int\limits_{\Delta t} \frac{F'_\alpha}{m} \frac{\partial \hat{f}}{\partial v_\alpha}\, dX\, dt\,. \tag{2.12}$$

Gathering all three terms together, we write the averaged Liouville equation in the form

$$\boxed{\frac{\partial f}{\partial t} + \mathbf{v} \cdot \nabla_{\mathbf{r}} f + \frac{\langle \mathbf{F} \rangle}{m} \cdot \nabla_{\mathbf{v}} f = \left(\frac{\partial \hat{f}}{\partial t} \right)_{\mathrm{c}},} \tag{2.13}$$

where

$$\boxed{\left(\frac{\partial \hat{f}}{\partial t} \right)_{\mathrm{c}} = - \frac{1}{\Delta X\, \Delta t} \int\limits_{\Delta X} \int\limits_{\Delta t} \frac{F'_\alpha}{m} \frac{\partial \hat{f}}{\partial v_\alpha}\, dX\, dt\,.} \tag{2.14}$$

Equation (2.13) and its right-hand side (2.14) are called the *kinetic equation* and the *collisional integral* (cf. definition (1.35)), respectively.

Therefore we have found the *most general* form of the kinetic equation with a collisional integral, which is nice but cannot be directly used in cosmic plasma physics, without making some additional simplifying assumptions. The main assumption, the binary character of collisions, will be taken into account in the next Section, see also Section 2.5.3.

2.2 A collisional integral and correlation functions

2.2.1 The exact distribution function

We shall distinguish different kinds of particles, for example, electrons and protons. With this aim we have to complicate the notation. Let $\hat{f}_k(\mathbf{r}, \mathbf{v}, t)$ be the exact distribution function (2.1) of particles of the *kind* k, i.e.

$$\hat{f}_k(\mathbf{r}, \mathbf{v}, t) = \sum_{i=1}^{N_k} \delta(\mathbf{r} - \mathbf{r}_{ki}(t))\, \delta(\mathbf{v} - \mathbf{v}_{ki}(t))\,, \tag{2.15}$$

the index i denoting the ith particle of kind k, N_k being the number of particles of kind k.

The force acting on a particle of kind k at a point $(\mathbf{r}, \mathbf{v})$ of the phase space X at a moment of time t, $\hat{F}_{k,\alpha}(\mathbf{r}, \mathbf{v}, t)$, is the sum of forces acting on this particle from all other particles. So the total force $\hat{F}_{k,\alpha}(\mathbf{r}, \mathbf{v}, t)$ depends upon the instant positions and velocities (generally with the time delay taken into account) of all the particles and can be written with the help of the exact distribution function as follows:

$$\hat{F}_{k,\alpha}(\mathbf{r}, \mathbf{v}, t) = \sum_{l} \sum_{i=1}^{N_l} \hat{F}^{(i)}_{kl,\alpha}(\mathbf{r}, \mathbf{v}, \mathbf{r}_{li}(t), \mathbf{v}_{li}(t)) =$$

$$= \sum_{l} \int_{X_1} \hat{F}_{kl,\alpha}(X, X_1)\, \hat{f}_l(X_1, t)\, dX_1 \,. \tag{2.16}$$

Here

$$\hat{f}_l(X, t) = \sum_{i=1}^{N_l} \delta(X - X_{li}(t))$$

is the exact distribution function of particles of kind l, the variable of integration is designated as $X_1 = \{\mathbf{r}_1, \mathbf{v}_1\}$ and $dX_1 = d^3\mathbf{r}_1\, d^3\mathbf{v}_1$.

Formula (2.16) takes into account that the forces considered are *binary* ones, i.e. they can be represented as a sum of interactions between two particles.

Making use of the representation (2.16), let us average the force term in the Liouville equation (2.2), as this has been done in (2.12). We have

$$\frac{1}{\Delta X\, \Delta t} \int_{\Delta X} \int_{\Delta t} \frac{1}{m_k} \hat{F}_{k,\alpha}(\mathbf{r}, \mathbf{v}, t)\, \frac{\partial \hat{f}_k}{\partial v_\alpha}\, dX\, dt =$$

$$= \frac{1}{\Delta X\, \Delta t} \int_{\Delta X} \int_{\Delta t} \sum_{l} \int_{X_1} \frac{1}{m_k} \hat{F}_{kl,\alpha}(X, X_1)\, \hat{f}_l(X_1, t)\, \frac{\partial}{\partial v_\alpha} \hat{f}_k(X, t)\, dX\, dX_1\, dt =$$

$$= \frac{1}{\Delta X} \int_{\Delta X} \sum_{l} \int_{X_1} \frac{1}{m_k} \hat{F}_{kl,\alpha}(X, X_1) \times$$

$$\times \frac{\partial}{\partial v_\alpha} \left[\frac{1}{\Delta t} \int_{\Delta t} \hat{f}_k(X, t)\, \hat{f}_l(X_1, t)\, dt \right] dX\, dX_1 \,. \tag{2.17}$$

Here we have taken into account that the exact distribution function $\hat{f}_l(X_1, t)$ is independent of the velocity $\mathbf{v}$, which is a part of the variable $X = \{\mathbf{r}, \mathbf{v}\}$ related to the particles of the kind k, and that the interaction law $\hat{F}_{kl,\alpha}(X, X_1)$ is explicitly independent of time t.

Formula (2.17) contains the *pair products* of exact distribution functions of different particle kinds, as is natural for the case of *binary interactions*.

2.2.2 Binary correlation

Represent the exact distribution function $\hat{f}_k$ as

$$\hat{f}_k(X, t) = f_k(X, t) + \hat{\varphi}_k(X, t), \tag{2.18}$$

where $f_k(X, t)$ is the *statistically averaged* distribution function, $\hat{\varphi}_k(X, t)$ is the deviation of the exact distribution function from the averaged one. In general the deviation is not small, of course. It is obvious that, according to definition (2.18),

$$\hat{\varphi}_k(X, t) = \hat{f}_k(X, t) - f_k(X, t);$$

hence

$$\langle \hat{\varphi}_k(X, t) \rangle = 0. \tag{2.19}$$

Consider the integrals of pair products, appearing in the averaged force term (2.17). In view of definition (2.18), they can be rewritten as

$$\frac{1}{\Delta t} \int_{\Delta t} \hat{f}_k(X, t)\, \hat{f}_l(X_1, t)\, dt = f_k(X, t)\, f_l(X_1, t) + f_{kl}(X, X_1, t), \tag{2.20}$$

where

$$f_{kl}(X, X_1, t) = \frac{1}{\Delta t} \int_{\Delta t} \hat{\varphi}_k(X, t)\, \hat{\varphi}_l(X_1, t)\, dt. \tag{2.21}$$

The function f_{kl} is referred to as the *correlation function* or, more exactly, the *binary* correlation function.

The physical meaning of the correlation function is clear from (2.20). The left-hand side of Equation (2.20) means the probability to find a particle of kind k at a point X of the phase space at a moment of time t *under condition* that a particle of kind l places at a point X_1 at the same time. In the right-hand side of (2.20) the distribution function $f_k(X, t)$ characterizes the probability that a particle of kind k stays at a point X at a moment of time t. The function $f_l(X_1, t)$ plays the analogous role for the particles of kind l.

If the particles of kind k did not interact with those of kind l their distributions would be independent, i.e. probability densities would simply multiply:

$$\langle \hat{f}_k (X,t) \, \hat{f}_l (X_1,t) \rangle = f_k (X,t) \, f_l (X_1,t) \,. \tag{2.22}$$

So in the left-hand side of Equation (2.20) there should be

$$f_{kl} (X, X_1, t) = 0 \,. \tag{2.23}$$

In other words there would be no correlation in the particle distribution.

With the proviso that the parameter characterizing the binary interaction, namely Coulomb collision considered below,

$$\zeta_{\rm i} \approx \frac{e^2}{\langle l \rangle} \Bigg/ \left\langle \frac{mv^2}{2} \right\rangle , \tag{2.24}$$

is small under conditions in a wide range, the correlation function must be *relatively small*:

if the interaction is weak, the second term in the right-hand side of (2.20) must be small in comparison with the first one.

We shall come back to the discussion of this property in Section 2.4. This fundamental property allows us to make a theory of cosmic plasma in many cases of astrophysical interest.

2.2.3 The collisional integral and binary correlation

Now let us substitute (2.20) in formula (2.17) for the averaged force term:

$$\frac{1}{\Delta X \, \Delta t} \int\limits_{\Delta X} \int\limits_{\Delta t} \frac{1}{m_k} \, \hat{F}_{k,\alpha} (X,t) \, \frac{\partial \hat{f}_k}{\partial v_\alpha} \, dX \, dt =$$

$$= \frac{1}{\Delta X} \int\limits_{\Delta X} \sum_l \int\limits_{X_1} \frac{1}{m_k} \, \hat{F}_{kl,\alpha} (X, X_1) \, \frac{\partial}{\partial v_\alpha} \, [f_k (X,t) \, f_l (X_1,t) \, +$$

$$+ f_{kl} (X, X_1, t)] \, dX \, dX_1 =$$

$$= \left[\frac{\partial}{\partial v_\alpha} f_k (X,t) \right] \left\{ \frac{1}{\Delta X} \int\limits_{\Delta X} \sum_l \int\limits_{X_1} \frac{1}{m_k} \hat{F}_{kl,\alpha} (X, X_1) \, f_l (X_1, t) \, dX \, dX_1 \right\} +$$

$$+\frac{1}{\Delta X}\int\limits_{\Delta X}\sum_{l}\int\limits_{X_1}\frac{1}{m_k}\hat{F}_{kl,\alpha}(X,X_1)\frac{\partial}{\partial v_\alpha}f_{kl}(X,X_1,t)\,dX\,dX_1 =$$

$$=\frac{1}{m_k}F_{k,\alpha}(X,t)\frac{\partial f_k(X,t)}{\partial v_\alpha}+$$

$$+\sum_{l}\int\limits_{X_1}\frac{1}{m_k}F_{kl,\alpha}(X,X_1)\frac{\partial f_{kl}(X,X_1,t)}{\partial v_\alpha}\,dX_1\,. \tag{2.25}$$

Here we have taken into account that the averaging of smooth functions does not change them, and the following definition of the *averaged force* is used:

$$F_{k,\alpha}(X,t)=\frac{1}{\Delta X}\int\limits_{\Delta X}\sum_{l}\int\limits_{X_1}\hat{F}_{kl,\alpha}(X,X_1)\;f_l(X_1,t)\,dX\,dX_1 =$$

$$=\sum_{l}\int\limits_{X_1}F_{kl,\alpha}(X,X_1)\;f_l(X_1,t)\,dX_1\,. \tag{2.26}$$

This definition coincides with the previous definition (2.12) of the average force, since

> all the deviations of the real force $\hat{\mathbf{F}}_k$ from the mean (smooth) force $\mathbf{F}_k$ are taken care of in the deviations $\hat{\varphi}_k$ and $\hat{\varphi}_l$ of the real distribution functions $\hat{f}_k$ and $\hat{f}_l$ from their mean values.

Thus the collisional integral can be represented in the form

$$\left(\frac{\partial\hat{f}_k}{\partial t}\right)_{\mathrm{c}}=-\sum_{l}\int\limits_{X_1}\frac{1}{m_k}F_{kl,\alpha}(X,X_1)\frac{\partial f_{kl}(X,X_1,t)}{\partial v_\alpha}\,dX_1\,. \tag{2.27}$$

Moreover, if in the last term of (2.25) the binary interactions can be represented by smooth functions of the type $e_k e_l\left(|\,\mathbf{r}_k-\mathbf{r}_l\,|\right)^{-2}$ with account of the Debye shielding (Sections 2.5 and 6.2), then formally the velocity dependence may be neglected.

Recall an important particular case considered in Section 1.2. For the Lorentz force (1.31) as well as for the gravitational one (1.18), the condition (1.25) is satisfied:

$$\frac{\partial}{\partial v_\alpha}\sum_{l}F_{kl,\alpha}(X,X_1)=0\,. \tag{2.28}$$

In fact this condition was tacitly assumed from the early beginning, from Equation (2.2). Anyway, in the case (2.28) we obtain from (2.27) the following expession

$$\left(\frac{\partial \hat{f}_k}{\partial t}\right)_c = -\frac{\partial}{\partial v_\alpha} \sum_l \int_{X_1} \frac{1}{m_k} F_{kl,\alpha}(X, X_1)\, f_{kl}(X, X_1, t)\, dX_1 . \qquad (2.29)$$

Hence the collisional integral, at least, for the Lorentz and gravity forces can be written in the divergent form in the velocity space $\mathbf{v}$:

$$\boxed{\left(\frac{\partial \hat{f}_k}{\partial t}\right)_c = -\frac{\partial}{\partial v_\alpha} j_{k,\alpha} ,} \qquad (2.30)$$

where the flux of particles of kind k in the velocity space (cf. Fig. 1.4b) is

$$j_{k,\alpha}(X, t) = \sum_l \int_{X_1} \frac{1}{m_k} F_{kl,\alpha}(X, X_1)\, f_{kl}(X, X_1, t)\, dX_1 . \qquad (2.31)$$

Therefore we arrive to conclusion that the averaged Liouville equation or **the kinetic equation for particles of kind k**

$$\frac{\partial f_k(X,t)}{\partial t} + v_\alpha \frac{\partial f_k(X,t)}{\partial r_\alpha} + \frac{F_{k,\alpha}(X,t)}{m_k} \frac{\partial f_k(X,t)}{\partial v_\alpha} =$$

$$= -\frac{\partial}{\partial v_\alpha} \sum_l \int_{X_1} \frac{1}{m_k} F_{kl,\alpha}(X, X_1)\, f_{kl}(X, X_1, t)\, dX_1 \qquad (2.32)$$

contains the *unknown* function f_{kl}. Hence the kinetic equation (2.32) for distribution function f_k is not closed. We have to find the equation for the correlation function f_{kl}. This will be done in the next Section.

2.3 Equations for correlation functions

To derive the equations for correlation functions (in the first place for the function of pair correlations f_{kl}), it is not necessary to introduce any new postulates or develop new formalisms. All the necessary equations and averaging procedures are at hand.

Looking at definition (2.21), we see that we need an equation which will describe the deviation of distribution function from its mean value, i.e. the function $\hat{\varphi}_k = \hat{f}_k - f_k$. In order to derive such equation, we simply have to subtract the averaged representation (2.32) from the exact Liouville equation (2.2). The result is

$$\frac{\partial \hat{\varphi}_k(X,t)}{\partial t} + v_\alpha \frac{\partial \hat{\varphi}_k(X,t)}{\partial r_\alpha} + \frac{\hat{F}_{k,\alpha}}{m_k}\frac{\partial \hat{f}_k}{\partial v_\alpha} - \frac{F_{k,\alpha}}{m_k}\frac{\partial f_k}{\partial v_\alpha} =$$
$$= \frac{\partial}{\partial v_\alpha}\sum_l \int_{X_1} \frac{1}{m_k} F_{kl,\alpha}(X,X_1)\, f_{kl}(X,X_1)\, dX_1 \,. \tag{2.33}$$

Here

$$\hat{F}_{k,\alpha}(X,t) = \sum_l \int_{X_1} F_{kl,\alpha}(X,X_1)\, \hat{f}_l(X_1,t)\, dX_1 \tag{2.34}$$

is the *exact* force (2.16) acting on a particle of the kind k at the point X of phase space, and

$$F_{k,\alpha}(X,t) = \sum_l \int_{X_1} F_{kl,\alpha}(X,X_1)\, f_l(X_1,t)\, dX_1 \tag{2.35}$$

is the statistically *averaged* force (2.26).

Thus the difference between the exact force and the averaged one is

$$\hat{F}_{k,\alpha} - F_{k,\alpha} = \sum_l \int_{X_1} F_{kl,\alpha}(X,X_1)\, \hat{\varphi}_l(X_1,t)\, dX_1 \,. \tag{2.36}$$

We substitute it in Equation (2.33) where the difference of force terms can be rewritten as follows:

$$\frac{\hat{F}_{k,\alpha}}{m_k}\frac{\partial \hat{f}_k}{\partial v_\alpha} - \frac{F_{k,\alpha}}{m_k}\frac{\partial f_k}{\partial v_\alpha} = \frac{\hat{F}_{k,\alpha} - F_{k,\alpha}}{m_k}\frac{\partial f_k}{\partial v_\alpha} + \frac{\hat{F}_{k,\alpha}}{m_k}\frac{\partial \hat{\varphi}_k}{\partial v_\alpha} \,.$$

The result of the substitution is

$$\frac{\hat{F}_{k,\alpha}}{m_k}\frac{\partial \hat{f}_k}{\partial v_\alpha} - \frac{F_{k,\alpha}}{m_k}\frac{\partial f_k}{\partial v_\alpha} =$$
$$= \sum_l \int_{X_1} \frac{1}{m_k} F_{kl,\alpha}(X,X_1)\, \hat{\varphi}_l(X_1,t)\, dX_1 \frac{\partial f_k}{\partial v_\alpha} + \frac{F_{k,\alpha}}{m_k}\frac{\partial \hat{\varphi}_k}{\partial v_\alpha} +$$
$$+ \sum_l \int_{X_1} \frac{1}{m_k} F_{kl,\alpha}(X,X_1)\, \hat{\varphi}_l(X_1,t)\, dX_1 \frac{\partial \hat{\varphi}_k}{\partial v_\alpha} \,. \tag{2.37}$$

On substituting (2.37) in Equation (2.33) we have the equation for the deviation $\hat{\varphi}_k$ of the exact distribution function $\hat{f}_k$ from its mean value f_k:

$$\frac{\partial \hat{\varphi}_k (X,t)}{\partial t} + v_\alpha \frac{\partial \hat{\varphi}_k (X,t)}{\partial r_\alpha} + \ldots = 0 . \tag{2.38}$$

Considering that we have to derive the equation for the pair correlation function

$$f_{kl} (X_1, X_2, t) = \langle \hat{\varphi}_k (X_1, t) \, \hat{\varphi}_l (X_2, t) \rangle ,$$

let us take two equations:
one for $\hat{\varphi}_k (X_1, t)$

$$\frac{\partial \hat{\varphi}_k (X_1,t)}{\partial t} + v_{1,\alpha} \frac{\partial \hat{\varphi}_k (X_1,t)}{\partial r_{1,\alpha}} + \ldots = 0 \tag{2.39}$$

and another for $\hat{\varphi}_l (X_2, t)$

$$\frac{\partial \hat{\varphi}_l (X_2,t)}{\partial t} + v_{2,\alpha} \frac{\partial \hat{\varphi}_l (X_2,t)}{\partial r_{2,\alpha}} + \ldots = 0 . \tag{2.40}$$

Add the equations resulting from (2.39) multiplied by $\hat{\varphi}_l$ and (2.40) multiplied by $\hat{\varphi}_k$. We obtain

$$\hat{\varphi}_l \frac{\partial \hat{\varphi}_k}{\partial t} + \hat{\varphi}_k \frac{\partial \hat{\varphi}_l}{\partial t} + v_{1,\alpha} \frac{\partial \hat{\varphi}_k}{\partial r_{1,\alpha}} \hat{\varphi}_l + v_{2,\alpha} \frac{\partial \hat{\varphi}_l}{\partial r_{2,\alpha}} \hat{\varphi}_k + \ldots = 0$$

or

$$\frac{\partial (\hat{\varphi}_k \hat{\varphi}_l)}{\partial t} + v_{1,\alpha} \frac{\partial (\hat{\varphi}_k \hat{\varphi}_l)}{\partial r_{1,\alpha}} + v_{2,\alpha} \frac{\partial (\hat{\varphi}_k \hat{\varphi}_l)}{\partial r_{2,\alpha}} + \ldots = 0 . \tag{2.41}$$

On averaging Equation (2.41) we finally have the equation for the *pair correlation* function in the following form:

$$\frac{\partial f_{kl}}{\partial t} + v_{1,\alpha} \frac{\partial f_{kl}}{\partial r_{1,\alpha}} + v_{2,\alpha} \frac{\partial f_{kl}}{\partial r_{2,\alpha}} +$$

$$+ \frac{F_{k,\alpha} (X_1)}{m_k} \frac{\partial f_{kl}}{\partial v_{1,\alpha}} + \frac{F_{k,\alpha} (X_2)}{m_k} \frac{\partial f_{kl}}{\partial v_{2,\alpha}} +$$

$$+ \frac{\partial f_k}{\partial v_{1,\alpha}} \sum_n \int_{X_3} \frac{1}{m_k} F_{kn,\alpha} (X_1, X_3) \, f_{nl} (X_3, X_2) \, dX_3 +$$

$$+ \frac{\partial f_l}{\partial v_{2,\alpha}} \sum_n \int_{X_3} \frac{1}{m_l} F_{ln,\alpha} (X_2, X_3) \, f_{nk} (X_3, X_1) \, dX_3 =$$

$$= -\frac{\partial}{\partial v_{1,\alpha}} \sum_n \int\limits_{X_3} \frac{1}{m_k} F_{kn,\alpha}(X_1, X_3)\, f_{kln}(X_1, X_2, X_3)\, dX_3 -$$

$$-\frac{\partial}{\partial v_{2,\alpha}} \sum_n \int\limits_{X_3} \frac{1}{m_l} F_{ln,\alpha}(X_2, X_3)\, f_{kln}(X_1, X_2, X_3)\, dX_3 . \qquad (2.42)$$

Here

$$f_{kln}(X_1, X_2, X_3, t) = \frac{1}{\Delta t} \int\limits_{\Delta t} \hat{\varphi}_k(X_1, t)\, \hat{\varphi}_l(X_2, t)\, \hat{\varphi}_n(X_3, t)\, dt \qquad (2.43)$$

is the function of *triple correlations* (see also Problem 2.1).

Thus Equation (2.42) for the pair correlation function contains the *unknown* function of triple correlations. In general,

> the chain of equations for correlation functions can be shown to be *unclosed*: the equation for the correlation function of sth order contains the function of the order $(s + 1)$.

2.4 Approximations for binary collisions

2.4.1 Small parameters of kinetic theory

In itself, the infinite chain of equations for the distribution function and correlation functions does not contain more information than the initial Liouville equation. Actually, the statistical mixing of trajectories in phase space with subsequent statistical smoothing over the physically infinitesimal volume allows to lose 'useless information' – the information about the exact motion of particles. The value of the chain also allows a direct introduction of new physical assumptions which make it possible to break the chain off at some term and to estimate the resulting error. We call this procedure a **well controlled approximation.**

There is no universal way of breaking the chain off. It is intimately related, in particular, to the physical state of a plasma. Different states (as well as different aims) require different approximations. In general, the physical state of a plasma can be characterized, at least partially, by **the ratio of the mean energy of two particle interaction to their mean kinetic energy** (parameter (2.24)). If the last one can be reasonably characterized by some temperature T (see Section 7.1), then this ratio

$$\zeta_{\rm i} \approx \frac{e^2}{\langle l \rangle} (k_{\rm B} T)^{-1} . \qquad (2.44)$$

As a mean distance between the particles we take $\langle l \rangle \approx n^{-1/3}$. Hence the ratio

$$\zeta_{\rm i} = \frac{e^2}{n^{-1/3}} \, (k_{\rm B} T)^{-1} = \frac{e^2}{k_{\rm B}} \times \frac{n^{1/3}}{T} \tag{2.45}$$

is termed the ***interaction parameter***. It is small for a sufficiently ***hot*** and ***rarefied*** plasma. So the thermal kinetic energy of plasma particles is much larger than their interaction energy. **The particles are almost free** or moving on definite trajectories in the external fields if the later are present.

We shall call this case the approximation of ***weak*** Coulomb interaction (see, for example, consideration of the Debye-shielding effect in Section 6.2). An existence of the small parameter allows us to have a complete description of this interaction by using the perturbation procedure. Moreover such a description is the simplest and the most exact one.

As a rule, the so-called ***plasma parameter***

$$\zeta_{\rm p} = \left(n \, r_{\rm D}^3 \right)^{-1} \tag{2.46}$$

is a small quantity as well as it can be expressed in terms of the small interaction parameter (Problem 2.2). The fact that $\zeta_{\rm p} \ll 1$ implies **a large number of plasma particles in a volume enclosed by the sphere of the Debye radius** (see Section 2.5.2). In many astrophysical applications the plasma parameter (2.46) is really small (e.g., Problem 2.3).

While constructing the kinetic theory, it is natural to use the perturbation theory with respect to the plasma parameter $\zeta_{\rm p}$. This means that

> the distribution function f_k must be taken to be of order unity, the pair correlation function f_{kl} of order $\zeta_{\rm p}$, the triple correlation function f_{kln} of order $\zeta_{\rm p}^2$, etc.

We shall see in what follows that this principle has a deep physical sense in kinetic theory of cosmic plasma.

2.4.2 The Vlasov kinetic equation

In the zeroth order with respect to the plasma parameter $\zeta_{\rm p}$, we obtain the Vlasov equation with the self-consistent electromagnetic field (Vlasov, 1945):

$$\frac{\partial f_k (X,t)}{\partial t} + v_\alpha \, \frac{\partial f_k (X,t)}{\partial r_\alpha} + \\ + \frac{e_k}{m_k} \left(\mathbf{E} + \frac{1}{c} \, \mathbf{v} \times \mathbf{B} \right)_\alpha \frac{\partial f_k (X,t)}{\partial v_\alpha} = 0 \, . \tag{2.47}$$

Here **E** and **B** are fields obeying Maxwell's equations

$$\operatorname{curl}\mathbf{E} = -\frac{1}{c}\frac{\partial \mathbf{B}}{\partial t}, \qquad \operatorname{div}\mathbf{E} = 4\pi\left(\rho^{0} + \rho^{\mathrm{q}}\right),$$
$$\operatorname{curl}\mathbf{B} = \frac{1}{c}\frac{\partial \mathbf{E}}{\partial t} + \frac{4\pi}{c}\left(\mathbf{j}^{0} + \mathbf{j}^{\mathrm{q}}\right), \qquad \operatorname{div}\mathbf{B} = 0, \tag{2.48}$$

where ρ^0 and $\mathbf{j}^0$ are the densities of external charges and currents, ρ^{q} and $\mathbf{j}^{\mathrm{q}}$ are the charge and current densities due to the plasma particles themselves:

$$\rho^{\mathrm{q}}(\mathbf{r}, t) = \sum_k e_k \int f_k(\mathbf{r}, \mathbf{v}, t)\, d^3\mathbf{v}, \tag{2.49}$$

$$\mathbf{j}^{\mathrm{q}}(\mathbf{r}, t) = \sum_k e_k \int \mathbf{v}\, f_k(\mathbf{r}, \mathbf{v}, t)\, d^3\mathbf{v}. \tag{2.50}$$

So, if we are considering processes which occur on a time scale much shorter than the time scale of collisions, we may use a description which includes the electric and magnetic fields arising from the plasma charge density and current density, but **neglects the microfields responsible for binary collisions**. This means that $\mathbf{F}' = 0$ in formula (2.11), therefore the collisional integral (2.14) is also equal to zero.

The Vlasov kinetic Equation (2.47) together with the definitions (2.49) and (2.50), and with Maxwell's Equations (2.48) serve as a classic basis for the theory of oscillations and waves in a plasma (e.g., Silin, 1971; Schmidt, 1979; Benz, 1993) with small parameter ζ_{p} and small correlational effects of higher orders. The Vlasov equation is also a proper basis for kinetic theory of wave-particle interactions in cosmic plasma (Chapter 5) and shock waves in collisionless plasma (Section 11.4).

One of the natural limitations of the Vlasov equation is that it will not make a plasma relax to a Maxwellian distribution (Section 7.2.4), since we effectively neglect collisions by neglecting the binary correlation function.

2.4.3 The Landau collisional integral

In the first approximation with respect to the small plasma parameter ζ_{p} we find the Maxwellian distribution function and the effect of Debye shielding. This is the subject of the Section 2.5.

Using the perturbation theory with respect to the small interaction parameter ζ_{i} we find the kinetic equation with the collisional integral given by

Landau (1937)

$$\left(\frac{\partial \hat{f}_k}{\partial t}\right)_c = -\frac{\partial}{\partial v_\alpha} \jmath_{k,\alpha}, \tag{2.51}$$

where the flux of particles of kind k in the velocity space (cf. formula (2.31)) is

$$\jmath_{k,\alpha} = \frac{\pi e_k^2 \ln\Lambda}{m_k} \sum_l e_l^2 \int_{v_l} \left\{ f_k \frac{\partial f_l}{m_l\, \partial v_{l,\beta}} - f_l \frac{\partial f_k}{m_k\, \partial v_{k,\beta}} \right\} \times$$

$$\times \frac{(u^2\, \delta_{\alpha\beta} - u_\alpha u_\beta)}{u^3}\, d^3\mathbf{v}_l\,. \tag{2.52}$$

Here $\mathbf{u} = \mathbf{v} - \mathbf{v}_l$ is the relative velocity, $d^3\mathbf{v}_l$ corresponds to the integration over the whole velocity space of 'field' particles l. $\ln\Lambda$ is the Coulomb logarithm which takes into account divergence of the Coulomb cross-section (see Section 6.1.4).

The date of publication of the Landau (1937) paper may be considered as the date of birth of the kinetic theory of collisional fully-ionized plasma. The theory of collisionless plasma begins with the classical paper of Vlasov (1938). In fact, **these two approaches correspond to different limiting cases.** Landau takes into account only the part of the particle interaction which determines dissipation while the Vlasov equation only allows for the average field, and is thus reversible. In the Vlasov theory the question of the role of collisions in the neighbourhood of resonances remains open.

Eight years later, the famous paper by Landau (1946) was devoted to this problem. Landau used the reversible Vlasov equation as the basis to study the dynamics of a small perturbation of the Maxwell distribution function. In order to solve the linearized Vlasov equation (Section 5.1.1), he made use of the Laplace transformation, and defined the rule to avoid a pole in the divergent integral (see Section 5.1.2) by the replacement $\omega \to \omega + i\,0$.

This technique for avoiding singularities may be replaced by a different procedure. Namely it is possible to add a small dissipative term $-\nu f^{(1)}(\mathbf{r}, \mathbf{v}, t)$ to the linearized Vlasov equation. In this way, the Fourier transform of the kinetic equation involves the complex frequency $\omega = \omega' + i\,\nu$, leading with $\nu \to 0$ to the same expression for the **Landau damping** (see Section 5.1.2).

Thus there are two different approaches to the description of plasma oscillation damping. The first is based on formal mathematical regularization of the Cauchy integral divergence. In this technique the physical nature of the damping is not considered, and the initial equation remains reversible.

The second approach reduces the reversible Vlasov equation to an irreversible one. Although the dissipation is assumed to be negligibly small, one cannot take the limit $\nu \to 0$ directly in the master equations: this can be done only in the final formulae. This second method of introducing the Landau damping is more natural. It shows that **collisions play the principal role in the physics of collisionless plasma**. It is this approach that has been adopted in Klimontovich (1986). A more comprehensive solution of this principal question, however, can only be obtained on the basis of the dissipative kinetic equation.

The example of the Landau resonance and Landau damping demonstrates that some fundamental problems still remain unsolved in the kinetic theory of plasma. They arise from inconsistent descriptions of the transition from the reversible equations of the mechanics of charge particles and fields to the irreversible equations for statistically averaged distribution functions (e.g., Klimontovich, 1998).

2.4.4 The Fokker-Plank equation

The smallness of the interaction parameter $\zeta_{\rm i}$ signifies that, in the Landau collisional integral, the distant Coulomb collisions are taken care of as the interactions with a **small momentum transfer** (Section 6.1). It comes as no surprise that the Landau integral is a particular case of the Fokker-Planck equation (Fokker, 1914; Plank, 1917). The latter generally describes the distribution function change due to **nonstop overlapping weak collisions** resulting in particle diffusion in velocity space:

$$\left(\frac{\partial \hat{f}}{\partial t} \right)_{\rm c} = - \frac{\partial}{\partial v_\alpha} \left[a_\alpha f \right] + \frac{\partial^2}{\partial v_\alpha \, \partial v_\beta} \left[b_{\alpha\beta} \, f \right] . \tag{2.53}$$

The Fokker-Planck equation formally coincides with the diffusion equation (which is irreversible of course) for some admixture with concentration f in an *external* field. The coefficient $b_{\alpha\beta}$ plays the role of the diffusion coefficient and is equal to

$$b_{\alpha\beta} = \frac{1}{2} \left(\delta v_{\alpha\beta} \right)_{\rm av} , \tag{2.54}$$

i.e. is expressed in terms of the averaged velocity changes in elementary acts – distant collisions:

$$\left(\delta v_{\alpha\beta} \right)_{\rm av} = \langle \, \delta v_\alpha \, \delta v_\beta \, \rangle . \tag{2.55}$$

The other coefficient is

$$a_\alpha = \left(\delta v_\alpha \right)_{\rm av} = \langle \, \delta v_\alpha \, \rangle . \tag{2.56}$$

In order to find the mean values appearing in the Fokker-Planck equation, we have to make clear the physical and mathematical sense of expressions (2.55) and (2.56).

The mean values of velocity changes are in fact statistically averaged and determined by the forces acting between a test particle and scatterers (field particles or waves). Because of this, these averaged quantities have to be expressed by the collisional integral with the corresponding cross-sections (Problem 2.4). For electrons and ions in a plasma, such calculations can be made and give the Landau integral. The kinetic equation found in this way will allow us to study the Coulomb interaction of accelerated particle beams with cosmic plasma (Chapter 3).

During the motion of a flux of accelerated particles in a plasma a *reverse* current of thermal electrons is generated, which tends to compensate the electric current of accelerated particles – the *direct* current. The electric field driving the reverse current makes a great impact on the particle flux kinetics. That is why, in order to solve the problem of accelerated particle propagation in, for example, the solar atmosphere, we inevitably have to apply a **combined approach**, which takes into account both the electric field influence on the accelerated particles (as in the Vlasov equation) and their scattering from the thermal particles of a plasma (as in the Landau equation; see Section 3.5).

The Landau collisional integral is effectively used in many problems. It permits a considerable simplification of the calculations of many quantities determined by collisions of charged particles, such as the viscosity coefficient, thermal conductivity, electric conductivity, etc. (Section 7.2.4).

2.5 The correlation function and Debye shielding

2.5.1 The Maxwellian distribution function

Consider the stationary ($\partial/\partial t = 0$) solution to the equations for correlation functions, assuming the plasma parameter $\zeta_{\rm p}$ to be small and using the **method of successive approximations** in the following form. We set $f_{kl} = 0$ in the averaged Liouville equation (2.32) for f_k, then assume that the triple correlation function f_{kln} is zero in Equation (2.42) for the correlation function f_{kl} etc.

The plasma is supposed to be stationary, *uniform* and in the thermodynamic equilibrium state, i.e. the particle velocity distribution is assumed to be a Maxwellian function

$$f_k(X) = f_k(v^2) = c_k \exp\left(-\frac{m_k v^2}{2k_{\rm B}T_k}\right). \tag{2.57}$$

The constant c_k is determined by the normalizing condition and equals

$$c_k = n_k \left(\frac{m_k}{2\pi\, k_{\rm B}T_k}\right)^{3/2}.$$

It is obvious that the Maxwellian function (2.57) satisfies the kinetic equation (2.32) under assumption made above if the average force is equal to zero:

$$F_{k,\alpha}(X,t) = F_{k,\alpha}(X) = 0. \tag{2.58}$$

Since we will need the same assumption in the next Section, we shall justify it there.

2.5.2 Pair correlations and the Debye radius

To a first approximation, i.e. with account of $f_{kl} \neq 0$, the distribution function is also uniform with respect to its space variables. Substitute (2.57) in Equation (2.42), neglecting all the interactions except Coulomb ones. For the latter, in circumstances where the averaged distribution functions for the components are *uniform*, we obtain the following expression for the averaged force (2.26):

$$\begin{aligned} F_{k,\alpha}(X_1) &= \sum_l \int\limits_{X_2} F_{kl,\alpha}(X_1, X_2)\, f_l(X_2)\, dX_2 = \\ &= \sum_l \int\limits_{\mathbf{r}_2} F_{kl,\alpha}(\mathbf{r}_1, \mathbf{r}_2)\, d^3\mathbf{r}_2 \int\limits_{\mathbf{v}_2} f_l(\mathbf{v}_2)\, d^3\mathbf{v}_2 = \\ &= -\int\limits_{\mathbf{r}_2} \sum_l \frac{\partial}{\partial r_{1,\alpha}} \left(\frac{e_k\, e_l}{|\,\mathbf{r}_1 - \mathbf{r}_2\,|}\right) n_l\, d^3\mathbf{r}_2 = \\ &= -\int\limits_{\mathbf{r}_2} \frac{\partial}{\partial r_{1,\alpha}} \left(\frac{e_k}{|\,\mathbf{r}_1 - \mathbf{r}_2\,|}\right) d^3\mathbf{r}_2 \cdot \sum_l n_l\, e_l = 0, \end{aligned} \tag{2.59}$$

if the plasma is assumed to be *quasi-neutral* (see also Section 6.2), or more exactly if here

$$\boxed{\sum_l n_l\, e_l = 0.} \tag{2.60}$$

Balanced charges of ions and electrons determine the name *plasma* according Langmuir (1928). So the average force (2.26) is equal to zero in the electrically neutral plasma but is not equal to zero in gravitational systems (Section 7.2.4).

As a first approximation, on putting the triple correlation function $f_{kln} = 0$, we obtain from Equation (2.42), in view of condition (2.59), the following equation for the binary or pair correlation function f_{kl}:

$$v_{1,\alpha} \frac{\partial f_{kl}}{\partial r_{1,\alpha}} + v_{2,\alpha} \frac{\partial f_{kl}}{\partial r_{2,\alpha}} =$$
$$= - \sum_n \int_{X_3} \left\{ \frac{1}{m_k} F_{kn,\alpha}(X_1, X_3)\, f_{nl}(X_3, X_2) \frac{\partial f_k}{\partial v_{1,\alpha}} + \right.$$
$$\left. + \frac{1}{m_l} F_{ln,\alpha}(X_2, X_3)\, f_{nk}(X_3, X_1) \frac{\partial f_l}{\partial v_{2,\alpha}} \right\} dX_3 . \tag{2.61}$$

Consider the particles of two kinds – electrons and ions, assuming the ions to be motionless and homogeneously distributed. Then the ions do not take part in any kinetic processes; hence $\hat{\varphi}_i \equiv 0$ for ions and the correlation functions associated with $\hat{\varphi}_i$ equal zero as well:

$$f_{ii} = 0 , \quad f_{ei} = 0 \quad \text{etc.} \tag{2.62}$$

Among the pair correlation functions, only one has a non-zero magnitude

$$f_{ee}(X_1, X_2) = f(X_1, X_2) . \tag{2.63}$$

Taking into account (2.62), (2.63), and (2.57), rewrite Equation (2.61) as follows

$$\mathbf{v}_1 \frac{\partial f}{\partial \mathbf{r}_1} + \mathbf{v}_2 \frac{\partial f}{\partial \mathbf{r}_2} =$$
$$= \frac{1}{k_B T} \int_{X_3} [\mathbf{v}_1 \cdot \mathbf{F}(X_1, X_3)\, f(X_3, X_2)\, f_e(\mathbf{v}_1) +$$
$$+ \mathbf{v}_2 \cdot \mathbf{F}(X_2, X_3)\, f(X_1, X_3)\, f_e(\mathbf{v}_2)]\, dX_3 . \tag{2.64}$$

Since $\mathbf{v}_1$ and $\mathbf{v}_2$ are arbitrary and refer to the same kind of particles (electrons), Equation (2.64) takes the form

$$\frac{\partial f}{\partial \mathbf{r}_1} = \frac{1}{k_B T} \int_{X_3} \mathbf{F}(X_1, X_3)\, f(X_3, X_2)\, f_e(\mathbf{v}_1)\, dX_3 . \tag{2.65}$$

Taking into account the character of Coulomb force in the same approximation as in formula (2.59) and assuming the correlation to exist only between the positions of the particles in space (rather than between velocities), let us integrate both sides of Equation (2.65) over $d^3\mathbf{v}_1\, d^3\mathbf{v}_2$. The result is

$$\frac{\partial g(\mathbf{r}_1,\mathbf{r}_2)}{\partial \mathbf{r}_1} = -\frac{ne^2}{k_{\mathrm{B}}T}\int_{\mathbf{r}_3} \nabla_{\mathbf{r}_1}\frac{1}{|\,\mathbf{r}_1-\mathbf{r}_3\,|}\, g(\mathbf{r}_2,\mathbf{r}_3)\, d^3\mathbf{r}_3\,. \tag{2.66}$$

Here the function

$$g(\mathbf{r}_1,\mathbf{r}_2) = \iint f(X_1,X_2)\, d^3\mathbf{v}_1\, d^3\mathbf{v}_2\,. \tag{2.67}$$

Integrate Equation (2.66) over $\mathbf{r}_1$ and designate the function

$$g(\mathbf{r}_1,\mathbf{r}_2) = g(r_{12}^2)\,,$$

where $r_{12} = |\,\mathbf{r}_1 - \mathbf{r}_2\,|$. We obtain the equation

$$g(r_{12}^2) = -\frac{ne^2}{k_{\mathrm{B}}T}\int \frac{g(r_{23}^2)}{r_{13}}\, d^3\mathbf{r}_3\,.$$

Its solution is

$$g(r) = \frac{c_0}{r}\exp\left(-\frac{r}{r_{\mathrm{D}}}\right), \tag{2.68}$$

where

$$\boxed{r_{\mathrm{D}} = \left(\frac{k_{\mathrm{B}}T}{4\pi ne^2}\right)^{1/2}} \tag{2.69}$$

is the Debye radius. It will be defined in just this way (see formula (6.28)) for the case when the shielding is due to the particles of one kind – due to electrons. A more general formula for the Debye radius will be derived in Section 6.2.

Thus the Debye radius is shown to be a characteristic length scale for the pair correlation function.

The constant of integration c_0 can be found, to complete this Section, by solving the Poisson equation for the potential φ (more justification is given in Section 6.2):

$$\Delta\varphi = -4\pi\, en\left\{1 - \left[1 + \frac{c_0}{r}\exp\left(-\frac{r}{r_{\mathrm{D}}}\right)\right]\right\} =$$

$$= n \, \frac{4\pi e \, c_0}{r} \, \exp\left(-\frac{r}{r_{\mathrm{D}}}\right). \tag{2.70}$$

Here it is taken into account that

$$\int\limits_{\mathbf{v}_1} \int\limits_{\mathbf{v}_2} \langle \hat{f}_k(X_1) \, \hat{f}_l(X_2) \rangle \, d^3\mathbf{v}_1 \, d^3\mathbf{v}_2 = n_k(\mathbf{r}_1) \, n_l(\mathbf{r}_2) + g_{kl}(\mathbf{r}_1, \mathbf{r}_2) \, .$$

The general solution of Equation (2.70) in the spherically symmetric case, i.e. the solution of equation

$$\frac{1}{r} \frac{d^2}{dr^2} (r\varphi) = \frac{4\pi e \, c_0}{r} \, \exp\left(-\frac{r}{r_{\mathrm{D}}}\right) n \, ,$$

is of the form

$$\varphi(r) = n \, \frac{4\pi e \, r_{\mathrm{D}}^2 \, c_0}{r} \, \exp\left(-\frac{r}{r_{\mathrm{D}}}\right) + c_1 + \frac{c_2}{r} \, .$$

Since, as $r \to 0$, the potential φ takes the form $(-e)/r$, $c_1 = c_2 = 0$, and the only non-zero constant is

$$c_0 = -\frac{1}{4\pi \, r_{\mathrm{D}}^2 \, n} \, . \tag{2.71}$$

Substituting (2.71) in solution (2.68) gives the sought-after pair correlation function, i.e. the velocity-integrated correlation function

$$g(r) = -\frac{1}{4\pi \, r_{\mathrm{D}}^2 \, n} \, \frac{1}{r} \, \exp\left(-\frac{r}{r_{\mathrm{D}}}\right) = -\frac{e^2}{k_{\mathrm{B}} T} \, \frac{1}{r} \, \exp\left(-\frac{r}{r_{\mathrm{D}}}\right). \tag{2.72}$$

Formula (2.72) shows that

> the Debye radius is a characteristic length scale of pair correlations in a fully-ionized equilibrium plasma:

$$\boxed{g(r) \sim \frac{1}{r} \, \exp\left(-\frac{r}{r_{\mathrm{D}}}\right).} \tag{2.73}$$

This result proves to be fair in the context of Section 6.2 where the Debye shielding will be considered in another approach. Comparison of formula (2.73) with (6.27) shows that the binary correlation function reproduces the shape of the shielded Coulomb potential.

It is known that cosmic plasma can exhibit *collective phenomena* arising out of mutual interactions of many charged particles. Since the Debye radius r_D is a characteristic length scale of pair correlations, the number $n\, r_D^3$ gives us a measure of the number of particles which can interact simultaneously. The inverse of this number is the plasma parameter (2.46).

2.5.3 Gravitational systems

There is a fundamental difference between plasma and the gravitational systems with potential (1.18), for example, the stars in a galaxy. This difference lies in the nature of the gravitational force: there is no shielding to vitiate this long-range $1/r^2$ force. The conventional wisdom of such system dynamics (e.g., Binney and Tremaine, 1987) asserts that the structure and evolution of a collection of N self-gravitating point masses can be described by the collisionless kinetic equation, the gravitational analog of the Vlasov equation (Problems 2.5 and 11.7). On the basis of what we have seen above,

the collisionless appoach in gravitational systems, i.e. the entire neglect of particle pair correlations, constitutes an **uncontrolled approximation.**

Unlike the case of plasma, we cannot derive the next order correction to the collisionless kinetic equation in the context of a systematic perturbation expansion.

Physically, this means that there is **no shielding**. This is manifested by the fact that the $1/r$ potential yields an infinite cross-section, so that, when evaluating the effects of collisions in the usual way (Section 6.1.4) for an infinite homogeneous system, we encounter logarithmic divergences in the limit of large impact parameter (formula (6.15)).

We may hope to circumvent this difficulty, the problematic Coulomb logarithm of gravitational dynamics, by *first* identifying the bulk mean field force $\langle \mathbf{F} \rangle$ in definition (2.11), acting at any given point in space and *then* treating fluctuations $\mathbf{F}'$ away from the mean field force. This splitting into a mean field plus fluctuations can be introduced formally (e.g., Kandrup, 1998) and allows one to write down the collisional integral of the type (2.14). However, this is difficult to implement concretely because of the apparent absence of a clean separation of time scales.

For the N-body problem with $N \gg 1$ we might expect that these fluctuations are small, so that their effects do in fact constitute a small perturbation. So it is assumed that, on long time scales, one must allow for discreteness effects, described by the Fokker-Plank equation (2.53) or the kinetic equation with the Landau collisional integral (2.51).

Given that theoretical analyses have as yet proven inconclusive, one might instead seek resource to numerical experiments. This, however, is also difficult for gravitational systems not characterized by a high degree of symmetry. There is in fact only one concrete setting where detailed computations have been done, namely the toy model of one-dimensional gravity.

In summary, even though a **mean gravitational field theory** based on the Vlasov equation may seem well motivated physically, there is as yet no rigorous proof of its validity and, in particular, no rigorous estimate as to the time scale on which it might be expected to fail.

Hydrodynamic description of gravitational systems has a difficulty of the same origin. The gravitational attraction cannot be screened (Section 7.2.5).

2.6 Comments on numerical simulations

At present, cosmic plasma processes are typically investigated in well developed and distinct approaches. One approach, described by the Vlasov equation, is the collisionless limit used when collective effects dominate. In cases where the plasma dynamics is determined by collisional processes in external fields and where the self-consistent fields can be neglected, the Fokker-Planck approach is used. At the same time, it is known that

> both collective kinetic effects and Coulomb collisions can play an essential role in a great variety of astrophysical phenomena

starting from the most simple one – propagation of fast particles in cosmic plasma (Chapter 3). Besides, as was mentioned in Section 2.4.3, **collisions play the principal role in the physics of collisionless plasma.** Taking collisions into account may lead not only to quantitative but also qualitative changes in the plasma behaviour, even if the collision frequency ν is much less than the electron plasma frequency.

It is known that, even in the collisionless limit, the kinetic equation is still too difficult for numerical simulations, and the particle methods are the most widely used algorithms. In these methods, instead of direct numerical solution of the kinetic equation, a set of ordinary differential equations (which are the characteristics of the Vlasov equation) for every 'macroparticle' is solved.

In the case of a collisional plasma, the position of a macroparticle satisfies the usual equation of the collisionless case

$$\dot{\mathbf{r}} \equiv \frac{d\,\mathbf{r}}{dt} = \mathbf{v}(t)\,, \tag{2.74}$$

but the momentum equation is modified owing to the Coulomb collisions. They are described by the Fokker-Planck operator (2.53) which introduces a friction (the coefficient a_α) and diffusion (the coefficient $b_{\alpha\beta}$) in velocity space. Thus it is necessary to find the effective collisional force $\mathbf{F}_c$ which acts on the macroparticles:

$$\dot{\mathbf{v}} \equiv \frac{d\,\mathbf{v}}{dt} = \frac{1}{m}\left(\mathbf{F}_{\mathrm{L}} + \mathbf{F}_{\mathrm{c}}\right)\,. \tag{2.75}$$

The collisional force can be introduced phenomenologically (e.g., Jones *et al.*, 1996) but a more mathematically correct approach can be constructed using the stochastic equivalence of the Fokker-Planck and Langevin equations (see Cadjan and Ivanov, 1999). So, **stochastic differential equations** can be regarded as an alternative to the description of a cosmic plasma in terms of distribution function.

The Langevin approach allows to overcome some difficulties related to the Fokker-Planck equation and to simulate plasma processes, taking account of both collective effects and Coulomb collisions. Generally, if we want to construct an effective method for the simulation of complex nonlinear processes in cosmic plasma, we have to satisfy the following obvious but conflicting conditions.

First, the method should be adequate for the task in hand. For a number of problems the application of simplified models of the collisional integral can provide a correct description and ensure good accuracy. The constructed model should describe collisional effects with the desired accuracy.

Second, the method should be computationally efficient. The algorithm should not be extremely time-consuming. In practice, some compromise between accuracy and complexity of the method should be achieved. Otherwise, we restrict ourselves either to a relatively simple setup of the problem or to a too-rough description of the phenomena.

A 'recipe': the choice of a particular collisional model (or a model of the collisional integral) is determined by the importance and particular features of the collisional processes in a given astrophysical problem.

2.7 Practice: Problems and Answers

Problem 2.1. By analogy with formula (2.20), show that

$$\langle \hat{f}_k(X_1,t)\,\hat{f}_l(X_2,t)\,\hat{f}_n(X_3,t)\rangle = \tag{2.76}$$

$$= f_k(X_1,t)\,f_l(X_2,t)\,f_n(X_3,t) +$$

$$+ f_k(X_1,t)\,f_{ln}(X_2,X_3,t) + f_l(X_2,t)\,f_{kn}(X_1,X_3,t) +$$

$$+ f_n(X_3,t)\,f_{kl}(X_1,X_2,t) + f_{kln}(X_1,X_2,X_3,t)\,.$$

Problem 2.2. Show that the interaction parameter

$$\zeta_{\rm i} = \frac{1}{4\pi}\,\zeta_{\rm p}^{2/3}\,, \tag{2.77}$$

if the Debye radius is given by formula (2.69). Discuss the difference between $\zeta_{\rm i}$ and $\zeta_{\rm p}$.

Problem 2.3. How many particles are inside the Debye sphere in the solar corona?

Answer. From formula (6.26) for the Debye radius in two-component equilibrium plasma (see also formula (6.75) in Problem 6.3) it follows that for electron-proton plasma with $T \approx 2\times 10^6$ K and $n \approx 2\times 10^8\ \text{cm}^{-3}$ the Debye radius

$$r_{\rm D} = \left(\frac{kT}{8\pi\,e^2\,n}\right)^{1/2} \approx 4.9\left(\frac{T}{n}\right)^{1/2} \approx 0.5\ \text{cm}\,. \tag{2.78}$$

The number of particles inside the Debye sphere

$$N_{\rm D} = n\,\frac{4}{3}\,\pi r_{\rm D}^3 \sim 10^8. \tag{2.79}$$

Hence the typical value of plasma parameter (2.46) in the corona is really small: $\zeta_{\rm p} \sim 10^{-8}$.

Problem 2.4. Express the collisional integral in terms of the differential cross-sections between particles (e.g., Smirnov, 1981).

Problem 2.5. Following Section 2.4.2, write and discuss the gravitational analog of the Vlasov equation.

Answer. The basic assumption underlying the Vlasov equation is that the gravitational N-body system can be described probabilistically in terms of a statistically smooth distribution function $f(X,t)$. The Vlasov equation manifests the idea that this function will stream freely in the self-consistent gravitational potential $\phi(\mathbf{r},t)$ (cf. (1.18)) associated with $f(X,t)$, so that

$$\frac{\partial f(X,t)}{\partial t} + v_\alpha \frac{\partial f(X,t)}{\partial r_\alpha} - \frac{\partial \phi}{\partial r_\alpha}\frac{\partial f(X,t)}{\partial v_\alpha} = 0\,. \tag{2.80}$$

Here

$$\Delta\phi = -\,4\pi G\rho(\mathbf{r},t) \tag{2.81}$$

and

$$\rho(\mathbf{r},t) = \int f(\mathbf{r},\mathbf{v},t)\,d^3\mathbf{v}\,. \tag{2.82}$$

Note that, in the context of the mean field theory, a distribution of particles over their masses has no effect.

Applying for example to the system of stars in a galaxy, Equation (2.80) indicates that the net gravitational force acting on a star is determined by the large-scale structure of the galaxy rather than by whether the star happens to lie close to some other star. The force on any star does not vary rapidly, and each star is supposed to accelerate smoothly through the force field generated by the galaxy as a whole.

In fact, encounters between stars may cause the acceleration $\dot{\mathbf{v}}$ to differ from the smoothed gravitational force $-\nabla\phi$ and therefore invalidate Equation (2.80). **Gravitational encounters are not screened**, they can be thought of as leading to an additional collisional term on the right side of the equation – a collisional integral. However, very little is known mathematically about such possibility as we can see in Section 2.5.3.

Chapter 3

Propagation of Accelerated Particles in Cosmic Plasma

Among a variety of kinetic phenomena related to fast particles in space, the simplest effect is Coulomb collisions under propagation of the particles in a plasma.

3.1 Derivation of the basic equation

3.1.1 Basic approximations

Among a rich variety of kinetic phenomena related to accelerated fast electrons and ions in space (e.g., Kivelson and Russell, 1995) let us consider in this Chapter only the simplest effect – Coulomb collisions under propagation of fast particle beams in a fully-ionized thermal plasma. We shall assume that there exists some external (background) magnetic field $\mathbf{B}_0$ which determines a direction of fast particle propagation and which can be locally considered as a uniform one.

Electric and magnetic fields, $\mathbf{E}$ and $\mathbf{B}$, related to a beam of fast particles will be discussed in Section 3.5. Heating of plasma will be considered, for example, in Section 6.3. So, untill this will be necessary,

> accelerated particles will be considered as 'test' particles that do not influence the background plasma and magnetic field $\mathbf{B}_0$.

Let $f = f\,(t, \mathbf{r}, \mathbf{v})$ be an unknown distribution function of test particles. In what follows, $q = Ze$ and $m = Am_{\text{p}}$ are electric charge and mass of a test particle, respectively.

Let us restrict a problem by consideration of fast but non-relativistic particles interacting with background plasma which consists of thermal electrons ($m_1 = m_e$ and $e_1 = -e$) and thermal protons ($m_2 = m_p$ and $e_2 = +e$). Both components of a plasma are in thermodynamic equilibrium. Using the kinetic equation with the Landau collisional integral (2.51) we obtain

$$\frac{\partial f}{\partial t} + v_\alpha \frac{\partial f}{\partial r_\alpha} + \frac{q}{m} \left\{ E_\alpha + \frac{1}{c} \left[\mathbf{v} \times (\mathbf{B} + \mathbf{B}_0) \right]_\alpha \right\} \frac{\partial f}{\partial v_\alpha} = - \frac{\partial}{\partial v_\alpha} j_\alpha , \tag{3.1}$$

with $\mathbf{E} = 0$ and $\mathbf{B} = 0$,

$$j_\alpha = \frac{\pi q^2 \ln \Lambda}{m} \sum_{l=1}^{2} e_l^2 \int_{v_l} \left\{ f \frac{\partial f_l}{m_l \, \partial v_{l,\beta}} - f_l \frac{\partial f}{m \, \partial v_\beta} \right\} \times$$

$$\times \frac{(u^2 \, \delta_{\alpha\beta} - u_\alpha u_\beta)}{u^3} \, d^3 \mathbf{v}_l \, . \tag{3.2}$$

Here $\mathbf{u} = \mathbf{v} - \mathbf{v}_l$ is the relative velocity, $d^3 \mathbf{v}_l$ corresponds to the integration over the whole velocity space of the plasma particles $l = 1, 2$. They are distributed by the Maxwellian function (2.57):

$$f_e(v) = n_e \left(\frac{m_e}{2\pi \, k_B T_e} \right)^{3/2} \exp \left(- \frac{m_e v^2}{2 k_B T_e} \right) \tag{3.3}$$

and

$$f_p(v) = n_p \left(\frac{m_p}{2\pi \, k_B T_p} \right)^{3/2} \exp \left(- \frac{m_p v^2}{2 k_B T_p} \right) . \tag{3.4}$$

For simplicity we assume $T_e = T_p = T$ as well as $n_e = n_p = n$. Also for simplicity we shall consider the stationary situation ($\partial / \partial t = 0$). Moreover we shall assume that the distribution function f depends on one spatial variable – the coordinate z measured along the field $\mathbf{B}_0$, on the value of velocity v and the angle θ between the velocity vector $\mathbf{v}$ and the axis z. Therefore

$$f = f(z, v, \theta) \, . \tag{3.5}$$

It is easy to see that, in this case, the term containing the Lorentz force, related to the external field $\mathbf{B}_0$, in Equation (3.1) is equal to zero. Under ussumptions made above, Equation (3.1) takes the following form:

$$v \cos\theta \frac{\partial f}{\partial z} = - \frac{1}{v^2} \frac{\partial}{\partial v} \left(v^2 j_v \right) - \frac{1}{v} \frac{1}{\sin\theta} \frac{\partial}{\partial \theta} \left(\sin\theta \, j_\theta \right) . \tag{3.6}$$

The distribution function f is *not* an isotropic one. So, the angular component j_θ of the particle flux is not equal to zero.

3.1.2 Dimensionless equation

Let us introduce the dimensionless non-relativistic energy of test particles

$$x = \frac{mv^2}{2k_{\rm B}T}\left(\frac{m_{\rm e}}{m}\right) \tag{3.7}$$

and the dimensionless column depth along the magnetic field

$$\zeta = \xi / \tilde{\xi}\,. \tag{3.8}$$

Here

$$\xi = \int_0^z n(z)\,dz\,, \ \ {\rm cm}^{-2}, \tag{3.9}$$

is the dimensional column depth passed by test particles along the z axis; the unit of its measurement is

$$\tilde{\xi} = \frac{k_{\rm B}^2 T^2}{\pi e^2 q^2 \ln \Lambda}\left(\frac{m}{m_{\rm e}}\right)^2, \ \ {\rm cm}^{-2}. \tag{3.10}$$

Equation (3.6) in the dimensionless variables (3.7) and (3.8) takes the following form (Somov, 1982):

$$\sqrt{x}\cos\theta\,\frac{\partial f}{\partial \zeta} = \frac{1}{\sqrt{x}}\frac{\partial}{\partial x}\left\{\sqrt{x}\,D_\gamma(x)\left[\frac{\partial f}{\partial x} + \left(\frac{m}{m_{\rm e}}\right) f\right]\right\} + D_\theta(x)\,\Delta_\theta f. \tag{3.11}$$

Here

$$D_\gamma(x) = \left[\frac{\operatorname{erf}(\sqrt{x})}{\sqrt{x}} - \frac{2}{\sqrt{\pi}}\exp(-x)\right] +$$

$$+\left(\frac{m_{\rm e}}{m_{\rm p}}\right)^{1/2}\left[\frac{\operatorname{erf}(\sqrt{X})}{\sqrt{X}} - \frac{2}{\sqrt{\pi}}\exp(-X)\right] \tag{3.12}$$

with

$$X = \frac{m_{\rm p}}{m_{\rm e}}\,x$$

and

$$\operatorname{erf}(w) = \frac{2}{\sqrt{\pi}}\int_0^w \exp(-t^2)\,dt\,,$$

which is the error function. The diffusion coefficient over the angle θ

$$D_\theta(x) = \frac{1}{8x^2}\left\{\left[\frac{\operatorname{erf}(\sqrt{x})}{\sqrt{x}}(2x-1) + \frac{2}{\sqrt{\pi}}\exp(-x)\right] +\right.$$

$$+ \left(\frac{m_e}{m_p} \right)^{1/2} \left[\frac{\operatorname{erf}(\sqrt{X})}{\sqrt{X}} (2X - 1) + \frac{2}{\sqrt{\pi}} \exp(-X) \right] \Bigg\} , \tag{3.13}$$

and

$$\Delta_\theta = \frac{1}{\sin\theta} \frac{\partial}{\partial\theta} \left(\sin\theta \frac{\partial}{\partial\theta} \right)$$

is the θ-dependent part of the Laplace operator.

To point out the similarity of the equation obtained with the Fokker-Plank equation (2.53), let us rewrite Equation (3.11) as follows:

$$\sqrt{x}\cos\theta \frac{\partial f}{\partial\zeta} = -\frac{\partial}{\partial x} [F(x) f] + \frac{1}{2} \frac{\partial^2}{\partial x^2} [D(x) f] + D_\theta(x) \Delta_\theta f . \tag{3.14}$$

Here the first coefficient

$$F(x) = \frac{dD_\gamma}{dx} - \left(\frac{m}{m_e} + \frac{1}{2x} \right) D_\gamma(x) \tag{3.15}$$

characterized the *regular losses* of energy when accelerated particles pass through the plasma. The second coefficient

$$D(x) = 2D_\gamma(x) \tag{3.16}$$

describes the *energy diffusion*. The third coefficient $D_\theta(x)$ corresponds to the fast particle diffusion over the angle θ.

Kudriavtsev (1958) derived the time-dependent equation which has the right-hand side similar to the one in our Equation (3.11) but for the isotropic distribution function $f = f(t, x)$ for fast ions in a thermal plasma. By using the Laplace transformation, Kudriavtsev solved the problem of maxwelization of fast ions that initially had the mono-energetic distribution $f(0, x) \sim \delta(x - x_0)$. The same problem has been solved numerically by MacDonald, Rosenbluth, and Chuck (1957). (Note that in formula (8) by Kudriavtsev for the 'radial' component j_v of the fast ion flow in the velocity space, the factor $\sqrt{\pi}$ must be in the nominator but not in the denominator.) Both solutions (analytical and numerical) show, of course, that the higher the ion energy, the longer the maxwellization process.

In the particular case when all the particles are the same ($m = m_e = m_p$), the right-hand side of Equation (3.11) can be found, for example, by using the formulae for the Fokker-Plank coefficients (2.56) and (2.54) from Balesku (1963).

3.2 A kinetic equation at high speeds

Bearing in mind particles accelerated to high speeds in a cosmic plasma, let us consider some approximations and some solutions of the kinetic Equation (3.11) that correspond to these approximations. First of all, we shall assume that the dimensionless energy (3.7) of the fast particles

$$x \gg 1 \, . \tag{3.17}$$

This means that speeds of the particles are much higher than the mean thermal velocity of plasma electrons (6.12). However, for simplicity, we restrict the problem by consideration of the fast but non-relativistic particles.

Under condition (3.17), we obtain from (3.12) and (3.13) the following simple formulae for the coefficients in the kinetic Equation (3.11):

$$D_\gamma(x) = \frac{1}{\sqrt{x}} \left(1 + \frac{m_{\rm e}}{m_{\rm p}} \right) , \tag{3.18}$$

$$D_\theta(x) = \frac{1}{2x\sqrt{x}} \, . \tag{3.19}$$

It is *not* taken into account here yet that $m_{\rm e} \ll m_{\rm p}$. The first term on the right-hand side of formula for D_γ (see the unit inside the brackets) is a contribution of collisions with the thermal electrons of a plasma, the second term (see the ratio $m_{\rm e}/m_{\rm p}$) comes from collisions with the thermal protons. However, the electrons and protons give equal contributions to the angular diffusion coefficient D_θ. This is important to see when we derive formula (3.19) from (3.13).

Under the same assumption, the Fokker-Plank type equation (3.14) has the following coefficients:

$$D(x) = \frac{2}{\sqrt{x}} \left(1 + \frac{m_{\rm e}}{m_{\rm p}} \right) , \tag{3.20}$$

$$F(x) = - \frac{m}{m_{\rm e}} \frac{1}{\sqrt{x}} \left(1 + \frac{m_{\rm e}}{m} \frac{1}{x} \right) , \tag{3.21}$$

and the same coefficient of angular diffusion $D_\theta(x)$ of course.

Formulae (3.18) and (3.20) demonstrate that

> energy diffusion due to collisions with thermal electrons is faster in $m_{\rm p}/m_{\rm e}$ times than that due to collisions with thermal protons.

However, the angular diffusion rate is equally determined by both electrons and protons in a plasma.

Since $x \gg 1$ and $m \geq m_e$, the second term on the right-hand side of the formula for $F(x)$ is always smaller than the first one. Taking into account that $m_e \ll m_p$ we also neglect the second term in formula for $D(x)$. Hence, in approximation under consideration,

$$F(x) = -\frac{m}{m_e} \frac{1}{\sqrt{x}} , \qquad D(x) = \frac{2}{\sqrt{x}} , \qquad D_\theta(x) = \frac{1}{2x\sqrt{x}} . \tag{3.22}$$

Let us estimate a relative role of the first and second terms on the right-hand side of Equation (3.14). Dividing the former by the last with account of (3.22) taken gives the ratio

$$\frac{2xF(x)}{D(x)} = \frac{m}{m_e} x , \tag{3.23}$$

which is always much greater than unity. So, for fast particles with speeds much greater than the thermal velocity of plasma electrons,

> the regular losses of energy due to Coulomb collisions always dominate the energy diffusion.

However, the energy diffusion may appear significant near the lower energy boundary $\mathcal{E}_1$ of the fast particle spectrum if $\mathcal{E}_1 \approx k_B T$. This seems to be the case of electron acceleration in high-temperature current sheets in solar flares (Chapters 17 and 18). However, this simply means that, near the lower energy $\mathcal{E}_1 \approx 10$ keV, the initial assumption (3.17) becomes invalid. Instead of (3.17), $x \to 1$; so we have to solve exactly Equation (3.11).

Compare the first and third terms on the right-hand side of (3.14). Dividing the former by the last with account of (3.22) taken gives the ratio

$$\frac{F(x)}{xD_\theta(x)} = 2\,\frac{m}{m_e} . \tag{3.24}$$

For fast protons and heavier ions, we neglect angular scattering in comparison with the regular losses of energy.

> For fast electons formula (3.24) shows that it is impossible to neglect the angular diffusion.

Since the case of fast electrons will be considered later on in more detail, let us rewrite the non-relativistic kinetic equation in the high-speed approximation as follows:

$$\cos\theta\,\frac{\partial f}{\partial \zeta} = \frac{1}{x}\frac{\partial f}{\partial x} + \frac{1}{2x^2}\,\Delta_\theta f. \tag{3.25}$$

Recall that the energy diffusion is neglected in (3.25) according to (3.23).

3.3 The classical thick-target model

We have just seen that, in the fast electron kinetic Equation (3.25), it is not reasonable to neglect the angular diffusion. Let us, however, consider the well-known and widely-used model of a thick target (Brown, 1971; Syrovatskii and Shmeleva, 1972). From Equation (3.25), by *neglecting* the angular diffusion, we obtain the following equation

$$\cos\theta\,\frac{\partial f}{\partial \zeta} = \frac{1}{x}\frac{\partial f}{\partial x}\,. \tag{3.26}$$

With a new variable $y = \zeta/\mu$, where $\mu = \cos\theta$, this equation becomes especially simple:

$$\frac{1}{x}\frac{\partial f}{\partial x} - \frac{\partial f}{\partial y} = 0\,. \tag{3.27}$$

General solution of this equation can be written as

$$f(x,y) = \mathcal{F}\left(\frac{x^2}{2} + y\right), \tag{3.28}$$

where $\mathcal{F}$ is an arbitrary function of its argument. Note that $\mu = \text{const}$, because we have neglected the angular diffusion; so the fast electrons move along straight lines $\theta = \text{const}$ without any scattering.

Consider the initial ($y = 0$) energy distribution of fast electrons – the *injection spectrum* – as a power law:

$$f(x,0) = c_0\, x^{-\gamma_0}\, \Theta(x - x_1)\, \Theta(x_2 - x)\, p_0(\mu)\,. \tag{3.29}$$

Here $\Theta(x)$ is the teta-function; $p_0(\mu)$ is the angular distribution of fast electrons, for example, for a beam of electrons injected parallel to the z axis

$$p_0(\mu) = \frac{1}{(1-\mu^2)^{1/2}}\,\delta\,(\mu - 1)\,. \tag{3.30}$$

According to (3.28) the general solution of the kinetic equation for the fast electrons at the column depth y has the following form:

$$f(x,y) = c_0\, 2^{-\gamma_0/2} \left(\frac{x^2}{2} + y \right)^{-\gamma_0/2} \Theta(x - x_1')\, \Theta(x_2' - x)\, p_0(\mu)\,, \qquad (3.31)$$

where $x_{1,2}' = Re \left(x_{1,2}^2 - 2y \right)^{1/2}$.

Let us consider the normalization condition for the distribution function, first, in the dimensional variables z, v, and θ (see definition (3.5)). If $n_b(z)$ is the **density of electrons in the beam** at distance z from the injection plane $z = 0$, then

$$n_b(z) = \int_0^{\infty} \int_0^{\pi} f\,(z,\, v,\, \theta)\, v^2 dv\, 2\pi\, \sin\theta\, d\theta, \;\; \text{cm}^{-3}. \qquad (3.32)$$

Now rewrite the same normalization condition in the dimensionless variable ζ, x, and μ:

$$n_b(\zeta) = \pi \left(\frac{2k_{\text{B}}T}{m_{\text{e}}} \right)^{3/2} \int_0^{\infty} \int_{-1}^{1} f\,(\zeta,\, x,\, \mu)\, \sqrt{x}\, dx\, d\mu\,, \;\; \text{cm}^{-3}. \qquad (3.33)$$

For initial energy distribution (3.29) and initial angular distribution (3.30), formula (3.33) gives

$$n_b(0) = \pi \left(\frac{2k_{\text{B}}T}{m_{\text{e}}} \right)^{3/2} c_0 \int_{x_1}^{x_2} x^{-\gamma_0+1/2}\, dx \equiv \int_{x_1}^{x_2} N(0,x)\, dx\,, \;\; \text{cm}^{-3}. \qquad (3.34)$$

Here

$$N(0,x) = \pi \left(\frac{2k_{\text{B}}T}{m_{\text{e}}} \right)^{3/2} c_0\, x^{-\gamma_0+1/2}\, \Theta(x - x_1)\, \Theta(x_2 - x) \qquad (3.35)$$

is the differential spectrum of the fast electron density at the boundary $\zeta = 0$ where they are injected.

To use the same notations as in Syrovatskii and Shmeleva (1972) let $\mathcal{E}$ be the kinetic energy of fast electrons measured in keV. Then we rewrite (3.35) as

$$N(0,\mathcal{E}) = K\, \mathcal{E}^{-(\gamma+1/2)}\, \Theta(\mathcal{E} - \mathcal{E}_1)\, \Theta(\mathcal{E}_2 - \mathcal{E})\,, \;\; \text{cm}^{-3}\, \text{keV}^{-1}, \qquad (3.36)$$

where the coefficient

$$K = \pi \left(\frac{2k_{\rm B}T}{m_{\rm e}} \right)^{3/2} c_0 \left(\frac{k_{\rm B}T}{\rm keV} \right)^{\gamma+1/2}, \quad {\rm cm}^{-3}\,{\rm keV}^{\,\gamma-1/2}, \tag{3.37}$$

and the spectral index

$$\boxed{\gamma = \gamma_0 - 1\,.} \tag{3.38}$$

Hence the *injection* spectrum of fast electrons is determined by parameters (3.37) and (3.38).

Substituting c_0 and γ_0 from (3.37) and (3.38) in (3.31) allows us to obtain the differential spectrum of the number density of fast electrons passed the coulomn depth ξ measured in cm^{-2} (see definition (3.9)):

$$N(\xi, \mathcal{E}) = K\,\mathcal{E}^{1/2} \left(\mathcal{E}^2 + \mathcal{E}_0^2 \right)^{-(\gamma+1/2)/2} \times \tag{3.39}$$

$$\times\, \Theta\left(\mathcal{E} - \mathcal{E}_1' \right)\, \Theta\left(\mathcal{E}_2' - \mathcal{E} \right), \quad {\rm cm}^{-3}\,{\rm keV}^{-1}.$$

Here

$$\mathcal{E}_0 = \left(2a_0 \xi \right)^{1/2} \tag{3.40}$$

is the minimal energy of electrons that can pass the depth ξ, the 'constant' a_0 (a slow function of energy $\mathcal{E}$) originates from the Coulomb logarithm and equals

$$a_0 = 2\pi e^4 \ln\Lambda \approx \tag{3.41}$$

$$\approx 1.3 \times 10^{-19} \times \left[\ln\left(\frac{\mathcal{E}}{mc^2} \right) - \frac{1}{2} \ln n + 38.7 \right], \quad {\rm keV}^2\,{\rm cm}^2.$$

In formula (3.39)

$$\mathcal{E}_{1,2}'(\xi) = \left(\mathcal{E}_{1,2}^2 - \mathcal{E}_0^2(\xi) \right)^{1/2} \tag{3.42}$$

are the new boundaries of energetic spectrum, when the fast electrons have passed the column depth ξ.

Solution (3.39) shows that

> the regular losses of energy shift the spectrum of fast electrons to lower energies and make it harder

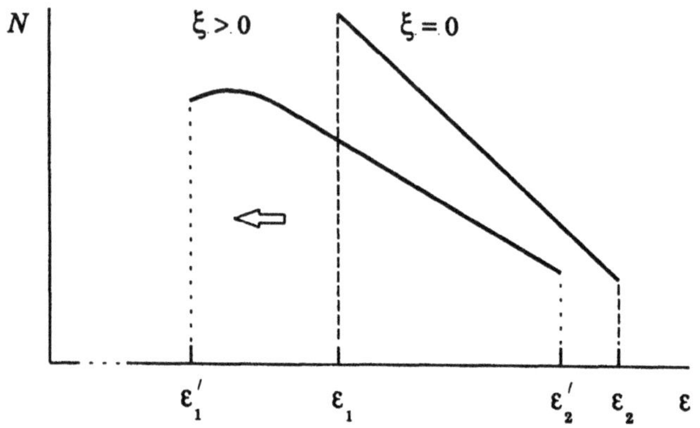

Figure 3.1: An injection spectrum ($\xi = 0$) and the spectrum of fast electrons that have passed the column depth ξ.

(see Figure 3.1). Both effects follow from the fact that, in Equation (3.26), we have taken into account **only the regular losses of energy** (3.22). For non-relativistic electrons $F(x) = -1/\sqrt{x}$.

Syrovatskii and Shmeleva (1972) used the solution (3.39) to calculate the hard X-ray bremsstrahlung which arises during inelastic collisions of accelerated electrons with thermal ions in the solar atmosphere during flares. Brown (1971), in the same approximation but using a different method, has found a similar formula for hard X-ray intensity but with the different numerical coefficient by factor π in Section 5 (see formulae (14) and (15)). Anyway, since that time,

> the simplest thick-target model is widely accepted as a likely mechanism and an appropriate mathematical tool to explain and describe the hard X-ray emission observed during flares

on the Sun and other stars or generally in cosmic plasma (see, however, Sections 3.4 and 3.5). In the classical formulation of the thick-target model, beams of accelerated electrons stream along the magnetic field lines and loose their energy by Coulomb collisions in denser layers of the solar atmosphere, mainly in the chromosphere.

3.4 An approximate account of scattering

As we have seen in Section 3.2, for fast electrons, we cannot neglect the angular scattering in comparison with the regular losses of energy in kinetic

Equation (3.14). Hence, in the classical thich-target model, we have to take the angular scattering into account at least approximately.

If, for example, the beam of fast electrons penetrates a plane parallel the stratified plasma such as the solar chromosphere, the scattering of an *average* beam of electrons may conveniently be described by the Chandrasekhar-Spitzer formulae (6.46) and (6.47) in terms of a coordinate z normal to the atmospheric strata and directed into the plasma. Then the *mean* electron energy $\mathcal{E}$ may be expressed as a function of z while the scattering is measured in terms of the angle $\theta(z)$ which the *mean* electron velocity $\mathbf{v}$ makes with the z axis at that point (Brown, 1972). So

$$v_{\parallel} \equiv v_z = v\mu \,, \quad \text{where} \quad \mu = \cos\theta. \tag{3.43}$$

A convenient change of the independent variable t to ξ is defined by

$$\xi = \int\limits_0^z n(z)\, dz \,, \ \text{cm}^{-2}, \tag{3.44}$$

i.e. ξ is the dimensional column depth passed by electrons along the z axis.

In terms of ξ, formulae (6.46) and (6.47) are:

$$\frac{d\mathcal{E}}{d\xi} = -\frac{a_0}{\mathcal{E}}\,\frac{v}{v_z} \tag{3.45}$$

and

$$\frac{dv_z}{d\xi} = -\frac{3}{2}\,\frac{a_0}{\mathcal{E}^2}\,v \,, \tag{3.46}$$

where $a_0 = 2\pi e^4 \ln\Lambda$ (see definition (3.41)). Thus

$$\frac{3}{2}\,\frac{1}{\mathcal{E}}\,\frac{d\mathcal{E}}{d\xi} = \frac{1}{v_z}\,\frac{dv_z}{d\xi}$$

with solution

$$\left(\frac{\mathcal{E}}{\mathcal{E}_0}\right)^{3/2} = \frac{v_z}{v_{z0}} \,, \tag{3.47}$$

where the suffix 0 refers to values at $\xi = 0$. Since

$$\frac{v_z}{v_{z0}} = \frac{\mu}{\mu_0}\left(\frac{\mathcal{E}}{\mathcal{E}_0}\right)^{1/2},$$

it follows from (3.47) that

$$\boxed{\frac{\mu}{\mu_0} = \frac{\mathcal{E}}{\mathcal{E}_0} .} \tag{3.48}$$

This nice formula (Brown, 1972) shows, in particular, that on average when an electron has suffered a 60° deflection by the column depth its energy has been reduced by 50 %.

Resubstituting (3.48) in (3.45) and (3.46) gives solutions for μ and $\mathcal{E}$:

$$\frac{\mu}{\mu_0} = \frac{\mathcal{E}}{\mathcal{E}_0} = \left(1 - \frac{3\, a_0 \xi}{\mu_0\, \mathcal{E}_0^2} \right)^{1/3} . \tag{3.49}$$

For small depth ξ

$$\frac{\mu}{\mu_0} = \frac{\mathcal{E}}{\mathcal{E}_0} \approx 1 - \frac{a_0}{\mu_0\, \mathcal{E}_0^2}\, \xi \, . \tag{3.50}$$

Compare this result with the general solution (3.28) obtained without account taken of scattering in the classical thick-target model. For electrons with initial energy $\mathcal{E}_0$ solution (3.28) gives us:

$$\frac{\mathcal{E}}{\mathcal{E}_0} = \left(1 - \frac{2a_0}{\mu_0\, \mathcal{E}_0^2}\, \xi \right)^{1/2} . \tag{3.51}$$

If

$$\xi \ll \xi_0 = \frac{\mathcal{E}_0^2}{2a_0} ,$$

then

$$\frac{\mathcal{E}}{\mathcal{E}_0} \approx 1 - \frac{a_0}{\mu_0\, \mathcal{E}_0^2}\, \xi \, . \tag{3.52}$$

Formula (3.52) coincides with (3.50). With increase of ξ, formula (3.49) predicts much faster losses of energy in comparison with the classical thick-target model which does not take collisional scattering into account.

In Figure 3.2, the dashed straight line (a) corresponds to the asymptotic formula (3.50) which is valid for small column depth ξ. The solid curve (b) represents the **classical thick-target model**; it takes only the collisional losses of energy into account. An approximate scattering model described above is presented by the curve (c) which demonstrates that

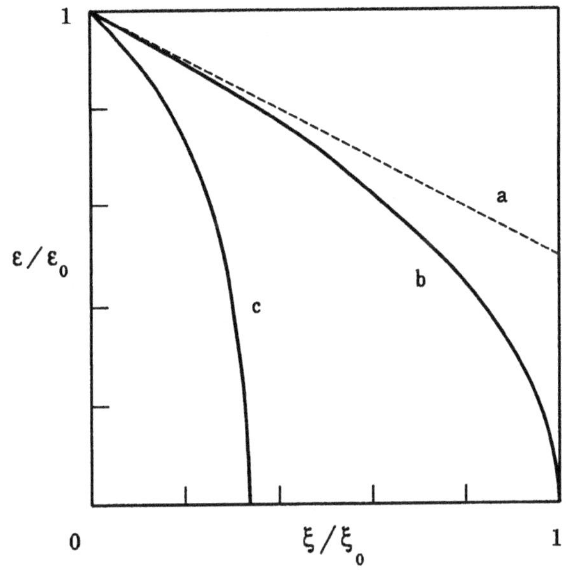

Figure 3.2: The mean energy $\mathcal{E}$ of fast electrons that have passed the column depth ξ (from Somov, 1982).

the collisional scattering and energy losses become very great in comparison with the classical thick-target model if the column depth ξ is not very small.

Brown (1972) used formula (3.49) to develop a solar flare model in which electrons are accelerated downward into the chromosphere. Here the electron distribution is greatly modified by collisions – not only by energy losses but also by scattering. **Directivity and polarization** of the hard X-ray bremsstrahlung emission have been calculated in the thick-target model in which the guiding field $\mathbf{B}_0$ is vertical. The model predicted that the degree of polarization should rise from zero to around 30 % near the solar limb.

Unfortunately the accuracy of the model decreases when the collisional scattering and energy losses become not small. The reason is that the mean rates (3.45) and (3.46) represent well the modification of the electron velocity distribution only at small depth ξ. A more accurate formulation of the kinetic problem will be given in the next Section with account taken of one more mechanism of the electron beam anisotropization. Generally, it seems true that the total absorption of the accelerated electrons in a thick target might result in *negligible* directivity and polarization of the hard X-ray emission.

3.5 The reverse-current electric-field effect

3.5.1 The necessity for a beam-neutralizing current

We assume that some external magnetic field $\mathbf{B}_0$ channels a fast particle propagation and can be locally considered as uniform. The electric and magnetic fields $\mathbf{E}$ and $\mathbf{B}$ related to a beam of fast electrons are superposed on this field. In this way, the beam will be considered as a real electric current $\mathbf{J}$ which influences the background plasma and magnetic field $\mathbf{B}_0$. In order not to obscure the essential physical points related to the electromagnetic field of the beam, we shall completely neglect all other processes like the radiative and hydrodynamic response of the background plasma to a fast heating by the electron beam (Section 6.3.2).

In the classical thick-target model for solar flares, if the fast electrons are supposed to have about the parallel velocities, then the number of injected beam particles per unit time has to be very large – in the order of $\geq 10^{36}$ electrons s^{-1} above 25 keV during the impulsive phase of a flare (Hoyng, Brown, and van Beek, 1976). Given the large electron fluxes implied by the hard X-ray observations, various authors realized that the beam electric current must be enormous – $J \geq 10^{17}$ A. This would imply the magnetic field of the beam $B \geq 10^5$ G. So, the magnetic energy contents of the coronal volume should be more than six orders of magnitude larger than the pre-flare contents for an average coronal field $B_0 \approx 100$ G. Such situation is not likely to occur because the electron beams are thought to be created by conversion of the magnetic energy available in the corona into kinetic energy (Chapter 18).

Apart from this energy problem there is another difficulty related to beams of $\sim 10^{36}$ electrons s^{-1}; they create an enormous charge displacement. For a typical coronal volume of 10^{28} cm^3 and an electron density 10^9 cm^{-3}, the total number of electrons is 10^{37}. A stream of 10^{36} electrons s^{-1} would evacuate all the electrons out of the volume in about 10 s. As a result an enormous charge difference between the corona and the chromosphere would be build up.

In reality the above mentioned problems will not occur, because the beam propagates in a **well-conducting background plasma**. The charge displacement by the beam will quickly create an electric field $\mathbf{E}_1$ which causes the plasma electrons to redistribute in such a way as to *neutralize* the local charge built:

$$\operatorname{div} \mathbf{E}_1 = 4\pi \rho^{\mathrm{q}} . \tag{3.53}$$

Because this electric field is caused by charge separation, it is frequently referred to as an *electrostatic* field.

The second effect is related to the inductive properties of a plasma. In a plasma the magnetic field will not vary considerably on a timescale shorter than the magnetic diffusion time. For beams with radii comparable to the radii of coronal flaring loops this scale is much longer than the duration of the impulsive phase. When the current varies in magnitude, immediately an *inductive* electric field $\mathbf{E}_2$ will be created. It drives a current $\mathbf{j}_2$ of plasma electrons in such a way to prevent magnetic field variations on a time scale shorter than the magnetic diffusion time. As a result the magnetic field will not vary much during the impulsive phase:

$$\mathrm{curl}\,\mathbf{B} \approx \mathrm{const} \approx 0 \approx \frac{4\pi}{c}\mathbf{j}_2 + \frac{1}{c}\frac{\partial}{\partial t}\mathbf{E}_2 . \qquad (3.54)$$

So, the electrostatic effect allows the plasma to 'absorb' the excess charge imposed by the beam of fast electrons; and the inductive effect prevents the magnetic field from changing faster than the allowed diffusion time.

> Both the electrostatic and the inductive electric field will effectively result in an electron plasma current which is in opposite direction of the beam current $\mathbf{J}$.

This electron plasma current is commonly referred to as the *reverse* or *return* current $\mathbf{J}_{rc}$.

Van den Oord (1990) has analyzed the electrostatic and inductive response of a plasma to a prescribed electron beam. By using the Maxwell equations together with the time-dependent Ohm's law (Section 7.3) and with the equation of motion for the plasma electrons in the hydrodynamic approximation (Section 7.1.3), he has shown that the non-linear terms are responsible for a coupling between the electrostatic (irrotational) and inductive (solenoidal) vector fields generated by the beam in a plasma. In order to obtain analytical solutions, van den Oord has decoupled the electrostatic and inductive fields, by ignoring the non-linear terms in the equation of motion, and has found solutions for a mono-energetic blunt beam.

An application of the model in conditions of the solar corona leads to the following results. Charge neutralization is accompanied by plasma oscillations (see formula (6.30)), that are present behind the beam front, and occurs on a time-scale of a few electron-ion collision times. This is also the time scale on which the plasma waves damp out. The net current in the system quickly becomes too low and therefore also the resulting magnetic field strength remains low ($B \ll B_0$).

Although the electric field near the beam front is locally strong, the oscillatory character prevents strong acceleration of the plasma electrons. According to the van den Oord model, all the beam energy is used initially to accelerate the plasma electrons from rest and later on to drive the reverse current against collisional losses. In what follows, we shall use these results and shall formulate an opposite problem in the kinetic approximation. We shall not consider the beam as prescribed. On the contrary, we consider **an influence of the electric field**, which drives the reverse current, **on the distribution function of fast electrons** in the thick-target plasma.

3.5.2 A formulation of a realistic kinetic problem

The *direct* electric current carried by the fast electrons is equal to

$$j_{dc}(z) = e \int f(v,\, \theta,\, z)\, v\, \cos\theta\, d^3\mathbf{v}\,. \tag{3.55}$$

We shall consider this current to be fully balanced by the reverse current of the thermal electrons in the ambient plasma,

$$j_{dc}(z) = j_{rc}(z) \equiv j(z)\,. \tag{3.56}$$

This means that here we do not consider a very fast process of the reverse current generation. The time-dependent process of current neutralization, with account of both electrostatic and inductive effects taken (Section 3.5.1), has been investigated in linear approximation by Van den Oord (1990). Instead of that we shall construct a self-consistent approach for solving the pure kinetic problem with a steady electric field $E = E(z)$ which drives the reverse current. So, using Ohm's law, we determine the reverse-current electric field to be equal to

$$E(z) = \frac{j(z)}{\sigma}\,. \tag{3.57}$$

Here σ is conductivity of the plasma; we assume that the conductivity is determined by Coulomb collisions (Section 7.3). The plasma turbulence effects are important, for example, in the heat conductive front between the high-temperature source of energy and cold plasma of the thick-target. Anyway, even though we expect the wave-particle interactions to have some effects on the fast electrons (Chapter 5), it is unlikely that such effects can change significantly the distribution function of fast electrons with energies far exceeding the energies of the particles in a background cold plasma.

What is really important is the **reverse-current electric field**, it results in an essential change of the fast electron behaviour in the plasma. That is why, to solve the thick-target problem, we develop a combined approach which takes into account the electric field (3.57) as in the Vlasov equation and Coulomb collisions as in the Landau equation. So the distribution function for the fast electrons in the target is described by the following equation (Diakonov and Somov, 1988):

$$v \cos\theta \, \frac{\partial f}{\partial z} - \frac{eE(z)}{m_{\rm e}} \cos\theta \, \frac{\partial f}{\partial v} - \frac{eE(z)}{m_{\rm e} v} \sin^2\theta \, \frac{\partial f}{\partial \cos\theta} = \left(\frac{\partial f}{\partial t} \right)_{\rm c} . \qquad (3.58)$$

Here the second and the third terms are the expression of the term

$$\frac{e_{\rm e}}{m_{\rm e}} \, \mathbf{E}\,(\mathbf{r}) \, \frac{\partial f}{\partial \mathbf{v}}$$

in the dimensional variables v and θ. On the right-hand side of Equation (3.58)

$$\left(\frac{\partial f}{\partial t} \right)_{\rm c} = \frac{1}{v^2} \frac{\partial}{\partial v} \left[v^2 \, \nu(v) \left(\frac{k_{\rm B} T_{\rm e}}{m_{\rm e}} \frac{\partial f}{\partial v} + v f \right) \right] +$$

$$+ \, \nu(v) \, \frac{\partial}{\partial \cos\theta} \left(\sin^2\theta \, \frac{\partial f}{\partial \cos\theta} \right) \qquad (3.59)$$

is the linearized collisional integral; $\nu(v)$ is the collisional rate for fast electrons in the cold plasma.

To set the mathematical problem in the simplest form, we assume that 'superhot' ($T_{\rm e,0} = T_0 \sim 10^8$ K) and 'cold' ($T_{\rm e,1} = T_1 \sim 10^4 - 10^6$ K $\ll T_0$) plasmas occupy the two half-spaces separated by the plane turbulent front ($z = 0$). The superhot region represents the source of energy, for example, the high-temperature current sheet in a solar flare. Let

$$f_{sh} = f_{sh}(v, \, \theta) \qquad (3.60)$$

be the electron distribution function in the source. f_{sh} is, for example, the Maxwellian function for the case of thermal electron runaway (Diakonov and Somov, 1988) or a superposition of thermal and nonthermal functions in the general case. To study the effect of the reverse-current electric field in the classical thick-target model, Litvinenko and Somov (1991) considered only accelerated electrons with an energetic power-law spectrum. Anyway, the function f_{sh} is normalized to the electron number density n_0 in the source:

$$\int f_{sh}(v, \, \theta) \, d^3\mathbf{v} = n_0 \, . \qquad (3.61)$$

Because the electron runaway in a turbulent plasma (Gurevich and Zhivlyuk, 1966) is similar to the ordinary collisional runaway effect (Section 6.4.3), the electrons with velocities

$$v_{\rm e} > v_{cr} \, , \tag{3.62}$$

where v_{cr} is some critical velocity, can freely penetrate through the turbulent front into the cold plasma. Electrons with lower velocities remain trapped in the source. In this Section, we are going to consider the distribution function for the fast electrons escaping into the cold plasma and propagating there. The boundary condition for the forward-flying (the suffix ff) fast electrons may be taken as

$$f_{ff}(v, \theta, 0) = f_{sh}(v, \theta)\, \Theta(v - v_{cr}) \, , \quad 0 \leq \theta \leq \pi/2 \, , \tag{3.63}$$

where Θ is the theta-function.

The distribution function for the back-flying electrons is determined from the solution of Equation (3.58) everywhere, including the boundary $z = 0$. Therefore the problem has been formulated. Note the obvious but important thing; Equation (3.58) contains **two unknown functions**: the fast electron distribution function $f(v, \theta, z)$ and the electric field $E(z)$. So, the kinetic Equation (3.58) must be solved together with Equations (3.55)–(3.57). This is the complete set of equations to be solved self-consistently.

3.5.3 Dimensionless parameters of the problem

In the dimensionless variables (3.7), (3.8) and $\mu = \cos\theta$, Equation (3.58) takes the form

$$\mu x^2 \frac{\partial f}{\partial \zeta} - 2\varepsilon \mu x^2 \frac{\partial f}{\partial x} - \varepsilon x \left(1 - \mu^2\right) \frac{\partial f}{\partial \mu} = x \frac{\partial f}{\partial x} + \tau x \frac{\partial^2 f}{\partial x^2} + \frac{1}{2} \Delta_\mu f. \tag{3.64}$$

Here the dimensionless electron energy

$$x = \frac{m_{\rm e} v^2}{2 k_{\rm B} T_0} \tag{3.65}$$

is normalized with the temperature T_0 of the superhot plasma; for example, $T_0 = T_{{\rm e},cs} \approx 100$ MK is an effective electron temperature of the high-temperature current sheet (Section 17.3). The ratio of the cold-to-superhot plasma temperature

$$\tau = \frac{T_1}{T_0} \approx 10^{-4}, \tag{3.66}$$

if we consider as example the injection of fast electrons into the solar chromosphere. The dimensionless column depth ζ (see definition (3.8)) equals the dimensional column depth passed by fast electrons

$$\xi = \int_0^z n(z)\, dz\,, \quad \text{cm}^{-2}, \tag{3.67}$$

divided by the unit of its measurement

$$\tilde{\xi} = \frac{k_{\text{B}}^2 T_0^2}{\pi e^4 \ln \Lambda}\,, \quad \text{cm}^{-2}. \tag{3.68}$$

The dimensionles electric field

$$\varepsilon = \frac{E}{E_{\text{D},1}} \, \frac{2}{\tau}\,, \tag{3.69}$$

where

$$E_{\text{D},1} = \frac{4\pi e^3 \ln \Lambda}{k_{\text{B}}} \, \frac{n_1}{T_1} \tag{3.70}$$

is the Dreicer field in the cold plasma of the target (cf. definition (6.65)).

The parameter ε can be found from the self-consistent solution of the complete set of equations and the boundary conditions as desribed in Section 3.5.2. The parameter ε is not small in a general case and, in particular, in the solar flare problem $\varepsilon \approx 2 - 20$ (see Fig. 4 in Diakonov and Somov, 1988). Therefore, from (3.69)

$$E = \varepsilon \, \frac{\tau}{2} \, E_{\text{D},1} \approx \left(10^{-4} - 10^{-3} \right) E_{\text{D},1}\,, \tag{3.71}$$

so Ohm's law (3.57) is well applicable in this case.

Let us set the specific form of the boundary distribution function (3.63). The processes of electron acceleration in cosmic plasma and their heating are always closely related. However, for the sake of contrast of them to each other, we consider separately two different functions.

(a) We shall suppose that the electron distribution in the superhot plasma is near to the Maxwellian one. So, the distribution function

$$f_{sh}(x, \mu) = n_0 \, c_0 \, \exp(-x) \, h(\mu)\,, \qquad \mu \geq 0, \tag{3.72}$$

with the constant

$$c_0 = \left(\frac{m_e}{2\pi k_{\text{B}} T_0} \right)^{3/2}.$$

(b) For accelerated electrons we shall use the power-law spectum as the boundary distribution function for the forward-flying electrons

$$f_{sh}(x,\,\mu) = n_0\, c_0\, x^{-\gamma}\, h(\mu)\,, \qquad \mu \geq 0, \tag{3.73}$$

with another normalization constant c_0. In principle, the function $h(\mu)$ is indefinite but should satisfy some additional conditions; at least the function $h(\mu)$ should be maximally smooth (Diakonov and Somov, 1988).

3.5.4 Coulomb energy losses

(a) Electric current in the thick target

In Equation (3.64), the term $\tau x\,(\partial^2 f/\partial x^2)$ describes the energy diffusion. As we know from Section 3.2, for fast electrons with velocities much greater than the thermal velocity of plasma electrons, the regular losses of energy due to collisions always dominate the energy diffusion. So, we neglect this term in comparison with the term $x\,(\partial f/\partial x)$.

However, as we also know from Section 3.2, we cannot neglect the term with the μ-dependent part $\Delta_\mu f$ of the differential operator Laplacian Δ. This term is responsible for the angular diffusion of electrons and is not small in comparison to the regular losses term $x\,(\partial f/\partial x)$.

Therefore we can ignore only the term with small parameter τ in Equation (3.64). After that we have

$$\mu x^2\,\frac{\partial f}{\partial \zeta} = 2\varepsilon\mu x^2\,\frac{\partial f}{\partial x} + \varepsilon x\,(1-\mu^2)\,\frac{\partial f}{\partial \mu} + x\,\frac{\partial f}{\partial x} + \frac{1}{2}\,\frac{\partial}{\partial \mu}\left[\left(1-\mu^2\right)\frac{\partial f}{\partial \mu}\right]. \tag{3.74}$$

By using this equation, we would like to obtain the equation which determines the behaviour of the direct electric current (3.55) carried by fast electrons in the target. It follows from definition (3.55) that

$$j_{dc}(\zeta) = 2\pi e\left(\frac{2k_{\rm B}T_0}{m_{\rm e}}\right)^2 \int\limits_0^\infty \int\limits_{-1}^{+1} f(x,\,\mu,\,\zeta)\, x\mu\, dx\, d\mu\,. \tag{3.75}$$

So we have to divide Equation (3.74) by x and to integrate it as in formula (3.75).

All terms on the right-hand side of (3.74), except one, give zero contributions. The only term $x\,(\partial f/\partial x)$, describing the regular energy losses due to Coulomb collisions, determines the changes of electric current

$$j(\zeta) = j_{dc}(\zeta) = j_{rc}(\zeta) \tag{3.76}$$

along the coulomn depth ζ into the target. It gives the right-hand side of the equation:

$$\frac{\mathrm{d}j}{\mathrm{d}\zeta} = -\, c_j \int_{-1}^{+1} f(x,\, \mu,\, \zeta)\, d\mu \tag{3.77}$$

with constant

$$c_j = \pi e \left(\frac{2k_{\mathrm{B}} T_0}{m_{\mathrm{e}}} \right)^2 . \tag{3.78}$$

The physical meaning of Equation (3.77) is that

> fast electrons lose their energy and mix with thermal particles of the ambient cold plasma due to Coulomb collisions.

Thus the self-consistent reverse-current problem demands to consider the term $x\,(\partial f / \partial x)$, describing the Coulomb energy losses.

(b) 2D versus 1D models

Equation (3.77) shows that the electric current $j(\zeta)$ decreases along the coulomn depth ζ into the target because of the 'falling out' of 'completely' stopped ($x = 0$) electrons from the distribution function owing to collisional losses of energy. From the electric current continuity equation it follows that a current change is possible only when there are electron 'sources' and/or 'sinks' in the target. In the energy region where Equation (3.64) is valid ($x \gg \tau$), the collisional friction force (Section 6.4.1) is inversely proportional to x. For this reason, the electrons with low energies quickly slow down to energies of the order of τ and thus mix with the thermal electrons in the ambient plasma. Since in Equation (3.74) formally $\tau = 0$, the 'falling out' takes place under $x = 0$ according to formula (3.77).

The models under consideration in this Chapter, except the classical thick-target model in Section 3.3, are *two-dimensional* (2D) in the velocity space (see definition (3.5)). This fact has an important consequence.

> Some electrons after injection into the target may make a curve trajectory and cross the boundary in the reverse direction without significant losses of energy.

These electrons come back to the source (the place of acceleration) without being stopped in the target; they determine the boundary distribution function for *back-flying* electrons and constitute a *significant* part (possibly the bulk) of all injected electrons.

Note that such a process is impossible in one-dimensional (1D) models because an electron cannot change the initial direction to the opposite one without being stopped to zeroth velocity and accelerated by the reverse-current electric field from the zeroth velocity in the reverse direction. So collisional energy losses are involved twice in the 1D dynamics of all fast electrons stopped in the target. In general, 1D kinetic models taking Coulomb collisions into account are non-physical approximations.

The other group of injected electrons considered in 2D models is composed of the fast electrons which, after moving in the target under electrostatic and friction forces, do not come back in the particle source. With suitable values of energy x and angle θ, they lose a lot of their initial energy and stop their motion in the target not far from the boundary. There seem to be small amounts of such particles. They determine the electric current change. Thus the current $j(\zeta)$ and, hence, the electric field $E(\zeta)$ can change slowly near the boundary.

Among the particles that determine the current, we may choose a small subgroup of fast electrons which penetrate to such a depth into the target where the electric field is very small ($\varepsilon \ll 1$) and further on they are moving affected only by collisions. Even for this small subgroup the 2D models are certainly more realistic in comparison with the 1D models which do not take into account the collisional scattering (Section 3.4)

3.5.5 Basic physical results

Usually to solve the 2D (in velocity space) kinetic equation one develops a complicated numerical method. Diakonov and Somov (1988) have developed a new technique to obtain an approximate analytical solution of Equation (3.58) taking the Coulomb collisions and the reverse-current field into account. They have applied this technique to the case of thermal runaway electrons in solar flares. It appears that the reverse-current electric field leads to a significant reduction of the convective heat flux carried by fast electrons escaping from the high-temperature plasma to the cold one.

> It is not justified to exclude the reverse-current electric-field effects in studies of convective heat transport by fast thermal electrons in cosmic plasma, for example, in solar flares.

Litvinenko and Somov (1991) have used the same technique to study the behaviour of the electrons accelerated inside a current sheet in the solar atmosphere during flares. They have shown that the reverse-current electric field results in an essential change of the fast electron behaviour in the thick target. The reverse-current field leads to a *quicker* decrease of the distribution function with the column depth in comparison with the classical thick-target model. It is worth mentioning here that both models (thermal and non-thermal) lead to practically the same value of the field near the boundary, ε_0, and this value is large: $\varepsilon_0 \gg 1$. So, the effects of the reverse-current field are not small.

The distribution function appears to be an *almost isotropic* one. The main part of the injected electrons returns into the source. As a result, the hard X-ray polarization appears much smaller than in the collisional thick-target model without taking account of the reverse current. In calculations by Litvinenko and Somov (1991), the maximum polarization was found to be of about 4 % only. So, a major conclusion of this section is that

> to have a more precise insight into the problem of electron acceleration in solar flares, we inevitably have to take into account the reverse-current electric-field effects.

They make the accelerated electron distribution to be almost isotropic and leads to a significant decrease of expected hard X-ray bremsstrahlung polarization (Somov and Tindo, 1978).

* * *

After all said above, it is rather surprising to conclude that the most of the above mentioned models, after the *classical* thick-target model (Section 3.3), are however not used to obtain realistic quantitative informaton on accelerated electrons in solar flares. The classical thick-target model is still very popular; for example, see Equation (16) in Park *et al.* (1997). Up to now we do not have a realistic time-dependent self-consistent thick-target model (which must be simple enough to be easily used) to interpret and analyze the hard X-ray emission so frequently detected in space.

Future models will incorporate such fine effects like a nonuniform ionization of chromospheric plasma in the thick-target (Brown *et al.*, 1998a), the time-of-flight effect (Brown *et al.*, 1998b), with account taken of the reverse-current electric field.

* * *

Let us clarify our plans. Before transition to the hydrodynamic description (Chapter 7) that is valid for systems containing a large number of colliding particles, we have to study two particular but interesting cases.

First, $N = 1$, a particle in a given force field. This simplest approximation gives us clear approach to several fundamental issues of collisionless plasma in space. In particular, it is necessary to outline the basis of kinetic theory for wave-particle interactions in cosmic plasmas (Chapter 5).

Second, $N = 2$, binary collisions of particles with the Coulomb potential of interaction. They are typical for collisional plasmas in space. We have to know the Coulomb collisions well to justify the hydrodynamic description of cosmic plasmas (Chapter 7) and the magnetohydrodynamics (Chapter 8) which is the simplest but sufficient approximation to describe many large-scale phenomena in space.

In the next Chapter we start from the former.

3.6 Practice: Problems and Answers

Problem 3.1. How deep can the accelerated electrons with the initial energy $\mathcal{E}_0 \approx 10$ keV penetrate from the solar corona into the chromosphere?

Answer. From formula (3.40) we find the simplest estimation for the column depth

$$\xi = \frac{\mathcal{E}_0^2}{2a_0}\,, \ \mathrm{cm}^{-2}. \tag{3.79}$$

Substituting $\mathcal{E}_0 \approx 10$ keV and $n \approx 10^{12}$ cm^{-3} in (3.41) gives $a_0 \approx 3 \times 10^{-18}$ keV^2 cm^2. With this value a_0 we find $\xi \approx 10^{19}$ cm^{-2}. At such depth in the chromosphere, the density of the plasma $n \approx 10^{12}$ cm^{-3} indeed.

Accelerated electrons with energies $\mathcal{E} > 10$ keV can penetrate deeper and contribute significantly to impulsive heating of the optical part of a solar flare (see a temperature enhancement at $\xi \approx 10^{20}$ cm^{-2} in Figure 6.2).

Problem 3.2. How strong can be the reverse-current electric field in the chromosphere during a solar flare?

Answer. According to (3.71), the electric field

$$E = \varepsilon \frac{\tau}{2} E_{\mathrm{D},1} \approx \left(10^{-4} - 10^{-3}\right) E_{\mathrm{D},1}\,. \tag{3.80}$$

In the chromosphere (Problem 6.4), the Dreicer field $E_D > 0.1\,\mathrm{V\,cm^{-1}}$. So, under injection of accelerated electrons into the chromosphere during the impulsive phase of a flare, the reverse-current field $E > 10^{-5} - 10^{-4}\,\mathrm{V\,cm^{-1}}$. With the length scale $l \sim 10^3\,\mathrm{km}$, this electric field gives rise to a potential $\phi \approx E\,l \sim 1 - 10\,\mathrm{keV}$.

Chapter 4

The Motion of a Particle in Given Fields

Surprisingly, the simplest situation – a single particle in given force fields – gives us clear approach to several facets of collisionless plasma in space.

4.1 A particle in constant homogeneous fields

In order to study the motion of a charged particle, let us consider the following basic equation:

$$\frac{d\mathbf{p}}{dt} = e\,\mathbf{E} + \frac{e}{c}\,\mathbf{v} \times \mathbf{B} + m\,\mathbf{g}\,. \tag{4.1}$$

By taking the scalar product of Equation (4.1) with the velocity vector $\mathbf{v}$ we obtain

$$\frac{d\mathcal{E}}{dt} = \mathbf{F} \cdot \mathbf{v}\,, \tag{4.2}$$

where

$$\mathbf{F} = e\,\mathbf{E} + m\,\mathbf{g}$$

is a *non-magnetic* force. The particle kinetic energy change during the time dt is $d\mathcal{E} = \mathbf{v} \cdot d\mathbf{p}$. Therefore, according to Equation (4.2), **the work on a particle is done by the non-magnetic force only**. In what follows we shall remember that magnetic fields are 'lazy' and do not work.

Consider the particle motion in *constant homogeneous* fields.

4.1.1 Constant non-magnetic forces

Let a non-magnetic force be parallel to the y axis, $\mathbf{F} = F\,\mathbf{e}_y$, and let the initial momentum of the particle be parallel to the x axis, $\mathbf{p}_0 = p_0\,\mathbf{e}_x$. Then

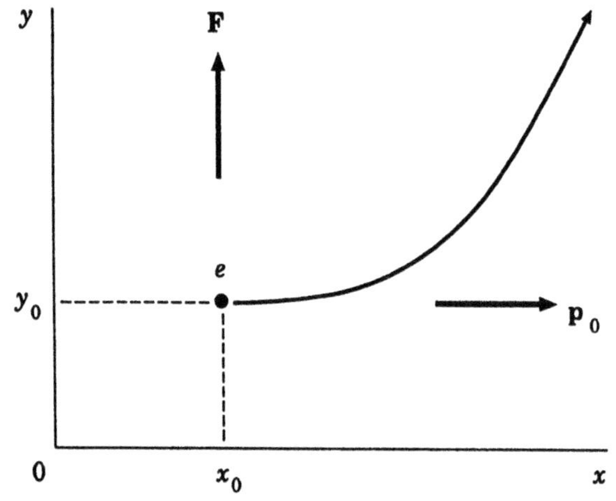

Figure 4.1: The trajectory of particle motion under the action of a constant non-magnetic force.

we integrate Equation (4.1) to find that the particle moves along the catenary shown in Figure 4.1:

$$y - y_0 = \frac{\mathcal{E}_0}{F}\left\{\cosh\left[\frac{F}{p_0 c}(x - x_0)\right] - 1\right\}. \tag{4.3}$$

Here $\mathcal{E}_0$ is an initial energy of the particle.

In the non-relativistic limit formula (4.3) is that of a parabola:

$$y - y_0 = \frac{F}{2mv_0^2}(x - x_0)^2 .$$

4.1.2 Constant homogeneous magnetic fields

The magnetic force in a constant and homogeneous field results in particle motions. Let us show that. From Equation (4.1) we have

$$\frac{d\mathbf{p}}{dt} = \frac{e}{c}\,\mathbf{v}\times\mathbf{B}. \tag{4.4}$$

In relativistic mechanics (e.g., Landau and Lifshitz, *Classical Theory of Field*, 1971, Ch. 2, § 9) the particle momentum and energy are

$$\mathbf{p} = \frac{m\,\mathbf{v}}{\sqrt{1 - v^2/c^2}} \quad \text{and} \quad \mathcal{E} = \frac{mc^2}{\sqrt{1 - v^2/c^2}}, \tag{4.5}$$

respectively. By using the Lorentz factor

$$\gamma_{\text{L}} = \frac{1}{\sqrt{1 - v^2/c^2}}, \tag{4.6}$$

we rewrite formulae (4.5) as

$$\mathbf{p} = \gamma_{\text{L}} m\,\mathbf{v} \quad \text{and} \quad \mathcal{E} = \gamma_{\text{L}} mc^2. \tag{4.7}$$

Hence

$$\mathbf{p} = \frac{\mathcal{E}}{c^2}\,\mathbf{v}\,. \tag{4.8}$$

We known by virtue of (4.2) that $\mathcal{E} = \text{const}$. Therefore $|\,\mathbf{v}\,| = \text{const}$, and from Equation (4.4)

$$\dot{\mathbf{v}} = \omega_{\text{B}}\,\mathbf{v} \times \mathbf{n}\,. \tag{4.9}$$

Here the overdot denotes the derivative with respect to time t, $\mathbf{n}$ is the unit vector along the field $\mathbf{B} = B\,\mathbf{n}$, and the constant

$$\omega_{\text{B}} = \frac{ecB}{\mathcal{E}} \tag{4.10}$$

is the *cyclotron* or Larmor frequency (or gyro-frequency sometimes in what follows). In the non-relativistic limit

$$\boxed{\omega_{\text{B}} = \frac{eB}{mc}\,.} \tag{4.11}$$

By integrating Equation (4.9) we find the linear differential equation

$$\dot{\mathbf{r}} = \omega_{\text{B}}\,\mathbf{r} \times \mathbf{n} + \mathbf{C}\,, \tag{4.12}$$

where vector $\mathbf{C} = \mathbf{const}$.

By taking the scalar product of Equation (4.12) with the unit vector $\mathbf{n}$ we have

$$\mathbf{n} \cdot \dot{\mathbf{r}} = C_{\|} \equiv v_{\|}\,(t = 0)\,.$$

The constant $\mathbf{C}_\perp$ can be removed from consideration by an appropriate choice of the moving reference system. $\mathbf{C}_\perp = 0$ in the reference system where $\mathbf{F} = 0$ (Section 4.1.3), and this choice is consistent with the initial Equation (4.4). Therefore

$$\dot{\mathbf{r}}_\perp = \omega_{\mathrm{B}}\, \mathbf{r}_\perp \times \mathbf{n}\,. \tag{4.13}$$

The vector $\mathbf{r}_\perp$ is changing with the velocity $\mathbf{v}_\perp$ which is perpendicular to $\mathbf{r}_\perp$ itself. Hence the change of vector $\mathbf{r}_\perp$ is a *rotation with the constant frequency* $\boldsymbol{\omega} = \omega_{\mathrm{B}}\, \mathbf{n}$. Thus we have

$$v_\perp = \omega_{\mathrm{B}}\, r_\perp = \mathrm{const} = v_\perp(0)\,,$$

and

$$r_\perp = \frac{v_\perp(0)}{\omega_{\mathrm{B}}} = \frac{\mathcal{E}\, v_\perp(0)}{ecB} = \frac{c p_\perp}{eB}\,,$$

since it follows from formula (4.8) that

$$\mathcal{E}\, v_\perp = c^2 p_\perp\,.$$

We have obtained the expression for the Larmor radius or gyro-radius

$$\boxed{r_{\mathrm{L}} = \frac{c p_\perp}{eB}\,.} \tag{4.14}$$

The term 'rigidity' is introduced in cosmic plasma physics:

$$\boldsymbol{\mathcal{R}} = \frac{c\mathbf{p}}{e}\,. \tag{4.15}$$

The rigidity of a particle is measured in Volts:

$$[\mathcal{R}] = \frac{[cp]}{[e]} = \frac{\mathrm{eV}}{\mathrm{e}} = \mathrm{V}.$$

Rigidity is usually used together with the term 'pitch-angle'

$$\theta = \left(\widehat{\mathbf{v}_0, \mathbf{B}}\right). \tag{4.16}$$

From (4.14) and (4.15) it follows that the particle's Larmor radius is

$$r_{\mathrm{L}} = \frac{\mathcal{R}_\perp}{B}\,. \tag{4.17}$$

That is why

> the particles with the same rigidity and pitch-angle move along the same trajectories in a magnetic field.

This fact is used in the physics of the magnetospheres of the Earth and other planets, as well as in cosmic ray physics.

4.1.3 Non-magnetic forces in a homogeneous magnetic field

Consider the case when a non-magnetic force $\mathbf{F}$ is perpendicular to the magnetic field $\mathbf{B}$ (see Figure 4.2). For the sake of simplicity consider the non-relativistic equation of motion:

$$m\dot{\mathbf{v}} = \mathbf{F} + \frac{e}{c}\,\mathbf{v} \times \mathbf{B}\,. \tag{4.18}$$

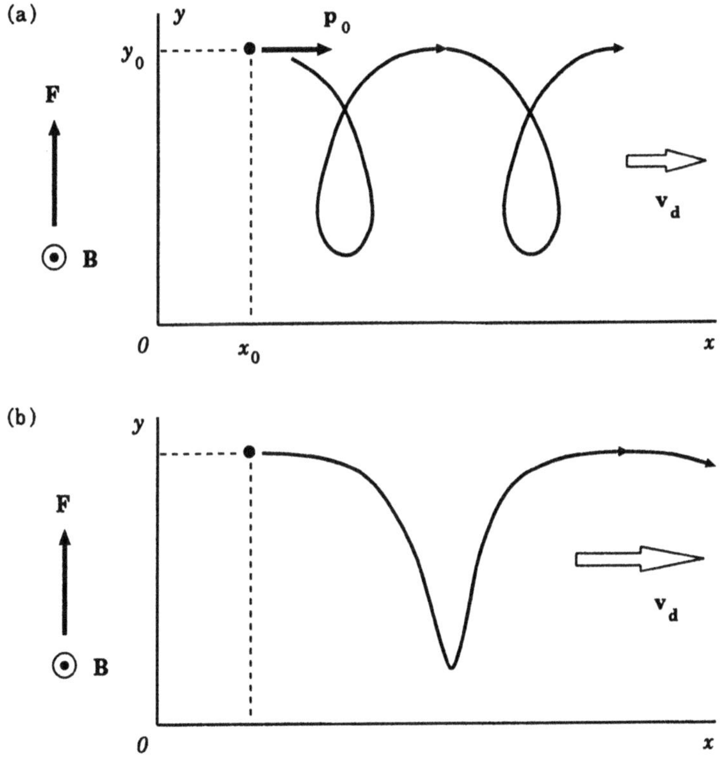

Figure 4.2: The trajectory of particle motion in a uniform magnetic field under the action of a non-magnetic force. Slow (a) and fast (b) drifts.

Let us try to find the solution of this equation in the form

$$\mathbf{v} = \mathbf{v}_{\mathrm{d}} + \mathbf{u}\,. \tag{4.19}$$

Here $\mathbf{v}_{\mathrm{d}}$ is some constant velocity, so that substituting (4.19) in Equation (4.18) gives

$$m\dot{\mathbf{u}} + \mathbf{0} = \frac{e}{c}\,\mathbf{u} \times \mathbf{B} + \mathbf{F} + \frac{e}{c}\,\mathbf{v}_{\mathrm{d}} \times \mathbf{B}\,.$$

We choose $\mathbf{v}_d$ in such a way that the two last terms vanish:

$$\mathbf{F} + \frac{e}{c}\,\mathbf{v}_d \times \mathbf{B} = 0\,.$$

This is the case if the following expression is chosen:

$$\boxed{\mathbf{v}_d = \frac{c}{e}\,\frac{\mathbf{F}\times\mathbf{B}}{B^2}\,.} \tag{4.20}$$

Actually, by using the known vector identity

$$\mathbf{a}\times(\mathbf{b}\times\mathbf{c}) = \mathbf{b}\,(\mathbf{a}\cdot\mathbf{c}) - \mathbf{c}\,(\mathbf{a}\cdot\mathbf{b})\,,$$

we infer

$$\frac{e}{c}\,\mathbf{v}_d \times \mathbf{B} = \mathbf{n}\,(\mathbf{n}\cdot\mathbf{F}) - \mathbf{F} = -\,\mathbf{F}\,,$$

since $\mathbf{F} \perp \mathbf{n} = \mathbf{B}/B$. So, formula (4.20) is correct.

Thus if a non-magnetic force $\mathbf{F}$ is perpendicular to the field $\mathbf{B}$, the particle motion is a sum of the *drift* with the velocity (4.20) called *drift velocity*, which is perpendicular to both $\mathbf{F}$ and $\mathbf{B}$, and the spiral motion round the field lines – the gyro-motion:

$$m\,\dot{\mathbf{u}} = \frac{e}{c}\,\mathbf{u}\times\mathbf{B}\,.$$

Depending on a relative speed of these two motions, we distinguish *slow* ($v_d < u$) and *fast* ($v_d > u$) drifts, see (a) and (b) in Figure 4.2.

As we have seen above, in collisionless plasma, any force $\mathbf{F}$, that is capable of accelerating or decelerating particles as they gyrate about the magnetic field $\mathbf{B}$, will result in a drift perpendicular to both the field and the force. If $\mathbf{F} = e\,\mathbf{E}$, then the drift is called *electric* drift, its velocity

$$\mathbf{v}_d = c\,\frac{\mathbf{E}\times\mathbf{B}}{B^2} \tag{4.21}$$

being independent of the particle charge and mass (Figure 4.3).

Since the drift velocity depends upon neither the charge nor the mass of the particle,

> the electric drift generates the motion of collisionless plasma as a whole with the velocity $\mathbf{v} = \mathbf{v}_d$ relative to a magnetic field.

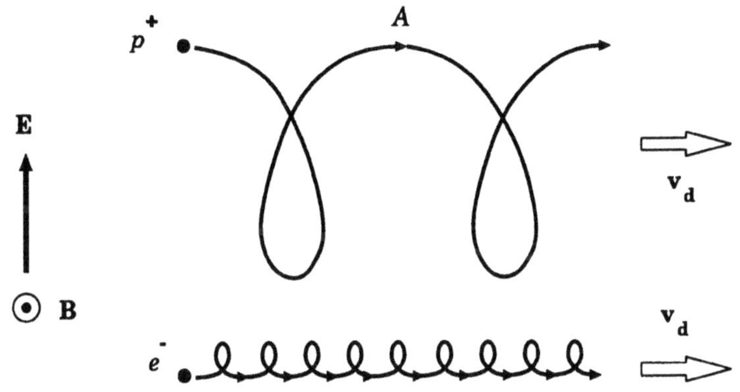

Figure 4.3: Electric drift. The energy of a positively charged particle is a maximum at the upper point A, hence the curvature radius of the trajectory is a maximum at this point.

We should not forget that formula (4.21) was obtained in the non-relativistic limit. In fact, formula (4.21) would formally result in $v_{\rm d} \geq c$ for $E \geq B$.

For the gravitational force $\mathbf{F} = m\,\mathbf{g}$ we have

$$\mathbf{v}_{\rm d} = \frac{mc}{e} \, \frac{\mathbf{g} \times \mathbf{B}}{B^2} \, . \tag{4.22}$$

The *gravitational* drift velocity is seen to depend upon the particle mass and charge. Positively charged particles drift in the direction coinciding with that of the product $\mathbf{g} \times \mathbf{B}$, while negatively charged particles drift in the opposite direction (Fig. 4.4). Therefore

> a gravitational field is capable of generating an electric current in a magnetized collisionless plasma.

4.2 Weakly inhomogeneous slowly changing fields

4.2.1 Small parameters in the motion equation

Let us take the non-relativistic Equation (4.18) for the motion of a charged particle and rewrite it as follows:

$$\frac{m}{e} \left(\ddot{\mathbf{r}} - \mathbf{g} \right) = \mathbf{E} + \frac{1}{c} \, \dot{\mathbf{r}} \times \mathbf{B} \, . \tag{4.23}$$

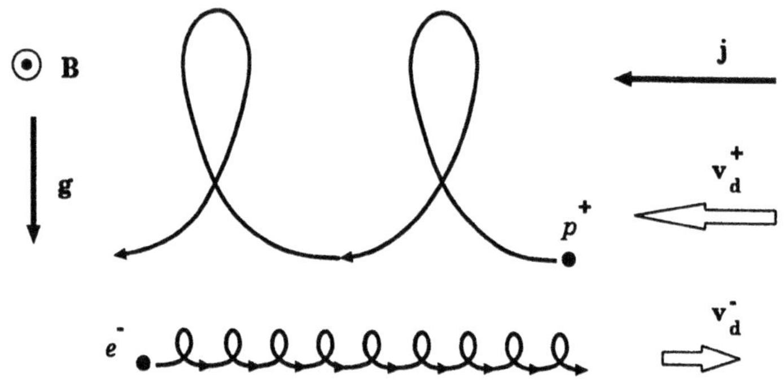

Figure 4.4: Gravitational drift. Initiation of an electric current by the action of the gravity force in a collisionless plasma with magnetic field.

On making this expression non-dimensional

$$\mathbf{r}^* = \frac{\mathbf{r}}{L}, \quad t^* = \frac{t}{\tau}, \quad \mathbf{v}^* = \frac{\mathbf{v}}{v_0}, \ldots, \quad \mathbf{B}^* = \frac{\mathbf{B}}{B_0},$$

we have

$$\frac{m}{e}\frac{L}{\tau^2}\left(\ddot{\mathbf{r}}^* - \mathbf{g}^*\right) = E_0\,\mathbf{E}^* + \frac{L}{c\tau}B_0\,\dot{\mathbf{r}}^* \times \mathbf{B}.$$

Normalize this equation with respect to the last term (the Lorentz force) by dividing the equation by $LB_0/c\tau$:

$$\frac{m}{e}\frac{c}{B_0}\frac{1}{\tau}\left(\ddot{\mathbf{r}}^* - \mathbf{g}^*\right) = \frac{E_0}{B_0}\frac{c\tau}{L}\,\mathbf{E}^* + \dot{\mathbf{r}}^* \times \mathbf{B}^*.$$

Introduce the dimensionless parameter

$$\alpha_{\mathrm{B}} = \frac{m}{e}\frac{c}{B_0}\frac{1}{\tau}.$$

Two situations are conceivable.

(a) Spatially homogeneous magnetic fields are slowly changing in time. The characteristic time $\tau = 1/\omega$, where ω is a characteristic field change frequency. Therefore the dimensionless parameter α_{B} is equal to

$$\alpha_{\mathrm{B}} = \frac{\omega}{\omega_{\mathrm{B}}}. \qquad (a)$$

(b) For the fields that are constant in time but weakly inhomogeneous, the characteristic time is to be defined as $\tau = L/v_0$, L and v_0 being the characteristic values of the field dimensions and the particle velocity, respectively.

In this case

$$\alpha_{\mathrm{B}} = \frac{r_{\mathrm{L}}}{L}\,. \tag{b}$$

Generally, a superposition of these two cases takes place. The field is called *weakly inhomogeneous and slowly changing*, if

$$\boxed{\alpha_{\mathrm{B}} \approx \frac{\omega}{\omega_{\mathrm{B}}} \approx \frac{r_{\mathrm{L}}}{L} \ll 1\,.} \tag{4.24}$$

The second parameter of the problem,

$$\alpha_{\mathrm{E}} = \frac{E_0}{B_0}\,\frac{c\tau}{L}\,,$$

characterizes the relative role of the electric field. We assume $\alpha_{\mathrm{E}} = 1$, because, if this parameter is small, this can be taken into account in the final result.

Thus we have

$$\alpha_{\mathrm{B}}\,(\ddot{\mathbf{r}}^* - \mathbf{g}^*) = \mathbf{E}^* + \dot{\mathbf{r}}^* \times \mathbf{B}^*\,, \tag{4.25}$$

the equation formally coinciding with the initial dimensional one. That is why it is possible to work with Equation (4.23), using as a *small parameter* the dimensional quantity m/e. This method is rather unusual but quite justified and widely used in plasma physics. The corresponding expansion in the Taylor series is termed the expansion in powers of m/e. We find such a solution of Equation (4.23).

4.2.2 Expansion in powers of m/e

Represent the solution of Equation (4.23) as a sum of two terms,

$$\mathbf{r}\,(t) = \mathbf{R}\,(t) + \mathbf{r}_{\mathrm{L}}(t)\,. \tag{4.26}$$

The first term $\mathbf{R}\,(t)$ describes the motion of the *guiding center* of the Larmor circle, the second term $\mathbf{r}_{\mathrm{L}}\,(t)$ corresponds to the rotational motion or gyromotion of the particle. The case of an electron e^- is shown in Figure 4.5.

Recall that for the constant homogeneous magnetic field (see (4.14))

$$r_{\mathrm{L}} = \frac{c p_{\perp}}{eB} = \frac{m}{e}\,\frac{c v_{\perp}}{B}\,,$$

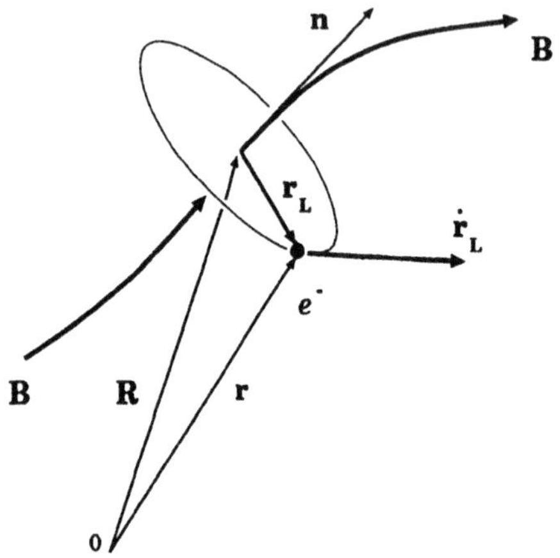

Figure 4.5: The Larmor motion of a negatively charged particle (an electron) in a weakly inhomogeneous slowly changing field.

i.e., the Larmor radius is proportional to the parameter m/e. It is natural to suppose that the dependence is the same for the weakly inhomogeneous slowly changing field, i.e.

$$|\mathbf{r}_L| \sim \frac{m}{e}.$$

Substitute (4.26) in Equation (4.23) and expand the fields $\mathbf{g}$, $\mathbf{E}$, and $\mathbf{B}$ in the Taylor series about the point $\mathbf{r} = \mathbf{R}$:

$$\begin{aligned} \mathbf{g}(\mathbf{r}) &= \mathbf{g}(\mathbf{R}) + (\mathbf{r}_L \cdot \nabla)\,\mathbf{g}(\mathbf{R}) + \dots, \\ \mathbf{E}(\mathbf{r}) &= \mathbf{E}(\mathbf{R}) + (\mathbf{r}_L \cdot \nabla)\,\mathbf{E}(\mathbf{R}) + \dots, \\ \mathbf{B}(\mathbf{r}) &= \mathbf{B}(\mathbf{R}) + (\mathbf{r}_L \cdot \nabla)\,\mathbf{B}(\mathbf{R}) + \dots. \end{aligned} \tag{4.27}$$

From Equation (4.23) we have

$$\ddot{\mathbf{r}} = \mathbf{g} + \left(\frac{m}{e}\right)^{-1} \left[\mathbf{E}(\mathbf{r}) + \frac{1}{\mathbf{c}}\,\dot{\mathbf{r}} \times \mathbf{B}(\mathbf{r})\right].$$

Hence the basic equation contains the small parameter m/e to the power (-1). By substituting (4.26) and (4.27) in this equation we obtain

$$\begin{aligned} \ddot{\mathbf{R}} + \ddot{\mathbf{r}}_L &= \mathbf{g}(\mathbf{R}) + (\mathbf{r}_L \cdot \nabla)\,\mathbf{g}(\mathbf{R}) + \\ &+ \left(\frac{m}{e}\right)^{-1} \{\mathbf{E}(\mathbf{R}) + (\mathbf{r}_L \cdot \nabla)\,\mathbf{E}(\mathbf{R})\} + \\ &+ \left(\frac{m}{e}\right)^{-1} \left\{\frac{1}{c}\left(\dot{\mathbf{R}} + \dot{\mathbf{r}}_L\right) \times [\mathbf{B}(\mathbf{R}) + (\mathbf{r}_L \cdot \nabla)\,\mathbf{B}(\mathbf{R})]\right\} + \dots. \end{aligned} \tag{4.28}$$

Note that we have to think carefully about smallness of different terms in Equation (4.28). For example, the magnitude of $\dot{\mathbf{r}}_L$ is not small:

$$|\,\dot{\mathbf{r}}_L\,| \sim \frac{|\,\mathbf{r}_L\,|}{\tau} \sim r_L\,\omega_B \sim \alpha_B\,\alpha_B^{-1} \sim 1\,.$$

The particle velocity is not small, although the Larmor radius is small. That is the physical reason for the term

$$\left(\frac{m}{e}\right)^{-1} \frac{1}{c}\,[\,\dot{\mathbf{r}}_L \times (\mathbf{r}_L \cdot \nabla)\,\mathbf{B}\,(\mathbf{R})\,]$$

having zero order with respect to the small parameter m/e.

The acceleration term $\ddot{\mathbf{r}}_L$ is not small either:

$$|\,\ddot{\mathbf{r}}_L\,| \sim \frac{|\,\mathbf{r}_L\,|}{\tau^2} \sim r_L\,\omega_B^2 \sim \alpha_B^{-1} \sim \left(\frac{m}{e}\right)^{-1}.$$

In the expansion (4.28) let us retain only the terms with the order of smallness less than one, that is

$$\begin{aligned}
\underbrace{\ddot{\mathbf{R}}}_{(0)} &= \underbrace{-\,\ddot{\mathbf{r}}_L}_{(-1)} + \underbrace{\mathbf{g}\,(\mathbf{R})}_{(0)} + \underbrace{\left(\frac{m}{e}\right)^{-1}\left[\mathbf{E}\,(\mathbf{R}) + \frac{1}{c}\,\dot{\mathbf{R}} \times \mathbf{B}\,(\mathbf{R})\right]}_{(-1)} + \\
&+ \underbrace{\left(\frac{m}{e}\right)^{-1}(\mathbf{r}_L \cdot \nabla)\,\mathbf{E}\,(\mathbf{R})}_{(0)} + \underbrace{\left(\frac{m}{e}\right)^{-1}\frac{1}{c}\,\dot{\mathbf{R}} \times [\,(\mathbf{r}_L \cdot \nabla)\,\mathbf{B}\,(\mathbf{R})\,]}_{(0)} + \\
&+ \underbrace{\left(\frac{m}{e}\right)^{-1}\frac{1}{c}\,\dot{\mathbf{r}}_L \times [\,(\mathbf{r}_L \cdot \nabla)\,\mathbf{B}\,(\mathbf{R})\,]}_{(0)} + O\left(\frac{m}{e}\right).
\end{aligned} \qquad (4.29)$$

Here the orders of smallness of the corresponding terms are given in brackets under the braces.

4.2.3 The averaging over gyro-motion

In order to obtain the equation for guiding center motion let us average Equation (4.29) over a small period of the Larmor rotation,

$$T_B = \frac{2\pi}{\omega_B}\,.$$

Since $\langle\, \mathbf{r}_{\text{L}} \,\rangle = \langle\, \dot{\mathbf{r}}_{\text{L}} \,\rangle = \langle\, \ddot{\mathbf{r}}_{\text{L}} \,\rangle = 0$, we infer the following equation

$$\ddot{\mathbf{R}} = \mathbf{g}\,(\mathbf{R}) + \frac{e}{m}\left[\mathbf{E}\,(\mathbf{R}) + \frac{1}{c}\,\dot{\mathbf{R}} \times \mathbf{B}\,(\mathbf{R})\right] + O\left(\frac{m}{e}\right) +$$

$$+ \frac{e}{mc}\langle\, \dot{\mathbf{r}}_{\text{L}} \times [\,(\mathbf{r}_{\text{L}} \cdot \nabla\,)\,\mathbf{B}\,(\mathbf{R})\,]\,\rangle\,. \tag{4.30}$$

Consider the last term which also has to be averaged. Here we may put

$$\dot{\mathbf{r}}_{\text{L}} = \omega_{\text{B}}\, \mathbf{r}_{\text{L}} \times \mathbf{n}\,.$$

We have to write down the following expression explicitly

$$(\,\mathbf{r}_{\text{L}} \times \mathbf{n}\,) \times [\,(\,\mathbf{r}_{\text{L}} \cdot \nabla\,)\,\mathbf{B}\,(\mathbf{R})\,]$$

and then to average it. It is a matter to do that, once we make use of the following tensor identities:

$$(\mathbf{a} \times \mathbf{b})_{\,\alpha} = e_{\alpha\beta\gamma}\, a_{\beta}\, b_{\gamma}\,.$$

Here $e_{\alpha\beta\gamma}$ is the unit antisymmetric tensor, and

$$e_{\alpha\beta\gamma}\, e_{\mu\nu\gamma} = \delta_{\alpha\mu}\, \delta_{\beta\nu} - \delta_{\alpha\nu}\, \delta_{\beta\mu}\,.$$

On rearrangement, we average the last term in Equation (4.30) to obtain

$$\frac{e}{mc}\,\langle\, \dot{\mathbf{r}}_{\text{L}} \times [\,(\,\mathbf{r}_{\text{L}} \cdot \nabla\,)\,\mathbf{B}\,(\mathbf{R})\,]\,\rangle = -\frac{\mathcal{M}}{m}\,\nabla B\,. \tag{4.31}$$

Here

$$\mathcal{M} = \frac{1}{c}\,\frac{e\,\omega_{\text{B}}}{2\pi}\left(\pi r_{\text{L}}^{2}\right) = \frac{1}{c}\,JS \tag{4.32}$$

is the *magnetic moment* of a particle on the Larmor orbit (Fig. 4.6). The case of electron e^- is shown. It is obvious why we call $\mathcal{M}$ a dipole magnetic moment, as that name usually refers to a property of a *current loop* defined as the current J flowing through the loop times the area S of the loop.

Note that, for any charge of a particle, positive or negative,

> the direction of the magnetic moment is opposite to the direction of the magnetic field.

So the **diamagnetic effect** has to occur.

Substituting the non-relativistic formula $\omega_{\text{B}} = eB/mc$ in (4.32) gives

$$\mathcal{M} = \frac{1}{2\pi}\,\frac{e^2}{mc^2}\,B\,\pi r_{\text{L}}^{2}\,. \tag{4.33}$$

Therefore

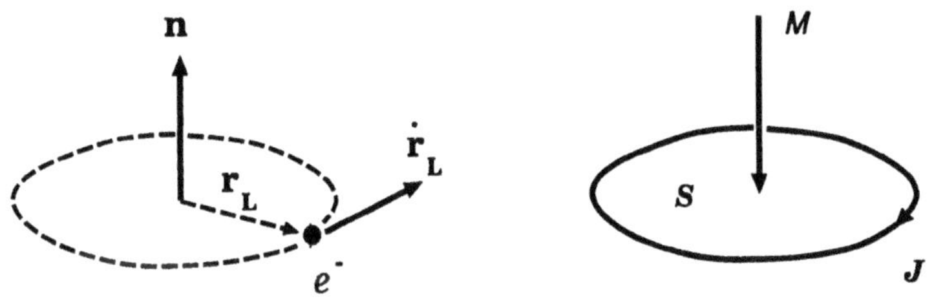

Figure 4.6: The motion of a negative charge particle on the Larmor orbit and its magnetic moment. The moment is antiparallel to the magnetic field.

the magnetic moment is proportional to the magnetic field flux through the surface covering the particle's Larmor orbit.

It is also obvious from (4.31) that we can use the following formula for the force acting on the magnetic moment:

$$\boxed{\mathbf{F} = -\mathcal{M}\,\nabla B\,.} \tag{4.34}$$

Let the field strength increase along the field direction. The force (4.34) is exerted along the field and **away from the direction of increase of the field**. We say that the particle experiences a *mirror force*, and we shall call the place where it turns around a *magnetic mirror* (Section 4.3).

Finally, from Equation (4.30), we obtain the equation of the guiding center motion:

$$\ddot{\mathbf{R}} = \mathbf{g}\,(\mathbf{R}) + \frac{e}{m}\left[\mathbf{E}\,(\mathbf{R}) + \frac{1}{c}\,\dot{\mathbf{R}} \times \mathbf{B}\,(\mathbf{R})\right] - \frac{\mathcal{M}}{m}\,\nabla B\,(\mathbf{R}) + O\left(\frac{m}{e}\right). \tag{4.35}$$

4.2.4 Spiral motion of the guiding center

Even without regarding the terms $O(m/e)$, Equation (4.35) is more difficult in comparison with (4.23). The term $\mathbf{g}\,(\mathbf{R})$, the term with electric field $\mathbf{E}\,(\mathbf{R})$, and the two last terms in Equation (4.35) apart, it is seen that

$$\ddot{\mathbf{R}} = \frac{e}{mc}\,\dot{\mathbf{R}} \times \mathbf{B}\,. \tag{4.36}$$

Therefore the guiding center *spirals*, as does the particle (cf. Equation (4.4)).

By analogy with formula (4.14), the guiding center spiral radius can be found

$$R_{\perp} = \frac{mc\dot{R}_{\perp}}{eB} . \tag{4.37}$$

So it is a small quantity of order

$$\frac{R_{\perp}}{r_{\rm L}} = \frac{\dot{R}_{\perp}}{v_{\perp}} \sim \frac{r_{\rm L}}{L}$$

as compared with the particle Larmor radius (4.14).

The radius of the guiding center spiral is of the order of m/e as compared with the particle Larmor radius. Consequently, this spiral has a higher order with respect to the small parameter m/e and can be neglected in the approximation under study.

4.2.5 Inertial and gradient drifts

Let us neglect the term $O(m/e)$ in Equation (4.35) and take the vector product of Equation (4.35) with the unit vector $\mathbf{n} = \mathbf{B}/B$:

$$\ddot{\mathbf{R}} \times \mathbf{n} = \mathbf{g} \times \mathbf{n} + \frac{e}{m} \mathbf{E} \times \mathbf{n} + \frac{eB}{mc} (\dot{\mathbf{R}} \times \mathbf{n}) \times \mathbf{n} + \frac{\mathcal{M}}{m} \, \mathbf{n} \times \nabla B .$$

From this we find the drift velocity across the magnetic field

$$\dot{\mathbf{R}}_{\perp} \equiv \mathbf{n} \times (\dot{\mathbf{R}} \times \mathbf{n}) = c \, \frac{\mathbf{E} \times \mathbf{n}}{B} + \frac{mc}{eB} \, \mathbf{g} \times \mathbf{n} +$$

$$+ \frac{\mathcal{M}c}{eB} \, \mathbf{n} \times \nabla B - \frac{mc}{eB} \, \ddot{\mathbf{R}} \times \mathbf{n} . \tag{4.38}$$

The first term on the right-hand side of Equation (4.38) corresponds to the *electric* drift (4.21), the second one presents the *gravitational* drift (4.22). The third term is new; it describes the *gradient* drift arising due to the magnetic field inhomogenity. The gradient drift velocity

$$\boxed{\mathbf{v}_{\rm d} = \frac{\mathcal{M}c}{eB} \, \mathbf{n} \times \nabla B .} \tag{4.39}$$

The same formula follows of course from (4.20) after substituting in it the formula (4.34) for the force acting on the magnetic moment $\mathcal{M}$ in the weakly inhomogeneous field.

The fourth term on the right-hand side of (4.38) corresponds to the *inertial* drift. Let us consider it in some detail. For calculating the inertial drift velocity, we have to know the guiding center *acceleration* $\ddot{\mathbf{R}}$. It will suffice for the calculation of $\ddot{\mathbf{R}}$ to consider Equation (4.38) in zeroth order, since the last term of (4.38) contains the small parameter m/e. We have

$$\ddot{\mathbf{R}} = \frac{d}{dt}\,\dot{\mathbf{R}} = \frac{d}{dt}\,(\dot{\mathbf{R}}_{\parallel} + \dot{\mathbf{R}}_{\perp}) = \frac{d}{dt}\left(v_{\parallel}\,\mathbf{n} + c\,\frac{\mathbf{E}\times\mathbf{n}}{B}\right).$$

(a) At first, consider the particular case assuming electric field $\mathbf{E} = 0$, magnetic field $\mathbf{B}$ being *time-independent* but weakly inhomogeneous. Under these conditions

$$\ddot{\mathbf{R}} = \frac{d}{dt}\,(v_{\parallel}\,\mathbf{n}) = \mathbf{n}\,\frac{dv_{\parallel}}{dt} + v_{\parallel}\,\frac{d\,\mathbf{n}}{dt}\,.$$

The first term on the right-hand side does not contribute to the drift velocity since $\mathbf{n}\times\mathbf{n} = \mathbf{0}$. Rewrite the second term as follows:

$$v_{\parallel}\,\frac{d\,\mathbf{n}}{dt} = v_{\parallel}\left(\frac{\partial\,\mathbf{n}}{\partial t} + v_{\parallel}\,(\mathbf{n}\cdot\nabla)\,\mathbf{n}\right).$$

The first term on the right equals zero for the time-independent field. The second one is equal to

$$v_{\parallel}^{2}\,(\mathbf{n}\cdot\nabla)\,\mathbf{n} = -v_{\parallel}^{2}\left(\frac{\mathbf{e}_{c}}{\mathcal{R}_{c}}\right).$$

Here $\mathcal{R}_c$ is a radius of *curvature* for the field line at a given point $\mathbf{R}$. At this point the unit vector $\mathbf{e}_c$ is directed from the curvature center 0_c (Figure 4.7).

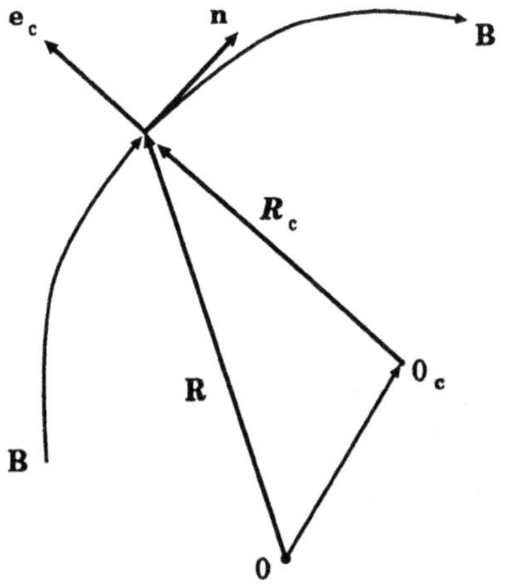

Figure 4.7: The frame of reference for derivation of the formula for the inertial drift in weakly inhomogeneous field.

Thus the dependence of the inertial drift on the curvature of the weakly inhomogeneous magnetic field is found

$$\dot{\mathbf{R}}_{\perp}\big|_{c} = \frac{1}{\mathcal{R}_{c}\,\omega_{B}}\, v_{\parallel}^{2}\, \mathbf{e}_{c} \times \mathbf{n}\,. \tag{4.40}$$

This is the drift of a particle under action of the *centrifugal* force

$$\mathbf{F}_{c} = \frac{m v_{\parallel}^{2}}{\mathcal{R}_{c}}\, \mathbf{e}_{c}\,.$$

(b) Note that the gradient drift in a time-independent weakly inhomogeneous magnetic field has an analogous structure, since

$$\nabla B = \frac{1}{2B}\, \nabla (\mathbf{B} \cdot \mathbf{B}) = \frac{1}{B}\, [\, (\mathbf{B} \cdot \nabla)\, \mathbf{B} + \mathbf{B} \times \operatorname{curl} \mathbf{B}\,]\,.$$

In a current-free region $\operatorname{curl} \mathbf{B} = 0$, and hence

$$\nabla B = \frac{1}{B}\, (\mathbf{B} \cdot \nabla)\, \mathbf{B} = (\mathbf{n} \cdot \nabla)\, \mathbf{B} = (\mathbf{n} \cdot \nabla)\, B\, \mathbf{n} = B\, (\mathbf{n} \cdot \nabla)\, \mathbf{n} +$$

$$+\, \mathbf{n}\, (\mathbf{n} \cdot \nabla B) = -B \left(\frac{\mathbf{e}_{c}}{\mathcal{R}_{c}} \right) + \mathbf{n}\, (\mathbf{n} \cdot \nabla B)\,.$$

The last term does not contribute to the gradient drift velocity (4.39). The contribution of the first term to the drift velocity is

$$\dot{\mathbf{R}}_{\perp} = \frac{\mathcal{M} c}{e B}\, \mathbf{n} \times \left((-B)\, \frac{\mathbf{e}_{c}}{\mathcal{R}_{c}} \right) = -\frac{\mathcal{M}}{e\, \mathcal{R}_{c}}\, \mathbf{n} \times \mathbf{e}_{c} = \frac{\mathcal{M}}{e\, \mathcal{R}_{c}}\, \mathbf{e}_{c} \times \mathbf{n}\,. \tag{4.41}$$

Here, according to definition (4.32) and formula (4.14), the magnetic moment

$$\mathcal{M} = \frac{1}{c}\, J S = \frac{e\, \omega_{B}\, r_{L}^{2}}{2c} = \frac{e\, v_{\perp}^{2}}{2c\, \omega_{B}}\,. \tag{4.42}$$

(c) On substituting formula (4.42) into (4.41) and uniting the two drifts (4.41) and (4.40), we arrive at

$$\dot{\mathbf{R}}_{\perp} = \frac{1}{\mathcal{R}_{c}\, \omega_{B}} \left(v_{\parallel}^{2} + \frac{1}{2}\, v_{\perp}^{2} \right) \mathbf{e}_{c} \times \mathbf{n}\,. \tag{4.43}$$

This formula unites the two drifts that depend on the field line curvature of a weakly inhomogeneous magnetic field.

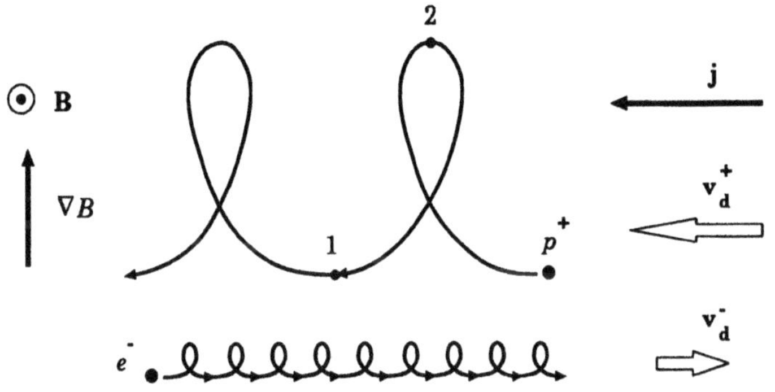

Figure 4.8: A simple interpretation of the gradient drift.

It is worth considering the part of the gradient drift that is independent of the field line curvature. Let the field lines be straight ($\mathcal{R}_c \to \infty$), their density increasing unidirectionally in Figure 4.8. The field strength B_2 at a point 2 is greater than at a point 1 and, according to (4.17),

$$r_L|_2 < r_L|_1 .$$

The particle moves in the manner indicated in Figure 4.8.

For comparison purposes, it is worth remembering another illustration. This is related to, on the contrary, the non-magnetic force **F** (Section 4.1.3). Under action of the force, the particle velocity at a point 1 in Figure 4.9, v_1,

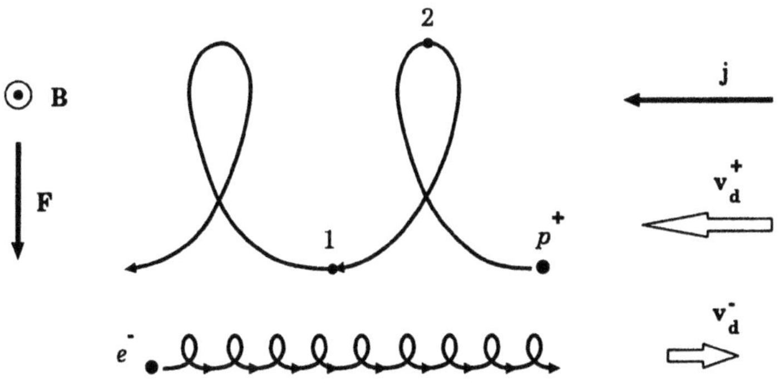

Figure 4.9: The physical nature of the drift under the action of a non-magnetic force **F**.

is greater than at a point 2. Hence the Larmor radius $r_L = cp_\perp/eB$ is greater at a point 1 than at a point 2 as well.

Figures 4.8 and 4.9 also demonstrate the validity of formula (4.34).

The drifts with velocity which depends on the particle charge and mass, like the gradient drift, can be important for the problem of element abundances or element fractionation (see Sections 21.2.5 and 21.3.3).

Recommended Reading: Sivukhin (1965), Morozov and Solov'ev (1966b)

4.3 Adiabatic invariants in cosmic plasmas

4.3.1 General definitions

As is known from mechanics (e.g., Landau and Lifshitz, *Mechanics*, 1960, Ch. 7, § 49), the so-called *adiabatic invariants* remain constant under changing conditions of motion, if these changes are slow. Recall that the system executing a *finite one-dimensional* motion is assumed to be characterized by a parameter λ that is slowly – adiabatically – changing with time:

$$\lambda \, / \, \dot{\lambda} \gg T \, . \tag{4.44}$$

Here T is a characteristic time for the system (e.g., a particle in given fields) motion.

More precisely, if the parameter λ did not change, the system would be closed and would execute a strictly periodic motion with the period T like a simple pendulum in gravitational field. In this case the energy of the system, $\mathcal{E}$, would be invariant.

Under the slowly changing parameter λ, if $\dot{\mathcal{E}} \sim \dot{\lambda}$, then the integral

$$I = \oint P \, dq \, , \tag{4.45}$$

rather than the energy $\mathcal{E}$, is conserved. Here P and q are the *generalized* momentum and coordinate, respectively. The integral is taken along the trajectory of motion under given $\mathcal{E}$ and λ. The integral I is referred to as the adiabatic invariant.

4.3.2 Three main invariants

(a) Motion in the Larmor plane

The motion of a charged particle in slowly changing weakly inhomogeneous fields has been considered in the previous section. Several types of periodic motion can be found. In particular, the particle's motion in the plane perpendicular to the magnetic field – the Larmor motion – is periodic. Let **P** be the generalized momentum. According to (4.45) for such a motion the adiabatic invariants are the integrals

$$I_1 = \oint P_1 \, dq_1 = \text{const} \quad \text{and} \quad I_2 = \oint P_2 \, dq_2 = \text{const} \,,$$

taken over a period of the motion of coordinates q_1 and q_2 in the plane of the Larmor orbit.

It is convenient to combine these integrals, that is simply to add them together:

$$I = \oint \mathbf{P}_\perp \cdot d\mathbf{q} = \text{const} \,. \tag{4.46}$$

(This is the same, of course, as $q = r_{\rm L} \phi$ in definition (4.45) with $0 \le \phi \le 2\pi$.) Here

$$\mathbf{P}_\perp = \mathbf{p}_\perp + \frac{e}{c} \mathbf{A}$$

is the generalized momentum (see Landau and Lifshitz, *Classical Theory of Field*, 1971, Ch. 3, § 16) projection onto the plane mentioned above. In this plane $\mathbf{q} = \mathbf{r}_{\rm L}$. The vector potential **A** is perpendicular to the vector **B** since $\mathbf{B} = \text{curl}\, \mathbf{A}$, and **p** is the ordinary kinetic momentum of a particle.

Now perform the integration in formula (4.46)

$$I = \oint \mathbf{P}_\perp \cdot d\mathbf{r}_{\rm L} = \oint \mathbf{p}_\perp \cdot d\mathbf{r}_{\rm L} + \frac{e}{c} \oint \mathbf{A} \cdot d\mathbf{r}_{\rm L} =$$

$$= 2\pi r_{\rm L} \, p_\perp - \frac{e}{c} \int_S \text{curl}\, \mathbf{A} \cdot d\mathbf{S} =$$

by virtue of the Stokes theorem

$$= 2\pi r_{\rm L} \, p_\perp - \frac{e}{c} \int_S \mathbf{B} \cdot d\mathbf{S} = 2\pi r_{\rm L} \, p_\perp - \frac{e}{c} B \, \pi r_{\rm L}^2 \,. \tag{4.47}$$

Substituting $r_{\rm L} = c p_\perp / eB$ (cf. formula (4.17)) into (4.47) gives

$$I = \frac{\pi c}{e} \frac{p_\perp^2}{B} = \text{const} \,.$$

Thus we come to the conclusion that the conserving quantity is

$$\boxed{\frac{p_\perp^2}{B} = \text{const} \,.} \tag{4.48}$$

This quantity is called the *first* or *transversal* adiabatic invariant.

According to definition (4.32), the particle magnetic moment for the Larmor orbit is

$$\mathcal{M} = \frac{1}{c}\, JS = \frac{p_\perp^2}{2mB} = \frac{\mathcal{K}_\perp}{B}\,. \tag{4.49}$$

Here use is made of the non-relativistic formula for the Larmor frequency (4.11) and the non-relativistic kinetic energy of the particle transversal motion is designated as

$$\mathcal{K}_\perp = \frac{p_\perp^2}{2m}\,.$$

When (4.48) is compared with (4.49), it is apparent that the particle *magnetic moment is conserved in the non-relativistic approximation.*

In the relativistic limit the particle magnetic moment (4.49) does not remain constant; however, the first adiabatic invariant can be interpreted to represent the magnetic field flux through the surface covering the particle Larmor orbit,

$$\Phi = B\,\pi r_{\rm L}^2 = \frac{\pi c^2}{e^2}\,\frac{p_\perp^2}{B} = \text{const}\,. \tag{4.50}$$

This also follows directly from (4.47), when we substitute the relativistic formula

$$p_\perp = r_{\rm L}\,\frac{eB}{c} \tag{4.51}$$

into the first term on the right-hand side of formula (4.47). We obtain

$$I = \frac{e}{c}\left(B\,\pi r_{\rm L}^2\right) = \frac{e}{c}\,\Phi\,. \tag{4.52}$$

Therefore

> in the *relativistic* case, the magnetic field flux Φ through the surface S covering the particle Larmor orbit is conserved.

(b) Magnetic mirrors and traps

Imagine the time-independent magnetic field, the field lines forming the convergent flux. As a rule, the field takes such a form in the vicinity of its sources, for instance, a sunspot S in the photosphere Ph in Figure 4.10.

The particle transversal momentum is

$$p_\perp = p\,\sin\theta\,, \tag{4.53}$$

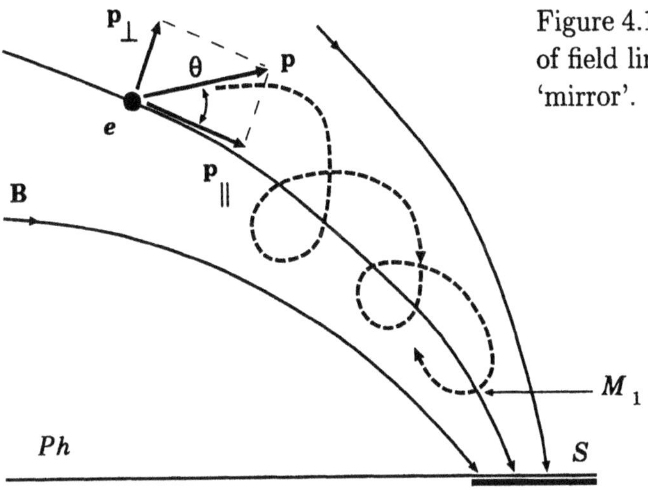

Figure 4.10: A converging flux of field lines forms a magnetic 'mirror'.

it being known that p = const, since by virtue of (4.2) we have $\mathcal{E}$ = const. Substituting (4.53) into (4.48) gives

$$\frac{\sin^2\theta}{B} = \text{const} = \frac{\sin^2\theta_0}{B_0}$$

or

$$\sin^2\theta = \frac{B}{B_0}\,\sin^2\theta_0\,. \tag{4.54}$$

This formula shows that, for the increasing B, a point M_1 must appear in which $\sin^2\theta_1 = 1$. The corresponding value of the field is equal to

$$B_1 = B_0\,/\,\sin^2\theta_0\,. \tag{4.55}$$

At this point the particle 'reflection' takes place:

$$p_{\parallel} = p\,\cos\theta_1 = 0\,.$$

The regions of convergent field lines are frequently referred to as magnetic 'mirrors' or 'corks'.

Such reflections constitute the principle of a *magnetic trap*. For example, magnetic fields create traps for fast particles in the solar atmosphere (Fig. 4.11). The particles are injected into the coronal magnetic tubes called flaring loops, during a flare. Suppose that occurs at the loop apex.

Let us also suppose that, having hit the chromosphere Ch, the particles 'die' because of collisions. The particles do not return to the coronal part of the trap, their energy being transferred to the chromospheric plasma, leading

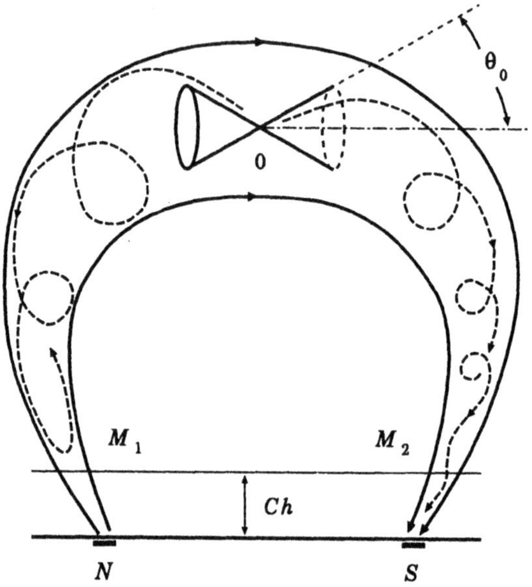

Figure 4.11: A coronal magnetic tube as a trap for particles accelerated in a solar flare. $\theta < \theta_0$ is the loss cone.

to its heating. Such particles are termed *precipitating* ones. Their pitch-angles have to be less than θ_0:

$$\theta < \theta_0 \tag{4.56}$$

with

$$\theta_0 = \arcsin \sqrt{B_0 \,/\, B_1} \tag{4.57}$$

in accordance with (4.55). Here B_0 is the magnetic field at the trap apex, B_1 is the field at the upper chromosphere level (points M_1 and M_2 in Fig. 4.11). The quantity $B_1 \,/\, B_0$ is called the *cork ratio*.

The angle region (4.56) is termed the *loss cone*. The particles with the initial momenta inside the loss cone precipitate from the trap. By contrast, the particles with $\theta > \theta_0$ at the loop apex experience reflection and do not reach the chromosphere. Such particles are termed *trapped* ones.

An interesting situation arises if the diffusion of the trapped particles into the loss cone is slower than their precipitation from the trap into the chromosphere. Then the distribution function becomes anisotropic (since the loss cone is 'eaten away') and non-equilibrium. The situation is quite analogous to the case of the distribution function formation with the positive derivative in some velocity region, like the *bump-on-tail* distribution (Figure 5.1). As a result, some *kinetic instabilities* (e.g., Silin, 1971; Schram, 1991; Shu, 1992) can be excited which lead to such plasma processes as wave excitation, anomalous particle transfer owing to the particles scattering off the waves, and *anomalous diffusion* into the loss cone (see also Chapter 5).

(c) Cyclical bounce motion, Fermi acceleration

Consider another example of a magnetic trap, namely that of a particle motion between two magnetic corks, the transversal drift being small during the period of longitudinal motion. In other words, the conditions of periodic longitudinal motion are changing adiabatically slowly. Then the *second* adiabatic invariant, referred to as the *longitudinal* one, is conserved:

$$I = \oint P_{\parallel}\, dl = p \oint \sqrt{1 - \sin^2\theta}\, dl = p \oint \sqrt{1 - \frac{B}{B_1}}\, dl\,. \tag{4.58}$$

Here account is taken of the facts that the vector **A** is perpendicular to the vector **B** and $p = |\,\mathbf{p}\,| = \text{const}$ since $\mathcal{E} = \text{const}$; the formula (4.54) for the first adiabatic invariant is used in the last equality.

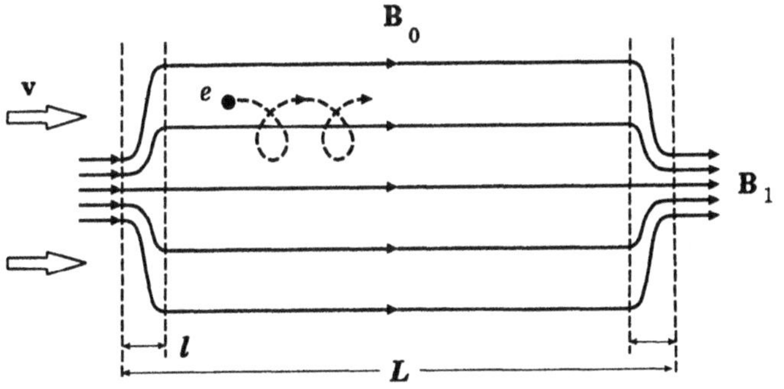

Figure 4.12: An idealized model of a long trap with a short moving cork.

Let us apply formula (4.58) to the case of a long trap with short corks: $l \ll L$ in Figure 4.12. The longitudinal invariant for such a trap is

$$I = \oint p_{\parallel}\, dl \approx 2p_{\parallel} L = \text{const}\,.$$

Therefore the second adiabatic invariant is associated with the cyclical bounce motion between two mirrors or corks and is equal to

$$\boxed{p_{\parallel} L = \text{const}\,.} \tag{4.59}$$

Suppose now that the distance between the corks is changing, that is the trap length $L = L(t)$. Then from (4.59) it follows that

$$p_{\parallel}(t) = p_{\parallel}(0)\,\frac{L(0)}{L(t)}\,. \tag{4.60}$$

It is evident from (4.60) that (a) increasing the distance between the corks decreases the longitudinal momentum and, consequently, the particle energy, and (b) particle acceleration takes place in the trap if two magnetic corks are approaching each other as is shown by vector **v** in Figure 4.12.

The former case can describe the so-called 'adiabatic cooling' of accelerated particles, for example, in a magnetic trap which is captured by the solar wind and is expanding into interplanetary space.

The latter case is more interesting. It corresponds to the Fermi mechanism (Fermi, 1949, 1954) of acceleration; for instance, by magnetic inhomogeneities in the solar wind or interplanetary medium. As a rule, the energy will increase or decrease according to whether the inhomogeneity of the magnetic field that causes the reflection moves toward the particle (head-on collision) or away from it (overtaken collision). It was shown by Fermi (1949) that,

> on the average, the energy tends to increase primarily because the head-on collisions are *more probable* than the overtaking collisions.

Through this *stochastic* mechanism the energy of the particle increases at a rate that, for relativistic particles, is proportional to their energy:

$$\frac{d\mathcal{E}}{dt} \propto \mathcal{E}\,. \tag{4.61}$$

That is why such a mechanism is called the *first-order* Fermi acceleration.

From formula (4.61) follows that the energy $\mathcal{E}$ increses exponentially with time:

$$\mathcal{E}(t) = \mathcal{E}_0 \exp\frac{t}{t_a}\,, \tag{4.62}$$

where t_a is the acceleration time scale.

Large-scale MHD turbulence is generally considered as a source of magnetic inhomogeneities accelerating particles in cosmic plasma. Acceleration of fast particles by MHD turbulence has long been recognized as a possible mechanism for solar and galactic cosmic rays (e.g., Davis, 1956).

Though the Fermi acceleration has been popular, it appears to be neither efficient nor selective. A mirror reflects particles on a nonselective basis: thermal particles may be reflected as well as suprathermal ones. Therefore one is faced with the conclusion (Eichler, 1979) that **most of the energy in the MHD turbulence goes into bulk heating** of the plasma rather than the selective acceleration of only a minority of particles. We shall come back to this question in Chapter 5.

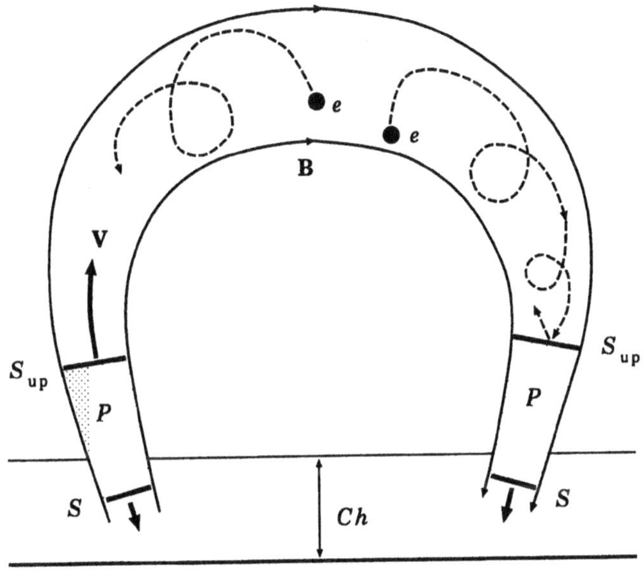

Figure 4.13: The flare-heated chromospheric plasma P rapidly expands into the corona. Particle acceleration may occur in a magnetic loop between two shock waves S_{up}.

Another example of a similar type seems to be the impulsive (with high rate of energy gain) acceleration between two approaching shocks S_{up} in a flaring loop as shown in Figure 4.13. To explain the hard X-ray and gamma-ray time profiles in solar flares, Bai *et al.* (1983) assumed that pre-accelerated electrons penetrate into the flare loop and heat the upper chromosphere to high-temperatures rapidly. As a consequence of the fast expansion of a high-temperature plasma into the corona – the process of chromospheric 'evaporation' –, two shock waves S_{up} move upward from both footpoints.

Energetic particles are to be reflected only by colliding with the shock fronts. In such a way, the **first-order Fermi acceleration** of particles between two shocks was suggested as a mechanism for the second-step acceleration of protons and electrons in flares. A similar example of the first-order Fermi-type acceleration also related to a collapsing ($L(t) \to 0$) magnetic trap in solar flares will be considered in Section 13.4.

(d) Drift in closed orbits, the flux invariant

Consider the axisymmetric trap which is modelled on, for example, the Earth's magnetic field. Three types of the particle's motion are shown in Figure 4.14.

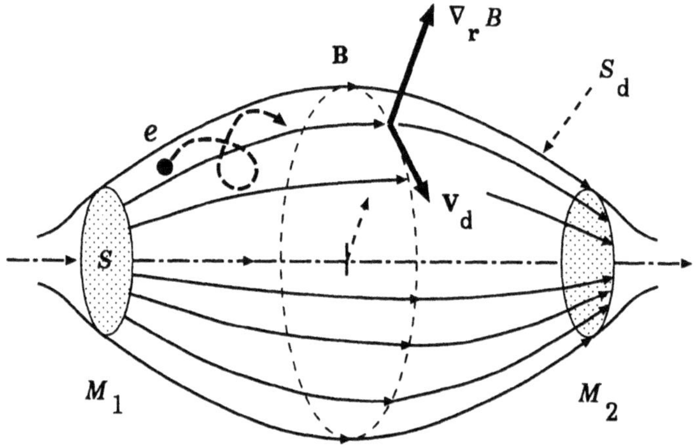

Figure 4.14: Particle drift in a trap, due to the radial gradient of field.

First, on the time scales of Larmor period, the particle spirals about a field line. Second, since there is a field-aligned gradient of the field strength, the particle oscillates between two mirrors M_1 and M_2. Third, if the guiding center does not lie on the trap's symmetry axis then **the radial gradient of field** (cf. Fig. 4.8) **causes the drift** around this axis. This drift (formula (4.39)) is superimposed on the particle's oscillatory of rotation.

As the particle bounces between the mirrors and also drifts from one field line to another one, it traces some magnetic surface S_d. The latter is called the *drift shell*. Let T_s be the period of particle motion on this surface.

If the magnetic field $\mathbf{B} = \mathbf{B}(t)$ is changing so slowly that $B \,/\, \dot{B} \gg T_s$, then a *third* adiabatic invariant, referred to as a flux one, is conserved:

$$\Phi = \int_S \mathbf{B} \cdot d\mathbf{S} = \text{const}\,. \qquad (4.63)$$

Thus the first adiabatic invariant implies conservation of the magnetic flux through the Larmor orbit, $B\,\pi r_L^2$, whereas

> the flux invariant implies conservation of the magnetic flux through the closed orbit of guiding center motion,

that is the flux through the shaded surface S in Figure 4.14.

4.3.3 Approximation accuracy. Exact solutions

Adiabatic invariants have been obtained in the approximation of weakly inhomogeneous slowly changing magnetic fields. The invariants are *approximate* integrals of motion, widely used in cosmic plasma physics. However, we should not forget two important facts. First, the adiabatic theory has a limited, though *exponential*, accuracy. Second, this theory has a limited, though wide, area of applicability. The following section presents a situation when the adiabatic theory does not apply.

Exact solutions to the equations of charged particle motion usually require numerical integration. The motion in the field of a magnetic dipole is a simple case that, nevertheless, is of practical significance. The reason for that is the possibility to approximate the Earth's magnetic field at moderately large distances by the dipole field. It was Störmer (1955) who contributed significantly to the solution of this problem.

Two types of trajectories are considered.

(a) The ones coming from infinity and returning there. These have been calculated in order to find out whether a particle can reach a given point along a given direction. An answer to this question is important for cosmic ray theories. For each point on the Earth and for each direction the so-called 'threshold rigidity' has been calculated. If a rigidity is greater than the threshold one, then the particle can reach the point. The vertical threshold rigidity is the most universally used one. This characterizes particle arrival in the direction of the smallest column depth of the Earth atmosphere.

(b) The orbits of trapped particles. Two radiation belts of the Earth, the inner and the outer, have been shown to exist. The mechanisms which generate trapped particles are not yet fully understood. They are presumably related to geomagnetic storms (see Tverskoy, 1969; Walt, 1994).

Recommended Reading: Northrop (1963), Kivelson and Russell (1995).

4.4 What is magnetic reconnection?

4.4.1 Neutral points of a magnetic field

The so-called *zeroth* or *neutral* points, lines and surfaces of magnetic field, which are the regions where magnetic field equals zero:

$$\mathbf{B} = 0\,, \tag{4.64}$$

are considered to be important for cosmic physics since Giovanelli (1946). They are of interest for the following reasons. First, **plasma behaviour is quite specific** in the vicinity of such regions (Dungey, 1958). Second, they predetermine a large number of astrophysical phenomena. We shall be primarily concerned with non-stationary phenomena in the solar atmosphere (such as flares, coronal transients, coronal mass ejections), accompanied by particle acceleration to high energies. Analogous phenomena take place on other stars, in planetary magnetospheres, and pulsars.

Neutral points most commonly appear as a result of the interaction of magnetic fluxes. The simplest way to recognize this is to consider the emerging flux in the solar atmosphere.

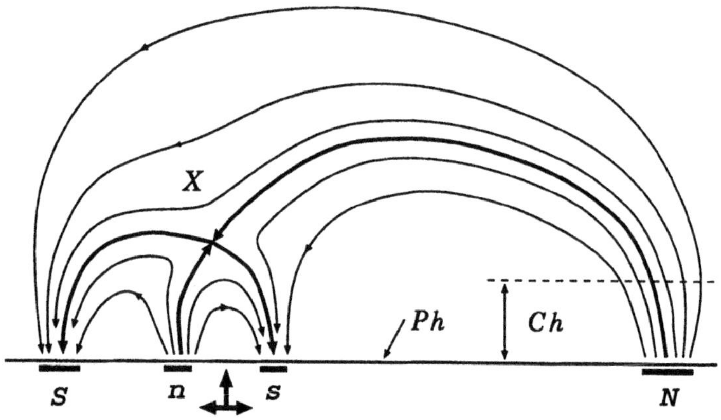

Figure 4.15: The emergence of a new magnetic flux (n, s) from under the photosphere Ph inside an active region whose magnetic field is determined by the sources S and N.

Figure 4.15 shows the sources N and S corresponding to the active region's magnetic field. The sources n and s play the role of a new flux emerging from under the photosphere Ph. The chromosphere is shown by the dashed line Ch. We consider an arrangement of the sources along a straight line, although the treatment can well be generalized (Section 16.5.2) to consider arbitrary configurations of the four sources in the photosphere.

Obviously a point can be found above the emerging flux, where oppositely directed but equal in magnitude magnetic fields 'meet'. Here the total field, that is the sum of the old and the new ones, is zero. Let us denote this point by X, bearing in mind that the field in its vicinity has the hyperbolic structure shown in Figure 4.16. In order to convince oneself that this is the case, we can consider the magnetic field in the simplest approach which is the *potential* approximation (Section 8.3.3).

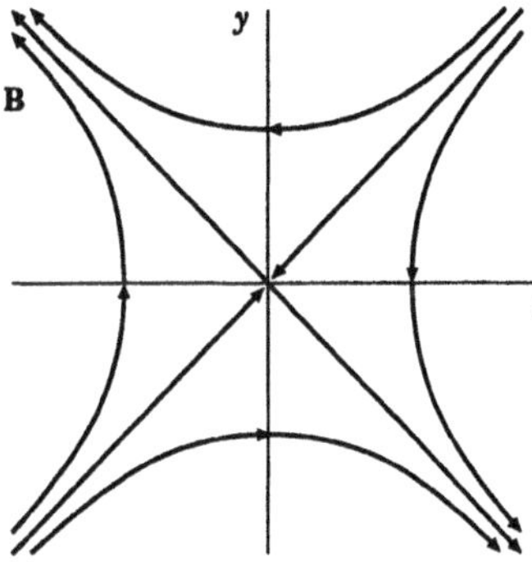

Figure 4.16: A hyperbolic zeroth point (line along the axis z) of a potential magnetic field.

4.4.2 Reconnection in vacuum

X-type points constitute a topological peculiarity of a magnetic field. They are the places where redistribution of magnetic fluxes occurs, which changes the connectivity of field lines. Let us illustrate such a process by the simplest example of two parallel electric currents of equal magnitude I in vacuum as shown in Figure 4.17. The magnetic field of these currents forms three

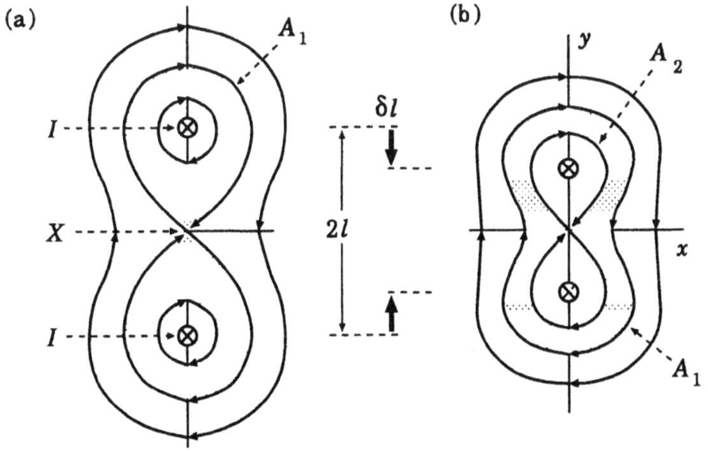

Figure 4.17: The potential field of two parallel currents: (a) the initial state, $2l$ is a distance between the currents; (b) the final state after they have been drawn nearer by a driven displacement δl.

different fluxes in the plane (x, y). Two of them belong to the upper and the lower currents, respectively, and are situated inside the *separatrix* field line A_1 which forms the figure of the eight-like curve with zeroth X-point. The third flux belongs to both currents and is situated outside the separatrix.

If the currents are displaced in the direction of each other, then the following magnetic flux redistribution will take place. The current's proper fluxes will diminish by the quantity δA, while their common flux will increase by the same quantity. So the field line A_2 will be the separatrix of the final state.

This process is realized as follows. Two field lines approach the X-point, merge there, forming a separatrix, and then they *reconnect* forming a field line which encloses both currents. Such a process is termed reconnection of field lines or *magnetic reconnection*. A_2 is the last reconnected field line.

Magnetic reconnection is of fundamental importance for the nature of many non-stationary phenomena in cosmic plasma. We shall discuss the physics of this process more fully in Chapters 16 to 22. Suffice it to note that **reconnection is inevitably associated with electric field generation.** This field is the inductive one, since

$$\mathbf{E} = -\frac{1}{c}\frac{\partial \mathbf{A}}{\partial t}, \tag{4.65}$$

where $\mathbf{A}$ is the vector potential of magnetic field,

$$\mathbf{B} = \operatorname{curl} \mathbf{A}. \tag{4.66}$$

In the above example the electric field is directed along the z axis. It is clear that, if δt is the characteristic time of the reconnection process shown in Figure 4.17, then according to (4.65)

$$E \approx \frac{1}{c}\frac{\delta A}{\delta t} \approx \frac{1}{c}\frac{A_2 - A_1}{\delta t}; \tag{4.67}$$

the last equality will be justified in Section 9.2.

Reconnection in vacuum is a real physical process: magnetic field lines move to the X-type neutral point and reconnect in it as well as

> the electric field is induced and can accelerate a charge particle or particles in the vicinity of the neutral point.

In this sense, a *collisionless* reconnection – the physical process in a high-temperature *rarefied* plasma such as the solar corona, geomagnetic tail, fusion plasmas, and so on – is simpler for understanding than reconnection in a highly-conducting collisional space plasma.

4.4.3 Reconnection in plasma

Let us try to predict plasma behaviour near the X-point as reconnection proceeds on the basis of our knowledge about the motion of a charge particle in given magnetic and electric fields.

The first obvious fact is that, given the non-zero electric field $\mathbf{E}$, the plasma begins to drift in the magnetic field $\mathbf{B}$, in a way shown in Figure 4.18a. The electric drift (cf. Fig. 4.3) velocity $\mathbf{v_d}$ is shown in four points. The magnetic field is considered as a uniform field in the vicinity of these points.

The second fact consists of the inapplicability of the adiabatic drift approximation near the zeroth point, since the Larmor radius (4.17) increases indefinitely as $B \to 0$. We have to solve the exact equations of motion. This will be done later on. However, we see at once that in this region an electric current $\mathbf{J}$ can flow along the z axis. The proper magnetic field of the current changes the initial field topology, so that there will be two symmetric zeroth points X_1 and X_2 on the x axis in Figure 4.18b instead of one X-point.

The same arguments concerning drift flows and X-point bifurcation are applicable to the new X-points. We can easily guess that the result of the interaction of line currents with the external hyperbolic field is a current sheet in the region of reconnection. The reconnecting current sheet is shown by thick solid straight line in Figure 4.18c. Note that the direction of the electric current can change at the external edges of the sheet. Here the currents can flow in the opposite direction (the *reverse* currents) with respect to the main current (the *direct* current) in the central part.

Reconnecting current sheets are *two-dimensional* and *two-scale* formations. The former means that one-dimensional models are in principle inadequate for describing the current sheets: both plasma inflow in the direction perpendicular to the sheet and plasma outflow along the sheet, along the x axis in Figure 4.19, have to be taken into account.

The existence of two scales implies that usually (for a sufficiently strong field and high conductivity like in the solar corona) the current sheet width $2b$ is much greater than its thickness $2a$. This is important since

> the wider the reconnecting current sheet, the larger the magnetic energy which can be accumulated

in the region of reconnecting fluxes interaction. On the other hand, a small thickness is responsible for the rate of accumulated energy dissipation, as well as for the possibility of non-stationary processes (for instance, tearing instability) in the current sheet.

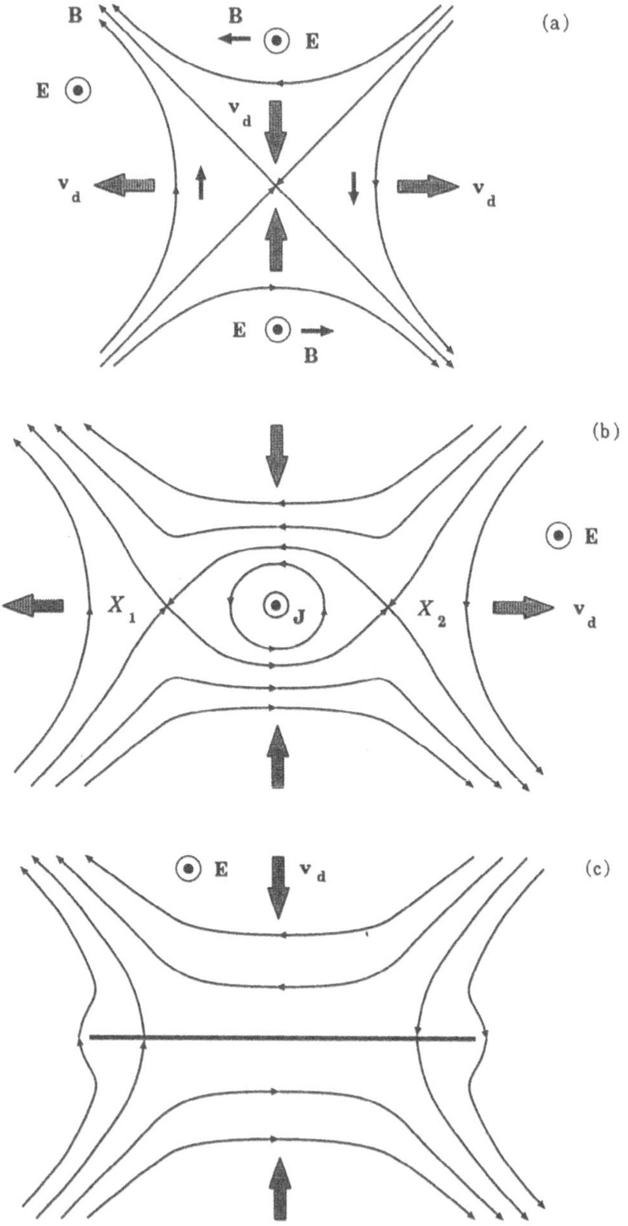

Figure 4.18: (a) Plasma flows owing to the electric drift in the vicinity of a zeroth point. (b) The appearance of secondary X-points – bifurcation of the initial zeroth line, given the current **J** flowing along it. (c) A thin current sheet.

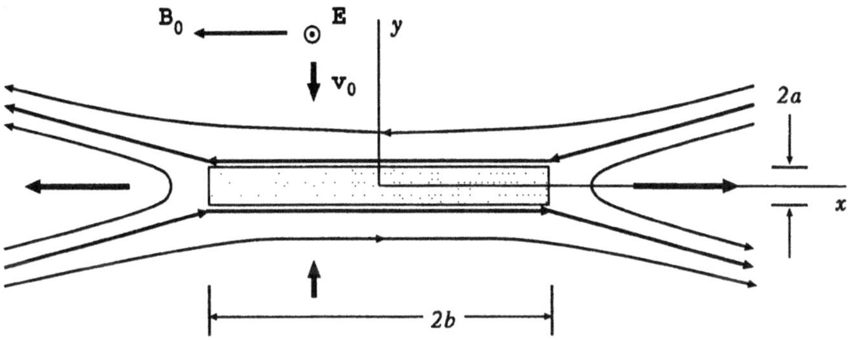

Figure 4.19: The simplest model of a current sheet – the neutral sheet.

4.4.4 Three stages in the reconnection process

Let us come back to an example considered in Section 4.4.2. Two parallel currents are displaced from the initial state (a) in Figure 4.20 to the final state (c) in plasma, which is the same as the state (b) in Figure 4.17.

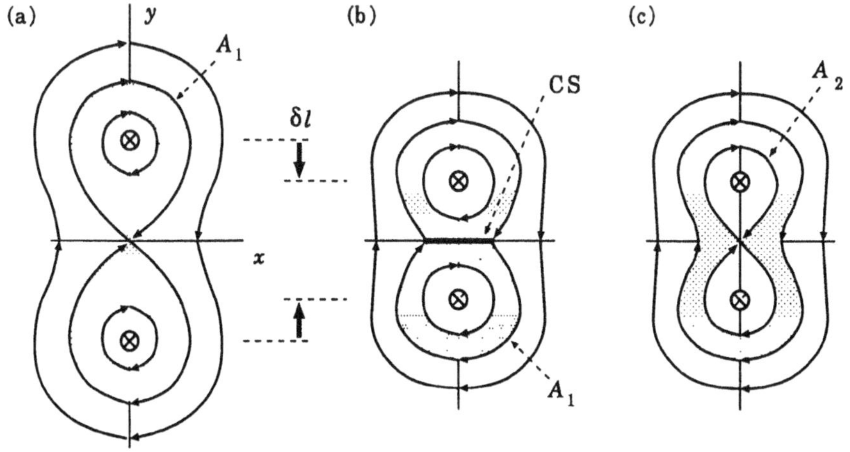

Figure 4.20: Three states of magnetic field: (a) the initial state; (b) the *pre-reconnection* state with a 'non-reconnecting' current sheet CS; (c) the final state after reconnection.

Contrary to the case of reconnection in vacuum, in cosmic plasma of *high conductivity* we have to add an intermediate state. Let us call it the *pre-reconnection state*. At this state, coming between the initial and final one, the electric currents have been displaced to the final positions, but the magnetic field lines have not started to reconnect yet, if the plasma conductivity can be considered as *infinite*. **The current sheet** along the X-type neutral line **protects the interacting fluxes from reconnection**. The energy of this interaction called the *free magnetic energy* is just the energy of the magnetic field of the current sheet.

Because of the *finite* conductivity, reconnection proceeds slow (or fast) depending on how high (or low) the conductivity is. Anyway, the final state (c) after reconnection is the same as the state (b) in Figure 4.17 with the line A_2 as the separatrix of the final state or the last reconnected line.

4.5 Acceleration in current sheets, why?

4.5.1 The origin of particle acceleration

The formation and properties of current sheets will be considered in Chapters 16 to 22 in different approximations. However, one property which is important from the standpoint of applications can be understood just now by considering the motion of a particle in given magnetic and electric fields. This property is particle acceleration.

In accordance with Figure 4.19, let the magnetic field **B** be directed along the x axis, changing the sign at $y = 0$ (the current sheet plane). That is why the $y = 0$ plane is called the *neutral* surface and the model under consideration is called the *neutral* current sheet. Certainly this simplest model is not well justified from physical point of view but mathematically convenient. Moreover, even being a strong idealization, the model allows us to understand why particles are accelerated in a reconnecting current sheet.

The electric field **E** is directed along the z axis in Figure 4.21, being constant and homogeneous. Thus

$$\mathbf{B} = \{ -hy, 0, 0 \}, \quad \mathbf{E} = \{ 0, 0, E \}, \tag{4.68}$$

where h and E are constants. We assume that the magnetic field changes its value gradually inside the current sheet with a gradient $h = | \nabla B |$.

Consider the particle motion in such crossed fields. One might argue that the motion is a sum of electric and gradient drifts (see formula (4.39) together with Figure 4.8). The electric drift makes a particle move to the neutral plane from both sides of this plane. So the electric drift creates some confinement of a particle near the neutral plane. The gradient drift drives a positively charged particle along the negative direction of the z axis in Figure 4.21, i.e. in the direction opposite to the electric field.

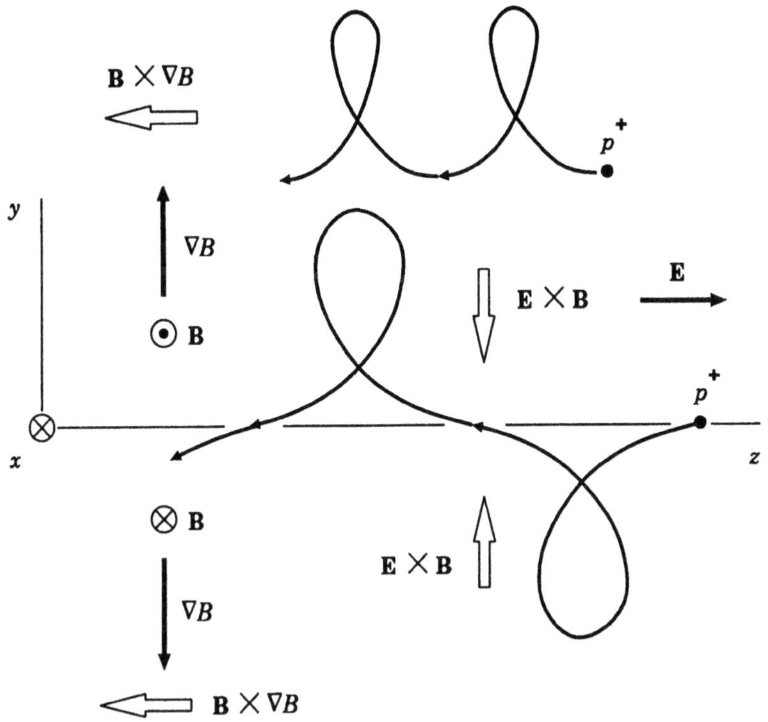

Figure 4.21: The drift motion of a particle near the neutral plane $y = 0$. The case of the slow drift of a positively charged particle is shown.

4.5.2 Acceleration in a neutral current sheet

As we have seen above, on the basis of drift consideration, one might think that the neutral current sheet is not a place for particle acceleration. However, this is not true. As the particle approaches the neutral plane, the Larmor radius $r_{\rm L} = \mathcal{R}_{\perp} / B$ increases indefinitely. The drift formalism is not applicable here. We have to solve the exact equations of particle motion. In the non-relativistic case, they are of the form:

$$\ddot{x} = 0 \, , \quad \ddot{y} = -\frac{eh}{mc} \, y \, \dot{z} \, , \quad \ddot{z} = \frac{e}{m} \left(E + \frac{h}{c} \, y \, \dot{y} \right) .$$

Rewrite these equations as follows:

$$\ddot{x} = 0 \, , \quad \ddot{y} + \frac{eh}{mc} \, \dot{z} \, y = 0 \, , \quad \ddot{z} = \frac{eE}{m} + \frac{eh}{mc} \, y \, \dot{y} \, . \tag{4.69}$$

The last equation is integrated to give

$$\dot{z} = \frac{eE}{m}\,t + \frac{eh}{2mc}\,y^2 + \text{const}\,. \tag{4.70}$$

The motion along the y axis is *finite*. This is a result of the above analysis of the character of motion in the drift approximation which applies when the particle is far enough from the neutral plane $y = 0$. That is the reason why, for large t (the ratio $y^2/t \to 0$), the first term on the right of Equation (4.70) plays a leading role. So we put asymptotically

$$\boxed{\dot{z} = \frac{eE}{m}\,t\,.} \tag{4.71}$$

As we shall see below, (4.71) is the *main* formula which desribes the effect of **acceleration by the electric field in the neutral sheet.**

After substituting (4.71) into the second equation of (4.69) we obtain

$$\ddot{y} + \frac{e^2 hE}{m^2 c}\,t\,y = 0\,.$$

Introducing the designation

$$\frac{e^2 hE}{m^2 c} = a^2\,,$$

we have

$$\boxed{\ddot{y} + \omega^2(t)\,y = 0,} \tag{4.72}$$

where $\omega^2(t) = a^2 t$.

Let us try to find the solution of Equation (4.72) in the form

$$y\,(t) = f(t)\,\cos\varphi(t)\,, \tag{4.73}$$

where $f(t)$ is a slowly changing function of the time t. Substituting (4.73) in Equation (4.72) results in

$$\ddot{f}\cos\varphi - 2\dot{f}\,\dot{\varphi}\,\sin\varphi - f\,\ddot{\varphi}\,\sin\varphi - f\,(\dot{\varphi})^2\cos\varphi + a^2 t\,f\cos\varphi = 0\,.$$

Since f is a slow function, the first term, containing the second derivative of f with respect to time, can be ignored. The remaining terms are regrouped in the following way:

$$f\left[-(\dot{\varphi})^2 + a^2 t\right] \cos\varphi - \left(2\dot{f}\,\dot{\varphi} + f\,\ddot{\varphi}\right)\sin\varphi = 0\,.$$

By the orthogonality of the functions $\sin\varphi$ and $\cos\varphi$, we have a set of two independent equations:

$$(\dot{\varphi})^2 = a^2\, t\,, \tag{4.74}$$

$$2\dot{f}\,\dot{\varphi} + f\,\ddot{\varphi} = 0\,. \tag{4.75}$$

The first equation is integrated, resulting in

$$\varphi = \frac{2}{3}\, a\, t^{3/2} + \varphi_0\,, \tag{4.76}$$

where φ_0 is a constant. Substitute this solution in Equation (4.75):

$$\frac{\dot{f}}{f} = -\frac{1}{2}\frac{\ddot{\varphi}}{\dot{\varphi}} = -\frac{1}{4}t^{-1}\,.$$

From this it follows that

$$f = C\, t^{-1/4}\,, \tag{4.77}$$

where C is a constant of integration.

On substituting (4.76) and (4.77) in (4.73), we obtain the sought-after description of the particle trajectory in a current sheet:

$$y\,(t) = C\, t^{-1/4} \cos\left(\frac{2}{3}\, a\, t^{3/2} + \varphi_0\right), \tag{4.78}$$

$$z\,(t) = \frac{eE}{m}\frac{t^2}{2} + z_0\,. \tag{4.79}$$

Eliminate the variable t between formulae (4.78) and (4.79). We have

$$y\,(z) = C\left[\frac{2m}{eE}\,(z - z_0)\right]^{-1/8} \cos\left\{\frac{2}{3}\, a\left[\frac{2m}{eE}\,(z - z_0)\right]^{3/4} + \varphi_0\right\}. \tag{4.80}$$

The amplitude of this function

$$A_y \sim z^{-1/8} \sim t^{-1/4} \tag{4.81}$$

slowly decreases as z increases.

Let us find the 'period' of the function (4.80): $\varphi \sim z^{3/4}$, hence $\delta\varphi \sim z^{-1/4}\,\delta z$. If $\delta \simeq 2\pi$, then

$$\delta z\,|_{2\pi} \sim z^{1/4}\,. \tag{4.82}$$

Thus the period of the function (4.80) is enhanced as shown in Figure 4.22.

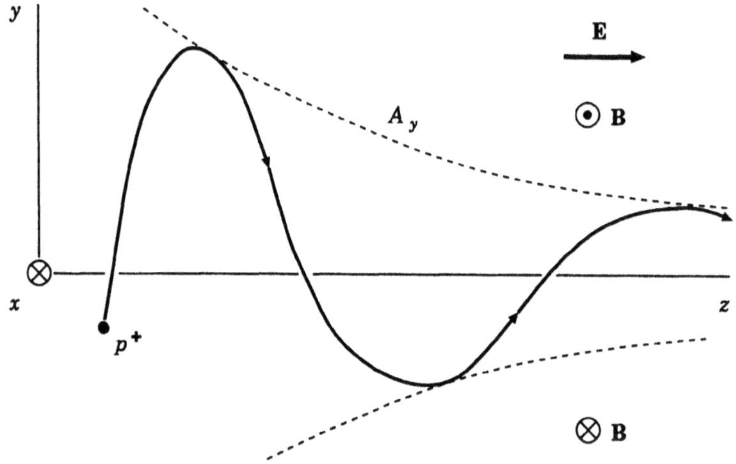

Figure 4.22: The trajectory of a particle accelerated by the electric field **E** in the neighbourhood of the neutral plane inside a neutral current sheet.

Note that the transversal velocity

$$\dot{y} \sim t^{-1/4}\,\dot{\varphi} \sim t^{1/4} \tag{4.83}$$

grows with time, but slower than the velocity component parallel to the electric field. From the main formula (4.71) it follows that

$$\dot{z} \sim t\,. \tag{4.84}$$

As a result, the particle is predominantly accelerated in the electric field direction along the current sheet.

An exact analytical solution to Equation (4.72) can be expressed as a linear combination of Bessel functions (Speiser, 1965). It has the same properties as (it asymptotically coincides with) the approximate solution. Equation (4.72) corresponds to the equation of a linear oscillator, with the spring constant becoming larger with time. In the neutral current sheet, the magnetic force returns the particle to the neutral plane: the larger the force, the higher the particle velocity.

The electric field provides particle acceleration along the reconnecting current sheet. This is the main effect.

Needless to say, the picture of acceleration in real current sheets is more complicated and interesting. In particular, acceleration efficiency depends strongly upon the small *transversal* component of the magnetic field which penetrates into the reconnecting current sheet and makes the accelerated particles be ejected from the sheet (Speiser, 1965). This effect, as well as the role of the *longitudinal* (along the z axis) component of a magnetic field inside the current sheet, will be considered in Chapters 18 and 20. Magnetical non-neutrality of the current sheet is of fundamental importance for acceleration of electrons, for example, in the solar atmosphere.

In fact, real current sheets are *non-neutral* not only in the sense of the magnetic field. They are also *electrically* non-neutral; they have an additional electric field directed towards the sheet plane from both sides. This electric field is important for ion acceleration and will be considered in Chapter 18.

4.6 Practice: Problems and Answers

Problem 4.1. Evaluate the Larmor frequency for thermal electrons and protons in the solar corona above a sunspot.

Answer. At typical temperature in the corona, $T \approx 2 \times 10^6$ K, from the non-relativistic formula (4.11), it follows that: the electron Larmor frequency

$$\omega_{\rm B}^{(e)} = 1.76 \times 10^7 \, B \, ({\rm G}) \, , \;\; {\rm rad \; s^{-1}} \, ; \tag{4.85}$$

the proton Larmor frequency

$$\omega_{\rm B}^{(p)} = 9.58 \times 10^3 \, B \, ({\rm G}) \, , \;\; {\rm rad \; s^{-1}} \, . \tag{4.86}$$

The Larmor frequency of electrons is $m_{\rm p} \, / \, m_{\rm e} \approx 1.84 \times 10^3$ times larger than that one of protons. Just above a sunspot the field strength can be as high as $B \approx 3000$ G. Here $\omega_{\rm B}^{(e)} \approx 5 \times 10^{10}$ rad s^{-1}. The emission of thermal electrons at this height in the corona can be observed at wavelength $\lambda \approx 4$ cm.

Problem 4.2. Under conditions of the corona (Problem 4.1) evaluate the *mean thermal velocity* and the Larmor radius of thermal electrons and protons.

Answer. The thermal velocity of particles with mass $m_{\rm i}$ and temperature $T_{\rm i}$ is

$$V_{\rm Ti} = \left(\frac{3 k_{\rm B} \, T_{\rm i}}{m_{\rm i}} \right)^{1/2} . \tag{4.87}$$

Respectively, for electrons and protons:

$$V_{\rm Te} = 6.74 \times 10^5 \sqrt{T_{\rm e}\,({\rm K})}\,, \quad {\rm cm\ s^{-1}}\,, \tag{4.88}$$

and

$$V_{\rm Tp} = 1.57 \times 10^4 \sqrt{T_{\rm p}\,({\rm K})}\,, \quad {\rm cm\ s^{-1}}\,. \tag{4.89}$$

At the coronal temperature $V_{\rm Te} \approx 9.5 \times 10^3$ km s^{-1} and $V_{\rm Tp} \approx 220$ km s^{-1}.

From (4.14) we find the following formulae for the Larmor radius:

$$r_{\rm L}^{(\rm e)} = \frac{V_{\rm Te}}{\omega_{\rm B}^{(\rm e)}} = 3.83 \times 10^{-2} \, \frac{\sqrt{T_{\rm e}\,({\rm K})}}{B\,({\rm G})}\,, \quad {\rm cm}\,, \tag{4.90}$$

and

$$r_{\rm L}^{(\rm p)} = \frac{V_{\rm Tp}}{\omega_{\rm B}^{(\rm p)}} = 1.64 \, \frac{\sqrt{T_{\rm p}\,({\rm K})}}{B\,({\rm G})}\,, \quad {\rm cm}\,. \tag{4.91}$$

At $T \approx 2 \times 10^6$ K and $B = 3000$ G we find $r_{\rm L}^{(\rm e)} \approx 0.2$ mm and $r_{\rm L}^{(\rm p)} \approx 1$ cm.

Problem 4.3. During solar flares electrons are accelerated to energies higher than 20–30 keV. These electrons produce the bremsstrahlung emission. The lower boundary of the spectrum of accelerated electrons is not known because the thermal X-ray emission of the high-temperature (superhot) plasma masks the lower boundary of the non-thermal X-ray spectrum. Assuming that the lower energy of accelerated electrons $\mathcal{K} \approx 30$ keV, find their velocity and the Larmor radius in the corona.

Answer. The kinetic energy of a particle

$$\mathcal{K} = \mathcal{E} - mc^2, \tag{4.92}$$

where $\mathcal{E}$ is the total energy (4.5), $mc^2 = 511$ keV for an electron. Since $\mathcal{K}/mc^2 \ll 1$, formula (4.92) can be used in the non-relativistic limit: $\mathcal{K} = mv^2/2$. From here the velocity of a 30 keV electron $v \approx 10^{10}$ cm s^{-1} $\approx 0.3\ c$.

The Larmor radius of a non-relativistic electron according to (4.14)

$$r_{\rm L}^{(\rm e)} = 5.69 \times 10^{-8} \, \frac{v_{\perp}\,({\rm cm\ s^{-1}})}{B\,({\rm G})}\,. \tag{4.93}$$

For a 30 keV electron

$$r_{\rm L}^{(\rm e)} \approx 5.6 \times 10^2 \, \frac{1}{B\,({\rm G})}\,. \tag{4.94}$$

Above a sunspot with $B \approx 3000$ G the Larmor radius $r_{\rm L}^{(e)} \approx 2$ mm. Inside a coronal magnetic trap with a field $B \approx 100$ G the electrons with kinetic energy $\mathcal{K} \approx 30$ keV have the Larmor radius $r_{\rm L}^{(e)} \approx 6$ cm.

Problem 4.4. Under conditions of the previous problem estimate the Larmor radius of a proton moving with the same velocity as a 30 keV electron.

Answer. For a non-relativistic proton it follows from formula (4.14) that

$$r_{\rm L}^{(p)} = 1.04 \times 10^{-4} \, \frac{v_{\perp}\,(\text{cm s}^{-1})}{B\,(\text{G})}\,, \quad \text{cm}\,. \tag{4.95}$$

Above a sunspot a proton with velocity $\approx 0.3\,c$ has the Larmor radius $\approx$ 3 m. Inside a coronal trap with magnetic field $\approx$ 100 G the Larmor radius $\approx 10^4$ cm. So

> non-relativistic protons (and ions) can be well trapped in coronal magnetic traps including collapsing ones

(Section 13.4). This is important for the problem of ion acceleration in solar flares.

Problem 4.5. The stronger magnetic field, the smaller is the Larmor radius $r_{\rm L}$ of an electron. Find the condition when $r_{\rm L}$ is so small as the *de Broglie* wavelength of the electron

$$\lambda_{\rm B} = \frac{h}{m_e v} = 1.22 \times 10^{-7} \, \frac{1}{\sqrt{\mathcal{K}(\text{eV})}}\,. \tag{4.96}$$

Here h is Planck's constant, $\mathcal{K}$ is the kinetic energy of the electron. If $\mathcal{K} = 1$ eV, the de Broglie wavelength $\lambda_{\rm B} \approx 10^{-7}$ cm $\approx$ 10 Angström.

Answer. In the non-relativistic limit, the electron with kinetic energy $\mathcal{K}$ has the Larmor radius

$$r_{\rm L} = 3.37 \, \frac{\sqrt{\mathcal{K}\,(\text{eV})}}{B\,(\text{G})}\,, \quad \text{cm}\,. \tag{4.97}$$

When the energy of the electron is 1 eV and the field has a strength of 1 G, the Larmor radius $r_{\rm L} \approx 3$ cm. However, for a field of 3×10^7 G, the Larmor radius is diminished to the de Broglie wavelength $\approx 10^{-7}$ cm. So, for white dwarfs which have $B > 10^7$ G, and especially for neutron stars, we have to take into account

> the *quantization* effect of the magnetic field: the Larmor radius is no longer arbitrary but can take only certain definite values.

We call a magnetic field the *superstrong* one, if $r_{\rm L} < \lambda_{\rm B}$. Substituting (4.97) and (4.96) into this condition, we rewrite it as follows

$$B > 3 \times 10^7 \, \mathcal{K} \, ({\rm eV}) \, , \ \ {\rm G} \, . \tag{4.98}$$

In superstrong fields the classic theory of particle motion, developed above, is no longer valid and certain *quantum effects* appear.

The energy difference between the levels of a non-relativistic electron in a superstrong field is

$$\delta \mathcal{E}_{B} \approx \frac{eB}{mc} \frac{h}{2\pi} \sim 10^{-8} \, B \, , \ \ {\rm eV} \, . \tag{4.99}$$

On the other hand, the difference between energy levels in an atom, for example a hydrogen atom, is of about 10 eV; this is comparable with $\delta \mathcal{E}_{B}$ in a superstrong field $B > 10^8 - 10^9$ G. In ordinary conditions B is not so large and does not affect the internal structure of atoms.

Inside and near neutron stars $B > 10^{11} - 10^{12}$ G. In such fields a lot of abnormal phenomena come into existence due to the profound influence of the external field on the interior of atoms. For example, the electron orbits around nuclei become very oblate. Two heavy atoms, e.g. iron atoms, combine into a molecule (Fe_2) and, moreover, these molecules form polymolecular substances, which are constituents of the hard surface of neutron stars (Ruderman, 1971; Rose, 1998). Exceedingly superstrong fields, $\sim 10^{15}$ G, are suggested in the magnetars, the highly magnetized neutron stars.

Chapter 5

Wave-Particle Interactions in Cosmic Plasma

The growth or damping of the waves, the emission of radiation, the scattering and acceleration of particles – all these processes result from wave-particle interaction in plasma.

5.1 The basis of kinetic theory

5.1.1 The linearized Vlasov equation

In this Chapter we shall only outline the physics and main methods used to describe the wave-particle interaction in collisionless cosmic plasmas as well as in Maxwellian plasmas where fast particles interact with electromagnetic waves. In the simplest – *linear* – approach, the idea is in the following. We assume the unperturbed plasma to be uniform and characterized by the distribution functions $f_k^{(0)}$ of its components k: electrons and ions. Let $\mathbf{B}^{(0)}$ be the unperturbed uniform magnetic field inside the plasma. We further assume that the only zero-order force is the Lorentz force with $\mathbf{E}^{(0)} = 0$.

The dynamics of individual particles is determined by the first-order forces related to the wave electric field $\mathbf{E}^{(1)}$ and wave magnetic field $\mathbf{B}^{(1)}$. To describe these particles we shall use the perturbation function $f_k^{(1)}$, which is linear in $\mathbf{E}^{(1)}$ and $\mathbf{B}^{(1)}$. Under the assumptions made, we see that the Vlasov equation (2.47) can be a proper basis for the kinetic theory of wave-particle interaction. For this reason we shall realize the following procedure.

(a) We linearize the Vlasov equation together with the Maxwell equations for the self-consistent wave field. Equation (2.47) becomes

$$\begin{aligned} &\frac{\partial f_k^{(1)}(X,t)}{\partial t} + v_\alpha \frac{\partial f_k^{(1)}(X,t)}{\partial r_\alpha} + \\ &+ \frac{e_k}{m_k} \left(\frac{1}{c}\, \mathbf{v} \times \mathbf{B}^{(0)} \right)_\alpha \frac{\partial f_k^{(1)}(X,t)}{\partial v_\alpha} = \\ &- \frac{e_k}{m_k} \left(\mathbf{E}^{(1)} + \frac{1}{c}\, \mathbf{v} \times \mathbf{B}^{(1)} \right)_\alpha \frac{\partial f_k^{(0)}(X,t)}{\partial v_\alpha} \,. \end{aligned} \tag{5.1}$$

The left-hand side of the linear Equation (5.1) is the Liouville operator (1.28) acting on the first-order distribution function for particles following ***unperturbed*** trajectories in phase space $X = \{\mathbf{r}, \mathbf{v}\}$.

This fact (together with the linear Lorentz force in the right-hand side of (5.1) and the linearized Maxwell equations) can be used to find the general solution of the problem. We are not going to do this here (see Problem 5.1). Instead, we shall make several simplifying assumptions to demonstrate the most important features of kinetic theory on the basis of Equation (5.1).

(b) Let us consider a harmonic perturbation varying as

$$f_k^{(1)}(t, \mathbf{r}, \mathbf{v}) = \tilde{f}_k(\mathbf{v}) \exp\left[-\mathrm{i}\,(\omega t - \mathbf{k}\cdot\mathbf{r})\right]. \tag{5.2}$$

Substituting (5.2) with a similar presentation of the perturbed electromagnetic field in Equation (5.1) gives the following linear equation:

$$\begin{aligned} \mathrm{i}\,(\omega - \mathbf{k}\cdot\mathbf{v})\, \tilde{f}_k - \frac{e_k}{m_k} \left(\frac{1}{c}\, \mathbf{v} \times \mathbf{B}^{(0)} \right)_\alpha \frac{\partial \tilde{f}_k}{\partial v_\alpha} = \\ = \frac{e_k}{m_k} \left[\tilde{\mathbf{E}} \left(1 - \frac{\mathbf{k}\cdot\mathbf{v}}{\omega} \right) + \mathbf{k} \left(\frac{\mathbf{v}\cdot\tilde{\mathbf{E}}}{\omega} \right) \right]_\alpha \cdot \frac{\partial f_k^{(0)}}{\partial v_\alpha} \,. \end{aligned} \tag{5.3}$$

Here the Faraday law (1.2) has been used to substitute for the wave magnetic field.

(c) We shall assume that the waves propagate parallel to the ambient field $\mathbf{B}^{(0)}$ which defines the z axis of a Cartesian system. From Section 4.1 it follows that in a uniform magnetic field there exist two constants of a particle's motion: the parallel velocity $\mathbf{v}_\parallel$ and the magnitude of the perpendicular velocity

$$v_\perp = |\,\mathbf{v}_\perp| = \left(v_x^2 + v_y^2 \right)^{1/2}.$$

Hence the unperturbed distribution function

$$f_k^{(0)} = f_k^{(0)}\left(v_{\parallel}, v_{\perp}\right), \tag{5.4}$$

as required by Jeans's theorem (see Problem 1.4). Therefore in what follows we can consider two cases of resonance, corresponding two variables in (5.4).

5.1.2 The Landau resonance and Landau damping

Let us consider *electrostatic waves* which have only a parallel electric field under the assumption of parallel propagation. In this case the linear Equation (5.3) reduces to

$$\mathrm{i}\left(\omega - k_{\parallel} v_{\parallel}\right) \tilde{f}_k - \frac{e_k}{m_k}\left(\frac{1}{c}\,\mathbf{v}\times\mathbf{B}^{(0)}\right)_{\alpha} \frac{\partial \tilde{f}_k}{\partial v_{\alpha}} = \frac{e_k}{m_k}\,\tilde{E}_{\parallel}\,\frac{\partial f_k^{(0)}}{\partial v_{\alpha}}\,. \tag{5.5}$$

Now let us find the charge density according to definition (2.49). The second term on the left-hand side of Equation (5.5) vanishes on integration over perpendicular velocity.

Therefore, for parallel propagating electrostatic waves, the harmonic perturbation of charge density is given by

$$\tilde{\rho}^{q} = -\,\mathrm{i}\,\tilde{\mathbf{E}}_{\parallel} \sum_k \frac{e_k^2}{m_k} \int \frac{\partial f_k^{(0)} / \partial v_{\parallel}}{\omega - k_{\parallel} v_{\parallel}}\, d v_{\parallel}\,. \tag{5.6}$$

Formula (5.6) shows that there is a *resonance* which occurs when

$$\boxed{\omega - k_{\parallel} v_{\parallel} = 0} \tag{5.7}$$

or when the particle velocity equals the parallel phase velocity of the wave, $\omega / k_{\parallel}$. This is the *Landau resonance*.

A physical picture of Landau resonance is simple.

> When condition (5.7) is satisfied the particle 'sees' the electric field of the wave as a static electric field in the particle's rest system

(see Problem 5.3).

Particles in resonance moving slightly faster than the wave will lose energy, while those moving slightly slower will gain energy. Since the Maxwellian distribution is decreasing with velocity,

> in a Maxwellian plasma, near the Landau resonance, there are more particles at lower velocities than at higher velocities. That is why **the plasma gains energy at the expense of the wave**.

This effect (see an illustration in Problem 5.6) is called the *Landau damping* (Landau, 1946) or collisionless damping. Landau damping is often the dominant **damping mechanism for waves**, such as ion-acoustic waves and Langmuir waves, in thermal plasma without a magnetic field. The absorption of longitudinal waves in cosmic plasma in the thermal equilibrium is often determined by **collisionless damping** (e.g., Zheleznyakov, 1996).

On the other hand, if a distribution function has more particles at higher velocities than at lower velocities in some region of phase space as shown in Figure 5.1, this distribution will be unstable to waves that are in resonance

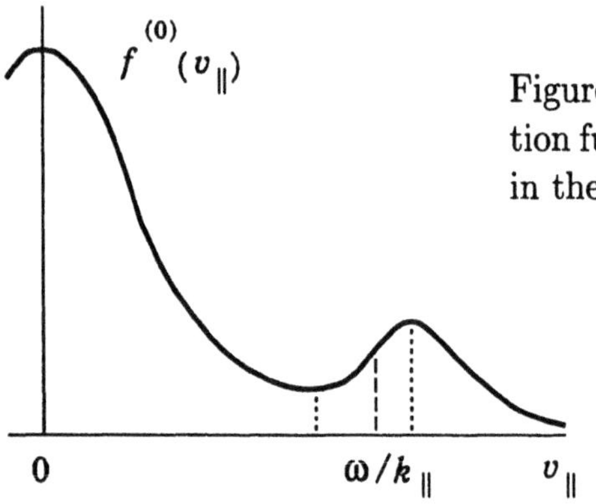

Figure 5.1: The 'bump-on-tail' distribution function with the resonance condition in the region of a positive slope.

with the particles. This is the known 'bump-on-tail' instability. Due to this type of instability, a beam of fast electrons (with velocities much higher than the thermal speed of electrons in the plasma) causes Langmuir waves to grow. Langmuir waves generated through the bump-on-tail instability play an essential role, for example, in solar radio bursts.

If we consider a stream of plasma with an average velocity impinding on a plasma at rest, we have just the same situation. The system has an instability such that the kinetic energy of the relative motion between the plasma steams is fed into a plasma wave of the appropriate phase velocity. So all the two-stream instabilities have, in fact, the same origin.

The above derivation emphasizes the close relation of the Landau damping with the Cherenkov effect (Problems 5.2–5). It has been definitely pointed out by Ginzburg and Zheleznyakov (1958) that

> the Landau damping and the Cherenkov absorption of plasma waves, the inverse Cherenkov effect, are the same phenomenon

initially described in two different ways.

The discussion hitherto has focused on the linear Landau damping, i.e. the behaviour of a small perturbation which satisfies the linearized Vlasov equation. However, this picture can be extended to finite amplitude perturbations. In the context of cosmic plasma physics, this means considering *nonlinear* Landau damping, which generalized the linear theory by incorporating the possibility of mode-mode couplings that allow energy transfer between different modes.

In fact, the linear theory illuminates only a narrow window out of the wealth of all effects related to wave-particle interactions. Mathematically, **linear theory uses a well-developed algorithm**. Few analytical methods are known to treat the much wider field of nonlinear effects, and most of these methods rely on approximations and lowest-order perturbation theory. The theory of *weak* wave-particle interaction or *weak turbulence* as well as the *quasi-linear* theory for different types of waves are still today the most important parts in cosmic physics applications (Benz, 1993; Treumann and Baumjohann, 1997).

5.1.3 Gyroresonance

As for the Landau resonance, we shall use the linear Equation (5.3) as a basis, assuming that a wave is propagating parallel to the ambient field $\mathbf{B}^{(0)}$. However, this time, we shall further assume that the wave electric field $\mathbf{E}^{(1)}$ and hence the wave magnetic field $\mathbf{B}^{(1)}$ are perpendicular to the ambient magnetic field.

Under the assumption of a harmonic perturbation (5.2) we shall make use of the so-called *polarized coordinates*:

$$\tilde{E}_l = \frac{\tilde{E}_x + \mathrm{i}\tilde{E}_y}{\sqrt{2}}\,, \qquad \tilde{E}_r = \frac{\tilde{E}_x - \mathrm{i}\tilde{E}_y}{\sqrt{2}}\,. \tag{5.8}$$

Subscripts l and r correspond to the waves with left- and right-hand circular polarizations, respectively.

If we multiply Equation (5.3) by velocity $\mathbf{v}_\perp$ and integrate over velocity space, we find the equation which determines (see definition (2.50)) the current density in the harmonic perturbation:

$$\tilde{\mathrm{j}}_l^{\mathrm{q}} = -\mathrm{i} \sum_{\mathrm{k}} \frac{e_{\mathrm{k}}^2}{m_{\mathrm{k}}} \tilde{E}_r \times \tag{5.9}$$

$$\times \int \left[\left(1 - \frac{k_\parallel v_\parallel}{\omega} \right) \frac{\partial f_k^{(0)}}{\partial v_\perp} + \frac{k_\parallel v_\perp}{\omega} \frac{\partial f_k^{(0)}}{\partial v_\parallel} \right] \frac{1}{\left(\omega - k_\parallel v_\parallel - s\, \omega_{\mathrm{B}}^{(\mathrm{k})} \right)}\, d^3\mathbf{v} \, .$$

Here $\omega_{\mathrm{B}}^{(\mathrm{k})}$ is the Larmor frequency of a particle of a kind k, the integer s can be positive or negative. The resonance condition in formula (5.9) for current density is the *gyroresonance*:

$$\boxed{\omega - k_\parallel v_\parallel - s\, \omega_{\mathrm{B}}^{(\mathrm{k})} = 0 \, .} \tag{5.10}$$

We see that a gyroresonant interaction occurs when the Doppler-shifted wave frequency $(\omega - k_\parallel v_\parallel)$ is an integer multiple of the Larmor frequency in the guiding center frame, i.e. $s\, \omega_{\mathrm{B}}^{(\mathrm{k})}$.

Depending upon the initial relative phase of the wave and particle, the particle will corotate with either an accelerating or decelerating electric field over a significant portion of its Larmor motion,

resulting in an appreciable gain or loss of energy, respectively.

If the particle and transversal electric field rotate in the same sense, the integer $s > 0$, whereas an opposite sense of rotation requires $s < 0$. However the strongest interaction usually occurs when the Doppler-shifted frequency exactly matches the particle Larmor frequency.

The gyroresonance is important for generating waves such as the *wistler* mode, which is polarized predominantly perpendicular to the ambient field.

For a wave to grow from gyroresonance, there should be a net decrease in particle energy as the particle diffuses down the phase-space density gradient defined by the numerator in (5.9)

i.e. by the expression enclosed in large square brackets under the integral in formula (5.9).

For the parallel propagation of a wave in plasma, the Landau resonance is associated with parallel electric fields. For perpendicular electric fields, particles and fields can be in gyroresonance. It is clear that the Landau resonance diffuses particles parallel to the ambient magnetic field, whereas **gyroresonance mainly causes diffusion in the pitch angle**.

As such, then, Landau-resonant instabilities are often driven by bump-on-tail distributions of particles, whereas gyroresonant instabilities are driven by *pitch-angle* anisotropy. Thus the gyroresonance-type instabilities can appear as soon as a 'tail' or beam is formed in the direction parallel to the background field $\mathbf{B}^{(0)}$. They excite waves that scatter the particles back to a nearly isotropic state.

5.2 Stochastic acceleration of particles by waves

5.2.1 The principles of particle acceleration by waves

In Section 5.1 we considered the resonant interaction between particles and one wave propagating parallel to the uniform magnetic field $\mathbf{B}^{(0)}$ in a uniform plasma without an external electric field: $\mathbf{E}^{(0)} = 0$. The dynamics of individual particles was determined by the first-order forces related to the wave electric field $\mathbf{E}^{(1)}$ and wave magnetic field $\mathbf{B}^{(1)}$. We described behavior of these particles by the linearized Vlasov equation (5.1) for the perturbation function $f_k^{(1)}$, which is linear in $\mathbf{E}^{(1)}$ and $\mathbf{B}^{(1)}$.

Under simplifing assumptions made, we saw that, in addition to the Landau resonance (5.7), other resonances (5.10) arise in wave-particle interaction. These are the gyroresonances which occur when the Doppler-shifted frequency, $\omega - k_{\|} v_{\|}$ (as observed by a particle moving with parallel velocity $v_{\|}$), is some integer multiple of the particle Larmor frequency $s\,\omega_{\mathrm{B}}^{(k)}$.

If a wave is, in general, oblique, its electric field has components transversal and parallel to $\mathbf{B}^{(0)}$, whereas if the wave is parallel, its electric field is transversal. Since the transversal field typically consists of left- and right-hand polarized components, the integer s may be either positive or negative. Anyway,

> the energy gain is severely limited due to the particle losing resonance with the wave.

Large gains of energy are possible, in principle, if a spectrum of waves is present. In this case, the resonant interaction of a particle with one wave can result in an energy change that brings this particle into resonance with a neighboring wave, which then changes the energy so as to allow the particle to resonate with another wave, and so on. Such an energy change can be diffusive, but over long time scales there is a net gain of energy, resulting in *stochastic* acceleration.

A traditional problem of the process under discussion is the so-called *injection* energy. The problem arises since for many waves in plasma their phase velocity along the ambient magnetic field, $\omega / k_{\|}$, is much greater than the mean thermal velocity of particles. Let us re-write the gyroresonance condition (5.10) as

$$\gamma_{\rm L}\left(\frac{\omega}{k_{\|}} - v_{\|}\right) = \frac{s\,\omega_{\rm B}^{(k)}}{k_{\|}}\,. \tag{5.11}$$

Here the relativistic Lorentz factor $\gamma_{\rm L}$ has been taken into account (Problem 5.3). Consider two opposite cases.

(a) For low thermal velocities we can neglect $v_{\|}$ in Equation (5.11) and see that, in order to resonate with a thermal particle, the waves must have very high frequencies $\omega \approx \omega_{\rm B}^{(k)}$ or very small $k_{\|}$.

For the case of thermal electrons and protons in the solar corona, their Larmor frequencies are very high (Problem 4.1). If we try to choose a minimal value of $k_{\|}$, we are strongly restricted by a maximal value of wavelenghts, which must be certainly smaller than the maximal size of an acceleration region. These difficulties naturally lead to much doubt about the viability of stochastic acceleration and to a search for *preacceleration* mechanisms.

(b) On the other hand, high energy particles need, according to the resonance condition (5.11), waves with very low frequencies: $\omega \ll \omega_{\rm B}^{(k)}$. Therefore

> a very *broad-band* spectrum of waves (extending from $\approx \omega_{\rm B}^{(k)}$ to very low frequencies) is necessary to accelerate particles from thermal to relativistic energies.

In principle, the so-called *wave cascading* from low to high frequencies can be a way of producing the necessary broad-band spectrum. The idea comes from the Kolmogorov theory of hydrodynamic turbulence (Kolmogorov, 1941). Here the evolution of turbulence can be described by the Kolmogorov-style dimensional analysis or by a **diffusion of energy in wavenumber space**. The idea was subsequently introduced to MHD by Zhou and Matthaeus (1990). They presented a general transfer equation for the wave spectral density. In Sections 5.2.2 and 5.2.3, we shall discuss briefly both approaches and their applications; see also Goldreich and Sridhar (1997).

Note that the stochastic acceleration of particles by waves is essentially the resonant form of Fermi acceleration (see Section 4.3.2 (c)). An important feature of stochastic acceleration is an isotropization process because

> the pitch-angle scattering increases the volume of wave phase space that can be sampled by the resonant particles (5.11).

So, if isotropization exists and keeps the distribution isotropic during an acceleration time, it increases the acceleration efficiency.

5.2.2 MHD turbulent cascading

In order not to obscure the essential physical point made in this section, we assume that a turbulence is isotropic and homogeneous. So we define a one-dimensional spectral density $W(k)$, which is the wave energy density per unit volume in the wave vector space $\mathbf{k}$.

First, we remind the Kolmogorov (1941) treatment of stationary fluid turbulence. The steady state assumption implies that the energy flux F through a sphere of radius k is independent of time. In the *inertial* range of wave numbers, for which supply and dissipation of energy are neglected, the flux F is also independent of the wave vector k. If $\mathcal{P}$ denotes the total rate of energy dissipation at the short wave ($k = k_{\rm max}$) edge of the inertial range, which equals the rate of energy supply at the long wave ($k = k_{\rm min}$) edge, then $F = \mathcal{P}$ and $dF/dk = 0$ in the inertial range in Figure 5.2.

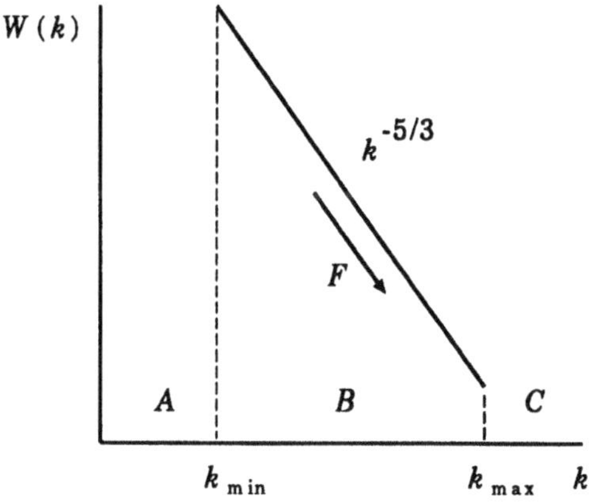

Figure 5.2: The energy per unit wave number in Kolmogorov's turbulence is plotted as a function of wavenumber in the inertial range B between the source A at small k and the sink C at large k.

Kolmogorov's theory adopts the hypothesis that with the above assumptions the flux F through a sphere of radius k in the *inertial* range depends only upon the energy in that sphere and upon the wave number. Thus by dimensional analysis we arrive at

$$F = \mathcal{P} \sim W^{3/2}\, k^{5/2}. \tag{5.12}$$

From here it follows that the one-dimensional spectral density

$$\boxed{W(k) = C_k\, \mathcal{P}^{\,2/3}\, k^{-5/3}.} \tag{5.13}$$

This is the famous *Kolmogorov spectrum* for the fluid isotropic turbulence, involving the Kolmogorov constant C_k.

Second, the Kolmogorov concept of independence of widely separated wave numbers in the inertial range of fluid turbulence was modified for the MHD case by Iroshnikov (1963) and Kraichnan (1965). When the magnetic energy in subinertial wave numbers exceeds the total energy in the inertial range, the predicted inertial range spectrum is proportional to $k^{-3/2}$, instead of $k^{-5/3}$. Note that the Kolmogorov spectrum is steeper than the Kraichnan spectrum ($5/3 > 3/2$).

Leith (1967) introduced a diffusion approximation for spectral transfer of energy in isotropic hydrodynamic turbulence. This approach may be viewed as an alternative to the straight-forward dimensional analysis discussed above. However, it is a very natural extension since this approach approximates the spectral transfer as a local process in wave number space, i.e. in accordance with the spirit of the Kolmogorov hypotheses that the total energy is conserved with respect to couplings between waves. Therefore

> just diffusion is a physically appealing framework for the simplest model to describe this kind of *local conservative* transfer.

So, if we assume that some waves, propagating to the uniform field $\mathbf{B}^{(0)}$, are injected at the longest wavelength $\lambda = \lambda_{\rm max}$ and that a Kolmogorov-like nonlinear cascade transfers the wave energy to smaller scales, then the diffusion equation in wave number space

$$\frac{\partial W}{\partial t} = \frac{\partial}{\partial k_{\|}} \left(D_{\|\|} \frac{\partial W}{\partial k_{\|}} \right) - \gamma \left(k_{\|} \right) W + S \tag{5.14}$$

can describe injection, cascading, and damping of the waves. Here $D_{\|\|}$ is a diffusion coefficient that depends on W and, in principle, can be determined for Kolmogorov-type cascading. $\gamma \left(k_{\|} \right)$ is the damping rate usually due to particle acceleration in high-temperature low-density cosmic plasma. The wave energy is dissipated by accelerating particles in smallest scales $\lambda \sim \lambda_{\rm min}$.

The source term S in Equation (5.14) is proportional to the injection rate Q of the wave energy. A mechanism by which the waves are generated is typically unknown but easily postulated. For example, MHD waves can be formed by a large-scale restructuring of the magnetic field in cosmic plasma, which presumably occurs in nonstationary phenomena with flare-like energy releases.

In summary, wave cascading and particle acceleration are described by one wave diffusion equation, in which the damping depends on the accelerating particle spectra, and by diffusion equations (one for each kind k of particles:

electrons, protons and other ions) for accelerating particles. The system is therefore highly coupled and generally nonlinear or *quasilinear* in the case of small-amplitude waves.

5.2.3 Stochastic acceleration of electrons

LaRosa *et al.* (1996), Miller *et al.* (1996) presented a model for the acceleration of electrons from thermal to relativistic energies in solar flares. They assume that Alfvénic outflows from the sites of reconnection generate a cascading MHD turbulence. The ratio of the gas pressure to the magnetic one is presumably small in this cascade. Thus the MHD turbulence has a small parameter β (our parameter γ^2) and mainly comprises of two low-amplitude wave modes: (a) Alfvén waves and (b) *fast* magnetoacoustic waves (Section 10.2.3). The authors do not consider a possible role of slow magnetoacoustic waves in the acceleration of protons.

LaRosa *et al.* assume that in the reconnection-driven turbulence there is an equipartition between these modes. About half of the energy of the turbulence resides in Alfvén waves and about half in fast magnetoacoustic waves (FMW). The threshold speed of the resonance determines the selectivity of the wave-particle interaction. Assuming $B^{(0)} \approx 500$ G, $T^{(0)} \approx 3 \times 10^6$ K, and $n^{(0)} \approx 10^{10}$ cm^{-3}, they found that the Alfvén speed $V_A \approx 0.036\, c$, the electron thermal speed $V_{Te} \approx 0.032\, c$, and the proton thermal speed $V_{Te} \approx 7.4 \times 10^{-4}\, c$. Therefore the threshold speed is far in the tail of the proton distribution, and a negligible number of protons could be accelerated by FMW or Alfvén waves. Consequently protons or other ions are a negligible dissipation source for these waves, but not for slow magnetoacoustic waves (SMW) ignored by LaRosa *et al.*

On the other hand, V_A is only slightly above V_{Te}, and a significant number of the ambient electrons can resonate with, and damp, the waves. Thus FMW almost exclusively accelerate electrons under the solar flare conditions accepted above. The process under consideration could be called *small-amplitude* Fermi acceleration or 'transit-time damping' (Miller *et al.*, 1996) to denote the *resonant character* of the wave-particle interaction. This is basically a resonant Fermi acceleration of second order.

> If we can ignore the gyroresonant part of the interaction, then only the parallel energy would systematically increase,

leading to a velocity-space anisotropy in the electron distribution function.

So, beyond the question of the origin of turbulence, the most significant question challenging electron energization by the Fermi process is pitch-angle scattering. In the absence of ancillary scattering, acceleration by FMW would lead to a systematic decrease of particle pitch-angles. Acceleration would then become less efficient, since only those waves with very high parallel phase speed would be able to resonate with the particles. However, as a tail is formed in the parallel direction, there would appear one or another instability which excites waves (for example, the fire-hose instability; see for a review Paesold and Benz, 1999) that can scatter the electrons back to a nearly isotropic state.

With the introduction of **isotropizing scattering of any origin**, we can avarage the momentum diffusion equation in spherical coordinates over the pitch-angle μ and obtain the isotropic momentum diffusion equation

$$\frac{\partial f}{\partial t} = \frac{1}{p^2} \frac{\partial}{\partial p} \left(p^2 D(p) \frac{\partial f}{\partial p} \right) . \tag{5.15}$$

Here

$$D(p) = \frac{1}{2} \int_{-1}^{+1} D_{pp} \, d\mu \, , \tag{5.16}$$

p is the magnitude of the momentum vector $\mathbf{p}$, and D_{pp} is the μ-dependent momentum diffusion coefficient (see Miller *et al.*, 1996). The quantity f is the phase-space distribution function, normalized such that $f(p,t)\, 4\pi p^2 dp$ equals the number of particles per unit volume with momentum in the interval dp about p.

Electron acceleration and wave evolution are thus described by the two coupled partial differential Equations (5.14) and (5.15). Their solution allows to evaluate the bulk energization of electrons by Fermi acceleration from the MHD turbulence expected in solar flares. LaRosa *et al.* (1996) has found that the Fermi acceleration acts fast enough to be the damping mechanism for the FMW turbulence. This means that Fermi acceleration becomes fast enough at short enough scales $\lambda \sim \lambda_{\text{min}}$ in the turbulent cascade of fast magnetoacoustic waves *to end* the cascade by dissipating the cascading turbulent energy into random-velocity kinetic energy of electrons. Practically all of the energy of the FMW turbulence is absorbed by the electrons while the protons get practically none.

5.2.4 Acceleration of protons and heavy ions

One of two 'sisters' of the Alfvén wave – the *fast* magnetoacoustic wave (FMW) – was discussed in the previous Section as a part of the turbulent cascade. The second 'sister' – the *slow* magnetoacoustic wave (SMW) – was forgotten forever from the early beginning. As we saw above, fast magnetoacoustic waves cascade to higher frequencies, eventually *Landau* resonate with and accelerate the thermal electrons by the small amplitude Fermi-type or 'transit-time' mechanism (LaRosa *et al.*, 1996; Miller *et al.*, 1996).

In this Section we shall discuss briefly the acceleration of protons and heavy ions by Alfvén waves that are a part of the same MHD turbulent cascade but *cyclotron* resonate with particles. More exactly, as the waves increase in frequency, they resonate with ions of progressively lower energies (Miller and Reames, 1996).

Let us consider for simplicity only the Alfvén waves with phase velocities parallel and antiparallel to the background field $\mathbf{B}^{(0)}$. These waves have left-hand circular polarization relative to $\mathbf{B}^{(0)}$ and occupy the frequency range below the cyclotron frequency (4.10) of Hydrogen (i.e., protons):

$$\omega < \omega_{\mathrm{B}}^{(\mathrm{H})} = \frac{ecB}{\mathcal{E}_{\mathrm{H}}} \, . \tag{5.17}$$

For simplicity we also take the low-frequency limit for the dispersion relation of the Alfvén waves under consideration:

$$\omega = V_{\mathrm{A}} \mid k_{\parallel} \mid . \tag{5.18}$$

In a multi-ion cosmic plasma, there are resonances and cutoffs in the dispersion relation corresponding to each kind i of ions. However, because of their small abundance, Fe and the Ne group do not affect the dispersion relation. The He group will produce a resonance at $\omega_{\mathrm{B}}^{(\mathrm{He})}$ and a cutoff at a slightly higher frequency. We shall take, however, the Alfvén wave dispersion relation (5.18) for all $\omega < \omega_{\mathrm{B}}^{(\mathrm{He})}$.

In general, a low-frequency Alfvén wave propagating obliquely with respect to the ambient field $\mathbf{B}^{(0)}$ has a linearly polarized magnetic field $\mathbf{B}^{(1)}$ normal to both $\mathbf{B}^{(0)}$ and $\mathbf{k}$ (see Fig. 10.1). The wave electric field $\mathbf{E}^{(1)}$ is normal to $\mathbf{B}^{(0)}$ and $\mathbf{B}^{(1)}$. A low-frequency FMW (Section 10.2.3) has a linearly polarized electric field $\mathbf{E}^{(1)}$ normal to both $\mathbf{B}^{(0)}$ and $\mathbf{k}$. In each case the electric field can be decomposed into left- and right-handed components. However, for parallel propagation, all Alfvén waves are left-handed, while all fast magnetoacoustic waves are right-handed.

Since we consider the Alfvén waves which phase velocities are strictly parallel and antiparallel to the background field, there is only one resonant wave and it is the backward-moving Alfvén wave (see Miller and Reames, 1996).

Applying the resonance condition (5.11) for this wave with $s = 1$, we find its wavenumber

$$k_{\parallel} = -\frac{\omega_{_{\rm B}}^{(i)}}{\gamma_{_{\rm L}}\left(V_{_{\rm A}} + v_{\parallel}\right)}\,. \tag{5.19}$$

Hence

> when the Alfvén wave frequency becomes close to the ion cyclotron frequency $\omega_{_{\rm B}}^{(i)}$, thermal ions of the kind i will be accelerated out of the background energies.

The first kind of ions encountered by the Alfvén waves will be the one with the lowest cyclotron frequency, namely Fe. (This is well visualized by Fig. 1 in Miller and Reames, 1996.) However, due to the low Fe abundance, the waves will not be completely damped and will continue to cascade up the group of ions with the next higher cyclotron frequency, namely Ne, Mg, and Si. These ions will be also accelerated but the waves will not be totally damped again. They encounter ^{4}He, C, N, and O. These ions do complitely dissipate the waves and halt the turbulent cascade.

Miller and Reames (1996) show that abundance ratios similar to those observed in interplanetary space after solar flares can result from the stochastic acceleration by cascading Alfvén waves in impulsive flares.

* * *

In general, stochastic acceleration is attractive on several points. One of them is that the stochastic interaction of particles with cascading waves in cosmic plasma offers, in principle, the opportunity to unify electron and ion acceleration within the context of a single model. Specifically, the picture that is emerging is one in which resonant wave-particle interactions are able to account for acceleration of particles out of the thermal background and to relativistic energies.

5.2.5 Electron-dominated solar flares

Hard X- and gamma-ray observations of solar flares have a wide range of energy from about 10 keV to about 10 GeV with relatively high spectral and temporal resolutions. Photon spectra over this wide range show significant deviations from the simple power law (e.g., Park, Petrosian, and Schwartz, 1997). The study of these deviations can provide important information about the acceleration mechanism. There is, however, some ambiguity in the analysis of the observational data because both accelerated electrons and protons contribute to the hard electromagnetic emission. Fortunately, there exist impulsive flares which have little or no evidence of nuclear excitation lines in the gamma-ray range. Such 'electron-dominated' events are uncontaminated by the proton processes and provide direct insights into the nature of the electron acceleration.

Park *et al.* use a model consisting of a finite-size region in the solar corona near the flare-loop top which contains a high-density of turbulence. Here the electrons are accelerated. Because of the rapid scattering by waves, the electrons trapped in this region have a nearly isotropic distribution. They emit bremsstrahlung photons which can be considered in a thin-target approximation. However, electrons eventually escape this region after an escape time of $\tau_{esc}(\mathcal{E})$ and lose most of their energy $\mathcal{E}$ in the chromosphere at the footpoints where they also emit hard X- and gamma-rays. This is called the thick-target source (Sections 3.3 to 3.5).

Instead of the simplified Equation (5.15), the Fokker-Planck equation (Section 2.4.4) re-written in energy space is used to describe the spectrum of electrons assuming isotropy and homogeneity:

$$\frac{\partial N}{\partial t} = -\frac{\partial}{\partial \mathcal{E}} \{[A(\mathcal{E}) - |B(\mathcal{E})|]\, N\} + \frac{\partial^2}{\partial \mathcal{E}^2} [D(\mathcal{E})\, N] - \frac{N}{\tau_{esc}(\mathcal{E})} + Q(\mathcal{E}). \quad (5.20)$$

Here $N(\mathcal{E}, t)\, d\mathcal{E}$ is the number of electrons per unit volume in the energy interval $d\mathcal{E}$, $A(\mathcal{E})$ is the systematic acceleration rate, $D(\mathcal{E})$ is the diffusion coefficient, $Q(\mathcal{E})$ is a source term. The energy loss term

$$B(\mathcal{E}) = \left(\frac{d\mathcal{E}}{dt}\right)_L \quad (5.21)$$

includes both Coulomb collision and synchrotron radiation losses.

Take the Maxwellian distribution as the source term

$$Q(\mathcal{E}) = Q_0 \frac{2}{\sqrt{\pi}} \left(\frac{\mathcal{E}}{k_{\rm B} T}\right)^{1/2} \exp\left(-\frac{\mathcal{E}}{k_{\rm B} T}\right), \quad (5.22)$$

where Q_0 is the rate at which the ambient plasma electrons of temperature T are accelerated. At steady state, the number of escaping particles is equal to the number of accelerated electrons:

$$\int \frac{N}{\tau_{esc}(\mathcal{E})}\, d\mathcal{E} = \int Q(\mathcal{E})\, d\mathcal{E} = Q_0 \, . \quad (5.23)$$

The temperature T of about 17 MK is taken. The coefficients $A(\mathcal{E})$, $D(\mathcal{E})$, and $\tau_{\rm esc}(\mathcal{E})$ of the Fokker-Planck equation are determined by the particle acceleration mechanism. They can be written in the form:

$$A(\mathcal{E}) = \mathcal{D}\,(q+2)\,(\gamma_{\rm L}\beta)^{\,q-1}, \tag{5.24}$$

$$D(\mathcal{E}) = \mathcal{D}\,\beta\,(\gamma_{\rm L}\beta)^{\,q}, \tag{5.25}$$

$$\tau_{esc}(\mathcal{E}) = \mathcal{T}_{esc}\,\frac{(\gamma_{\rm L}\beta)^{\,s}}{\beta}\,. \tag{5.26}$$

Here $\mathcal{D}$, $\mathcal{T}_{esc}$, q, and s are independent of the kinetic energy $\mathcal{E} = \gamma_{\rm L} - 1$ measured in units of $m_e c^2$, and βc is the velocity of electrons.

The acceleration timescale τ_a, which is also the timescale for reaching the steady state in Equation (5.20), can be estimated as

$$\tau_a(\mathcal{E}) \approx \tau_{_D}(\mathcal{E}) \approx \frac{\mathcal{E}^2}{D(\mathcal{E})}\,. \tag{5.27}$$

This should be less than the rise time of a flare. For three of four flares described by Park *et al.* (1997), the overall rise time τ_r of the hard X-rays is about 10 s and the total duration of the flare τ_f is about 100 s. For the most impulsive flare $\tau_r < 2$ s and $\tau_f \approx 8$ s. Hence the steady state approximation is justified. After setting $\partial/\partial t = 0$, we can divide Equation (5.20) by one of the parameters, say the diffusion coefficient $\mathcal{D}$, without changing the steady state solution.

The acceleration timescale for an electron with energy $\mathcal{E} = 1$ is approximately $\mathcal{D}^{-1}$. Therefore, for the three flares with the rise time $\tau_r \approx 10$ s, we estimate $\mathcal{D} \approx 0.15\ \mathrm{s}^{-1}$. For the shortest flare $\mathcal{D} \approx 1\ \mathrm{s}^{-1}$. Shorter rise times are possible, but these require higher values of the turbulence energy density and the magnetic field. With $\mathcal{D}$ fixed, the number of free parameters in the general stochastic model described above is reduced by one.

The numerical solutions show that the wistler wave resonant acceleration of electrons fits the observed spectra over the entire range of energy in all four flares. The high-energy cutoff in the two flares can be attributed to synchrotron radiation losses in the presence of a 500 G magnetic field at the acceleration site. The observed break in the photon spectra of all four flares around 1 MeV can be attributed to a combination of the energy dependence of the escape time $\tau_{esc}(\mathcal{E})$ of particles out of the acceleration region and the change in the energy dependence of the bremsstrahlung cross-section between the nonrelativistic and relativistic regimes. Futher steepening of the spectrum at even lower energies is caused by Coulomb losses (Park *et al.*, 1997).

5.3 The relativistic electron-positron plasma

According to present views, in a number of astrophysical objects there is a relativistic plasma that mainly consists of electrons and positrons. Among these objects are pulsar magnetospheres (Ruderman and Sutherland, 1975; Michel, 1991), accretion discs in close binary systems (e.g., Takahara and Kusunose, 1985; Rose, 1998), relativistic jets from active galactic nuclei (Begelman *et al.*, 1984; Peacock, 1999), and magnetospheres of rotating black holes in active galactic nuclei (e.g., Hirotani and Okamoto, 1998).

Because of synchrotron losses, the relativistic collisionless plasma in a strong magnetic field should be strongly anisotropic: its particle momenta should have a virtually one-dimensional distribution distended along the field. The transversal (with respect to the field) momentum of a particle is small compared with the longitudinal momentum. In accordance with Ruderman and Sutherland (1975), such a particle distribution is formed near the pulsar surface under the action of a strong longitudinal electric field and synchrotron radiation. What equations can be used as starting ones for a description of the electron-positron plasma? The answer depends upon a property of the plasma, which we would like to describe.

It is known that the anisotropy can result in various types of instabilities, for example, the fire-hose instability of the relativistic electron-positron plasma (e.g., Mikhailovskii, 1979). Behaviour of Alfvén waves in the isotropic and anisotropic plasmas can be essentially different (e.g., Mikhailovskii *et al.*, 1985).

We can suppose that the anisotropic relativistic approach of a type of the CGL approximation (see Section 7.3.5) can be used to consider the problem of Alfvén waves of finite amplitude. However the dispersion effects are important for such waves and are not taken into account in the CGL approximation. The problem can be analysed on the basis of the standard kinetic approach with use of the Vlasov equation (Section 2.4.2). As we saw above, such a procedure is sufficiently effective in the case of linear problems but is complicated in study of nonlinear processes when one must deal with parts of the distribution function square and cubic to the wave amplitude.

More effective kinetic approaches are reviewed and demonstrated in Mikhailovskii *et al.* (1985). One of them is based on expansion in the series of the inverse power of the background magnetic field (cf. Section 4.2) and allowance for the cyclotron effects as a small corrections. Using this approach, Mikhailovskii *et al.* consider the nonlinear Alfvén waves both in the case of an almost one-dimensional momentum particle distribution (the case of a pulsar plasma) and in the case of an isotropic plasma. The later case is interesting, in particular, for the reason that it has been also analysed by means of the hydrodynamic equations (Section 15.1.4). Two types of Alfvén solitons (the moving-wave type and the nonlinear wave-packet type) can exist in relativistic collisionless electron-positron plasma.

Recommended Reading: Lifshitz and Pitaevskii, *Physical Kinetics* (1981) Ch. 3 and 5.

5.4 Practice: Problems and Answers

Problem 5.1. Write the general solution of the linear Equation (5.1).

Answer. Since the left-hand side of (5.1) is the time derivative (more exactly, the Liouville operator (1.28) acting on the first-order distribution function for particles following *unperturbed* trajectories), the solution of (5.1) is formally the integral over time

$$f_k^{(1)}(\mathbf{r}, \mathbf{v}, t) = -\frac{e_k}{m_k} \int_{-\infty}^{t} \left(\mathbf{E}^{(1)} + \frac{1}{c} \mathbf{v} \times \mathbf{B}^{(1)} \right)_{\alpha} \frac{\partial f_k^{(0)}(\mathbf{r}, \mathbf{v}, \tau)}{\partial v_{\alpha}} \, d\tau. \quad (5.28)$$

Here the integration follows an unperturbed-particle trajectory to the point $(\mathbf{r}, \mathbf{v})$ in phase space X.

In principle, substitution of (5.28) into the Poisson law for electrostatic waves gives a perturbation of electric charge density (2.49). Similarly, one can determine a perturbation of current density (2.50) by substitution of (5.28) into the Ampér law in the case of electromagnetic waves. In practice, solving (5.28) is fairly complicated.

Problem 5.2. Show that, for a particle with velocity $\mathbf{v}$ in a plasma without an magnetic field, the resonance condition correspondes to:

$$\omega - \mathbf{k} \cdot \mathbf{v} = 0. \quad (5.29)$$

This is usually called the *Cherenkov condition.*

Problem 5.3. Consider a wave that has frequency ω and wave vector $\mathbf{k}$ in the laboratory frame. Show that in the rest frame of the particle the frequency of the wave is

$$\omega_0 = \gamma_{\mathrm{L}} (\omega - \mathbf{k} \cdot \mathbf{v}), \quad (5.30)$$

where

$$\gamma_{\mathrm{L}} = \left(1 - \frac{v^2}{c^2}\right)^{-1/2} \tag{5.31}$$

is the Lorentz factor of the particle. Therefore the Cherenkov resonance condition (5.29) corresponds to $\omega_0 = 0$, which means that the fields appear static in the rest frame of the particle.

Answer. Apply the Lorentz transformation (see Landau and Lifshitz, *Classical Theory of Field*, 1971, Ch. 1) to the four-vector $\{\mathbf{k}, i\omega/c\}$.

Problem 5.4. In a transparent medium with a refraction index n, greater than unity, the Cherenkov condition (5.29) can be satiesfied for fast particles with

$$\beta = \frac{v}{c} \geq \frac{1}{n}. \tag{5.32}$$

Let χ be the angle between the particle's velocity $\mathbf{v}$ and the wave vector $\mathbf{k}$ of appearing emission, which is called *Cherenkov emission*. Show that Cherenkov emission is confined to the surface of a cone with the cone half-angle

$$\chi_{\max} = \arccos\frac{1}{n}. \tag{5.33}$$

Problem 5.5. Discuss an analogy between the Cherenkov emission pattern and the bow wave of a ship or a supersonic aircraft.

Problem 5.6. Consider the Landau resonance for electrons in a Maxwellian plasma. It is clear that electrons moving much slower or much faster than the wave tend to see the electric field that averages to zero. So we have to consider only the particles in some small part of velocity space close to the phase velocity as shown in Figure 5.3.

Since the slope of the initial distribution function is negative, there are more electrons at lower velocity than at higher velocity near the resonance (5.7). Estimate a difference.

Problem 5.7. Show that for a purely transversal wave in which both the wave electric field and magnetic field are perpendicular to the wave vector, the net energy change for a particle moving with the phase speed of the wave is zero.

Answer. Transform the numerator of the formula (5.9) to the wave frame, i.e. to a frame in which the parallel phase velocity $\omega/v_{\|} = 0$. In

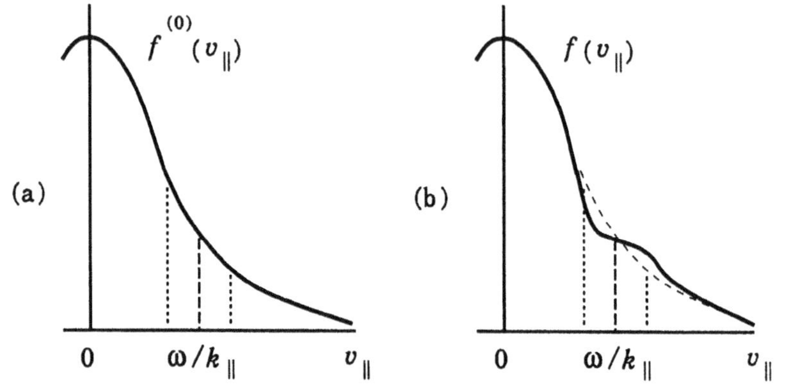

Figure 5.3: The Landau damping. (a) The initial distribution function of thermal electrons with some narrow region centered at the resonance with the wave. (b) The distribution function after an evolution due to interaction of the electrons with the wave.

this frame the gradient in velocity space appears to be equal to a gradient with respect to pitch angle. Hence the effect of the gyroresonance is to cause particles to change only pitch angle but not their energy.

Problem 5.8. Show that the Landau damping prevents plasma waves from escaping the region where $\omega = \omega_{pl}^{(e)}$ (see definition (6.76)) into rarefied plasma, for example, from the solar corona to interplanetary medium (see Zheleznyakov, 1996).

Hint. Consider the dispersion equation for electromagnetic waves in a homogeneous equilibrium plasma without a magnetic field.

Chapter 6

Coulomb Collisions of Particles in Cosmic Plasma

Binary collisions of particles with the Coulomb potential of interaction are typical for physics of collisional plasmas in space as well as for stellar dynamics.

6.1 Close and distant collisions

6.1.1 The Rutherford formula and collision parameters

Binary interactions of particles, described by the Coulomb potential

$$\varphi(r) = \frac{e}{r}\,, \tag{6.1}$$

have been studied in mechanics (see Landau and Lifshitz, *Mechanics*, 1960. Ch. 4, § 19). Considering binary interactions as *collisions*, we are interested only in their final result, the duration of the interaction and the actual form of particle trajectories being neglected. Thus in the centre of mass system, each particle is deflected through an angle χ defined by the relation

$$\tan\frac{\chi}{2} = \frac{e_1 e_2}{m v^2 l} \tag{6.2}$$

or

$$l\,(\chi) = \frac{e_1 e_2}{m v^2}\cot\frac{\chi}{2}\,. \tag{6.3}$$

Here

$$m = \frac{m_1 m_2}{m_1 + m_2}$$

is the reduced mass, v is the relative particle velocity at infinity, l is the 'impact parameter'. The last is the closest distance of the particle's approach, were it not for their interaction as shown in Figure 6.1.

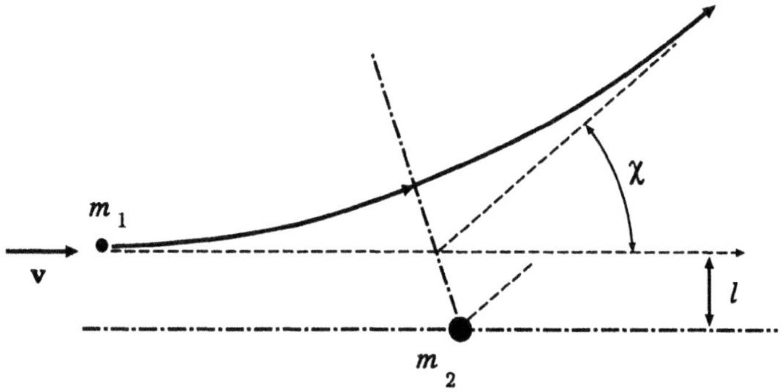

Figure 6.1: The trajectory of a light particle with mass m_1 near a heavy particle with mass m_2.

For particles deflected through a right angle

$$l\left(\frac{\pi}{2}\right) \equiv l_\perp = \frac{e_1 e_2}{m v^2}\,, \tag{6.4}$$

so the initial formula (6.2) is conveniently rewritten as

$$\boxed{\tan\frac{\chi}{2} = \frac{l_\perp}{l}\,.}$$

The collisions are called *close* if

$$\pi/2 \le \chi \le \pi\,, \quad \text{i.e.} \quad 0 \le l \le l_\perp\,. \tag{6.5}$$

Correspondingly, for *distant* collisions $l > l_\perp$.

The average characteristics of the Coulomb collisions are obtained with the aid of the formula for the *differential* cross-section. It is called the Rutherford formula and is derived from (6.3) as follows:

$$d\sigma = 2\pi\, l(\chi)\, dl = 2\pi\, l(\chi) \left| \frac{dl}{d\chi} \right| d\chi =$$

$$= \frac{\pi e_1^2 e_2^2}{m^2 v^4}\, \frac{\cos(\chi/2)}{\sin^3(\chi/2)}\, d\chi = \left(\frac{e_1 e_2}{2 m v^2}\right)^2 \frac{d\Omega}{\sin^4(\chi/2)}\,. \tag{6.6}$$

Here the modulus bars indicate the absolute value of the derivative $dl/d\chi$ because it has a negative sign: with increase of the impact parameter l, the scattering angle χ decreases; $d\Omega = 2\pi \sin\chi\, d\chi$.

By integrating (6.6) over the back hemisphere (6.5), we find the *total* cross-section of close collisions

$$\sigma_{cl} = \frac{\pi e_1^2 e_2^2}{m^2 v^4} = \pi l_\perp^2 . \tag{6.7}$$

This formula follows directly from definition (6.4), of course, without integrating the differential cross-section (6.6).

6.1.2 The test particle concept

By analogy with the usual gas, the concept of a 'test' particle is introduced to analyse the collisions in plasma. For instance the frequency of test particle (m_1, e_1) collisions with 'field' particles (m_2, e_2) is introduced:

$$\nu_{cl} = n_2\, v_1\, \sigma_{cl} = \frac{\pi e_1^2 e_2^2\, n_2}{m_1^2 v_1^3} . \tag{6.8}$$

Here, for simplicity's sake, it is assumed that $m_2 \gg m_1 \approx m$ and $v_2 \ll v_1$. So this is, for example, the case of an electron colliding with ions.

The length of *mean free path* λ of a test particle in a gas consisting of field particles is, by definition, the distance along which the particle suffers one collision,

$$\lambda = v_1\, \nu^{-1} .$$

From this it follows for *close* collisions that

$$\lambda_{cl} = \frac{1}{n_2\, \sigma_{cl}} . \tag{6.9}$$

Hence the time between two consecutive collisions is

$$\tau_{cl} = \frac{\lambda_{cl}}{v_1} = \frac{m_1^2 v_1^3}{\pi e_1^2 e_2^2\, n_2} \sim \frac{v_1^3}{n_2} , \tag{6.10}$$

or the frequency of close collisions

$$\nu_{cl} = \frac{\pi e_1^2 e_2^2\, n_2}{m_1^2 v_1^3} , \tag{6.11}$$

which is the same as (6.8) of course.

6.1.3 Particles in a magnetic trap

Formulae (6.8) and (6.10) are frequently used in order to find out what approximation we have to use to consider the plasma. For example, if the length of mean free path λ of the test particles inside a magnetic trap (Section 4.3.2) in the solar atmosphere is greater than the trap's size, then such particles can be considered in the collisionless approximation. Here *charge separation* may be found to be essential, as well as the electric field resulting from it (e.g., Alfvén and Fälthammar, 1963).

> While the magnetic mirror is the primary trapping mechanism, the electrostatic potential also traps electrons

with energies low to overcome the electrostatic potential. It is noted (Spicer and Emslie, 1988) that the potential produced, in general, has an energy equivalent of the average electron energy and that the number and energy fluxes of the particles that escape from the trap can be reduced by as much as ~ 50 or more depending on the magnetic mirror ratio of the flare loop and the ratio of the ion and electron anisotropy factors.

Some other effects due to non-collisional particles in traps are mentioned in Section 4.3.2; they will be discussed in Sections 13.3 and 13.4.

On the other hand, if the length of the mean free path of the test particles is much less than the trap's size, the collisions play an important role. As a rule they maxwellise the plasma (the gas of test particles), making it an equilibrium one. In such a plasma the notion of *temperature* is meaningful, as we shall see in Chapter 7. For example, while considering thermal electrons (having the density $n_{\rm e}$ and the temperature $T_{\rm e}$) in the trap, an electron with the *mean thermal velocity* (see definition (4.87))

$$V_{\rm Te} = \sqrt{\frac{3k_{_{\rm B}} T_{\rm e}}{m_{\rm e}}} \tag{6.12}$$

should be taken as the test particle. Then we obtain the known 'T to the 3/2 power' law for the time of the Coulomb collisions (6.10):

$$\boxed{\tau \sim \frac{T_{\rm e}^{3/2}}{n_{\rm e}}\,.} \tag{6.13}$$

> The hotter the plasma is, the more non-collisional is it with respect to some physical phenomenon or another.

The characteristic time τ of the Coulomb collisions has to be compared with the characteristic times of other physical processes: the time of particle motion between magnetic corks in the trap, the period of the Larmor rotation, the time of heating or cooling, etc.

6.1.4 The role of distant collisions

Because for small angles χ the differential cross-section (6.6) is

$$d\sigma \sim \frac{d\chi}{\chi^3}\,, \tag{6.14}$$

the total cross-section diverges.

Such divergence of the cross-section always occurs, once the interaction potential has no restricting factor,

or, to put it another way, if the interaction forces do not break off at some distance, as in the case of hard balls. This fact is of fundamental importance, for example, in *stellar dynamics* (Jeans, 1929; Chandrasekhar, 1943) or, more exactly, in any astrophysical system governed by gravitational force (say a gravitational system), see Sections 2.5.3 and 7.2.5.

Although each distant collision causes only a small deflection of the test particle trajectory, they are present in such large numbers that their total action upon the particle is *greater* or much greater than that of relatively rare close collisions. Let us convince ourselves that this is true.

Each collision causes a small change in momentum perpendicular to the initial direction of the particle's motion:

$$\delta p_{\perp} = p\,\sin\chi = m_1 v_1\,\frac{2\tan(\chi/2)}{1+\tan^2(\chi/2)} = \frac{2\,m_1 v_1\,(l_{\perp}/l)}{1+(l_{\perp}/l)^2} = 2\,m_1 v_1\,\frac{x}{1+x^2}\,.$$

Here $x = l_{\perp}/l$, and $0 \leq x \leq 1$.

Since distant collisions occur chaotically, one is usually interested in the mean rate of change in the quantity $p_{\perp}^2$:

$$\frac{d}{dt}\,p_{\perp}^2 = \int\limits_{\chi=\pi/2}^{\chi=0} (\delta p_{\perp})^2\, n_2\, v_1\, d\sigma =$$

$$= 8\pi\, n_2\, m_1^2 v_1^3\, l_{\perp}^2 \int\limits_1^0 \frac{dx}{(1+x^2)^2\, x} \sim \ln x \,\Big|_1^0 . \tag{6.15}$$

The integral diverges logarithmically on the upper limit. Restrict it to some maximal value of the impact parameter $\Lambda = l_{\max}/l_{\perp}$. Then the integral is approximately equal to

$$\frac{d}{dt}\, p_{\perp}^{2} = 8\pi\, n_2\, m_1^2 v_1^3\, l_{\perp}^2 \,\ln\Lambda = 8\pi e_1^2 e_2^2\, \frac{n_2}{v_1}\, \ln\Lambda\,. \tag{6.16}$$

The factor $\ln\Lambda$ is referred to as the Coulomb logarithm.

Introduce the characteristic time $\tau_{\perp}$ during which the perpendicular component of the momentum acquires a value equal to the initial momentum $m_1 v_1$:

$$\tau_{\perp} = (m_1 v_1)^2 \left(\frac{d}{dt}\, p_{\perp}^2\right)^{-1} = \frac{m_1^2 v_1^3}{8\pi e_1^2 e_2^2\, n_2\, \ln\Lambda}\,. \tag{6.17}$$

In other words, the mean resulting deflection becomes comparable with the quantity $\pi/2$ in a time $\tau_{\perp}$. Recall that this deflection through a large angle is a result of many distant collisions.

The effective frequency of distant collisions that corresponds to the time $\tau_{\perp}$ is

$$\nu_{\perp} = \frac{1}{\tau_{\perp}} = \frac{8\pi e_1^2 e_2^2\, n_2}{m_1^2 v_1^3}\, \ln\Lambda\,, \tag{6.18}$$

which is $8\ln\Lambda$ larger than the close collisions frequency (6.11):

$$\nu_{\perp} = 8\ln\Lambda \cdot \nu_{cl}\,. \tag{6.19}$$

The factor $8\ln\Lambda$ is usually much greater than unity; its typical value is $\sim 10^2$ under physical definition of $\ln\Lambda$ given in Section 6.2.

> The influence of close collisions on many kinetic processes in cosmic plasma is, as a rule, negligibly small in comparison to the action of distant collisions.

For example, the distant collisions determine an evolution of the distribution function of fast electrons injected into the thermal plasma in the solar atmosphere. However **this does not mean that the close collisions do never play any role** in cosmic plasma physics. Just in the same example, the close collisions of fast electrons with thermal ions create hard X-ray bremsstrahlung emission in the range 10–100 keV, because the close collisions are responsible for large exchange of the particle momentum. For typical flare parameters ($h\nu \approx 20$ keV, $\ln\Lambda \approx 20$) the efficiency of the bremsstrahlung process is $\sim 3\times 10^{-6}$ (e.g., Brown, 1971; Korchak, 1971).

6.2 Debye shielding and plasma oscillations

While considering the distant collisions, we have removed the divergence of the integral (6.15) which describes the mean rate of change of the test particle transversal momentum, purely formally – by artificially restricting the radius of action of the Coulomb forces at some maximal distance $l_{\max}$. Meanwhile this maximal distance may be chosen quite justifiably, based on the following reasoning. In a plasma,

> each charged particle attracts oppositely charged particles and, at the same time, repels the particles of the same charge.

As a consequence, the oppositely charged particles tend to gather around the particle, thus weakening its Coulomb field. As a result of such 'shielding' the action of the field extends over a distance no greater than some quantity $r_{\rm D}$, the so-called *Debye radius*.

The concept of Debye shielding has a clear meaning; let us assume that a plasma contains an immovable charge which then creates the electrostatic field in its vicinity. Some *equilibrium* distribution of *two components*: positive and negative plasma particles is established in this field. Its electrostatic potential φ is related to the densities of ions $n_{\rm i}$ and electrons $n_{\rm e}$ via the Poisson equation

$$\Delta \varphi = -4\pi e \left(Z n_{\rm i} - n_{\rm e} \right) , \tag{6.20}$$

where Ze is the ion charge.

In the thermodynamic equilibrium state the ion and electron densities in the electrostatic field with potential $\varphi(r)$ are to be distributed according to Boltzmann's law

$$n_{\rm i} = n_{\rm i}^{0} \exp \left(- \frac{Ze\varphi}{k_{\rm B} T_{\rm i}} \right) , \qquad n_{\rm e} = n_{\rm e}^{0} \exp \left(\frac{e\varphi}{k_{\rm B} T_{\rm e}} \right) . \tag{6.21}$$

The constant coefficients are set equal to the mean densities $n_{\rm i}^{0}$ and $n_{\rm e}^{0}$ of plasma particles, since $\varphi \rightarrow 0$ far from the particle considered.

Supposing that the Coulomb interaction is so weak that

$$Ze\varphi \ll k_{\rm B} T_{\rm i} \quad \text{and} \quad e\varphi \ll k_{\rm B} T_{\rm e}, \tag{6.22}$$

or restricting our consideration to the approximate solutions applicable at large distances from the shielded charge, expand both exponents (6.21) in a series and substitute in Equation (6.20). We obtain

$$\frac{1}{r^2} \frac{d}{dr} \left(r^2 \frac{d\varphi}{dr} \right) =$$

$$= -4\pi e \left[Z n_{\rm i}^0 \left(1 - \frac{Ze\varphi}{k_{\rm B} T_{\rm i}}\right) - n_{\rm e}^0 \left(1 + \frac{e\varphi}{k_{\rm B} T_{\rm e}}\right)\right] = \frac{\varphi}{r_{\rm D}^2}\,. \tag{6.23}$$

As usual the actual plasma is *quasi-neutral on average* (see below); instead of this let us assume here (like in Section 2.5.2) that the plasma is *ideally neutral*:

$$Z n_{\rm i}^0 = n_{\rm e}^0\,. \tag{6.24}$$

Then on the right-hand side of Equation (6.23) we have two terms for two-component plasma:

$$\frac{1}{r_{\rm D}^2} = \frac{1}{r_{\rm D}^{(\rm e)\,2}} + \frac{1}{r_{\rm D}^{(\rm i)\,2}} = \frac{4\pi e^2 n_{\rm e}^0}{k_{\rm B} T_{\rm e}} \left(1 + Z\,\frac{T_{\rm e}}{T_{\rm i}}\right) \tag{6.25}$$

or

$$r_{\rm D} = \left[\frac{k_{\rm B}}{4\pi e^2 n_{\rm e}^0}\,\frac{T_{\rm e} T_{\rm i}}{Z T_{\rm e} + T_{\rm i}}\right]^{1/2} \tag{6.26}$$

is known as the Debye radius, being first derived by Debye and Hückel (1923) in the theory of electrolytes.

The solution of Equation (6.23) corresponding to the charge e situated at the origin of the coordinates is the potential

$$\boxed{\varphi = \frac{e}{r}\,\exp\left(-\frac{r}{r_{\rm D}}\right).} \tag{6.27}$$

At distances greater than $r_{\rm D}$, the electrostatic interaction is exponentially small.

> The Debye length is an effective range for collisions, the potential between charged particles being the *shielded* Coulomb potential (6.27) rather than the Coulomb one (6.1) which would apply in a vacuum.

That is why the Debye radius $r_{\rm D}$ is substituted in the Coulomb logarithm (6.16) in place of $l_{\rm max}$.

A formula that is simpler than (6.26) is frequently used for the Debye radius, namely

$$r_{\rm D}^{(\rm e)} = \left(\frac{k_{\rm B} T}{4\pi e^2 n}\right)^{1/2}. \tag{6.28}$$

This variant of the formula for the Debye radius implies that the shielding is due to just the particles of one sign, more exactly, *electrons*, i.e. in the formula (6.25) we have $T_i = 0$ (the approximation of *cold ions*) and $T_e = T$ (see Problem 7.2). This is the *electron* Debye radius. The corresponding formula for the Coulomb logarithm is

$$\ln \Lambda = \ln \frac{3}{2e^3} \left(\frac{k_B^3 \, T^3}{\pi n_e} \right)^{1/2} . \tag{6.29}$$

Its values typical of the solar atmosphere are around 20 (Problem 6.1).

Formula (6.28) shows that the electron Debye radius increases with an increase of temperature, since electrons with higher kinetic energy can withstand the attraction of the positive ion charge Ze up to larger distances. It decreases with an increase of density n_0, since a larger number of electrons and ions can be accommodated in shorter distances to screen the electric field of charge Ze.

Debye shielding length is fundamental to the nature of a plasma. That is why this important characteristic appears again and again in cosmic plasma physics.

The first point to note is that a plasma maintains *approximate charge neutrality* (see Sections 7.3.6 and 2.5.2). The reason for this is simply that any significant imbalance of positive and negative charge could only be maintained by a huge electric field (e.g., Boyd and Sanderson, 1969). The movement of electrons to neutralize a charge inhomogeneity would be followed by an oscillatory motion (e.g., Alfvén and Fälthammar, 1963, Ch. 4).

This brings us to a second characteristic of plasmas called the *plasma frequency* or, more exactly, the electron plasma frequency:

$$\omega_{pl}^{(e)} = \left(\frac{4\pi e^2 n_e}{m_e} \right)^{1/2} . \tag{6.30}$$

A charge density disturbance oscillates with this frequency (Problem 7.3). These oscillations are called *Langmuir waves* or *plasma waves*. Therefore, under most circumstances,

> plasma cannot sustain electric fields for lengths in excess of the Debye radius or times in excess of a plasma period $T_{pl}^{(e)} = 2\pi / \omega_{pl}^{(e)}$.

However one cannot talk of plasma oscillations unless a large number of thermal particles are involved in the motion. It is the Debye shielding length which determines the spatial range of the field set up by the charge inequality:

$$r_{\rm D} = \frac{1}{\sqrt{3}} \, \frac{V_{\rm Te}}{\omega_{pl}^{(e)}} \, . \tag{6.31}$$

Here $V_{\rm Te}$ is the mean thermal velocity of electrons. Therefore the Debye length

$$\boxed{r_{\rm D} \approx \frac{V_{\rm Te}}{\omega_{pl}^{(e)}} \, .} \tag{6.32}$$

So a fully-ionized plasma in the termodynamic equilibrium is a quasi-neutral medium. The *space* and *time* scales of charge separation in such plasma are the Debye radius and the inverse plasma frequency. Therefore the plasma oscillations are a typical example of **collective phenomena** (Section 2.5.2). The Coulomb collisions, of course, damp the amplitude of the plasma oscillations with the rate which is proportional to the frequency $\nu_{\rm ei}$ of electron-ion collisions.

6.3 Collisional relaxations in cosmic plasma

6.3.1 Some exact solutions

It was shown in Section 6.1 that, as a result of the Coulomb collisions, a particle deflects through an angle comparable with $\pi/2$ in a characteristic time given by (6.17). Some more exact calculations of the Coulomb collisions times, that take into account the motion of field particles, have been carried out by Spitzer (1940) and Chandrasekhar (1943). These calculations are cumbersome, so we give only their final results.

Consider the electron component of a plasma. Suppose that the test particles likewise are electrons moving with mean thermal velocity. Then the exact calculation gives instead of the formula (6.17) the time

$$\tau_{\rm ee} = \frac{m_{\rm e}^2 \, (3k_{\rm B} T_{\rm e}/m_{\rm e})^{3/2}}{e_{\rm e}^4 \, n_{\rm e} \, 8\pi \, \ln \Lambda} \cdot \frac{1}{0.714} \, . \tag{6.33}$$

This is called the time of mutual electron collisions or simply the *electron collisional time*. Comparison of formula (6.33) with (6.17) shows that the difference (the last factor in (6.33)) is not large. So the consideration of binary collisions in the approximation used in Section 6.1 is accurate enough, at least for astrophysical applications.

The analogous time of mutual collisions for ions, having mass $m_{\rm i}$, charge $e_{\rm i}$, temperature $T_{\rm i}$ and density $n_{\rm i}$, is equal to

$$\tau_{\rm ii} = \frac{m_{\rm i}^2 \, (3k_{\rm B} T_{\rm i}/m_{\rm i})^{3/2}}{e_{\rm i}^4 \, n_{\rm i} \, 8\pi \, \ln \Lambda} \cdot \frac{1}{0.714} \, . \tag{6.34}$$

If a plasma is quasi-neutral: $e_{\rm i} \, n_{\rm i} \approx -e_{\rm e} \, n_{\rm e} = en$, where $e_{\rm i} = -Z e_{\rm e}$, and if $T_{\rm e} \approx T_{\rm i}$, then the ratio

$$\boxed{\frac{\tau_{\rm ii}}{\tau_{\rm ee}} \approx \left(\frac{m_{\rm i}}{m_{\rm e}} \right)^{1/2} \frac{1}{Z^3} \, .} \tag{6.35}$$

Collisions between ions occur much more rarely than those between electrons.

However, it is not the time of collisions between ions $\tau_{\rm ii}$ – the *ion collisional time*, but rather the time of electron-ion collisions that is the greatest. This characterizes, in particular, the process of temperature equalizing between the electron and ion components in a plasma. The rate of temperature equalizing can be determined from the equation

$$\frac{dT_{\rm e}}{dt} = \frac{T_{\rm i} - T_{\rm e}}{\tau_{\rm ei}(\mathcal{E})} \, , \tag{6.36}$$

where $\tau_{\rm ei}(\mathcal{E})$ is the time of equilibrium establishment between the electron and ion plasma components. It characterizes the rate of exchange of energy $\mathcal{E}$ between the components and equals (see Spitzer, 1962)

$$\tau_{\rm ei}(\mathcal{E}) = \frac{m_{\rm e} m_{\rm i} \left[3k_{\rm B} \left(T_{\rm e}/m_{\rm e} + T_{\rm i}/m_{\rm i} \right) \right]^{3/2}}{e_{\rm e}^2 \, e_{\rm i}^2 \, (6\pi)^{1/2} \, 8 \, \ln \Lambda} \, . \tag{6.37}$$

For comparison with formula (6.35) let us put $T_{\rm i} = T_{\rm e}$. Then

$$\tau_{\rm ei}(\mathcal{E}) = 0.517 \, \frac{e_{\rm i}^2}{e_{\rm e}^2} \left(\frac{m_{\rm i}}{m_{\rm e}} \right)^{1/2} \tau_{\rm ii} \, . \tag{6.38}$$

Thus the time of energy exchange between electrons and ions is much greater than the time of mutual ion collisions.

In a plasma consisting of electrons and protons with equal temperatures we have

$$\tau_{\rm ep}(\mathcal{E}) \approx 22 \, \tau_{\rm pp} \approx 950 \, \tau_{\rm ee} \, . \tag{6.39}$$

The energy exchange between electron and ion components occurs so slowly that for each component a distribution may be set up which is close to Maxwellian with the proper temperature.

That is the reason why we often deal with a *two-temperature* plasma. Moreover, the so-called adiabatic model for two-temperature plasma (Section 6.3.3) is often used in astrophysics.

6.3.2 Two-temperature plasma in solar flares

(a) Impulsive heating by accelerated electrons

Let us illustrate the situation, discussed above, by two examples from the physics of flares. The first is the impulsive heating of the solar atmosphere by a powerful beam of accelerated electrons (Somov, Syrovatskii, and Spektor, 1981). The beam impinges on the chromosphere from the coronal part of a flare along the magnetic field tubes. The maximal energy flux is $F_{\max} \approx 10^{11}$ erg cm^{-2} s^{-1}. The time profile with the maximum at $t \approx 5$ s of the energy flux at the upper boundary of the chromosphere has been used for numerical solution of the two-temperature dissipative hydrodynamic equations (Chapter 2 in Somov 1992).

Yohkoh observations, made using three of the instruments on board – the Hard X-ray Telescope (HXT), the Soft X-ray Telescope (SXT), and the Bragg Crystal Spectrometer (BCS) – show that the nonthermal electron energy flux can be even larger, for example, in the flare of 16 December 1991 (see Fig. 6a in McDonald *et al.*, 1999), the maximal energy flux is $F_{\max} \approx 2.5 \times 10^{29}$ erg s^{-1}/2×10^{17} cm^2 $\sim 10^{12}$ erg cm^{-2} s^{-1}. Weak beams do not produce a significant response of the chromosphere (see Fig. 6b in McDonald *et al.*, 1999), of course, just hard X-ray bremsstrahlung.

In the chromosphere, beam electrons lose their energy by mainly Coulomb collisions.

The fastest process is the primary one, namely that of energy transfer from the beam electrons to the thermal electrons

of chromospheric plasma (Figure 6.2). As a result, plasma electrons are rapidly heated to high temperatures: in a matter of seconds the electron temperature reaches values of the order of ten million degrees. At the same time, the ion temperature lags considerably, by one order of magnitude, be-

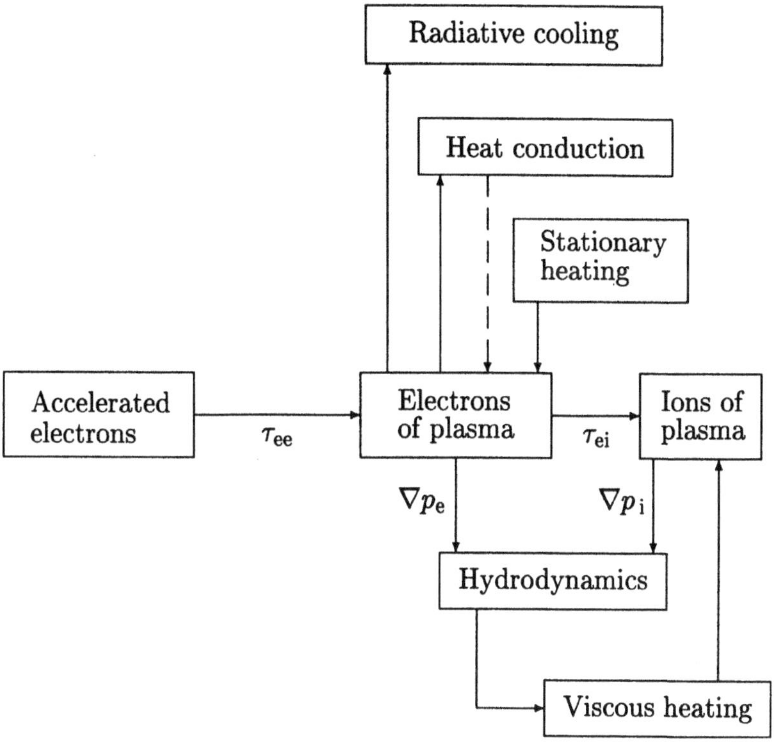

Figure 6.2: A scheme of the energy exchange in the two-temperature model of hydrodynamic response of the solar atmosphere to impulsive heating by an electron beam.

hind the electron temperature (Figure 6.3). Here the Lagrange variable

$$\xi = - \int_{z_{\max}}^{z} n(z)\, dz + \xi_{\min}\,, \ \ \text{cm}^{-2}\,, \tag{6.40}$$

z is the height above the photosphere, $z_{\max}$ corresponds to the transition layer between the chromosphere and corona before an impulsive heating. Therefore ξ is the column depth – the number of atoms and ions in a column (of the unit cross-section) measured down into the chromosphere from its upper boundary, the transition layer.

The column depth $\xi_{\min} = n_c l_r$ is the number of ions inside a flaring loop which is the coronal part of a reconnected field line tube (see Sections 16.5, 17.4, 13.4); l_r is the length of the reconnected field line, n_c is the plasma density inside the tube above the transition layer between the chromosphere and corona before an impulsive heating. Let us assume, for simplicity, that

$$\xi_{\min} \ll \xi_1 = \frac{\mathcal{E}_1^2}{2a_1} \text{ cm}^{-2} . \tag{6.41}$$

Here ξ_1 is the column thickness that the accelerated electrons with the minimal energy $\mathcal{E}_1$ measured in keV can pass in a plasma before they stop (see formula (3.40)). The assumption (6.41) means that we can neglect the energy losses of the electrons in the coronal part of the loop. In this way, we consider direct impulsive heating of the chromosphere by an electron beam. Accelerated electrons penetrate into the chromosphere to significant depth; for this reason a significant fraction of the beam energy is lost as radiation in optical and EUV lines. The column depth of evaporated plasma $\xi \approx 2 \times 10^{19}$ cm^{-2} but its temperature does not exceed $T_{\max} \approx 10^7$ K.

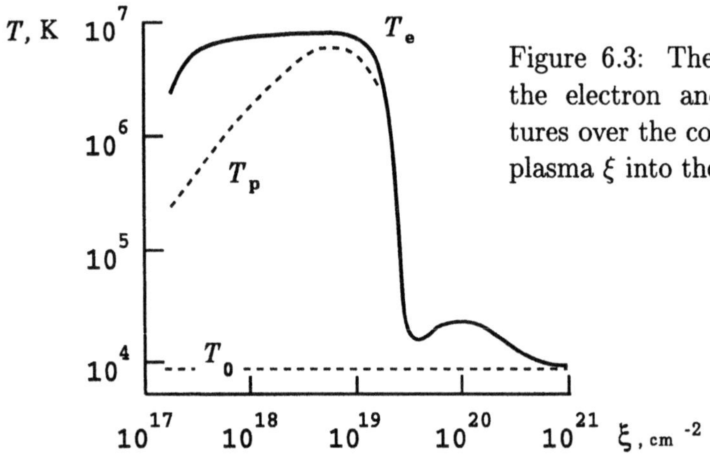

Figure 6.3: The distribution of the electron and ion temperatures over the column depth of a plasma ξ into the chromosphere.

The difference between the electron and ion temperatures is essential, at first, for the dynamics of high-temperature plasma

which absorbs the main part ($\geq 90\,\%$) of the beam energy flux. Imagine that only the electrons have a chance to be heated, while the ion heating can be neglected. In this case the electron temperature is twice as large as it would be in the case of equal heating of the electrons and ions,

$$(T_e)_1 \simeq 2\,(T_e)_2 \,.$$

The rate of high-temperature plasma cooling is mainly determined by heat fluxes into colder plasma. These can be evaluated by the formula for the classical heat flux

$$F_c = \kappa_e \nabla T_e \,. \tag{6.42}$$

Here $\kappa_e = \kappa_0 T_e^{5/2}$ is the classical conductivity due to the Coulomb collisions of plasma electrons. From formula (6.42) we see that the heat flux is proportional to $T_e^{7/2}$. Therefore the real heat flux

$$F_c(T_e)_1 \simeq 2^{7/2} F_c(T_e)_2 \tag{6.43}$$

can be an order of magnitude ($2^{7/2} \sim 10$) larger than the flux calculated in one-temperature ($T_e = T_i$) models. Because of this, the one-temperature models are much less dynamic than one would expect.

The effect becomes even more important if the accelerated electrons heat a preliminary (before a flare) evaporated 'warm' plasma. This formally means that, in formula (17.24), the column depth $\xi_{\min} = n_c l_r$ is not small in comparison with ξ_1. So we have to take into account the direct impulsive heating of the plasma inside the coronal part of the flaring loop. Such process (Duijveman et al., 1983; MacNeice et al., 1984) can very efficiently produce a 'superhot' plasma which has an electron temperature T_e much higher than in the case considered above.

(b) Heating by high-temperature current sheet

The difference between the electron and ion temperatures is known to be critical for a wide variety of kinetic effects, in particular for the generation of some turbulence (for example, ion-acoustic or ion-cyclotron) in the impulsively heated plasma. The turbulence, in its turn, has a great impact on the efficiency of heating and particle acceleration in a plasma.

The second example, when the electron component of a plasma has a temperature that is considerably different from the ion temperature, is supplied by the high-temperature turbulent-current sheets (Somov, 1981 and 1986; Somov and Titov, 1983) in the regions of reconnection. Since the sheet thickness, $2a$, is small in comparison with its width, $2b$ (Figure 4.19), the plasma inflow quickly enters the region of the Joule dissipation of reconnecting magnetic field components. Here the impulsively **fast heating of the electrons and ions takes place, resulting in considerably different temperatures**. The conditions in reconnecting current sheets in the solar corona, especially, in flares (Chapter 17) are such that

> the Coulomb exchange of energy between the impulsively heated electrons and ions can be entirely neglected.

One of distinctive features of reconnection in current sheets, proposed as the primary energy source in flares, is the presence of fast plasma outflows,

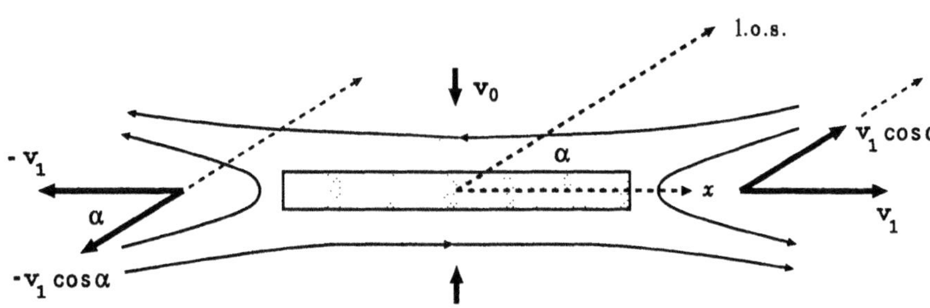

Figure 6.4: High-temperature plasma velocities near a reconnecting current sheet.

or jets, whose velocities are nearly equal to the Alfvén velocity, see definition (10.30). Outflows can give origin to plasma velocity distributions with equal and opposite components along the x axis in Figure 6.4 and, as a consequence, along the line-of sight (l.o.s.) to an observer. Therefore they can, in this way, create a **symmetric supra-thermal broadening in the soft X-ray lines** observed during solar flares. The broadening mainly depends on the electron and ion temperatures inside the current sheet (Antonucci and Somov, 1992).

A comparison of the supra-thermal profiles of the Fe XXV emission lines observed at flare onset with the predictions of the high-temperature turbulent-current sheet model suggests that the observed supra-thermal broadenings are consistent with the presence in the flare region of several small-scale or one (a few) curved large-scale reconnecting current sheets (Antonucci, Benna, and Somov, 1996).

The energy release by reconnection has been invoked to explain both large-scale events, such as solar flares and coronal mass ejections (CMEs), and small-scale phenomena, such as the coronal and chromospheric microflares that probably heat the corona (Section 21.4) and accelerate the solar wind. Ultraviolet observations of the so-called explosive events in the solar chromosphere by SUMER (the Solar Ultraviolet Measurements of Emitted Radiation instrument) on the spacecraft SOHO (the Solar and Heliospheric Observatory) reveal the presence of **bi-directional plasma jets** ejected from small sites above the solar surface (Innes *et al.*, 1997; cf. Antonucci and Somov, 1992). The structure of these jets evolves in the manner predicted by theoretical models of reconnection (see Fig. 1 in Somov and Syrovatskii, 1976b), thereby leading strong support to the view that reconnection is the fundamental process for accelerating plasma on the Sun.

6.3.3 An adiabatic model for two-temperature plasma

As we saw in Section 6.3.1, equilibrium in an electron-proton plasma is achieved in three stages. First, the electrons reach a Maxwellian distribution with temperature T_e on a timescale τ_{ee}. Then, on a longer timescale, $\tau_{pp} \sim (m_p/m_e)^{1/2}\,\tau_{ee}$, the protons reach a Maxwellian distribution with temperature T_p. Finally, the two temperatures equalize on the longest timescale of order $\tau_{ep} \sim (m_p/m_e)\,\tau_{ee}$.

Suppose that a two-temperature plasma is created by a strong shock wave in an electron-proton plasma. **The shock primarily heats ions** because the kinetic energy of a particle is proportional to the particle mass. In the postshock region, the protons reach thermal equilibrium on a timescale of τ_{pp} after they are heated through the shock (Zel'dovich and Raizer, 1966). Within this time the proton temperature is significantly higher than the electron one. Subsequently the protons share their thermal energy with the electrons through Coulomb collisions.

> In cosmic plasma, sometimes, a difference between electron and ion temperatures can be observed at huge linear scales.

For example, the so-called X-ray clusters, or clusters of galaxies in the Universe, with X-ray temperatures $(4 - 10) \times 10^7$ K show noticeable differences between their electron and ion temperatures at radii greater than 2 Mpc.

Note that the clusters of galaxies are the largest objects in the Universe, containing galaxies and dark matter, collisionless particles and a diffuse gas component. The last one is called the *intracluster medium* and has a temperature of about 10^8 K, thus emitting X-rays mainly through the thermal bremsstrahlung of the electrons. In the outer parts of the clusters, the free-free cooling time is much longer than the Hubble time. So we neglect radiative cooling in such plasma which is supposed to be heated by the shock in the accretion flow (see a review in Takizawa, 1998).

If we could also neglect heat conduction then the electrons would be considered as an adiabatic gas. It would be very convenient to calculate the electron and ion temperature profiles by using the *adiabatic model* of a two-temperature plasma by Fox and Loeb (1997). This is the case if tangled magnetic fields, for example of turbulent origin, can suppress heat conduction in high-temperature plasma. So we assume that there exists

> a chaotic magnetic field that is sufficiently strong to suppress heat conduction in high-temperature plasma, *yet small enough* to have negligible dynamical and dissipative effects including Joule heating.

These conditions seem to be approximatelly satiesfied in cluster environments; for more detail see Fox and Loeb (1997).

The general case of a strong shock in a fully ionized plasma with heat conduction is complicated by the fact that the electron thermal speed exceeds the shock speed, allowing the electrons to preheat the plasma ahead of the shoch (see Zel'dovich and Raizer, 1966, and references there). Usually **heat conduction determines internal scales of the problem** being in competition with the thermal instability driven by radiative cooling (Field, 1965; see also Somov and Syrovatskii, 1976b). Radiation emitted by the high-temperature plasma behind the shock also may heat a preshock region. So we have to be very careful when we apply the adiabatic model of two-temperature plasma to astrophysical conditions.

6.3.4 Two-temperature accretion flows

Magnetized accretion discs have become the most convincing physical paradigm to explain emission from the central engines of active galactic nuclei and X-ray binary sources (Section 8.4). The observed radiation comes from the energy dissipation required to maintain steady accretion of plasma on to the central object. In the standard model of the optically-thin accreation disc, the heat energy released by viscous dissipation is radiated almost immediatelly by the accreating plasma. So

> the net luminosity must be equal to ($\approx$ one-half) the gravitational energy released as the mass falls onto the central object.

In a few of binary stellar systems, the mass of the primary star has been measured and found to be consistent with the mass of a neutron star, $\sim 1.4\, M_{\odot}$. In several other systems, however, the mass of the primary is found to be greater than $3\ M_{\odot}$, which makes these stars too massive to be neutron stars. These are considered as black hole candidates.

Although neutron stars and black holes have been distinguished on the basis of their masses, the real physical distinction between the two is that black holes must have a horizon (a surface through which the matter and energy fall in but from which nothing escapes) while neutron stars are normal stars with surfaces. This basic difference provides an opportunity to test the reality of black holes (e.g., Narayan *et al.*, 1997).

Two-temperature advection-dominated accretion flows have received much attention in an effort to explain low-luminosity stellar and galactic accreting sources (e.g., Blackman, 1999; Wiita, 1999). Here the ions are assumed to receive the energy dissipated by the steady accretion **without having enough**

time to transfer their energy to the cooler electrons before falling on to the central object. While the electrons can almost always radiate efficiently, the protons will not, as long as Coulomb processes are the only thing that share energy between electrons and protons. So some or most of the dissipated energy is *advected* (Section 8.4.2), not radiated, as it would have been if the electrons received all of the dissipated energy.

> When the central object is a black hole, the advected energy is lost forever rather than reradiated as it would be for a neutron star.

Precisely such observed differences between corresponding X-ray binary systems have been purported to provide evidence for black hole horizons (Narayan *et al.*, 1997; see also Chakrabarti, 1999).

6.4 Dynamic friction in cosmic plasma

6.4.1 The collisional drag force and energy losses

(a) Chandrasekhar-Spitzer's formulae

As in Sections 6.1 and 6.3, we use the concept of a test particle to illustrate the effects of the collisional drag force in cosmic plasma. A test particle of mass m_1 and charge e_1 is incident with velocity $\mathbf{v}$ in a gas containing field particles of mass m_2, charge e_2 and density n_2. In what follows, $v_{\parallel}$ will be the component of the test particle velocity parallel to the original direction of its motion.

First, for simplicity, let us consider the field particles *at rest*. As in Section 6.1, **integration over all possible values of the impact parameter up to the upper cut-off** at $l = l_{\max}$ yields the following formulae describing the *mean* rates of energy losses and of scattering for the incident particle (cf. Spitzer, 1962):

$$\frac{d\mathcal{E}}{dt} = -\frac{2\pi e_1^2 e_2^2 \ln \Lambda}{\mathcal{E}} \frac{m_1}{m_2} n_2 v \tag{6.44}$$

and

$$\frac{d}{dt} v_{\parallel} = -\frac{\pi e_1^2 e_2^2 \ln \Lambda}{\mathcal{E}^2} \left(1 + \frac{m_1}{m_2} \right) n_2 v^2 . \tag{6.45}$$

Here $\mathcal{E}$ is the energy of the incident particle (see definition (4.5)).

If we consider a beam of accelerated electrons in cosmic ionized plasma, the most important are interactions with electrons and protons. So

$$\frac{d\mathcal{E}}{dt} = -\frac{2\pi e^4 \ln \Lambda}{\mathcal{E}} \left(1 + \frac{m_e}{m_p} \right) n_e v \tag{6.46}$$

and

$$\frac{d}{dt} v_{\parallel} = -\frac{\pi e^4 \ln \Lambda}{\mathcal{E}^2} \left(3 + \frac{m_e}{m_p} \right) n_e v^2 \,. \tag{6.47}$$

Thus

> both ambient electrons and protons produce scattering (6.47) of the incident electrons but **only ambient electrons contribute significantly to the energy losses**;

the contribution of protons in the rate of energy losses (6.46) is proportional to the small ratio m_e/m_p. This is consistent, of course, with what we have concluded in Section 3.2 for fast particles propagating in thermal plasma.

Note that we neglect collective effects due to interaction of the plasma and the electron beam as a whole without any justification here. It must be emphasized also at this point that formulae (6.46) and (6.47) describe the *mean* rates of change of $\mathcal{E}$ and $v_{\parallel}$ for the electrons of an incident beam but neglect the dispersions about these means. The accuracy of such procedure decreases as the scattering and energy losses become not small. These ristrictions will be discussed somewhere else (Section 3.4). Now we recall that we have neglected the proper motions of the plasma particles. Let us take them into account.

(b) Energy losses in cosmic plasma

The most general non-relativistic formula for Coulomb losses in the many-component thermal plasma is given, for example, in Trubnikov (1965), Syvukhin (1966) and can be expressed as follows:

$$P \equiv \frac{d\mathcal{E}}{dt} = \sum_k \left(\frac{d\mathcal{E}}{dt} \right)_k = -\sum_k \frac{4\pi e^4 \ln \Lambda}{m_k} \frac{Z^2 Z_k^2 n_k}{v_k} \mathcal{P}_k \left(\frac{v}{v_k}, \frac{m_k}{M} \right) . \tag{6.48}$$

Here Z_k, m_k, n_k and v_k are the charge, mass, density and thermal velocity of the plasma particles of the kind k; they have a temperature T_k. Z, $M = Am_p$ and v are the charge, mass and velocity of the incident particles; their kinetic energy $\mathcal{E} = Mv^2/2$. Note that, contrary to definition (6.12) of the mean thermal velocity, in formula (6.48) the thermal velocity is equal to the *most probable* velocity of thermal particles (e.g., Syvukhin, 1966):

$$v_k = \left(\frac{2k_{\rm B} T_k}{m_k} \right)^{1/2} . \tag{6.49}$$

It is convenient to determine the dimensionless variable

$$x_k = \frac{v}{v_k} = \left(\frac{m_k}{M} \frac{\mathcal{E}}{k_{\rm B} T_k} \right)^{1/2} \tag{6.50}$$

and to rewrite the dimensionless function $\mathcal{P}_k$ as follows

$$\mathcal{P}_k \left(x_k, \frac{m_k}{M} \right) = \frac{1}{x_k} \, \mathrm{erf}\,(x_k) - \left(1 + \frac{m_k}{M} \right) \frac{2}{\sqrt{\pi}} \exp\left(-x_k^2 \right) . \tag{6.51}$$

Here

$$\mathrm{erf}\,(x_k) = \frac{2}{\sqrt{\pi}} \int\limits_0^{x_k^2} \exp\left(-t^2 \right) dt \tag{6.52}$$

is the probability integral.

Consider the low-energy limit. Note that

$$\mathcal{P}_k \left(x_k, \frac{m_k}{M} \right) \approx \frac{2}{\sqrt{\pi}} \left[-\frac{m_k}{M} + \frac{2}{3} \left(1 + \frac{m_k}{M} \right) x_k^2 \right] \quad \text{if} \quad x_k \ll 1 . \tag{6.53}$$

Hence the dimensionless function

$$\mathcal{P}_k \left(0, \frac{m_k}{M} \right) = -\frac{2}{\sqrt{\pi}} \frac{m_k}{M} < 0 \tag{6.54}$$

and, according to formula (6.48), the energy losses rate

$$P_k \equiv \left(\frac{d\mathcal{E}}{dt} \right)_k = \frac{8\sqrt{\pi} e^4 \ln \Lambda}{M} \frac{Z^2 Z_k^2 n_k}{v_k} > 0 . \tag{6.55}$$

This means that a test particle with zeroth (or very small) velocity takes energy from the field particles having the temperature T_k. **The hot field particles heat a cold test particle.**

Consider an opposite limiting case. If $x_k \gg 1$, then, being positive, the function

$$\mathcal{P}_k \left(x_k, \frac{m_k}{M} \right) \approx \frac{1}{x_k} \to 0 \quad \text{when} \quad x_k \gg 1 . \tag{6.56}$$

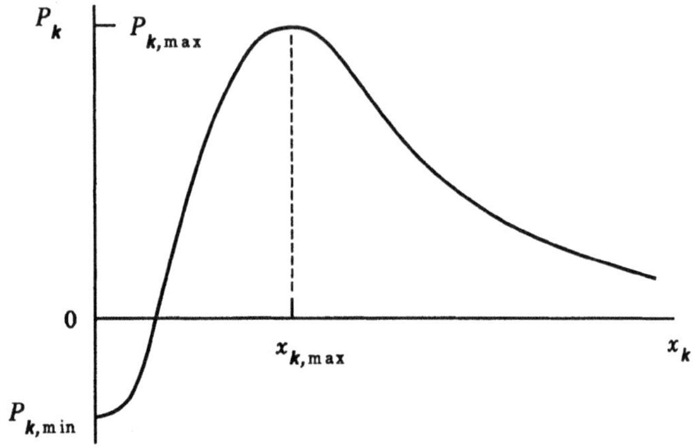

Figure 6.5: The Coulomb losses (with the sign *minus* in formula (6.48)) of a test particle as a function of its velocity measured in the most probable velocity of the field thermal particles.

So the higher the energy of a test particle, the smaller are the Coulomb losses.

The maximum of the dimensionless function $\mathcal{P}_k$ is reached at $x_{k,\,\max} \approx 1.52$, see schematical Figure 6.5.

Cosmic plasma consists of many components. To obtain the total losses it is necessary to sum over all of them in formula (6.48). However, two components – electrons and protons – give the largest contribution. In a plasma consisting of electrons and protons with $n_{\rm e} = n_{\rm p} = n$ and temperatures $T_{\rm e}$ and $T_{\rm p}$ we have (see Korchak, 1980):

$$P = -c_{\mathcal{E}}\,\frac{Z^2}{A}\,\frac{n\,\ln\Lambda}{\sqrt{k_{\rm B}T_{\rm e}}}\left[\mathcal{P}_{\rm e}\left(x_{\rm e},\frac{m_{\rm e}}{M}\right)+\left(\frac{m_{\rm e}T_{\rm e}}{m_{\rm p}T_{\rm p}}\right)^{1/2}\mathcal{P}_{\rm p}\left(x_{\rm p},\frac{m_{\rm p}}{M}\right)\right], \quad (6.57)$$

where the constant $c_{\mathcal{E}} \approx 1.6 \times 10^{-23}$.

The location of both maxima of the function (6.57) is determined by conditions:

$$x_1 = x_{\rm p} \approx 1.52 \quad \text{and} \quad x_2 = x_{\rm e} \approx 1.52\,. \quad (6.58)$$

As follows from formula (6.57), the ratio of losses in the maxima

$$\frac{P_{\max,\,{\rm p}}}{P_{\max,\,{\rm e}}} = \left(\frac{m_{\rm e}}{m_{\rm p}}\,\frac{T_{\rm e}}{T_{\rm p}}\right)^{1/2} \approx \frac{1}{43}\left(\frac{T_{\rm e}}{T_{\rm p}}\right)^{1/2}. \quad (6.59)$$

The electron Coulomb losses maximum is the main energy threshold of the particle acceleration from low energies. The proton barrier is considerably lower than the electron one.

The energy loss contribution of the proton component of cosmic plasma does not seem to be important. This is not always true, however. First of all, formula (6.59) shows that the Coulomb losses on thermal protons increase with the growth of the ratio T_e/T_p. This may be an important case if particles of low energies are accelerated in high-temperature current sheets (Section 17.3). The second argument comes from a consideration of very low energies of accelerated particles. In this region, the efficiency of acceleration is low for the majority of accelerating mechanisms. However, just in this region of low energies,

> the Coulomb losses can strongly influence the nuclear composition and the charge-state of accelerated particles

(e.g., Korchak, 1980; Holman, 1995).

(c) Dynamic friction in cosmic plasma

The collisional drag force F_f acts on a test particle (mass M, charge Ze) moving through the many-component plasma with the Maxwellian distribution of field particles:

$$M \frac{d}{dt} v_{\parallel} = -F_f = -\sum_k F_k \left(v_{\parallel} \right) . \tag{6.60}$$

Here the velocity component $v_{\parallel}$ is parallel to the vector of the initial velocity of an incident test particle.

For a test particle with a velocity v much below the thermal velocity (6.49) of the field particles with the mass m_k, temperature T_k, and number density n_k,

$$F_f \approx \sum_k \frac{4\pi e^4 \ln \Lambda}{k_{\text{B}}} \frac{Z^2 Z_k^2 n_k}{T_k} \left(1 + \frac{m_k}{M} \right) \frac{2}{3\sqrt{\pi}} \frac{v_{\parallel}}{v_k} \sim v_{\parallel} . \tag{6.61}$$

Therefore at small velocities the collisional drag force is proportional to the component $v_{\parallel}$ (cf. (1.32)).

When the test particle velocity exceeds the thermal velocity of the field particles, the drag force decreases with $v_{\parallel}$ as follows:

$$F_f = \sum_k F_k \approx \sum_k \frac{2\pi e^4 \ln \Lambda}{k_{\text{B}}} \frac{Z^2 Z_k^2 n_k}{T_k} \left(1 + \frac{m_k}{M} \right) \left(\frac{v_{\parallel}}{v_k} \right)^{-2} \sim v_{\parallel}^{-2} . \tag{6.62}$$

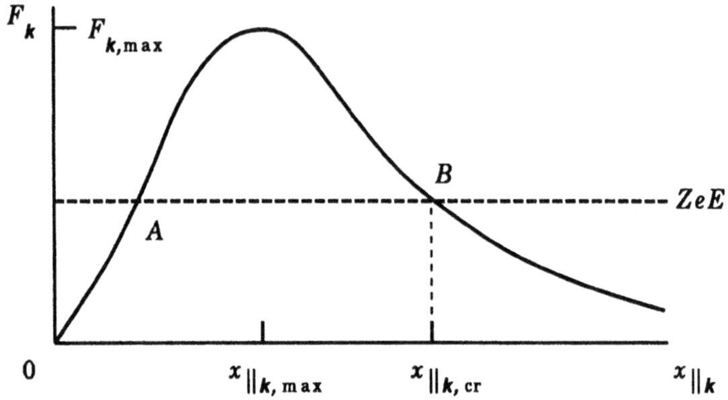

Figure 6.6: The collisional drag force F_k (with the sign *minus* in formula (6.60)) on a test particle as a function of its velocity $v_{\parallel}$ measured in the most probable velocity v_k of the field particles of the kind k.

The general formula for collisional drag force is given, for example, in Sivukhin (1966) and is illustrated by schematical Figure 6.6; here the dimensionless variable $x_{\parallel k} = v_{\parallel}/v_k$. The drag force vanishes when $x_{\parallel k} = 0$; it linearly increases with increasing $x_{\parallel k}$, becoming a maximum when

$$x_{\parallel k} = x_{\parallel k,\max} \approx 0.97, \tag{6.63}$$

and then falls off, approaching zero asymptotically as $x_{\parallel k} \to \infty$. This behaviour of the drag force has important consequences discussed below.

6.4.2 Electric runaway

It has been assumed above that the plasma is characterized by the Maxwellian distribution and that there are no external fields. Let us now assume that a uniform electric field $\mathbf{E}$ is switched on at some instant of time, the velocity distribution being assumed to be Maxwellian at this time. At least, at the beginning of the process when the velocity distribution has not yet changed appreciably, the time variation of the test-particle momentum $M\mathbf{v}$ due to Coulomb collisions with plasma particles will still be given by formulae (6.61) and (6.62) supplemented by the electric force $Ze\mathbf{E}$ in Equation (6.60). Thus, considering the component $v_{\parallel}$ as a component of the test-particle velocity $\mathbf{v}$ which is parallel to the electric field $\mathbf{E}$, we rewrite Equation (6.60) as follows:

$$M \frac{d}{dt} v_{\parallel} = -F_f + ZeE = -\sum_k F_k + ZeE\,. \tag{6.64}$$

If the test-particle velocity is not small in comparison with the thermal velocity v_k, then the collisional drag force on a test particle falls off with increasing velocity v, according to formula (6.62), while the electric force is velocity independent. Therefore

> for all particles with high enough velocities the electric force exceeds the collisional drag force, and the particles are able to *run away* from the thermal distribution.

Equating the electric and collisional drag forces allows us to see the critical velocity $v_{\rm cr}$ above which runaway will occur for a given electric field strength E, see point B in Figure 6.6. Runaway in cosmic plasma can occur as long as there is a component of the electric field along the magnetic field. Before the acceleration of the heavy ions becomes significant, the acceleration of the light electrons gives rise to the *electron runaway* effect which was first predicted by Giovanelli (1949). He has shown that

- as the electric field applied to a highly ionized gas is increased, **the current**, which is initially limited by elastic collisions between electrons and positive ions, **increases** rapidly as the field strength reaches a critical value;
- this is due to a reduction in the cross-section of positive ions for scattering of electrons with increasing electron velocity.

In a very strong electric field (or in a plasma of sufficiently low density and high temperature) all the electrons are accelerated by the field, i.e. become the runaway electrons. The Dreicer field (Dreicer, 1959):

$$E_{\rm Dr} = \frac{4\pi e^3 \ln \Lambda}{k_{\rm B}} \, \frac{n_{\rm e}}{T_{\rm e}} \tag{6.65}$$

approximately correspondes to the electric field strength for which $v_{\rm cr} = v_{\rm e}$. Here $v_{\rm e}$ is the most probable velocity of thermal electrons (6.49).

In a weak field only very fast electrons will run away, i.e. those velocity $v_{\parallel} > v_{\rm cr}$. The velocity $v_{\rm cr}$ depends in an essential manner on the magnitude of electric field. In a weak field, the velocity $v_{\rm cr}$ is naturally much larger than the thermal velocity of electrons in the plasma. Therefore the number of runaway electrons is not large. In order to determine it we must know the way in which the density of electrons having a velocity $v_{\parallel} \sim v_{\rm cr}$ varies. This means that we must know the velocity distribution for the electrons for $v_{\parallel} \sim v_{\rm cr}$. To consider this problem (e.g., Gurevich, 1961) it is necessary to solve the kinetic equation taking collisions and the electric field into account (Section 3.5).

To have an idea of the magnitude of the Dreicer field (see Problem 6.4), let us substitute the definition of the Debye radius (6.26) in formula (6.65) and assume that $T_e = T_p = T$ and $n_e = n_p = n$. We find

$$\boxed{E_{Dr} = \frac{e}{r_D^2}\,\frac{\ln\Lambda}{2} \sim \frac{e}{r_D^2}\,.} \tag{6.66}$$

So the Dreicer field is approximatelly equal to the electric field of a positive charge at a distance slightly smaller than the Debye radius.

6.4.3 Thermal runaway in cosmic plasma

Consider a plasma with a non-uniform distribution of electron temperature T_e. Let l_T be the characteristic length of the temperature profile and λ_e be the mean free path of thermal electrons. For the classical heat conductivity to be applicable, it is necessary to satisfy a condition (Section 7.2.4):

$$\lambda_e \ll l_T \equiv \frac{T_e}{|\nabla T_e|}\,. \tag{6.67}$$

The mean free path of a particle increases with its velocity. This can be seen from formula (6.10) which gives us the mean free path

$$\lambda = \tau v_1 \sim v_1^4\,. \tag{6.68}$$

That is why

> a number of fast electrons can penetrate from a hot plasma into cold one even if the gradient of temperature is very small.

In such a way, the hot plasma can lose some part of its thermal energy transfered by fast thermal escaping electrons. In addition to the usual heat flux (6.42), which is determined *locally* by the Coulomb collisions of plasma electrons, there appears a *non-local* energy flux carried by the fast electrons practically without collisions. A classical diffusive heat transfer and a convective one, determined by *thermal runaway* electrons, are always present in plasma. It is interesting for astrophysical applications that, at not too small temperature gradients, the convective transfer of thermal energy can play a principal role. Gurevich and Istomin (1979) have examined the case of a small temperature gradient. By using a perturbation analysis for the high-speed kinetic equation (Section 3.2), they have shown that the fast growth of the mean free path with increasing velocity gives an abrupt growth of the number of fast electrons in the cold plasma.

The opposite case of a **large temperature gradient** in the narrow transition layer between a high-temperature plasma and a cold one was investigated by many authors with applications to the problem of energy transfer in the solar atmosphere. For example, Shoub (1983) has solved numerically the boundary value problem for the Fokker-Plank equation in the model of the transition layer between the corona and the chromosphere in quiet conditions. An excess of fast electrons has been found in the low transion layer region. As for solar flares, the prevailing view is that the high-temperature plasma can lose energy efficiently by the convective heat transfer by runaway electrons (see Somov, 1992).

In both cases, however, it is important to take into account that **the fast runaway electrons**, similar to any beam of fast particles, **generate the electric field which drives the reverse current of thermal electrons**. Diakonov and Somov (1988) have found an analytical solution to the self-consistent kinetic problem on the beam of escaping thermal electrons and its associated reverse current (Sections 3.5.2–3.5.5). They have shown that the reverse-current electric field in solar flares leads to a significant reduction of the convective heat flux carried by fast electrons escaping from the high-temperature plasma to the cold one.

Recommended Reading: Sivukhin (1966), Somov (1992).

6.5 Practice: Problems and Answers

Problem 6.1. For an electron, which moves in the solar corona with a mean thermal velocity (Problem 4.2), evaluate the characteristic time of *close* and *distant* collisions with thermal protons.

Answer. Characteristic time of close electron-proton collisions follows from formula (6.10) and is equal to

$$\tau_{cl,\,\mathrm{ep}} = \frac{m_{\mathrm{e}}^2}{\pi e^4}\, \frac{V_{\mathrm{Te}}^3}{n_{\mathrm{p}}} \approx 4.96 \times 10^{-18}\, \frac{V_{\mathrm{Te}}^3}{n_{\mathrm{p}}}\,, \ \mathrm{s}. \qquad (6.69)$$

At typical temperatures of electrons in the corona $T_{\mathrm{e}} \approx 2 \times 10^6$ K, their thermal velocity (4.88) $V_{\mathrm{Te}} \approx 9.5 \times 10^8$ cm s^{-1}. Substituting this value in (6.69) and assuming $n_{\mathrm{p}} \approx n_{\mathrm{e}} \approx 2 \times 10^8\ \mathrm{cm}^{-3}$, we find that $\tau_{cl,\mathrm{ep}} \approx 22$ s.

According to (6.17) the characteristic time of distant collisions is 8 ln Λ shorter than the close collision time (6.69). Hence, first, we have to find the value of the Coulomb logarithm (6.29):

$$\ln \Lambda = \ln \left[\left(\frac{3 k_{\mathrm{B}}^{3/2}}{2 \pi^{1/2}\, e^3} \right) \left(\frac{T_{\mathrm{e}}^3}{n_{\mathrm{e}}} \right)^{1/2} \right] \approx \ln \left[1.25 \times 10^4 \left(\frac{T_{\mathrm{e}}^3}{n_{\mathrm{e}}} \right)^{1/2} \right] . \tag{6.70}$$

At typical coronal temperature and density, formula (6.70) gives $\ln \Lambda \approx 22$.

With this value of $\ln \Lambda$ formula (6.17) gives

$$\tau_{\perp,\mathrm{ep}} = \frac{m_{\mathrm{e}}^2}{\pi e^4} \frac{1}{8 \ln \Lambda} \frac{V_{\mathrm{Te}}^3}{n_{\mathrm{p}}} \approx 2.87 \times 10^{-20} \frac{V_{\mathrm{Te}}^3}{n_{\mathrm{p}}}\,,\ \mathrm{s}\,. \tag{6.71}$$

In the solar corona $\tau_{\perp,\mathrm{ep}} \approx 0.1$ s. Therefore *distant* collisions of thermal electrons with thermal protons in the corona are really much more frequent in comparison with close collisions.

Problem 6.2. Under conditions of Problem 6.1 evaluate the *exact* (determined by formulae (6.33) and (6.34)) *collisional times* between thermal electrons and between thermal protons, respectively. Compare these times with the characteristic time of energy exchange between electrons and protons in the coronal plasma.

Answer. By substituting $\ln \Lambda$ in (6.33), we have the following expression for the thermal electron collisional time

$$\tau_{\mathrm{ee}} = \frac{m_{\mathrm{e}}^2}{0.714\, e^4\, 8\pi \ln \Lambda} \frac{V_{\mathrm{Te}}^3}{n_{\mathrm{e}}} \approx 4.04 \times 10^{-20} \frac{V_{\mathrm{Te}}^3}{n_{\mathrm{e}}}\,,\ \mathrm{s}\,. \tag{6.72}$$

In the solar corona $\tau_{\mathrm{ee}} \approx 0.2$ s. For thermal protons formula (6.34) gives

$$\tau_{\mathrm{pp}} = \frac{m_{\mathrm{p}}^2}{0.714\, e^4\, 8\pi \ln \Lambda} \frac{V_{\mathrm{Tp}}^3}{n_{\mathrm{p}}} \approx 1.36 \times 10^{-13} \frac{V_{\mathrm{Tp}}^3}{n_{\mathrm{p}}}\,,\ \mathrm{s}\,. \tag{6.73}$$

Assuming $T_{\mathrm{p}} = T_{\mathrm{e}}$ and $n_{\mathrm{p}} = n_{\mathrm{e}}$, we find the proton collisional time in the solar corona $\tau_{\mathrm{pp}} \approx 7$ s; this is in a good agreement with formula (6.35), of course.

Finally, let us find the time of energy exchange between electrons and protons. By using formula (6.39), we have

$$\tau_{\mathrm{ep}}(\mathcal{E}) \approx 22\, \tau_{\mathrm{pp}} \approx 164\ \mathrm{s}\,. \tag{6.74}$$

So the energy exchange between electron and proton components in the coronal plasma is the slowest process determined by Coulomb collisions.

Problem 6.3. Evaluate the Debye radius and the plasma frequency in the solar corona.

Answer. From (6.26) it follows that for electron-proton plasma with $T_e = T_p = T$ and $n_e = n_p = n$ the Debye radius

$$r_D = \left(\frac{k_B T}{8\pi e^2 n} \right)^{1/2} \approx 4.9 \left(\frac{T}{n} \right)^{1/2}, \text{ cm}. \tag{6.75}$$

Under conditions in the solar corona $r_D \approx 0.5$ cm.

The electron plasma frequency (6.30)

$$\omega_{pl}^{(e)} \approx 5.64 \times 10^4 \sqrt{n_e}, \text{ rad s}^{-1}, \tag{6.76}$$

or

$$\nu_{pl}^{(e)} = \omega_{pl}^{(e)} / 2\pi \approx 10^4 \sqrt{n_e}, \text{ Hz}. \tag{6.77}$$

In the solar corona $\omega_{pl}^{(e)} \sim 10^9$ rad s^{-1}.

Problem 6.4. Evaluate and compare Dreicer's electric fields in the solar corona and in the chromosphere.

Answer. From (6.65) it follows that

$$E_{Dr} = \frac{4\pi e^3 \ln \Lambda}{k_B} \frac{n_e}{T_e} \approx 6.54 \times 10^{-8} \frac{n_e\,(\text{cm}^{-3})}{T_e\,(\text{K})}, \text{ V cm}^{-1}. \tag{6.78}$$

Here it was taken $\ln \Lambda \approx 21.6$ according to Problem 6.1.

At typical temperature and number density of electrons in the solar corona $T_e \approx 2 \times 10^6$ K and $n_e \approx 2 \times 10^8$ cm^{-3}, we find that the Dreicer electric field $E_{Dr} \approx 7 \times 10^{-6}$ V cm^{-1}. The same value follows, of course, from formula (6.66) with $r_D \approx 0.5$ cm (see Problem 6.3).

In the solar chromosphere $n_e > 2 \times 10^{10}$ cm^{-3} and $T_e < 10^4$ K. According to formula (6.78), the Dreicer electric field $E_{Dr} > 0.1$ V cm^{-1} in the chromosphere is, at least, 10^4 times stronger than the coronal one.

Problem 6.5. Define the dynamic friction by gravitational force as momentum loss by a massive moving object, for example a star in a galaxy, due to its gravitational interaction with its own gravitationally induced wake. Discuss two possibilities: the background medium consists of collisionless matter (other stars in the galaxy), the medium is entirely gaseous (e.g., Ostriker, 1999). The first case, the **gravitational drag** in collisionless systems (Chandrasekhar, 1943b), has widespread theoretical application in modern astrophysics.

Chapter 7

A Hydrodynamic Description of Cosmic Plasma

In this Chapter we are not concerned with individual particles but we will treat individual kinds of particles as continuous mediums interacting between themselves and with an electromagnetic field.

7.1 Transition to macroscopic transfer equations

The averaged Liouville equation or kinetic equation gives a *microscopic* (though averaged in a statistical sense) description of the plasma state's evolution. Consider the way of transition to a less comprehensive *macroscopic* description of a plasma. We start from the kinetic equation for particles of kind k, in the form derived in Section 2.2:

$$\frac{\partial f_k(X,t)}{\partial t} + v_\alpha \frac{\partial f_k(X,t)}{\partial r_\alpha} + \frac{F_{k,\alpha}(X,t)}{m_k} \frac{\partial f_k(X,t)}{\partial v_\alpha} = \left(\frac{\partial \hat{f}_k}{\partial t} \right)_{\rm c} . \tag{7.1}$$

Here the statistically averaged force is

$$F_{k,\alpha}(X,t) = \sum_l \int_{X_1} F_{kl,\alpha}(X,X_1)\, f_l(X_1,t)\, dX_1$$

and the collisional integral

$$\left(\frac{\partial \hat{f}_k}{\partial t} \right)_{\rm c} = - \frac{\partial}{\partial v_\alpha} j_{k,\alpha}(X,t) ,$$

where the flux of particles of kind k

$$j_{k,\alpha}(X,t) = \sum_l \int_{X_1} \frac{1}{m_k} F_{kl,\alpha}(X, X_1)\, f_{kl}(X, X_1, t)\, dX_1 \tag{7.2}$$

in the six-dimensional phase space $X = \{\mathbf{r}, \mathbf{v}\}$.

7.1.1 Distribution function moments

Before turning our attention to the deduction of equations for the macroscopic quantities or macroscopic *transfer* equations, let us define the following *moments* of the distribution function.

(a) The zeroth moment (without multiplying the distribution function f_k by the velocity)

$$\int_v f_k(\mathbf{r}, \mathbf{v}, t)\, d^3\mathbf{v} = n_k(\mathbf{r}, t) \tag{7.3}$$

is obviously the number of particles of kind k in a unit volume, i.e. the *number density* of particles of kind k. It is related to the *mass density* in a natural way:

$$\rho_k(\mathbf{r}, t) = m_k\, n_k(\mathbf{r}, t)\,.$$

The plasma mass density is accordingly

$$\rho(\mathbf{r}, t) = \sum_k \rho_k(\mathbf{r}, t) = \sum_k m_k\, n_k(\mathbf{r}, t)\,. \tag{7.4}$$

(b) The first moment of the distribution function, i.e. the integral of the product of the velocity to the first power and the distribution function f_k,

$$\int_v v_\alpha\, f_k(\mathbf{r}, \mathbf{v}, t)\, d^3\mathbf{v} = n_k\, u_{k,\alpha} \tag{7.5}$$

is the product of the number density of particles of kind k by their *mean velocity*

$$u_{k,\alpha}(\mathbf{r}, t) = \frac{1}{n_k} \int_v v_\alpha\, f_k(\mathbf{r}, \mathbf{v}, t)\, d^3\mathbf{v}\,. \tag{7.6}$$

Consequently, the *mean momentum* of particles of kind k in a unit volume is expressed in terms of the first moment of the distribution function as follows

$$m_k\, n_k\, u_{k,\alpha} = m_k \int_v v_\alpha\, f_k(\mathbf{r}, \mathbf{v}, t)\, d^3\mathbf{v}\,. \tag{7.7}$$

(c) The second moment of the distribution function is defined to be

$$m_k \int_v v_\alpha v_\beta \, f_k \left(\mathbf{r}, \mathbf{v}, t\right) d^3\mathbf{v} = m_k n_k \, u_{k,\alpha} u_{k,\beta} + p_{\alpha\beta}^{(k)} = \Pi_{\alpha\beta}^{(k)} \left(\mathbf{r}, t\right). \qquad (7.8)$$

Here

$$p_{\alpha\beta}^{(k)} = m_k \int_v v'_\alpha v'_\beta \, f_k \left(\mathbf{r}, \mathbf{v}, t\right) d^3\mathbf{v}\,, \qquad (7.9)$$

$v'_\alpha = v_\alpha - u_{k,\alpha}$ is the deviation of the particle velocity from its mean value $u_{k,\alpha} = \langle\, v_{k,\alpha} \,\rangle_v$ in the sense of the definition (7.6); so that $\langle\, v'_\alpha \,\rangle = 0$.

$\Pi_{\alpha\beta}^{(k)}$ is the *tensor of momentum flux density* for particles of kind k. Its component $\Pi_{\alpha\beta}^{(k)}$ is the αth component of the momentum transported by the particles of kind k, in a unit time, across the unit area perpendicular to the axis r_β. $p_{\alpha\beta}^{(k)}$ is termed the *pressure tensor*.

7.1.2 Equations for moments

Higher moments will be introduced as needed. Note, for now, that the deduction of macroscopic equations is nothing but the derivation of the equations for the distribution function moments.

(a) Let us calculate the *zeroth* moment of the kinetic Equation (7.1):

$$\int_v \frac{\partial f_k}{\partial t} \, d^3\mathbf{v} + \int_v v_\alpha \frac{\partial f_k}{\partial r_\alpha} \, d^3\mathbf{v} + \int_v \frac{F_{k,\alpha}}{m_k} \frac{\partial f_k}{\partial v_\alpha} \, d^3\mathbf{v} = \int_v \left(\frac{\partial \hat{f}_k}{\partial t} \right)_{\mathrm{c}} d^3\mathbf{v}\,. \qquad (7.10)$$

We interchange the order of integration over velocities and the differentiation with respect to time t in the first term and with respect to coordinates r_α in the second one. Under the second integral

$$v_\alpha \frac{\partial f_k}{\partial r_\alpha} = \frac{\partial}{\partial r_\alpha} \left(v_\alpha f_k \right) - f_k \frac{\partial v_\alpha}{\partial r_\alpha} = \frac{\partial}{\partial r_\alpha} \left(v_\alpha f_k \right) - 0\,,$$

since $\mathbf{r}$ and $\mathbf{v}$ are independent variables in phase space X.

Taking into account that the distribution function quickly approaches zero as $v \to \infty$, the integral of the third term is taken by parts and is equal to zero (Problem 7.1).

Finally, the integral of the right-hand side of (7.10) describes the change in the number of particles of kind k in a unit volume, in a unit time, as a result of collisions with particles of other kinds. If the processes of transformation, during which the particle kind can be changed (such as ionization, recombination, charge exchange, dissociation etc., see Problem 7.2), are not allowed for, then the last integral is zero as well.

Thus, by integration of (7.10), the following equation is found to result from (7.1)

$$\boxed{\frac{\partial n_k}{\partial t} + \frac{\partial}{\partial r_\alpha} n_k u_{k,\alpha} = 0 .} \tag{7.11}$$

This is the usual *continuity equation* expressing the conservation of particles of kind k.

Equation (7.11) for the zeroth moment n_k depends on the unknown first moment $u_{k,\alpha}$. This is important for what follows.

(b) Calculate the *first* moment of the kinetic Equation (7.1) multiplied by mass m_k:

$$m_k \int\limits_v \frac{\partial f_k}{\partial t} v_\alpha \, d^3\mathbf{v} + m_k \int\limits_v v_\alpha v_\beta \frac{\partial f_k}{\partial r_\beta} d^3\mathbf{v} + \int\limits_v v_\alpha F_{k,\beta} \frac{\partial f_k}{\partial v_\beta} d^3\mathbf{v} =$$

$$= m_k \int\limits_v v_\alpha \left(\frac{\partial \hat{f}_k}{\partial t} \right)_{\rm c} d^3\mathbf{v} .$$

With allowance made for the definitions (7.5) and (7.8), we obtain the *momentum conservation law*

$$\frac{\partial}{\partial t} (m_k n_k \, u_{k,\alpha}) + \frac{\partial}{\partial r_\beta} \left(m_k n_k \, u_{k,\alpha} u_{k,\beta} + p_{\alpha\beta}^{(k)} \right) -$$

$$- \langle F_{k,\alpha}(\mathbf{r},t) \rangle_v = \langle F_{k,\alpha}^{(\rm c)} (\mathbf{r},t) \rangle_v . \tag{7.12}$$

The *mean force* acting on the particles of kind k in a unit volume (the mean force *per unit volume*) is

$$\langle F_{k,\alpha} (\mathbf{r},t) \rangle_v = \int\limits_v F_{k,\alpha} (\mathbf{r},\mathbf{v},t) \; f_k (\mathbf{r},\mathbf{v},t) \; d^3\mathbf{v} . \tag{7.13}$$

This should not be confused with the *statistical mean* force acting on a single particle (see definition (2.26))

$$F_{k,\alpha} (X,t) = \sum_l \int\limits_{X_1} F_{kl,\alpha} (X,X_1) \; f_l (X_1,t) \; dX_1 .$$

In the particular case of the Lorentz force, we rewrite the mean force per unit volume as follows:

$$\langle F_{k,\alpha}(\mathbf{r},t)\rangle_v = n_k e_k \left[E_\alpha + \frac{1}{c}(\mathbf{u}_k \times \mathbf{B})_\alpha \right]$$

or

$$\boxed{\langle F_{k,\alpha}(\mathbf{r},t)\rangle_v = \rho_k^{\mathrm{q}} E_\alpha + \frac{1}{c}(\mathbf{j}_k^{\mathrm{q}} \times \mathbf{B})_\alpha .} \tag{7.14}$$

Here ρ_k^{q} and $\mathbf{j}_k^{\mathrm{q}}$ are the densities of electric charge and current, produced by the particles of kind k.

The right-hand side of Equation (7.12) contains the mean force resulting from collisions, i.e. the *mean collisional force*

$$\langle F_{k,\alpha}^{(c)}(\mathbf{r},t)\rangle_v = m_k \int\limits_v v_\alpha \left(\frac{\partial \hat{f}_k}{\partial t}\right)_{\mathrm{c}} d^3\mathbf{v} =$$

$$= -m_k \int\limits_v v_\alpha \frac{\partial}{\partial v_\beta} j_{k,\beta}\, d^3\mathbf{v} . \tag{7.15}$$

Integrate (7.15) by parts. For this purpose, find the derivative

$$\frac{\partial}{\partial v_\beta}(v_\alpha j_{k,\beta}) = j_{k,\beta}\frac{\partial v_\alpha}{\partial v_\beta} + v_\alpha \frac{\partial}{\partial v_\beta} j_{k,\beta} .$$

From this it follows that

$$v_\alpha \frac{\partial}{\partial v_\beta} j_{k,\beta} = -j_{k,\beta}\,\delta_{\alpha\beta} + \frac{\partial}{\partial v_\beta}(v_\alpha j_{k,\beta}) =$$

$$= -j_{k,\alpha} + \frac{\partial}{\partial v_\beta}(v_\alpha j_{k,\beta}) . \tag{7.16}$$

On substituting (7.16) and (7.2) in (7.15) and integrating, we obtain the most general formula for the mean collisional force

$$\langle F_{k,\alpha}^{(c)}(\mathbf{r},t)\rangle_v = m_k \int\limits_v j_{k,\alpha}(\mathbf{r},\mathbf{v},t)\, d^3\mathbf{v} = \tag{7.17}$$

$$= \sum_{l \neq k} \int\limits_v \int\limits_{v_1} \int\limits_{r_1} F_{kl,\alpha}(\mathbf{r},\mathbf{v},\mathbf{r}_1,\mathbf{v}_1)\, f_{kl}(\mathbf{r},\mathbf{v},\mathbf{r}_1,\mathbf{v}_1,t)\, d^3\mathbf{r}_1\, d^3\mathbf{v}_1\, d^3\mathbf{v} .$$

Note that,

for the particles of the same kind, the elastic collisions cannot change the total particle momentum per unit volume.

That is why $l \neq k$ in the sum (7.17).

If there are several kinds of particles, the mean collisional force can be expressed in terms of the *mean momentum loss* during the collisions of a particle of kind k with the particles of other kinds:

$$\langle F_{k,\alpha}^{(c)}(\mathbf{r}, t) \rangle = - \sum_{l \neq k} \frac{m_k n_k (u_{k,\alpha} - u_{l,\alpha})}{\tau_{kl}}, \tag{7.18}$$

where $\tau_{kl}^{-1} = \nu_{kl}$ is the mean frequency of collisions between the particles of kinds k and l ($l \neq k$). This force is zero, once the particles of all kinds have identical velocities. If $u_{l,\alpha} < u_{k,\alpha}$ then the force is negative:

the fast particles of kind k slow down by dint of collisions with the slowly moving particles of other kinds.

(c) In order to derive the *energy conservation law*, we multiply Equation (7.1) by the particle's kinetic energy $m_k v_\alpha^2/2$ and integrate over velocities, taking into account that

$$v_\alpha^2 = u_{k,\alpha}^2 + (v'_\alpha)^2 + 2\, u_{k,\alpha}\, v'_\alpha .$$

On rearrangement, we obtain the following general equation:

$$\frac{\partial}{\partial t}\left(\frac{\rho_k u_k^2}{2} + \rho_k\, \varepsilon_k\right) + \frac{\partial}{\partial r_\alpha}\left[\rho_k u_{k,\alpha}\left(\frac{u_k^2}{2} + w_k\right) + \pi_{\alpha\beta}^{(k)}\, u_{k,\beta} + q_{k,\alpha}\right] =$$

$$= \rho_k^{\mathrm{q}}\,(\mathbf{E} \cdot \mathbf{u}_k) + \left(\mathbf{F}_k^{(c)} \cdot \mathbf{u}_k\right) + Q_k^{(c)}(\mathbf{r}, t) . \tag{7.19}$$

Here

$$m_k\, \varepsilon_k(\mathbf{r}, t) = \frac{1}{n_k} \int\limits_v \frac{m_k\, (v'_\alpha)^2}{2}\, f_k(\mathbf{r}, \mathbf{v}, t)\, d^3\mathbf{v} =$$

$$= \frac{m_k}{2n_k} \int\limits_v (v'_\alpha)^2\, f_k(\mathbf{r}, \mathbf{v}, t)\, d^3\mathbf{v} \tag{7.20}$$

is the *mean kinetic energy* of chaotic (non-directed) motion per single particle of kind k. The term

$$\rho_k u_{k,\alpha}\left(\frac{u_k^2}{2} + w_k\right)$$

can be called the 'advection' flux of energy.

* * *

If the particles of the kth kind are in the *thermodynamic equilibrium* state, then f_k is the Maxwellian function with the *temperature* T_k and the mean kinetic energy of chaotic motion per single particle of kind k

$$m_k \, \varepsilon_k = \frac{3}{2} \, k_{\rm B} T_k \, . \tag{7.21}$$

In this case the pressure tensor (7.9) is isotropic:

$$p_{\alpha\beta}^{(k)} = p_k \, \delta_{\alpha\beta} \, , \tag{7.22}$$

where

$$p_k = n_k \, k_{\rm B} T_k \tag{7.23}$$

is the gas pressure of the particles of kind k. This is also the equation of state for the *perfect* gas. The heat function per unit mass or, more exactly, the *specific enthalpy* is

$$w_k = \varepsilon_k + \frac{p_k}{\rho_k} = \frac{5}{2} \, \frac{k_{\rm B} T_k}{m_k} \, . \tag{7.24}$$

This is a particular case of the thermodynamic equilibrium state; it will be discussed in Section 7.2.4.

* * *

Generally, without supposing thermodynamic equilibrium, the part associated with the deviation of the distribution function from the isotropic one (which does not need to be a Maxwellian function) is distinguished in the pressure tensor:

$$\pi_{\alpha\beta}^{(k)} = p_{\alpha\beta}^{(k)} - p_k \, \delta_{\alpha\beta} \, , \tag{7.25}$$

where $\pi_{\alpha\beta}^{(k)}$ is called the *viscous stress tensor*. The vector

$$q_{k,\alpha} = \int\limits_v \frac{m_k (v')^2}{2} \, v'_\alpha \, f_k \, ({\bf r}, {\bf v}, t) \; d^3 {\bf v} \tag{7.26}$$

is the *heat flux density* due to the particles of kind k in a system of coordinates, in which the gas of these particles is immovable at a given point of space.

The right-hand side of the energy conservation law (7.19) contains the following terms:

$$\rho_k^{\mathrm{q}} \left(\mathbf{E} \cdot \mathbf{u}_k \right) = n_k e_k \, E_\alpha \, u_{k,\alpha} \,, \tag{7.27}$$

which is the work done by the Lorentz force (without the magnetic field, of course) in unit time on unit volume, and the second term

$$\left(\mathbf{F}_k^{(\mathrm{c})} \cdot \mathbf{u}_k \right) = u_{k,\alpha} \int\limits_v m_k \, v_\alpha' \left(\frac{\partial \hat{f}_k}{\partial t} \right)_{\mathrm{c}} d^3\mathbf{v} \,, \tag{7.28}$$

which is the work done by the force of friction of particles of kind k with all other particles in unit time on unit volume, i.e.

> the work resulting from the mean momentum change of particles of kind k (moving with the mean velocity $\mathbf{u}_k$) owing to collisions with all other particles.

The last term

$$Q_k^{(c)} \left(\mathbf{r}, t \right) = \int\limits_v \frac{m_k \left(v' \right)^2}{2} \left(\frac{\partial \hat{f}_k}{\partial t} \right)_{\mathrm{c}} d^3\mathbf{v} \tag{7.29}$$

is the rate of thermal energy release (heating) in a gas of the particles of kind k due to collisions with other particles.

7.1.3 General properties of the transfer equations

Equations (7.11), (7.12), and (7.19) are referred to as the equations of particle, momentum and energy *transfer*, respectively; and the approximation in which they have been obtained may be called the model of *mutually penetrating* charged gases. These gases are not assumed to be in the thermodynamic equilibrium.

The equations of mass, momentum and energy transfer are written in the 'divergent' form. Other forms are sometimes convenient. For instance, the equation of momentum transfer or simply the equation of motion (7.12) can be brought into the frequently used form, with the aid of the continuity Equation (7.11):

$$\rho_k \left(\frac{\partial \, u_{k,\alpha}}{\partial t} + u_{k,\beta} \, \frac{\partial \, u_{k,\alpha}}{\partial r_\beta} \right) = - \frac{\partial}{\partial r_\beta} \, p_{\alpha\beta}^{(k)} +$$

$$+ \left\langle F_{k,\alpha} \left(\mathbf{r}, t \right) \right\rangle_v + \left\langle F_{k,\alpha}^{(\mathrm{c})} \left(\mathbf{r}, t \right) \right\rangle_v . \tag{7.30}$$

The so-called ***substantial derivative*** appears on the left-hand side of this equation:

$$\boxed{\frac{d}{dt} = \frac{\partial}{\partial t} + u_\beta \frac{\partial}{\partial r_\beta} = \frac{\partial}{\partial t} + \mathbf{u} \cdot \nabla_{\mathbf{r}} \, .} \tag{7.31}$$

This derivative – the total time derivative following a *fluid element* – is typical of hydrodynamic equations, to which the equation of motion (7.30) belongs. For the case of the Lorentz force (7.14), it can be rewritten as follows:

$$\rho_k \frac{d\, u_{k,\alpha}}{dt} = -\frac{\partial}{\partial r_\beta} p_{\alpha\beta}^{(k)} + \rho_k^{\mathrm{q}} E_\alpha + \frac{1}{c} \left(\mathbf{j}_k^{\mathrm{q}} \times \mathbf{B} \right)_\alpha + \langle F_{k,\alpha}^{(\mathrm{c})} (\mathbf{r}, t) \rangle_v \, . \tag{7.32}$$

Therefore, when we treat a plasma as one or several continuous media (mutually penetrating charged gases), the average properties (quantities like density, velocity, and pressure) are governed by the basic conservation laws for mass, momentum, and energy in the media.

Unfortunately, the transfer equations for local macroscopic quantities are as much unclosed as the initial kinetic Equation (7.1), which is the first equation of the chain for correlational functions. For example, formula (7.17) contains the unknown binary correlation function f_{kl}.

It is more important, however, that the transfer equations are unclosed in 'orthogonal' direction: the Equation (7.11) for the zeroth moment, density n_k, depends on the unknown first moment, velocity $u_{k,\alpha}$, and so on.

> The transfer equations represent the first three links in the chain of the equations for the distribution function moments.

We will come back to this critical point in the next Section.

7.2 Hydrodynamic equations for cosmic plasma

7.2.1 The continuity equation

In order to consider cosmic plasma as a *single* hydrodynamic medium, we have to sum each of the three transfer equations over all kinds of particles. Let us start from the continuity Equation (7.11). With allowance for the definition of the plasma mass density (7.4), we have

$$\frac{\partial \rho}{\partial t} + \operatorname{div} \left(\sum_k \rho_k \mathbf{u}_k \right) = 0 \, . \tag{7.33}$$

The mean velocities of motion for all kinds of particles are supposed to be equal to the plasma hydrodynamic velocity:

$$\mathbf{u}_1(\mathbf{r},t) = \mathbf{u}_2(\mathbf{r},t) = \cdots = \mathbf{u}(\mathbf{r},t).$$

However this is not a general case. In general, the mean velocities are not the same, and a frame of reference can be chosen in which

$$\rho\,\mathbf{u} = \sum_k \rho_k \mathbf{u}_k . \tag{7.34}$$

Then from (7.33) and (7.34) we obtain the usual *continuity equation*

$$\frac{\partial \rho}{\partial t} + \operatorname{div} \rho\,\mathbf{u} = 0 . \tag{7.35}$$

7.2.2 The momentum conservation law in cosmic plasma

In much the same way, we handle the momentum transfer Equation (7.32). On summing over all kinds of particles, we obtain the following equation:

$$\rho\,\frac{d\,u_\alpha}{dt} = -\,\frac{\partial}{\partial r_\beta}\,p_{\alpha\beta} + \rho^{\mathrm{q}}\,E_\alpha + \frac{1}{c}\,(\,\mathbf{j}\times\mathbf{B}\,)_\alpha + \sum_k \langle\, F^{(\mathrm{c})}_{k,\alpha}(\mathbf{r},t)\,\rangle_v . \tag{7.36}$$

Here the *volume charge* density in plasma is

$$\rho^{\mathrm{q}} = \sum_k n_k e_k = \frac{1}{4\pi}\operatorname{div}\mathbf{E}, \tag{7.37}$$

and the electric current density is

$$\mathbf{j} = \sum_k n_k e_k\,\mathbf{u}_k = \frac{c}{4\pi}\operatorname{curl}\mathbf{B} - \frac{1}{4\pi}\frac{\partial \mathbf{E}}{\partial t} . \tag{7.38}$$

Maxwell's equations (1.4) and (1.1) have been used on the right-hand side of Equations (7.37) and (7.38).

Since elastic collisions do not change the total momentum, we have

$$\sum_k \langle\, F^{(\mathrm{c})}_{k,\alpha}(\mathbf{r},t)\,\rangle_v = 0 . \tag{7.39}$$

On substituting (7.37) – (7.39) in Equation (7.36), the latter can be rearranged to give the *momentum conservation law* for plasma

$$\boxed{\rho\,\frac{d\,u_\alpha}{dt} = -\,\frac{\partial}{\partial r_\beta}\,p_{\alpha\beta} + F_\alpha(\mathbf{E},\mathbf{B}) .} \tag{7.40}$$

Here the Lorentz force is written in terms of the electric and magnetic field vectors:

$$F_\alpha(\mathbf{E},\mathbf{B}) = -\frac{\partial}{\partial t}\,\frac{(\mathbf{E}\times\mathbf{B})_\alpha}{4\pi c} - \frac{\partial}{\partial r_\beta}\,T_{\alpha\beta}\,. \tag{7.41}$$

The tensor

$$T_{\alpha\beta} = \frac{1}{4\pi}\left[-E_\alpha E_\beta - B_\alpha B_\beta + \frac{1}{2}\,\delta_{\alpha\beta}\,(E^2 + B^2)\right] \tag{7.42}$$

is called the *Maxwellian tensor* of stresses.

The divergent form of the *momentum conservation law* is

$$\boxed{\frac{\partial}{\partial t}\left[\rho\,u_\alpha + \frac{(\mathbf{E}\times\mathbf{B})_\alpha}{4\pi c}\right] + \frac{\partial}{\partial r_\beta}\,(\Pi_{\alpha\beta} + T_{\alpha\beta}) = 0\,,} \tag{7.43}$$

where

$$\Pi_{\alpha\beta} = p_{\alpha\beta} + \rho\,u_\alpha u_\beta \tag{7.44}$$

is the *momentum flux density* tensor analogous to (7.8). Recall that

$$p_{\alpha\beta} = p\,\delta_{\alpha\beta} + \pi_{\alpha\beta}\,,$$

where

$$\pi_{\alpha\beta} = \sum_k \pi_{\alpha\beta}^{(k)}$$

is the *viscous stress* tensor (see definition (7.25)) and

$$p = \sum_k p_k \tag{7.45}$$

is the total plasma pressure, the sum of *partial pressures*.

The momentum conservation law in the form (7.40) or (7.43) is applied for a wide range of conditions in cosmic plasmas like **fluid relativistic flows**, for example, astrophysical jets (Section 8.4). The assumption that the astrophysical plasma behaves as a continuum medium, which is essential if these forms of the momentum conservation law are to be applied, is excellent in the cases in which we are often interested: the Debye length and the particle Larmor radii are much smaller than the plasma flow scales.

7.2.3 The energy conservation law

In a similar manner, the energy conservation law is derived: we sum the general Equation (7.19) over k and then substitute in the resulting equation the electric current density (7.38) expressed in terms of the electric field $\mathbf{E}$ and magnetic field $\mathbf{B}$. On rearrangement, the following divergent form of the energy conservation law (cf. the simplified Equation (1.43) for electromagnetic field energy and kinetic energy of charged particles) is obtained:

$$\frac{\partial}{\partial t}\left(\frac{\rho u^2}{2} + \rho\varepsilon + \frac{E^2+B^2}{8\pi}\right) + \\ + \frac{\partial}{\partial r_\alpha}\left[\rho u_\alpha\left(\frac{u^2}{2} + w\right) + \frac{c}{4\pi}(\mathbf{E}\times\mathbf{B})_\alpha + \pi_{\alpha\beta}u_\beta + q_\alpha\right] = \\ = \left(u_\alpha F_\alpha^{(c)}\right)_{ff}. \tag{7.46}$$

The right-hand side of Equation (7.46) contains the total work of *friction forces* (7.28) in unit time on unit volume

$$\left(u_\alpha F_\alpha^{(c)}\right)_{ff} = \sum_k \left(F_{k,\alpha}^{(c)}\, u_{k,\alpha}\right) = \sum_k u_{k,\alpha}\int_v m_k\, v'_\alpha \left(\frac{\partial f_k}{\partial t}\right)_{\mathrm{c}} d^3\mathbf{v}\,. \tag{7.47}$$

This work is not zero. By contrast, the total heat release under elastic collisions between particles of different kinds (see definition (7.29)) is

$$\sum_k Q_k^{(c)}(\mathbf{r},t) = \sum_k \int_v \frac{m_k\,(v')^2}{2}\left(\frac{\partial f_k}{\partial t}\right)_{\mathrm{c}} d^3\mathbf{v} = 0\,. \tag{7.48}$$

Elastic collisions in a plasma conserve both the total momentum (see Equation (7.39)) and the total energy (see Equation (7.48)).

7.2.4 The equation of state and transfer coefficients

The hydrodynamic Equations (7.35), (7.43), and (7.46) for a plasma would be closed with respect to the three unknown terms ρ, $\mathbf{u}$ and T, if it were possible to express the *unknown* quantities p, $\pi_{\alpha\beta}$, q_α, and $(u_\alpha F_\alpha^{(c)})_{ff}$ in terms of these three variables. For this purpose, we have to know the equation of state and the so-called *transfer coefficients*. How can we find them?

According to the general principles of statistical physics, any distribution function tends, by virtue of collisions, to assume the Maxwellian form. In this case the equation of state is that of the perfect gas.

The Maxwellian distribution is the kinetic equation solution for a stationary homogeneous plasma in the thermal equilibrium state, i.e. for a plasma in **thermodynamic equilibrium**. Then spatial gradients and derivatives with respect to time are zero. In fact they are always nonzero. However,

> if the gradients and derivatives are *small*, then the real distribution function differs *little* from the local Maxwellian one, the difference being proportional to the small gradients or derivatives.

Thus if we are interested in the processes occurring in a time t, which is much greater than the characteristic collision time τ, and at a distance L, which is much larger than the particle mean free path λ,

$$t \gg \tau \,, \qquad L \gg \lambda \,, \tag{7.49}$$

then the particle distribution function $f_k(\mathbf{r}, \mathbf{v}, t)$ can be thought of as a sum of the local Maxwellian distribution

$$f_k^{(0)}(\mathbf{r}, \mathbf{v}, t) = n_k(\mathbf{r}, t) \left[\frac{m_k}{2\pi k_{\text{B}} T_k(\mathbf{r}, t)} \right]^{3/2} \times$$

$$\times \exp \left\{ - \frac{m_k \left[\mathbf{v} - \mathbf{u}_k(\mathbf{r}, t) \right]^2}{2k_{\text{B}} T_k(\mathbf{r}, t)} \right\} \tag{7.50}$$

and some **small additional term** $f_k^{(1)}(\mathbf{r}, \mathbf{v}, t)$. Therefore

$$f_k(\mathbf{r}, \mathbf{v}, t) = f_k^{(0)}(\mathbf{r}, \mathbf{v}, t) + f_k^{(1)}(\mathbf{r}, \mathbf{v}, t) \,, \qquad \left| f_k^{(1)} \right| < f_k^{(0)} . \tag{7.51}$$

We seek the additional term $f_k^{(1)}$ in the *linear* approximation with respect to the factors disturbing the Maxwellian distribution, such as gradients of physical parameters, electric fields etc. The quantities $\pi_{\alpha\beta}$, q_α etc., which in their turn are proportional to the same factors, can be expressed in terms of $f_k^{(1)}$. (For example, in the case of the heat flux q_α, both the additional term $f_k^{(1)}$ and the flux q_α are chosen to be proportional to the temperature gradient.) The proportionality coefficients are the sought-after transfer coefficients.

Since the Maxwellian function (7.50) and its derivatives are uniquely determined by the parameters n_k, $\mathbf{u}_k$, and T_k, the transfer coefficients are expressed in terms of the same quantities. This procedure makes it possible to close the set of hydrodynamic equations for a cosmic plasma under the conditions (7.49).

Recommended Reading: Braginskii (1965), Hollweg (1986), Bobrova and Sasorov (1993).

7.2.5 Gravitational systems

Remember that there is a big difference between cosmic plasmas and cosmic gravitational systems (Section 2.5.3). The gravitational attraction cannot be screened. A large-scale gravitational field always exists over a system. This follows from the formula (2.59) which shows that the averaged gravitational force cannot be equal to zero because the neutrality condition (2.60) cannot be satisfied if all the particles have the same charge sign.

The large-scale gravitational field makes an overall thermodynamic equilibrium impossible. On the contrary, the electric force in a plasma is screened beyond the Debye radius and does not come in the way of the plasma having a proper thermodynamic equilibrium. Therefore

> those results of cosmic plasma physics which explicitly depend upon the plasma being in thermodynamic equilibrium do not hold for gravitational systems.

For gravitational systems, like the stars in a galaxy, we may hope that the final distribution function reflects something about the initial conditions rather than just reflecting the relaxation mechanism. The random motions of the stars may be not only non-Maxwellian but even direction dependent within the system. So galaxies may be providing us with clues on how they were formed (e.g., Palmer, 1994; Bertin, 1999; Peacock, 1999).

If we assume that the stars form a collisionless system (see, however, Section 2.5.3), they do not exert pressure. Such a pressureless gravitating system is unstable (Jean's instability). Presumably a real galaxy should possess something akin to pressure to withstand the collapsing action of its gravity. This 'pressure' is associated with the random motion of stars. So the role of sound speed is assumed to be played by the root mean speed of the stars.

Another justification for treating a galaxy in the hydrodynamic approximation is that we consider processes on a spatial scale which is large enough to contain a large number of stars – one of the two requirements of the continuum mechanics. Anyway, several aspects of the structure of a galaxy can be understood by assuming that it is made up of a continuum medium.

More often than not, hydrodynamics provides a first level description of an astrophysical phenomenon governed predominantly by the gravitational force. Magnetic fields are usually included later on in order to address additional issues. For example, the early stages of star formation during which an interstellar cloud of low density collapses under the action of its own gravity can be modeled in the hydrodynamic approximation. However, when we want to explain the difference between the angular momentum of the cloud and that of the born star, we have to include the effect of a magnetic field.

7.3 The generalized Ohm's law in cosmic plasma

The Ohm's law, $\mathbf{j} = \sigma \mathbf{E}$, relates the current to the electric field. The coefficient σ is electric conductivity. As we have seen in Chapter 1 (see also Sections 4.4, 18.1, and 18.2), the electric field determines the electron and ion acceleration, rather than their velocity. That is why, generally, such a simple relation is absent. Moreover, while considering a plasma, it is necessary to take into account the presence of a magnetic field and the motion of plasma as a whole, and as a medium consisting of several moving components.

7.3.1 Basic equations

Recall the way of deriving the usual Ohm's law. The electric current is determined by the relative motion of electrons and ions. Considering the processes in which all quantities vary only slightly in a time between the electron-ion collisions, electron inertia can be neglected. An equilibrium is set up between the electric field action and electrons-on-ions friction (see point A in Figure 6.6). The condition for this equilibrium with respect to the electron gas results in Ohm's law.

It is therefore concluded that, in order to deduce the generalized Ohm's law for the plasma with magnetic field, we have to consider at least two equations of motion – for the electron and ion components. A crude theory of conductivity in a fully-ionized plasma can be given in terms of a two-fluid approximation. The more general case, with the motion of neutrals taken into account, has been considered by Schlüter (1951), Alfvén and Fälthammar (1963); see also different applications of the generalized Ohm's law in the *three-component* cosmic plasma (Shabansky, 1971; Kunkel, 1984; Hénoux and Somov, 1991 and 1997; Murata, 1991).

Let us write the transfer Equations (7.12) for the electrons and ions, taking proper account of the Lorentz force (7.14) and the friction force (7.18).

We have two following equations:

$$m_e \frac{\partial}{\partial t} (n_e u_{e,\alpha}) = - \frac{\partial \Pi_{\alpha\beta}^{(e)}}{\partial r_\beta} - e n_e \left[\mathbf{E} + \frac{1}{c} (\mathbf{u}_e \times \mathbf{B}) \right]_\alpha + $$
$$+ m_e n_e \nu_{ei} (u_{i,\alpha} - u_{e,\alpha}) , \tag{7.52}$$
$$m_i \frac{\partial}{\partial t} (n_i u_{i,\alpha}) = - \frac{\partial \Pi_{\alpha\beta}^{(i)}}{\partial r_\beta} + Z_i e n_i \left[\mathbf{E} + \frac{1}{c} (\mathbf{u}_i \times \mathbf{B}) \right]_\alpha + $$
$$+ m_e n_i \nu_{ei} (u_{e,\alpha} - u_{i,\alpha}) . \tag{7.53}$$

The last term in (7.52) represents the mean momentum transmitted, because of collisions (formula (7.18)), from ions to electrons. It is equal, with opposite sign, to the last term in (7.53). It is assumed that there are just two kinds of particles, their total momentum remaining constant under the action of elastic collisions.

Suppose that the ions are protons ($Z_i = 1$) and electrical *neutrality* is observed:

$$n_i = n_e = n .$$

Multiply Equation (7.52) by $-e/m_e$ and add it to Equation (7.53) multiplied by e/m_i. The result is

$$\frac{\partial}{\partial t} [e n (u_{i,\alpha} - u_{e,\alpha})] = \left[\frac{e}{m_i} F_{i,\alpha} - \frac{e}{m_e} F_{e,\alpha} \right] +$$
$$+ e^2 n \left(\frac{1}{m_e} + \frac{1}{m_i} \right) E_\alpha + \frac{e^2 n}{c} \left[\left(\frac{\mathbf{u}_e}{m_e} \times \mathbf{B} \right)_\alpha + \left(\frac{\mathbf{u}_i}{m_i} \times \mathbf{B} \right)_\alpha \right] -$$
$$- \nu_{ei} e n \left[(u_{i,\alpha} - u_{e,\alpha}) + \frac{m_e}{m_i} (u_{i,\alpha} - u_{e,\alpha}) \right] . \tag{7.54}$$

Here

$$F_{e,\alpha} = - \frac{\partial \Pi_{\alpha\beta}^{(e)}}{\partial r_\beta} \quad \text{and} \quad F_{i,\alpha} = \frac{\partial \Pi_{\alpha\beta}^{(i)}}{\partial r_\beta} . \tag{7.55}$$

Let us introduce the velocity of the centre-of-mass system

$$\mathbf{u} = \frac{m_i \mathbf{u}_i + m_e \mathbf{u}_e}{m_i + m_e} . \tag{7.56}$$

Since $m_i \gg m_e$,

$$\mathbf{u} = \mathbf{u}_i + \frac{m_e}{m_i} \mathbf{u}_e \approx \mathbf{u}_i . \tag{7.57}$$

On treating Equation (7.54), we neglect the small terms of the order of the ratio m_e/m_i. On rearrangement, we obtain the equation for the current

$$\mathbf{j} = en\,(\mathbf{u}_i - \mathbf{u}_e) \tag{7.58}$$

in the system of coordinates (7.57). This equation is

$$\frac{\partial \mathbf{j}'}{\partial t} = \frac{e^2 n}{m_e}\left[\mathbf{E} + \frac{1}{c}(\mathbf{u}\times\mathbf{B})\right] - \frac{e}{m_e c}(\mathbf{j}'\times\mathbf{B}) - - \nu_{ei}\,\mathbf{j}' + \frac{e}{m_i}\mathbf{F}_i - \frac{e}{m_e}\mathbf{F}_e\,. \tag{7.59}$$

The prime designates the electric current in the system of moving plasma, i.e. in the rest-frame of the plasma. Let $\mathbf{E}_u$ denote the electric field in this frame of reference, i.e.

$$\mathbf{E}_u = \mathbf{E} + \frac{1}{c}\,\mathbf{u}\times\mathbf{B}\,. \tag{7.60}$$

Divide Equation (7.59) by ν_{ei} and represent it in the form

$$\mathbf{j}' = \frac{e^2 n}{m_e \nu_{ei}}\,\mathbf{E}_u - \frac{\omega_B^{(e)}}{\nu_{ei}}\,\mathbf{j}'\times\mathbf{n} - \frac{1}{\nu_{ei}}\frac{\partial \mathbf{j}'}{\partial t} + \frac{1}{\nu_{ei}}\left(\frac{e}{m_i}\mathbf{F}_i - \frac{e}{m_e}\mathbf{F}_e\right), \tag{7.61}$$

where $\mathbf{n} = \mathbf{B}/B$ and $\omega_B^{(e)} = eB/mc$. This is a differential equation for $\mathbf{j}'$.

Replace the third and the fourth terms on the right by some effective electric field such that

$$\sigma \mathbf{E}_{eff} = -\frac{1}{\nu_{ei}}\frac{\partial \mathbf{j}'}{\partial t} + \frac{e}{\nu_{ei}}\left(\frac{1}{m_i}\mathbf{F}_i - \frac{1}{m_e}\mathbf{F}_e\right), \tag{7.62}$$

where

$$\boxed{\sigma = \frac{e^2 n}{m_e\,\nu_{ei}}} \tag{7.63}$$

is the *plasma conductivity* in the absence of magnetic field. Combine the fields (7.60) and (7.62),

$$\mathbf{E}' = \mathbf{E}_u + \mathbf{E}_{eff}\,,$$

in order to rewrite (7.61) in the form

$$\mathbf{j}' = \sigma \mathbf{E}' - \frac{\omega_B^{(e)}}{\nu_{ei}}\,\mathbf{j}'\times\mathbf{n}\,. \tag{7.64}$$

We will consider (7.64) as an *algebraic* equation in $\mathbf{j}'$, neglecting the $\partial \mathbf{j}'/\partial t$ dependence of the field (7.62). Note, however, that the term $\partial \mathbf{j}'/\partial t$ is by no means small in the problem of the particle acceleration by a strong electric field in cosmic plasma.

Collisionless reconnection is an example in which **particle inertia** (usually combined with anomalous resistivity, see Sections 16.4.3 and 17.3) of the current replaces classical resistivity in allowing fast reconnection to occur (e.g., Drake and Kleva, 1991; Horiuchi and Sato, 1994, 1997).

7.3.2 The general solution

Let us find the solution to (7.64) as a sum of three currents

$$\boxed{\mathbf{j}' = \sigma_{\parallel}\,\mathbf{E}'_{\parallel} + \sigma_{\perp}\,\mathbf{E}'_{\perp} + \sigma_{\rm H}\,\mathbf{n}\times\mathbf{E}'_{\perp}\,.} \tag{7.65}$$

Substituting formula (7.65) in Equation (7.64) gives

$$\sigma_{\parallel} = \sigma = \frac{e^2 n}{m_{\rm e}\nu_{\rm ei}}\,, \tag{7.66}$$

$$\sigma_{\perp} = \sigma\,\frac{1}{1+\left(\omega_{\rm B}^{(\rm e)}\tau_{\rm ei}\right)^2}\,, \qquad \tau_{\rm ei} = \frac{1}{\nu_{\rm ei}}\,; \tag{7.67}$$

$$\sigma_{\rm H} = \sigma_{\perp}\left(\omega_{\rm B}^{(\rm e)}\tau_{\rm ei}\right) = \sigma\,\frac{\omega_{\rm B}^{(\rm e)}\tau_{\rm ei}}{1+\left(\omega_{\rm B}^{(\rm e)}\tau_{\rm ei}\right)^2}\,. \tag{7.68}$$

Formula (7.65) is called the *generalized* Ohm's law. It shows that the presence of a magnetic field in a plasma not only changes the magnitude of the conductivity, but the form of Ohm's law as well: generally, the electric field and the resulting current are not parallel, since $\sigma_{\perp} \neq \sigma_{\parallel}$. Therefore the electric conductivity of a plasma in a magnetic field is *anisotropic*. Moreover, the current component, which is perpendicular to both the magnetic and electric fields, appears in the plasma. This component is the so-called Hall current (Figure 7.1).

7.3.3 The conductivity of magnetized plasma

The magnetic field influence on the conductivity of the 'direct' current $\sigma_{\perp}$ and that of Hall's current $\sigma_{\rm H}$ is determined by the parameter $\omega_{\rm B}^{(\rm e)}\tau_{\rm ei}$, which

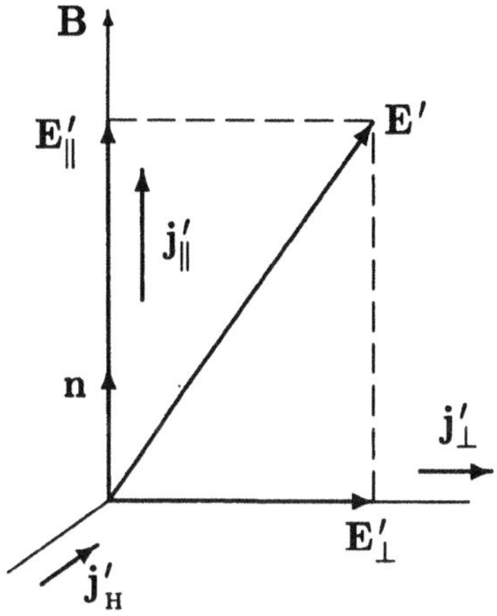

Figure 7.1: Direct ($\mathbf{j}'_{\parallel}$ and $\mathbf{j}'_{\perp}$) and Hall's ($\mathbf{j}'_{H}$) currents in a plasma with electric ($\mathbf{E}'$) and magnetic ($\mathbf{B}$) fields.

is nothing other than the *turning angle* of an electron on the Larmor circle in the intercollisional time. Consider **two limiting cases**.

(a) Let the turning angle be small:

$$\omega_{B}^{(e)} \tau_{ei} \ll 1 \,. \tag{7.69}$$

Obviously this inequality corresponds to the *weak* magnetic field or *dense cool* plasma, so that the current is scarcely affected by the field:

$$\sigma_{\perp} \approx \sigma_{\parallel} = \sigma \,, \qquad \frac{\sigma_{H}}{\sigma} \approx \omega_{B}^{(e)} \tau_{ei} \ll 1 \,. \tag{7.70}$$

Thus in a frame of reference associated with the plasma, the usual Ohm's law with *isotropic* conductivity holds.

(b) The opposite case, when the electrons **spiral freely between collisions**:

$$\omega_{B}^{(e)} \tau_{ei} \gg 1 \,, \tag{7.71}$$

corresponds to the *strong* magnetic field and hot rarefied plasma. This plasma is termed the *magnetized* one. It is frequently encountered under astrophysical conditions. In this case

$$\sigma_{\parallel} = \sigma \,, \quad \sigma_{\perp} \approx \sigma \left(\omega_{B}^{(e)} \tau_{ei} \right)^{-2} , \quad \sigma_{H} \approx \sigma \left(\omega_{B}^{(e)} \tau_{ei} \right)^{-1} , \tag{7.72}$$

or

$$\sigma_{\parallel} \approx \left(\omega_{B}^{(e)} \tau_{ei} \right) \sigma_{H} \approx \left(\omega_{B}^{(e)} \tau_{ei} \right)^{2} \sigma_{\perp} \,.$$

Hence in the magnetized plasma, for example in the solar corona (see Problems 7.6 and 7.7),

$$\boxed{\sigma_{\parallel} \gg \sigma_{H} \gg \sigma_{\perp} \,.} \tag{7.73}$$

In other words, the impact of the magnetic field on the direct current is especially strong for the component resulting from the electric field $\mathbf{E}'_{\perp}$. The current in the $\mathbf{E}'_{\perp}$ direction is considerably weaker than it would be in the absence of a magnetic field. Why is this so?

7.3.4 The physical interpretation

The physical mechanism of this phenomenon is shown in Figure 7.2.

> The primary effect of the electric field $\mathbf{E}'_{\perp}$ in the presence of the magnetic field $\mathbf{B}$ is not the current in the direction $\mathbf{E}'_{\perp}$, but rather the electric *drift* in the direction perpendicular to both $\mathbf{B}$ and $\mathbf{E}'_{\perp}$.

The electric drift velocity (4.21) is independent of the particle's mass and charge. The electric drift of electrons and ions generates the motion of the plasma as a whole with the velocity $\mathbf{v} = \mathbf{v}_{\rm d}$. This would be the case if there were no collisions at all (Figure 4.3).

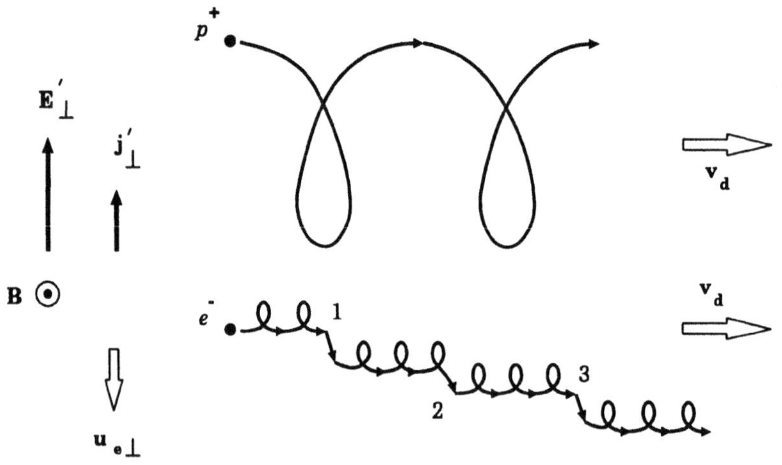

Figure 7.2: Initiation of the current in the direction of the perpendicular field $\mathbf{E}'_{\perp}$ as the result of collisions (1, 2, ...) against a background of the electric drift. Only collisions of electrons are shown.

Collisions, even infrequent ones, result in a disturbance of the particle's Larmor motion, leading to a displacement of the ions (not shown in Figure 7.2) along $\mathbf{E}'_{\perp}$, and the electrons in the opposite direction (see Figure 7.2). The small electric current (a factor of $\omega_{\rm B}^{(e)} \tau_{\rm ei}$ smaller than the drift one) appears in the direction $\mathbf{E}'_{\perp}$.

To ensure the current across the magnetic field, the so-called Hall field is necessary, that is the electric field component perpendicular to both the current $\mathbf{j}'_{\perp}$ and the field $\mathbf{B}$ (e.g., Braginskii, 1965; Sivukhin, 1977, Ch. 7, § 98). This is the secondary effect but it is not small in a strong magnetic field.

> The Hall field balances the forces acting on the perpendicular current carriers due to the presence of a magnetic field,

i.e.

$$\mathbf{F}\,(\,\mathbf{j}'_{\perp}) = \frac{en}{c}\,\mathbf{u}_{i\perp} \times \mathbf{B} - \frac{en}{c}\,\mathbf{u}_{e\perp} \times \mathbf{B}\,.$$

Hence the magnitude of the Hall field is

$$\mathbf{E}'_{\mathrm{H}} = \frac{1}{enc}\,\mathbf{j}'_{\perp} \times \mathbf{B}\,. \tag{7.74}$$

The Hall field in plasma is frequently set up automatically, as a consequence of small charge separation within the limits of quasineutrality. In this case the 'external' field, which has to be applied to the plasma, is determined by the expressions

$$E'_{\|} = j'_{\|}\,/\sigma_{\|} \quad \text{and} \quad E'_{\perp} = j'_{\perp}\,/\sigma_{\perp}\,.$$

We shall not discuss here the dissipation process under the conditions of anisotropic conductivity. In general, the symmetric highest component of the *conductivity tensor* can play the most important role (e.g., Landau, Lifshitz, and Pitaevskii, 1984, Ch. 3) in this process of fundamental significance for the flare energy release problem. In the particular case of a fully-ionized plasma, the tendency for a particle to spiral round the magnetic field lines insures the great reduction in the transversal conductivity (7.67). However, since the dissipation of the energy of the electric current into Joule heat is due solely to collisions between particles, the reduced conductivity does not lead to increased dissipation (Problem 7.9).

7.3.5 Cosmic plasma conductivity

Note another property of the generalized Ohm's law. As a rule, under laboratory conditions, one cannot neglect the gradient forces (7.55). On the contrary, these forces usually play no part in cosmic plasma. We shall ignore them. Moreover, let us restrict our consideration to very *slow* particle motions. These motions are supposed to be so slow that the following two conditions are fulfilled.

First, we neglect the electron inertia in comparison with that of the ions and make use of (7.57). This condition is usually written in the form

$$\omega = \frac{1}{t} \ll \omega_{\rm B}^{(\rm i)} = \frac{eB}{m_{\rm i} c} \, ,$$

where t is a characteristic time of the motion. To put it another way, the plasma motions have to be so slow that their frequency is smaller than the lowest gyro-frequency of the particles in the magnetic field.

Second, it is supposed that

$$\nu_{\rm ei} \gg \omega \quad \text{and} \quad \nu_{\rm ei} \gg \omega_{\rm B}^{(\rm e)} \, , \tag{7.75}$$

i.e.

$$\nu_{\rm ei} \, t \gg 1 \quad \text{and} \quad \omega_{\rm B}^{(\rm e)} \tau_{\rm ei} \ll 1 \, .$$

Hence the hydrodynamic approximation can be used, the conductivity being isotropic. The generalized Ohm's law assumes the following form, which is specific to magnetohydrodynamics (MHD):

$$\mathbf{j}' = \sigma \left(\mathbf{E} + \frac{1}{c} \, \mathbf{u} \times \mathbf{B} \right) . \tag{7.76}$$

The MHD approximation is the subject of the next chapter. Numerous applications of MHD to various phenomena in cosmic plasma will be considered in the remainder of the book.

Note that in the opposite case (7.71), i.e. even when the parameter $\omega_{\rm B}^{(\rm e)} \tau_{\rm ei}$ tends to infinity (like in the solar corona, see Problem 7.6) and collisions are negligible, the *quasi-hydrodynamic* description of plasma, the Chew-Goldberger-Low approximation (Chew, Goldberger, and Low, 1956) is possible and quite useful. This is because the strong magnetic field makes the plasma, even a non-collisional one, more 'connected', so to speak, more hydrodynamic in the directions perpendicular to the field. That allows one to write down a set of two-dimensional MHD equations for the almost non-collisional plasma in a magnetic field (e.g., Volkov, 1966).

7.3.6 Volume charge and quasi-neutrality

One more remark concerning the generalized Ohm's law is important for the following. While deriving the law, the *exact* charge neutrality of plasma or exact electric neutrality has been assumed:

$$Z_{\rm i} n_{\rm i} = n_{\rm e} = n \, ,$$

i.e. the absolute absence of the *volume charge* in plasma $\rho^{\mathrm{q}} = 0$. The same assumption was also used in Sections 6.2 and 2.5. However, there is no need for this. It is sufficient to require *quasi-neutrality*, i.e. the numbers of ions (with account of their charge taken) and electrons per unit volume are very nearly equal:

$$\frac{Z_{\mathrm{i}} n_{\mathrm{i}} - n_{\mathrm{e}}}{n_{\mathrm{e}}} \ll 1 .$$

So

> the volume charge density has to be small in comparison to the plasma density.

Once the volume charge density

$$\rho^{\mathrm{q}} \neq 0 ,$$

yet another term must be taken into account in the generalized Ohm's law:

$$\mathbf{j}_{\mathrm{u}}^{\mathrm{q}} = \rho^{\mathrm{q}} \mathbf{u} . \tag{7.77}$$

This is the so-called *convective current.* It is caused by the volume charge transfer and must be added to the conduction current (7.65).

The volume charge, the associated electric force and the convection current are of great importance in electrodynamics of relativistic objects such as black holes (see Novikov and Frolov, 1989) and pulsars (see Michel, 1991). Charge-separated plasmas originate in magnetospheres of rotating black holes in active galactic nuclei. The shortage of charge leads to the emergence of a strong electric field along the magnetic field lines. The parallel electric field accelerates migratory electrons and/or positrons to ultrarelativistic energies (e.g., Hirotani and Okamoto, 1998).

Charge density oscillations in a plasma, the Langmuir waves, are considered in Problems 7.3 and 7.4.

* * *

The volume charge can be evaluated in the following manner. On the one hand, from Maxwell's equation $\operatorname{div} \mathbf{E} = 4\pi \rho^{\mathrm{q}}$ we get

$$\rho^{\mathrm{q}} \approx \frac{E}{4\pi L} . \tag{7.78}$$

On the other hand, the non-relativistic equation of plasma motion yields

$$e n_{\mathrm{e}} E \approx \frac{p}{L} \approx \frac{n_{\mathrm{e}} k_{\mathrm{B}} T}{L} ,$$

so that

$$E \approx \frac{k_{\mathrm{B}} T}{eL} \,. \tag{7.79}$$

On substituting (7.79) in (7.78), we find the following estimate

$$\frac{\rho^{\mathrm{q}}}{en_{\mathrm{e}}} \approx \frac{k_{\mathrm{B}} T}{eL} \, \frac{1}{4\pi L} \, \frac{1}{en_{\mathrm{e}}}$$

or

$$\boxed{\frac{\rho^{\mathrm{q}}}{en_{\mathrm{e}}} \approx \frac{r_{\mathrm{D}}^{2}}{L^2} \,.} \tag{7.80}$$

Since the usual *concept of plasma* implies that the Debye radius $r_{\mathrm{D}} \ll L$, the volume charge density is small in comparison with the plasma density.

7.4 Practice: Problems and Answers

Problem 7.1. Show that the third integral in Equation (7.10) equals zero.

Answer. Find the derivative

$$\frac{\partial}{\partial v_\alpha} \left(\frac{F_{k,\alpha}}{m_k} \, f_k \right) = \frac{F_{k,\alpha}}{m_k} \frac{\partial f_k}{\partial v_\alpha} + \frac{f_k}{m_k} \frac{\partial F_{k,\alpha}}{\partial v_\alpha} = \frac{F_{k,\alpha}}{m_k} \frac{\partial f_k}{\partial v_\alpha} \,.$$

The condition (1.25) has been used on the right-hand side as the condition

$$\frac{\partial F_{k,\alpha}}{\partial v_\alpha} = 0 \,. \tag{7.81}$$

Hence

$$\int\limits_{v} \frac{F_{k,\alpha}}{m_k} \frac{\partial f_k}{\partial v_\alpha} \, d^3 \mathbf{v} = \frac{F_{k,\alpha}}{m_k} \, f_k(\mathbf{r}, \mathbf{v}, t) \, \Bigg|_{\mathbf{v} \to -\infty}^{\mathbf{v} \to +\infty} = 0 \,,$$

if the distribution function f_k quickly approaches zero as $v \to \infty$. Q.E.D.

Problem 7.2. Write the continuity equation with account of ionization and recombination.

Answer. The continuity equation including source/sink terms related to ionization/recombination or charge exchange reads

$$\frac{\partial n_k}{\partial t} + \frac{\partial}{\partial r_\alpha} \, n_k \, u_{k,\alpha} = \sum_{l} \left(\gamma_{lk} \, n_l - \gamma_{kl} \, n_k \right) , \tag{7.82}$$

where n_k denotes the particle density of species k, either neutral or ionized. The right-hand side is the change of n_k due to collisions. The γ_{kl} and γ_{lk} denote the rate of transformation of species k into species l and vice versa. These rates must obey the relation

$$\sum_k \sum_l (\gamma_{lk} n_l - \gamma_{kl} n_k) = 0, \tag{7.83}$$

which ensures the total particle number density conservation.

Problem 7.3. Justify formula (6.30) for the frequency of the charge density oscillations in a *cold* two-component plasma.

Answer. Let us assume that the ions do not move at all (they are infinitely massive) are uniformly distributed in space. So they have a fixed density n_0. Let us neglect all magnetic fields. However, we shall assume that any variations of electron density n_e, electron velocity u_e, and related electric field $\mathbf{E}$ occur only in one dimension – the x axis. Then we are left with a set of three equations: (a) the continuity equation (7.11) for electrons

$$\frac{\partial n_e}{\partial t} + \frac{\partial}{\partial x} n_e u_e = 0, \tag{7.84}$$

(b) the motion equation (7.30)

$$m_e n_e \left(\frac{\partial u_e}{\partial t} + u_e \frac{\partial u_e}{\partial x} \right) = - \frac{\partial p_e}{\partial x} - e n_e E_x, \tag{7.85}$$

(c) the electric field equation

$$\frac{\partial E_x}{\partial x} = 4\pi e (n_0 - n_e). \tag{7.86}$$

We cannot solve these nonlinear equations exactly, except for very special cases. One of them is trivial:

$$n_e = n_0, \quad u_e = 0, \quad p_e = \text{const}, \quad E_x = 0. \tag{7.87}$$

This solution corresponds to a *stationary* electron gas of uniform density.

Let us linearize Equations (7.84)–(7.86) with respect to the state (7.87). This yields the following set of *linear* equations:

$$\frac{\partial n_1}{\partial t} + n_0 \frac{\partial u_1}{\partial x} = 0, \tag{7.88}$$

$$m_e n_0 \frac{\partial u_1}{\partial t} = - \frac{\partial p_1}{\partial x} - e n_e E_1 \,, \tag{7.89}$$

$$\frac{\partial E_1}{\partial x} = 4\pi e \, n_1 \,. \tag{7.90}$$

Consider the special case of *cold* electrons ($p_e = 0$). We eliminate u_1 and E_1 from the set by taking the time derivative of Equation (7.88) to obtain the oscillator equation

$$\frac{\partial^2 n_1}{\partial t^2} + \left(\frac{4\pi e^2 n_0}{m_e} \right) n_1 = 0 \,. \tag{7.91}$$

If we displace some electrons to produce an initial perturbation, we create a positive-charge density at the position where they started. This positive-charge perturbation attracts the electrons, which will tend to move back to their original position, but will overshoot it. They come back again, overshoot it, and so on. Without any damping, the energy put into the plasma to create the perturbation will remain in the plasma. So the oscillation will continue forever with the frequency

$$\omega_{pl}^{(e)} = \pm \left(\frac{4\pi e^2 n_e}{m_e} \right)^{1/2} \tag{7.92}$$

called the *electron plasma frequency*.

Therefore, in a two-component cold plasma, there exist the **oscillations of charge density** – *Langmuir waves* which frequency is independent of the wave vector **k**; so the group velocity is zero. Thus

> in a cold plasma, Langmuir waves are spatially localized oscillations of electric charge density which do not propagate at all.

Note that there is no equivalent to these oscillations in gasdynamics or gravitational dynamics, for which there is no electric force and charge separation.

Problem 7.4. What happens with a Langmuir wave, if we take the electron temperature into account?

Answer. Let us drop the assumption of zero pressure in the linear equations (7.88) – (7.90). We then must include the perturbation of pressure

$$\frac{\partial p_1}{\partial x} = n_0 k_{\rm B} \frac{\partial T_1}{\partial x} + k_{\rm B} T_0 \frac{\partial n_1}{\partial x} \tag{7.93}$$

in Equation (7.89). We must now relate n_1 to T_1 and vice versa. For example, we could argue that for long-wavelength waves the compression is the one-dimensional ($N = 1$) adiabatic process with $\gamma = (N+2)/N = 3$. In this case, the perturbation of electron pressure becomes

$$\frac{\partial p_1}{\partial x} = 3k_{\rm B}T_0\,\frac{\partial n_1}{\partial x}\,. \tag{7.94}$$

We expect now an initial perturbation to propagate through the plasma as a wave. Thus a wave solution of the form

$$f_1(x,t) = \tilde{f}_1 \exp\left(-\mathrm{i}\,\omega t + \mathrm{i}\,kx\right) \tag{7.95}$$

should satisfy the linear differential equations. The quantities with tildes are the complex amplitudes. They obey three linear algebraic equations:

$$-\mathrm{i}\,\omega\,\tilde{n}_1 + \mathrm{i}\,k\,n_0\,\tilde{u}_1 = 0\,,$$

$$-\mathrm{i}\,\omega\,m_{\rm e}n_0\,\tilde{u}_1 + \mathrm{i}\,k\,3k_{\rm B}T_0\,\tilde{n}_1 + en_0\,\tilde{E}_1 = 0\,,$$

$$\mathrm{i}\,k\,\tilde{E}_1 + 4\pi e\,\tilde{n}_1 = 0\,.$$

To have a nontrivial solution, the determinant must be zero. Its solution is

$$\boxed{\omega = \pm\,\omega_{pl}^{(\rm e)}\left(1 + 3k^2 r_{\rm D}^2\right)^{1/2},} \tag{7.96}$$

where $r_{\rm D}$ is the electron Debye radius. This dispersion relation is shown in Figure 7.3.

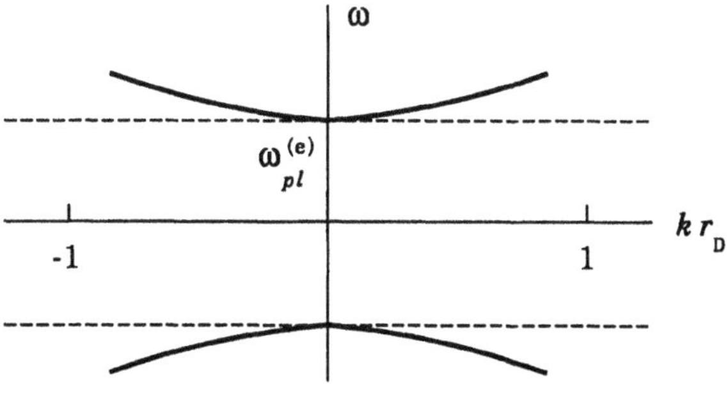

Figure 7.3: A dispersion diagram (solid curves) for Langmuir waves in a warm plasma.

Therefore the frequency ω of Langmuir waves in a plasma with warm electrons depends on the wave vector k. So the group velocity, $\partial\omega/\partial k$, of Langmuir waves in a warm plasma without magnetic field is not equal to zero.

They oscillate at the plasma frequency $\omega_{pl}^{(e)}$ and propagate in a warm plasma.

Problem 7.5. Show that, when the ions are allowed to move, ion contributions are important only for slow variations or *low-frequency* waves because the ions cannot react quickly enough.

Answer. Linearizing the continuity equations for electrons and ions, the motion equations for electrons and ions, as well as the electric field equation, let us assume that the electrons and ions both obey the adiabatic equation (7.94). Then we again use the wave solution (7.95) to reduce the linearized differential equations to algebraic ones and to obtain the determinant. Because $m_i/m_e \gg 1$, we neglect the term $m_e\,\omega^2$ in this determinant as compared with the term $m_i\,\omega^2$. By so doing, we obtain the relation

$$\omega^2 = k^2\left(\frac{\gamma_i\,k_B T_i}{m_i} + \frac{\gamma_e\,k_B T_e}{m_i}\,\frac{1}{1+\gamma_e\,k^2 r_D^2}\right). \tag{7.97}$$

This dispersion relation is shown in Figure 7.4.

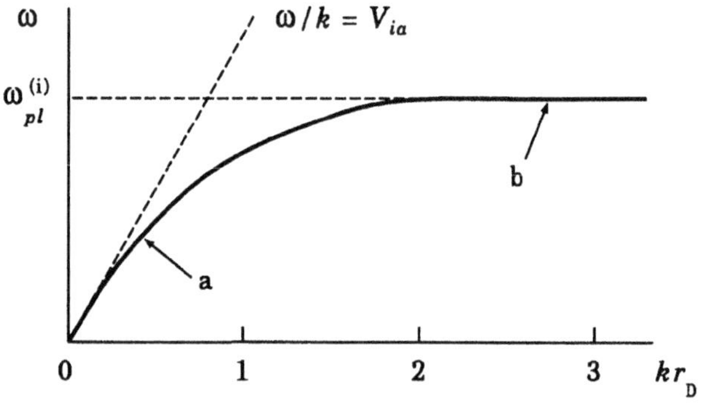

Figure 7.4: A dispersion diagram for ion-acoustic waves (part a) and for ion plasma waves (part b) in a warm plasma without magnetic field.

In the limit of small kr_D

$$\omega^2 = k^2\left(\frac{\gamma_i\,k_B T_i}{m_i} + \frac{\gamma_e\,k_B T_e}{m_i}\right) = k^2 V_{ia}^2.$$

This is the so-called *ion-acoustic wave*. They are shown by a curve part (a) in Figure 7.4. The group velocity of the wave is independent of k:

$$V_{gr} = V_{ia} = \left(\frac{\gamma_i\,k_B T_i + \gamma_e\,k_B T_e}{m_i}\right)^{1/2}. \tag{7.98}$$

An opposite limit is obtained for cold ions. If ion temperature $T_i \to 0$, then $kr_D \gg 1$, i.e., short wavelengths are under consideration. In this case, shown by the curve part (*b*) in Figure 7.4 the cold ions oscillate with a frequency

$$\omega_{pl}^{(i)} = \pm \left(\frac{4\pi e^2 n_e}{m_i} \right)^{1/2} \tag{7.99}$$

called the *ion plasma frequency*.

Ion-acoustic waves are observed in many cases. They were registred by *Voyager 1* in the upstream side of the Jovian bow shock. Ion-acoustic waves play an important role in solar flares (Section 17.3.5).

Problem 7.6. Estimate the parameter $\omega_B^{(e)} \tau_{ei}$ in the solar corona above a sunspot.

Answer. Just above a sunspot the field strength can be as high as $B \approx 3000\,\mathrm{G}$. Here $\omega_B^{(e)} \approx 5 \times 10^{10}\,\mathrm{rad\ s^{-1}}$ (Problem 4.1). Characteristic time of *close* electron-proton collisions $\tau_{ep} \approx 22$ s (Problem 6.1). Therefore $\omega_B^{(e)} \tau_{ei} \sim 10^{12}\,\mathrm{rad} \gg 1$.

Distant collisions are much more frequent (Problem 6.1). However even with $\tau_{ep} \approx 0.1$ s we obtain $\omega_B^{(e)} \tau_{ei} \sim 10^{10}\,\mathrm{rad} \gg 1$. So, for anisoptropic conductivity in the corona, the approximate formulae (7.72) can be well used.

Problem 7.7. Evaluate the characteristic value of the parallel conductivity (7.66) in the solar corona.

Answer. It follows from formula (7.66) that

$$\sigma_{\parallel} = \frac{e^2 n}{m_e} \, \tau_{ei} = 2.53 \times 10^8 \, n \, \tau_{ei} \sim 10^{16} - 10^{17}\,, \;\; \mathrm{s^{-1}}\,, \tag{7.100}$$

if we take $\tau_{ep} \sim 0.2 - 2.0$ s (Problem 6.1).

Problem 7.8. Show that in the solar corona a viscosity coefficient can be given by formula (Hollweg, 1986):

$$\eta = 10^{-16} \, T_p^{5/2}\,, \;\; \mathrm{g\ cm^{-1}\,s^{-1}}\,, \tag{7.101}$$

where T_p is the proton temperature, and the Coulomb logarithm has been taken to be 22. So, with $T_p \approx 2 \times 10^6$ K, the viscosity coefficient in the corona $\eta \sim 1\,\mathrm{g\,cm^{-1}\,s^{-1}}$.

Why does the viscosity grow with the proton temperature? Why does it grow so quickly?

Problem 7.9. Consider the generalized Ohm's law in the case when the electric field is perpendicular to the magnetic field $\mathbf{B} = B\,\mathbf{n}$. So

$$\mathbf{j}' = \sigma_{\perp}\,\mathbf{E}'_{\perp} + \sigma_{\rm H}\,\mathbf{n} \times \mathbf{E}'_{\perp}\,, \qquad (7.102)$$

where

$$\sigma_{\perp} = \sigma\,\frac{1}{1 + \left(\omega_{\rm B}^{(\rm e)}\tau_{\rm ei}\right)^2} \quad \text{and} \quad \sigma_{\rm H} = \sigma\,\frac{\omega_{\rm B}^{(\rm e)}\tau_{\rm ei}}{1 + \left(\omega_{\rm B}^{(\rm e)}\tau_{\rm ei}\right)^2}\,. \qquad (7.103)$$

This indicates that a current in the direction of $\mathbf{E}'_{\perp}$ is reduced in the ratio $1/1+\left(\omega_{\rm B}^{(\rm e)}\tau_{\rm ei}\right)^2$ by the magnetic field. In addition, the other current $\left(\omega_{\rm B}^{(\rm e)}\tau_{\rm ei}\right)^2$ times as large flows in the direction perpendicular to both $\mathbf{B}$ and $\mathbf{E}'_{\perp}$; this is the Hall current.

Show that the reduction in the 'direct' conductivity (cf. Fig. 7.1) does not increase the rate of dissipation of current energy (see Cowling, 1976, § 6.2).

Chapter 8

Magnetohydrodynamics of Cosmic Plasma

Magnetohydrodynamics (MHD) is the simplest but sufficient approximation to describe many large-scale phenomena in space.

8.1 Basic assumptions and the MHD equations

8.1.1 Old and new simplifying assumptions

As we have seen in the previous chapter, the set of kinetic equations (the averaged Liouville equation and the chain of equations for correlation functions) gives us a set of transfer equations for local macroscopic quantities. The set is generally infinite as well. It determines the behaviour of different kinds of particles, such as electrons and ions, once two conditions are complied with:

(a) many collisions occur in a characteristic time τ of the process:

$$\tau \gg \tau_c \,, \tag{8.1}$$

(b) the particle's path between two collisions – the particle's free path – is significantly smaller than the distance, L, over which macroscopic quantities change considerably:

$$L \gg \lambda_c \,. \tag{8.2}$$

Here τ_c and λ_c are the collisional time and the collisional mean free path, respectively. Once these conditions are satisfied, we can close the set of hydrodynamic *transfer* equations, as has been discussed in Section 7.2.4.

While considering the generalized Ohm's law, two other assumptions have been made. The first is complementary to the restriction (8.1) on the characteristic time τ of the process and can be written in the form (7.75):

$$\omega = \frac{1}{\tau} \ll \omega_{\mathrm{B}}^{(\mathrm{i})} = \frac{eB}{m_{\mathrm{i}}c} . \tag{8.3}$$

This allows us to neglect electron inertia in comparison to that of ions.

The second condition,

$$\omega_{\mathrm{B}}^{(\mathrm{e})} \tau_{\mathrm{ei}} \ll 1 , \tag{8.4}$$

is necessary to write down Ohm's law in the form

$$\mathbf{j} = \sigma \left(\mathbf{E} + \frac{1}{c} \mathbf{v} \times \mathbf{B} \right) + \rho^{\mathrm{q}} \mathbf{v} . \tag{8.5}$$

Here $\mathbf{v}$ is the macroscopic velocity of plasma considered as a continuous medium, $\mathbf{E}$ and $\mathbf{B}$ are the electric and magnetic fields in the 'laboratory' system of coordinates, where we measure $\mathbf{v}$. Accordingly,

$$\mathbf{E}_{\mathrm{v}} = \mathbf{E} + \frac{1}{c} \mathbf{v} \times \mathbf{B} \tag{8.6}$$

is the electric field in a frame of reference related to the plasma. The isotropic conductivity is (formula (7.63)):

$$\sigma = \frac{e^2 n}{m_{\mathrm{e}}\, \nu_{\mathrm{ei}}} . \tag{8.7}$$

Under the conditions listed above, we use the general hydrodynamic-type equations which are the conservation laws for mass (7.35), momentum (7.36) and energy (7.46).

> These equations have a much wider area of applicability than the equations of ordinary magnetohydrodynamics derived below.

The latter are much simpler than the equations derived in Section 7.2. Therefore **new additional simplifying assumptions** are necessary. Let us introduce them. There are two.

* * *

First, the plasma conductivity σ is assumed to be large, the electromagnetic processes being not very fast. Then, in Maxwell's equation (1.1)

$$\operatorname{curl} \mathbf{B} = \frac{4\pi}{c} \mathbf{j} + \frac{1}{c} \frac{\partial \mathbf{E}}{\partial t} ,$$

we ignore the *displacement current* in comparison to the *conductive* one. The corresponding condition is found by evaluating the currents as follows

$$\frac{4\pi}{c} j \gg \frac{1}{c}\frac{E}{\tau} \quad \text{or} \quad 4\pi\sigma E \gg \omega E .$$

Thus we suppose that

$$\boxed{\omega \ll 4\pi\sigma .} \tag{8.8}$$

In the same order with reference to the small parameter ω/σ (or, more exactly, $\omega/4\pi\sigma$), we can neglect the convective current in comparison with the conductive one in Ohm's law (8.5). Actually,

$$\rho^{q} v \approx v \operatorname{div} \mathbf{E} \frac{1}{4\pi} \approx \frac{L}{\tau}\frac{E}{L}\frac{1}{4\pi} \approx \frac{\omega}{4\pi} E \ll \sigma E ,$$

once the condition (8.8) is satisfied.

The conductivity of cosmic plasma, which is often treated in the MHD approximation, is very high. This is the reason why condition (8.8) is satisfied up to frequencies close to optical ones.

Neglecting the displacement current and the convective current, Maxwell's equations and Ohm's law result in the following relations:

$$\mathbf{j} = \frac{c}{4\pi} \operatorname{curl} \mathbf{B} , \tag{8.9}$$

$$\mathbf{E} = -\frac{1}{c} \mathbf{v} \times \mathbf{B} + \frac{c}{4\pi\sigma} \operatorname{curl} \mathbf{B} , \tag{8.10}$$

$$\rho^{q} = -\frac{1}{4\pi c} \operatorname{div} (\mathbf{v} \times \mathbf{B}) , \tag{8.11}$$

$$\operatorname{div} \mathbf{B} = 0 , \tag{8.12}$$

$$\frac{\partial \mathbf{B}}{\partial t} = \operatorname{curl} (\mathbf{v} \times \mathbf{B}) + \frac{c^2}{4\pi\sigma} \Delta \mathbf{B} . \tag{8.13}$$

Once two vectors **B** and **v** are given, the current density **j**, the electric field **E**, and the charge density ρ^{q} are completely determined by formulae (8.9)—(8.11). Thus

> the problem is reduced to finding the interaction of the magnetic field **B** and the hydrodynamic velocity field **v**.

As a consequence, the approach under discussion has come to be known as *magnetohydrodynamics*.

The corresponding equation of plasma motion is obtained by substitution of formulae (8.9)–(8.11) in the equation of momentum transfer (7.36). With due regard for the manner in which viscous forces are usually written in hydrodynamics, we have

$$\rho \frac{d\mathbf{v}}{dt} = -\nabla p + \rho^{q}\,\mathbf{E} - \frac{1}{4\pi}\,\mathbf{B} \times \operatorname{curl}\mathbf{B} +$$

$$+\eta\,\Delta\,\mathbf{v} + \left(\zeta + \frac{\eta}{3}\right)\nabla \operatorname{div}\mathbf{v}\,. \tag{8.14}$$

Here η is the first viscosity coefficient, ζ is the second viscosity coefficient (see Landau and Lifshitz, *Fluid Mechanics*, 1959, Ch. 2, § 15).

* * *

A second additional simplifying assumption has to be introduced now. Treating Equation (8.14), the electric force $\rho^{q}\,\mathbf{E}$ can be ignored in comparison to the magnetic one if

$$v^2 \ll c^2\,, \tag{8.15}$$

that is in the non-relativistic approximation. To make certain that this is true, evaluate the electric force using (8.11) and (8.10):

$$\rho^{q}\,E \approx \frac{1}{4\pi c}\,\frac{vB}{L}\,\frac{vB}{c} \approx \frac{B^2}{4\pi}\,\frac{1}{L}\,\frac{v^2}{L^2}\,, \tag{8.16}$$

the magnetic force being proportional to

$$\frac{1}{4\pi}\,|\,\mathbf{B} \times \operatorname{curl}\mathbf{B}\,| \approx \frac{B^2}{4\pi}\,\frac{1}{L}\,. \tag{8.17}$$

Comparing (8.16) with (8.17), we see that the electric force is a factor of v^2/c^2 short of the magnetic one.

In a great number of astrophysical applications of MHD, the plasma velocities fall far short of the speed of light. The Sun is a good case in point. Here the largest velocities observed, for example, in coronal transients do not exceed several thousands of km/s, i.e. $\leq 10^8$ cm/s. Under these conditions, **we neglect the electric force acting upon the volume charge** in comparison with the magnetic force.

Note however that relativistic objects such as accretion discs near black holes (see Chapter 7 in Novikov and Frolov, 1989), and pulsar magnetospheres are at the other extreme (e.g., Michel, 1991; Rose, 1998). The electric force acting on the volume charge plays a crucial role in the electrodynamics of relativistic objects.

8.1.2 Non-relativistic magnetohydrodynamics

With the assumptions made above, we write the following set of equations of non-relativistic MHD

$$\frac{d\mathbf{v}}{dt} = -\frac{\nabla p}{\rho} - \frac{1}{4\pi\rho}\mathbf{B}\times\operatorname{curl}\mathbf{B} + \frac{\eta}{\rho}\Delta\mathbf{v} + \frac{1}{\rho}\left(\zeta + \frac{\eta}{3}\right)\nabla\operatorname{div}\mathbf{v}\,, \tag{8.18}$$

$$\frac{\partial\mathbf{B}}{\partial t} = \operatorname{curl}\left(\mathbf{v}\times\mathbf{B}\right) + \nu_{\mathrm{m}}\,\Delta\mathbf{B}\,, \tag{8.19}$$

$$\operatorname{div}\mathbf{B} = 0\,, \tag{8.20}$$

$$\frac{\partial\rho}{\partial t} + \operatorname{div}\rho\mathbf{v} = 0\,, \tag{8.21}$$

$$\frac{\partial}{\partial t}\left(\frac{\rho v^2}{2} + \rho\varepsilon + \frac{B^2}{8\pi}\right) = -\operatorname{div}\mathbf{G}\,, \tag{8.22}$$

$$p = p\,(\rho, T)\,. \tag{8.23}$$

Here

$$\nu_{\mathrm{m}} = \frac{c^2}{4\pi\sigma} \tag{8.24}$$

is the *magnetic diffusivity* (or viscosity). It plays the same role in Equation (8.19) as the kinematic viscosity $\nu = \eta/\rho$ in the equation of plasma motion (8.18). The vector $\mathbf{G}$ is defined as the energy flux density (cf. Equation (7.46))

$$G_\alpha = \rho v_\alpha\left(\frac{v^2}{2} + w\right) + \frac{1}{4\pi}\left[\mathbf{B}\times(\mathbf{v}\times\mathbf{B})\right]_\alpha -$$

$$-\frac{\nu_{\mathrm{m}}}{4\pi}\left(\mathbf{B}\times\operatorname{curl}\mathbf{B}\right)_\alpha - \sigma^{\mathrm{v}}_{\alpha\beta}\,v_\beta - \kappa\,\nabla_\alpha T, \tag{8.25}$$

where the *specific enthalpy* is

$$w = \varepsilon + \frac{p}{\rho} \tag{8.26}$$

(see definition (7.24)).

The Poynting vector appearing as a part in expression (8.25) is rewritten using formula (8.10):

$$\mathbf{G}_{\mathrm{P}} = \frac{c}{4\pi}\,\mathbf{E}\times\mathbf{B} = \frac{1}{4\pi}\,\mathbf{B}\times(\,\mathbf{v}\times\mathbf{B}\,) - \frac{\nu_{\mathrm{m}}}{4\pi}\,\mathbf{B}\times\mathrm{curl}\,\mathbf{B}\,. \tag{8.27}$$

As usually in electrodynamics, the flux of electromagnetic energy disappeares when electric field **E** is parallel to magnetic field **B**.

The energy flux density due to friction processes is written as the contraction of the velocity vector **v** and the viscous stress tensor

$$\sigma^{\mathrm{v}}_{\alpha\beta} = \nu\left(\frac{\partial v_\alpha}{\partial r_\beta} + \frac{\partial v_\beta}{\partial r_\alpha} - \frac{2}{3}\,\delta_{\alpha\beta}\,\frac{\partial v_\gamma}{\partial r_\gamma}\right) + \zeta\,\delta_{\alpha\beta}\,\frac{\partial v_\gamma}{\partial r_\gamma} \tag{8.28}$$

(see Landau and Lifshitz, *Fluid Mechanics*, 1959, Ch. 2, § 15). Finally, the heat flux density is $-\kappa\nabla T$, where κ is the plasma thermal conductivity.

* * *

The equation of state (8.23) can be rewritten in other thermodynamic variables. It is convenient to transform the energy conservation law (8.22) from the divergent form to the hydrodynamic one containing the substantial derivative (7.31). To do this, we have to make use of Equations (8.18)–(8.21) and the thermodynamic identities

$$d\varepsilon = T\,ds + \frac{p}{\rho^2}\,d\rho \quad \text{and} \quad dw = T\,ds + \frac{1}{\rho}\,dp\,.$$

Here s is the entropy per unit mass. On rearrangement, Equation (7.19) results in the heat trasfer equation

$$\rho T\,\frac{ds}{dt} = \frac{\nu_{\mathrm{m}}}{4\pi}\,(\mathrm{curl}\,\mathbf{B})^2 + \sigma^{\mathrm{v}}_{\alpha\beta}\,\frac{\partial v_\alpha}{\partial r_\beta} + \mathrm{div}\,\kappa\nabla T\,. \tag{8.29}$$

It shows that

> the heat abundance change $dQ = \rho T\,ds$ in a moving element of unit volume is a sum of the Joule and viscous heating and conductive heat redistribution to neighbour elements.

The equation of motion (8.18) is written as the Navier-Stokes equation. However, it can be recast into the momentum conservation law in the divergent form:

$$\frac{\partial}{\partial t}\,\rho v_\alpha = -\frac{\partial}{\partial r_\beta}\,\Pi^*_{\alpha\beta}\,. \tag{8.30}$$

Here the asterisk refers to the total (unlike (7.44)) momentum flux density tensor $\Pi^*_{\alpha\beta}$, which is equal to

$$\Pi^*_{\alpha\beta} = p\,\delta_{\alpha\beta} + \rho v_\alpha v_\beta + \frac{1}{4\pi}\left(\frac{B^2}{2}\,\delta_{\alpha\beta} - B_\alpha B_\beta\right) - \sigma^{\text{v}}_{\alpha\beta}\,. \tag{8.31}$$

In contrast to Equation (7.43), the momentum of electromagnetic field does not appear in the non-relativistic Equation (8.30). It is negligibly small in comparison to the plasma stream momentum ρv_α. Formally this fact is a consequence of neglecting the displacement current in Maxwell's equations.

8.1.3 Relativistic magnetohydrodynamics

Relativistic MHD models are of considerable interest in several areas of astrophysics. The theory of gravitational collapse and models of supernova explosions are based on relativistic hydrodynamic models for a star. In most models a key feature is the occurrence of a relativistic shock, for example, to expel the bulk of the star. The effects of deviations from spherical symmetry due to an initial angular momentum and magnetic field require the use of relativistic MHD models.

In the theories of galaxy formation, relativistic fluid models have been used, for example, in order to describe the evolution of perturbations of the baryon and radiation components of the cosmic medium. Theories of relativistic stars are also based on relativistic fluid model (Zel'dovich and Novikov, 1978; Rose, 1998).

When the medium interacts electromagnetically and is highly conducting, the simplest description is in terms of relativistic MHD. From the mathematical viewpoint, the relativistic MHD was mainly treated in the framework of *general relativity*. This means that the MHD equations were studied in conjunction with Einstein's equations. Lichnerowicz (1967) has made a thorough and deep investigation of the initial value problem.

In many applications, however, we neglect the gravitational field generated by the conducting medium in comparison with the background gravitational field as well as in many cases we simply use *special relativity*. Mathematically this amounts to taking into account only the **conservation laws for matter and the electromagnetic field**, neglecting Einstein's equations. Such relativistic MHD theory is much simpler than the full general relativistic theory. So more detailed results can be obtained (e.g., Anile, 1989; Novikov and Frolov, 1989; Koide *et al.*, 1999).

8.2 Magnetic flux conservation. Ideal MHD

8.2.1 Integral and differential forms of the law

Equations (8.21), (8.18), and (8.22) are the conservation laws for mass, momentum, and energy, respectively. Let us show that, with the proviso that $\nu_{\mathrm{m}} = 0$, Equation (8.19) is the magnetic flux conservation law.

Consider the derivative of the vector $\mathbf{B}$ flux through a surface S moving with the plasma (Figure 8.1).

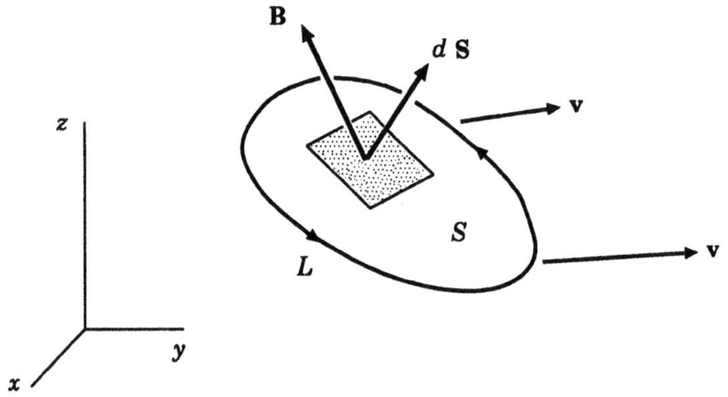

Figure 8.1: The magnetic field $\mathbf{B}$ flux through the surface S moving with a plasma with velocity $\mathbf{v}$.

According to the known formula of vector analysis (e.g., Smirnov, 1965), we have

$$\frac{d}{dt} \int_S \mathbf{B} \cdot d\mathbf{S} = \int_S \left(\frac{\partial \mathbf{B}}{\partial t} + \mathbf{v} \operatorname{div} \mathbf{B} + \operatorname{curl} (\mathbf{B} \times \mathbf{v}) \right) \cdot d\mathbf{S} \, . \tag{8.32}$$

By virtue of Equation (8.20), formula (8.32) is rewritten as follows

$$\frac{d}{dt} \int_S \mathbf{B} \cdot d\mathbf{S} = \int_S \left(\frac{\partial \mathbf{B}}{\partial t} - \operatorname{curl} (\mathbf{v} \times \mathbf{B}) \right) \cdot d\mathbf{S} \, ,$$

or, making use of Equation (8.19),

$$\frac{d}{dt} \int_S \mathbf{B} \cdot d\mathbf{S} = \nu_{\mathrm{m}} \int_S \Delta \mathbf{B} \cdot d\mathbf{S} \, .$$

The right-hand side of this formula can be rewritten with the help of the Stokes theorem and Equation (8.9):

$$\frac{d}{dt}\int\limits_S \mathbf{B}\cdot d\mathbf{S} = -\,\nu_{\rm m}\oint\limits_L \operatorname{curl}\mathbf{B}\cdot d\mathbf{l} = -\,\frac{c}{\sigma}\oint\limits_L \mathbf{j}\cdot d\mathbf{l}\,. \tag{8.33}$$

Here L is the 'liquid' contour bounding the surface S. We have also used here that

$$\Delta\mathbf{B} = -\operatorname{curl}\operatorname{curl}\mathbf{B}\,.$$

Equation (8.33) is equivalent to (8.19) and presents an integral form of the magnetic flux conservation law.

> The magnetic flux through any surface moving with the plasma is conserved, once the magnetic diffusivity $\nu_{\rm m}$ can be ignored.

If we cannot neglect magnetic diffusivity, then the rate of flux change through a surface connected with the plasma is proportional to the electric resistivity σ^{-1} of the plasma. Let us clarify the conditions when it is possible to neglect magnetic diffusivity.

* * *

The relative role of the dissipation processes in the *differential* Equation (8.19) can be evaluated by proceeding as follows. In a spirit similar to that of Section 4.2, we pass on to the dimensionless variables

$$\mathbf{r}^* = \frac{\mathbf{r}}{L}\,, \quad t^* = \frac{t}{\tau}\,, \quad \mathbf{v}^* = \frac{\mathbf{v}}{v}\,, \quad \mathbf{B}^* = \frac{\mathbf{B}}{B_0}\,. \tag{8.34}$$

On substituting definition (8.34) into Equation (8.19) we obtain

$$\frac{B_0}{\tau}\,\frac{\partial\,\mathbf{B}^*}{\partial t^*} = \frac{vB_0}{L}\,\operatorname{curl}^*(\,\mathbf{v}^*\times\mathbf{B}^*\,) + \nu_{\rm m}\,\frac{B_0}{L^2}\,\Delta^*\,\mathbf{B}^*\,.$$

Normalize this equation with respect to its left-hand side, i.e.

$$\frac{\partial\,\mathbf{B}^*}{\partial t^*} = \frac{v\tau}{L}\,\operatorname{curl}^*(\,\mathbf{v}^*\times\mathbf{B}^*\,) + \frac{\nu_{\rm m}\tau}{L^2}\,\Delta^*\,\mathbf{B}^*\,. \tag{8.35}$$

Note that the dimensionless Equation (8.35) contains two dimensionless parameters. The first one,

$$\delta = \frac{v\tau}{L}\,,$$

will be discussed in the next Section. Here, for simplicity, we assume $\delta = 1$. The second parameter,

$$\boxed{\mathrm{Re_m} = \frac{L^2}{\nu_{\mathrm{m}}\, \tau} = \frac{vL}{\nu_{\mathrm{m}}}\,,} \tag{8.36}$$

is termed the *magnetic* Reynolds number, by analogy with the *hydrodynamic* Reynolds number $\mathrm{Re} = vL/\nu$. This parameter characterizes the ratio of the first term on the right-hand side of (8.35) to the second one. Omitting the asterisk, we write Equation (8.35) in the *dimensionless* form

$$\frac{\partial \mathbf{B}}{\partial t} = \mathrm{curl}\,(\,\mathbf{v} \times \mathbf{B}\,) + \frac{1}{\mathrm{Re_m}}\,\Delta \mathbf{B}\,. \tag{8.37}$$

The larger the magnetic Reynolds number, the smaller the role played by magnetic diffusivity.

If $\mathrm{Re_m} \gg 1$, we neglect the plasma resistivity and associated Joule heating and magnetic field dissipation, just as one neglects viscosity effects under large Reynolds numbers in ordinary hydrodynamics.

In laboratory experiments, for example in devices for studying the processes of current sheet formation and rupture during reconnection (e.g., Altyntsev, 1977; Bogdanov *et al.*, 1986), the magnetic Reynolds number is usually not large: $\mathrm{Re_m} \sim 1 - 3$. In this case the resistivity has a dominant role, and Joule dissipation is important.

8.2.2 An approximation and the equations of ideal MHD

Under astrophysical conditions, owing to the high conductivity and the enormous length scales usually considered, the magnetic Reynolds number is very large: $\mathrm{Re_m} > 10^{10}$ (e.g., Problem 8.1). Furthermore the usual Reynolds number is, as a rule, large as well. Therefore, in a great number of problems, it is sufficient to consider an ideal medium with infinite conductivity:

$$\mathrm{Re_m} \gg 1\,, \qquad \mathrm{Re} \gg 1\,. \tag{8.38}$$

We also assume the heat exchange to be of minor importance. This assumption is not universally true. Sometimes the thermal conductivity (due to electrons or radiation) is so effective that the plasma motion must be considered as isothermal, rather than adiabatic. However, while treating the 'ideal medium', all dissipative transfer coefficients as well as the thermal conductivity are set equal to zero in the non-relativistic MHD equations (8.18)–(8.24):

$$\boxed{\nu_{\mathrm{m}} = 0\,, \quad \eta = \zeta = 0\,, \quad \kappa = 0\,.} \tag{8.39}$$

The complete set of MHD equations for the ideal medium has two different (but equivalent) forms. The first is the *transfer equations* form:

$$\frac{\partial \mathbf{v}}{\partial t} + (\mathbf{v}\cdot\nabla)\mathbf{v} = -\frac{\nabla p}{\rho} - \frac{1}{4\pi\rho}\,\mathbf{B}\times \operatorname{curl}\mathbf{B}\,,$$

$$\frac{\partial \mathbf{B}}{\partial t} = \operatorname{curl}(\mathbf{v}\times\mathbf{B})\,, \qquad \operatorname{div}\mathbf{B} = 0\,, \tag{8.40}$$

$$\frac{\partial \rho}{\partial t} + \operatorname{div}\rho\mathbf{v} = 0\,, \quad \frac{\partial s}{\partial t} + (\mathbf{v}\cdot\nabla)s = 0\,, \quad p = p(\rho, s)\,.$$

The other form of ideal MHD equations is the *divergent* form which also corresponds to the *conservation laws* for energy, momentum, mass and magnetic flux:

$$\frac{\partial}{\partial t}\left(\frac{\rho v^2}{2} + \rho\varepsilon + \frac{B^2}{8\pi}\right) = -\operatorname{div}\mathbf{G}\,, \tag{8.41}$$

$$\frac{\partial}{\partial t}(\rho v_\alpha) = -\frac{\partial}{\partial r_\beta}\,\Pi^*_{\alpha\beta}\,, \tag{8.42}$$

$$\frac{\partial \rho}{\partial t} = -\operatorname{div}\rho\mathbf{v}\,, \tag{8.43}$$

$$\frac{\partial \mathbf{B}}{\partial t} = \operatorname{curl}(\mathbf{v}\times\mathbf{B})\,, \tag{8.44}$$

$$\operatorname{div}\mathbf{B} = 0\,, \tag{8.45}$$

$$p = p(\rho, s)\,. \tag{8.46}$$

Here the energy flux density and the momentum flux density tensor are, respectively, equal to (cf. (8.25) and (8.31))

$$\mathbf{G} = \rho\mathbf{v}\left(\frac{v^2}{2} + w\right) + \frac{1}{4\pi}\Big(B^2\,\mathbf{v} - (\mathbf{B}\cdot\mathbf{v})\,\mathbf{B}\Big)\,, \tag{8.47}$$

$$\Pi^*_{\alpha\beta} = p\,\delta_{\alpha\beta} + \rho v_\alpha v_\beta + \frac{1}{4\pi}\left(\frac{B^2}{2}\,\delta_{\alpha\beta} - B_\alpha B_\beta\right). \tag{8.48}$$

The magnetic flux conservation law (8.44) written in the integral form

$$\frac{d}{dt} \int\limits_S \mathbf{B} \cdot d\mathbf{S} = 0 \,, \tag{8.49}$$

where the integral is taken over an arbitrary surface moving with the plasma, is quite characteristic of ideal MHD. It allows us to clearly represent the magnetic field as a set of field lines attached to the medium, as if they were 'frozen into' it. For this reason, Equation (8.44) is frequently referred to as the 'freezing-in' equation.

The freezing-in property converts the notion of magnetic field line from the purely geometric (as for the field in vacuum) to the material sphere. In the ideally conducting medium, the field lines move together with the plasma. The medium displacement conserves not only the magnetic flux but each of the field lines as well. To convince oneself that this is the case, one has to consider a thin tube of magnetic field lines.

In general, the field intensity **B** is a *local* quantity. However, the magnetic field lines are *integral* characteristics of the field. Their analysis becomes more complicated. Nonetheless, a large number of actual fields have been studied because the general features of the morphology – an investigation of *non-local* structures – of magnetic fields are fairly important in cosmic plasma physics.

The geometry of the field lines appears in different ways in the equilibrium criteria for cosmic plasma. For example, much depends on whether the field lines are concave or convex, on the value of the gradient of the so-called *specific volume* of magnetic flux tubes (Chapter 14), on the presence of X-type points (Section 9.3) as well as on a number of other *topological* characteristics, e.g. magnetic *helicity* (Chapter 21).

8.3 The main approximations in ideal MHD

8.3.1 Dimensionless equations

The equations of MHD, even the ideal ones, constitute a set of nonlinear differential equations in partial derivatives. The order of the set is rather high, while its structure is complicated. To formulate a problem in the context of MHD, we have to know the initial and boundary conditions admissible by this set of equations. To do this, in turn, we have to know the type of these equations, in the sense adopted in mathematical physics (e.g., Vladimirov, 1967).

To formulate a problem, one usually uses one or another approximation, which makes it possible to isolate the main effect – the essence of the phenomenon. For instance, if the magnetic Reynolds number is small, then the plasma moves comparatively easily with respect to the magnetic field. This is the case in MHD generators and other laboratory and technical devices (see Sutton and Sherman, 1965, § 1.3; Shercliff, 1965, § 6.5).

The opposite approximation is that of large magnetic Reynolds numbers, when magnetic field 'freezing in' takes place in the plasma (the previous Section). Obviously, the transversal (with respect to the magnetic field) plasma flows are implied. Along the field, Equation (8.44) holds. This approximation is quite characteristic of the cosmic plasma dynamics.

How can we isolate the main effect in a physical phenomenon and correctly formulate the problem? From the above examples concerning the magnetic Reynolds number, the following rule suggests itself:

> take the dimensional parameters characterizing the phenomenon at hand, combine them into dimensionless combinations and then, on calculating their numerical values, make use of the corresponding approximation in the set of *dimensionless* equations.

Such an approach is effective in hydrodynamics (Sedov, 1973, Vol. 1, Ch. 7).

Let us start with the set of MHD equations for an ideal medium (8.40):

$$\frac{\partial \mathbf{v}}{\partial t} + (\mathbf{v} \cdot \nabla)\, \mathbf{v} = -\frac{\nabla p}{\rho} - \frac{1}{4\pi\rho}\, \mathbf{B} \times \operatorname{curl} \mathbf{B}\,, \tag{8.50}$$

$$\frac{\partial \mathbf{B}}{\partial t} = \operatorname{curl}\,(\mathbf{v} \times \mathbf{B})\,, \tag{8.51}$$

$$\frac{\partial \rho}{\partial t} + \operatorname{div} \rho \mathbf{v} = 0\,, \tag{8.52}$$

$$\frac{\partial s}{\partial t} + (\mathbf{v} \cdot \nabla)\, s = 0\,, \tag{8.53}$$

$$\operatorname{div} \mathbf{B} = 0\,, \tag{8.54}$$

$$p = p\,(\rho, s)\,. \tag{8.55}$$

Let the quantities L, τ, v, ρ_0, p_0, s_0, and B_0 be the characteristic values of length, time, velocity, density, pressure, entropy and field strength, respectively. Rewrite Equations (8.50)–(8.55) in the dimensionless variables

$$\mathbf{r}^* = \frac{\mathbf{r}}{L}\,, \quad t^* = \frac{t}{\tau}\,, \;\dots \quad \mathbf{B}^* = \frac{\mathbf{B}}{B_0}\,.$$

Omitting the asterisk, we obtain the equations in dimensionless variables (Somov and Syrovatskii, 1976a):

$$\varepsilon^2 \left\{ \frac{1}{\delta} \frac{\partial \mathbf{v}}{\partial t} + (\mathbf{v} \cdot \nabla)\, \mathbf{v} \right\} = -\gamma^2 \frac{\nabla p}{\rho} - \frac{1}{\rho}\, \mathbf{B} \times \operatorname{curl} \mathbf{B}\,, \tag{8.56}$$

$$\frac{\partial \mathbf{B}}{\partial t} = \delta \operatorname{curl}\,(\mathbf{v} \times \mathbf{B})\,, \tag{8.57}$$

$$\frac{\partial \rho}{\partial t} + \delta \operatorname{div} \rho \mathbf{v} = 0\,, \tag{8.58}$$

$$\frac{\partial s}{\partial t} + \delta\,(\mathbf{v} \cdot \nabla)\, s = 0\,, \tag{8.59}$$

$$\operatorname{div} \mathbf{B} = 0\,, \tag{8.60}$$

$$p = p\,(\rho, s)\,. \tag{8.61}$$

Here

$$\boxed{\delta = \frac{v\tau}{L}\,, \quad \varepsilon^2 = \frac{v^2}{V_{\rm A}^2}\,, \quad \gamma^2 = \frac{p_0}{\rho_0 V_{\rm A}^{\,2}}} \tag{8.62}$$

are three dimensionless parameters characterizing the problem;

$$V_{\rm A} = \frac{B_0}{\sqrt{4\pi\rho_0}} \tag{8.63}$$

is the characteristic value of the Alfvén speed. If the gravitational force were taken into account in (8.50), Equation (8.56) would contain another dimensionless parameter, $gL/V_{\rm A}^{\,2}$, where g is the gravitational acceleration. The analysis of these parameters allows us to gain an understanding of the approximations which are possible in the ideal MHD.

8.3.2 Weak magnetic fields in cosmic plasma

We begin with the assumption that

$$\varepsilon^2 \gg 1 \quad \text{and} \quad \gamma^2 \gg 1\,. \tag{8.64}$$

Then, as is seen from Equation (8.56), in the zero-order approximation relative to the small parameters ε^{-2} and γ^{-2}, we can neglect the magnetic force as compared to the inertia force and the gas pressure gradient. In subsequent approximations, the magnetic effects are treated as a small correction to the hydrodynamic ones.

A lot of problems of cosmic physics are solved in this approximation, termed the *weak* magnetic field approximation. Among the simplest of them are the ones concerning the weak field's influence on hydrostatic equilibrium. An example is the problem of the influence of poloidal and toroidal magnetic fields on the equilibrium of a self-gravitating plasma ball (a star, magnetoid of quasar's kernel etc., see examples in Section 14.1.3).

Some other problems are in fact analogous to the previously mentioned ones. They are called *kinematic* problems, since

> they treat the influence of a given plasma flow on the magnetic field; the reverse influence is considered to be negligible.

Such problems are reduced to finding the magnetic field distribution resulting from the known velocity field. An example is the problem of magnetic field amplification and support by stationary plasma flows (magnetic dynamo) or turbulent amplification. The simplest example is the problem of magnetic field amplification by plasma differential rotation (Elsasser, 1956; Moffat, 1978; Parker, 1979; Rüdiger and von Rekowski, 1998).

8.3.3 Strong magnetic fields in cosmic plasma

The opposite approximation – that of the *strong* magnetic field – has been less well studied. It reflects the specificity of MHD to a greater extent than the weak field approximation. The strong field approximation is valid when the **magnetic force dominates all the others** (inertia force, gas pressure gradient, etc.). Within the framework of Equation (8.56), the magnetic field is referred to as a strong one if in some region under consideration

$$\varepsilon^2 \ll 1 \quad \text{and} \quad \gamma^2 \ll 1\,, \tag{8.65}$$

i.e. if the magnetic energy density greatly exceeds that of the kinetic and thermal energies:

$$\frac{B_0^2}{8\pi} \gg \frac{\rho_0 v^2}{2} \quad \text{and} \quad \frac{B_0^2}{8\pi} \gg 2n_0 k_{\rm B} T_0\,.$$

From Equation (8.56) it follows that, in the zeroth order with respect to the small parameters (8.65), the magnetic field is *force-free*, i.e. it obeys the equation

$$\mathbf{B} \times \operatorname{curl} \mathbf{B} = 0\,. \tag{8.66}$$

This conclusion is quite natural:

> if the magnetic force dominates all the others, then the magnetic field must balance itself in the region under consideration.

If, in addition, electric currents are absent in some region (in the zeroth approximation relative to the small parameters ε^2 and γ^2), then the strong field is simply *potential* in this region:

$$\operatorname{curl} \mathbf{B} = 0\,, \qquad \mathbf{B} = \nabla \Psi\,, \qquad \Delta \Psi = 0\,. \tag{8.67}$$

In principle, the magnetic field can be force-free or even potential for another reason: due to the equilibrium of non-magnetic forces. However, this does not happen frequently.

Consider the first order in the small parameters (8.65). There are two possibilities.

(a) Suppose that

$$\varepsilon^2 \ll \gamma^2 \ll 1\,. \tag{8.68}$$

Then, we neglect the inertia force in Equation (8.56) as compared to the gas pressure gradient. Decomposing the magnetic force into a *magnetic tension* force and a *magnetic pressure* gradient force,

$$\mathbf{F}_m = -\frac{1}{4\pi}\,\mathbf{B} \times \operatorname{curl} \mathbf{B} = \frac{1}{4\pi}\,(\mathbf{B} \cdot \nabla)\,\mathbf{B} - \nabla\,\frac{B^2}{8\pi}\,, \tag{8.69}$$

we obtain the following dimensionless equation:

$$(\mathbf{B} \cdot \nabla)\,\mathbf{B} = \nabla \left(\frac{B^2}{2} + \gamma^2 p \right). \tag{8.70}$$

Owing to the presence of the gas pressure gradient, the magnetic field differs from the force-free one at any moment of time:

> the magnetic tension force $(\mathbf{B} \cdot \nabla)\,\mathbf{B}/4\pi$ must balance not only the magnetic pressure gradient but that of the gas pressure as well.

Obviously the effect is proportional to the small parameter γ^2.

This approximation can be naturally called the *magnetostatic* one since $\mathbf{v} = 0$. It effectively works in regions of a strong magnetic field where the gas pressure gradients are large, for example, in coronal loops and current sheets in the solar corona (Problem 17.2).

(b) The inertia force also causes the magnetic field to deviate from the force-free one:

$$\varepsilon^2 \left\{ \frac{1}{\delta}\,\frac{\partial \mathbf{v}}{\partial t} + (\mathbf{v} \cdot \nabla)\,\mathbf{v} \right\} = -\,\frac{1}{\rho}\,\mathbf{B} \times \operatorname{curl} \mathbf{B}\,. \tag{8.71}$$

Here we ignored (in the first order) the gas pressure gradient as compared with the inertia force. Thus it is not the relation (8.68) between the small parameters (8.65), but rather its converse, that should be obeyed, i.e.

$$\gamma^2 \ll \varepsilon^2 \ll 1 \, . \tag{8.72}$$

The problems on plasma flows in a strong field are of considerable interest in cosmic electrodynamics. To solve them, inequalities (8.72) can be assumed to hold. Then we can use (8.71) as the MHD equation of motion. The approximation corresponding inequalities (8.72) is naturally termed the approximation of *strong* field and *cold* plasma.

The main applications of the strong-field-cold-plasma approximation are concerned with the solar atmosphere and the Earth's magnetosphere. Both phenomena are sufficiently well studied from the observational viewpoint. We can proceed with confidence from qualitative interpretation to the construction of quantitative models. The presence of a sufficiently strong magnetic field and a comparatively rarefied plasma is common for both phenomena. This justifies the applicability of the approximation at hand. Analogous conditions are reproduced under laboratory modelling of these phenomena. Some other astrophysical applications of the strong-field-cold-plasma approximation will be discussed in the next Section and next Chapter.

* * *

In closing, consider the dimensionless parameter $\delta = v\tau / L$. As is seen from Equation (8.71), it characterizes the relative role of the local $\partial / \partial t$ and transport $(\mathbf{v} \cdot \nabla)$ terms in the substantial derivative d/dt.

If $\delta \gg 1$ then, in the zeroth approximation relative to the small parameter δ^{-1}, the plasma flow can be considered to be *stationary*

$$\varepsilon^2 \, (\mathbf{v} \cdot \nabla) \, \mathbf{v} = - \frac{1}{\rho} \, \mathbf{B} \times \operatorname{curl} \mathbf{B} \, . \tag{8.73}$$

If $\delta \ll 1$, i.e. plasma displacement is small during the magnetic field change, then the transport term $(\mathbf{v} \cdot \nabla)$ can be ignored in the substantial derivative and the equation of motion in the strong-field-cold-plasma approximation takes the form

$$\varepsilon^2 \, \frac{\partial \mathbf{v}}{\partial t} = - \frac{1}{\rho} \, \mathbf{B} \times \operatorname{curl} \mathbf{B} \, , \tag{8.74}$$

other equations becoming linear. This case corresponds to small plasma displacements from the equilibrial state, i.e. small perturbations. (If need be, the right-hand side of Equation (8.74) can be linearized in the usual way.)

Generally the parameter $\delta \approx 1$ and the set of MHD equations in the approximation of strong field and cold plasma for ideal medium assumes the following dimensionless form:

$$\varepsilon^2 \frac{d\mathbf{v}}{dt} = -\frac{1}{\rho}\, \mathbf{B} \times \operatorname{curl} \mathbf{B}\,, \tag{8.75}$$

$$\frac{\partial \mathbf{B}}{\partial t} = \operatorname{curl}\,(\,\mathbf{v} \times \mathbf{B}\,)\,, \tag{8.76}$$

$$\frac{\partial \rho}{\partial t} + \operatorname{div} \rho \mathbf{v} = 0\,. \tag{8.77}$$

In the next chapter we shall consider some continuous plasma flows in a strong magnetic field, which are described by Equations (8.75)–(8.77).

8.4 Accretion discs and relativistic jets

8.4.1 Angular momentum transfer in binary stars

Magnetic fields were discussed as a possible means of angular transport in the development of *accretion disc* theory in the early seventies (Shakura and Sunyaev, 1973; Novikov and Thorne, 1973). Interest in the role of magnetic fields in binary stars steadily increased after the discovery of the nature of AM Herculis. It appeared that the optical counterpart of the soft X-ray source has linear and circular polarization in the V and I spectral bands, of a strength an order of magnitude larger than previously observed in any object. This suggested the presence of a very strong field, with $B \sim 10^8$ G, assuming the fundamental cyclotron frequency to be observed.

Similar systems were soon discovered. Evidence for strong magnetic fields was subsequently found in the X-ray binary pulsars and the intermediate polar binaries, both believed to include accretion discs. A magnetically channelled wind from the main sequence star has been invoked to explain the higher rates of mass transfer observed in binaries above the period gap, and in an explanation of the gap. The winds from accretion discs have been suggested as contributing to the inflow by removing angular momentum.

Magnetohydrodynamics in binary stars is now an area of central importance in stellar astrophysics (Campbell, 1997; Rose, 1998). Magnetic fields are believed to play a role even in apparently non-magnetic binaries. They provide the most viable means, through the so-called shear instabilities, of generating the MHD turbulence in an accretion disc necessary to drive the plasma inflow via the resulting magnetic and viscous stresses.

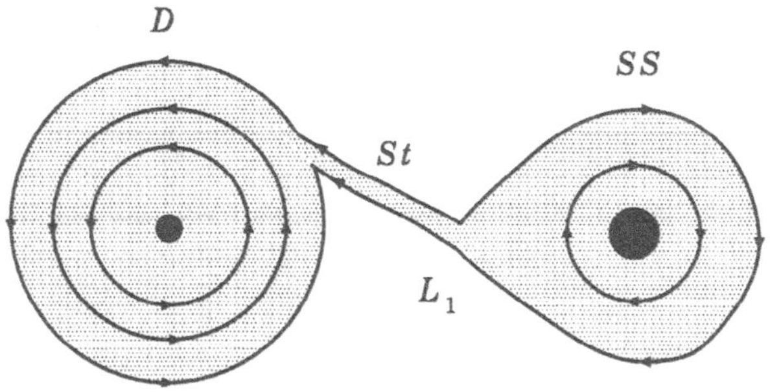

Figure 8.2: The standard model of a binary system viewed down the rotational pole. The tidally and rotationally distorted secondary star SS loses plasma from the unstable L_1 point. The resulting plasma stream St feeds an accretion disc D, centred on the primary star.

The fundamental problem is the role of magnetic fields in redistributing angular momentum in binary stars. The disc is fed by the plasma stream originated in the $L1$ region (Figure 8.2) of the secondary star. In a steady state,

> plasma is transported through the disc at the rate it is supplied by the stream and the angular momentum will be advected outwards.

Angular momentum avdection requires coupling between rings of rotating plasma; ordinary hydrodynamic viscosity is too weak to provide this. Hence some form of **anomalous viscosity** must be invoked to explain the plasma flow through the disc.

The key point is the recognition that a simple linear instability generates MHD turbulence. This turbulence transports angular momentum outward through the disc, allowing accretion to proceed. Although turbulence seems like a natural and straightforward transport mechanism, it turns out that the **magnetic fields are essential**. Purely hydrodynamic turbulence is not self-sustaining and does not produce sustained outward transport of angular momentum (see Hawley and Balbus, 1999).

It is most probable that the accretion discs have turbulent motions generated by the shear instabilities. The turbulence and strong radial shear lead to the generation and maintenance of a large scale magnetic field.

> Viscous and magnetic stresses cause radial advection of the angular momentum via the azimutal forces.

Provided these forces oppose the large-scale azimutal motion, plasma will spiral in through the disc as angular momentum flows outwards. Presumably, the approximation of a weak field (Section 8.3.2) can be used inside the disc to model these effects. Most models to date involve a vertically averaged structure. The future aim is to find 3D solutions which self-consistently incorporate the magnetic shear instabilities and vertical structure.

The stellar spin dynamics and stability are also important, of course. For example, in spin evolution calculations, a compact white draft, or neutron star, is usually treated as a rigid body. This is valid provided the dynamic time-scale for adjustments in the stellar structure is short compared to the spin evolution time scale. In general, however, a strongly-magnetic primary star may experience significant distortions from spherical symmetry due to non-radial internal magnetic forces. This fact can be demonstrated by the tensor virial theorem in MHD (Section 14.1.3).

8.4.2 Accretion discs near black holes

In interacting binary stars there is an abundance of evidence for the presence of accretion discs: (a) double-peaked emission lines are observed; (b) eclipses of an extended light source centered on the primary occur, and (c) in some cases eclipses of the secondary star by the disc are also detected. The case for the presence of accretion discs in active galactic nuclei is less clear. Nonetheless, the disc-fed accreation onto a super-massive black hole is the commonly accepted model for these astronomical objects. In fact, active galactic nuclei also exhibit the classical double-peaked, broad emission lines which are considered to be characteristic for a rotating disc.

As the plasma accretes in the gravitational potential of the central mass, magnetic field lines are convected inwards, amplified and finally deposited on **the horizon of the black hole** (Section 6.3.4). As long as a magnetic field is confined by the disc, a differential rotation causes the field to wrap up tightly, becoming highly sheared and predominantly azimuthal in orientation. A dynamo in the disc may be responsible for the maintenance and amplification of the magnetic field.

In the standard model of an accretion disc (Shakura and Sunyaev, 1973; Novikov and Thorne, 1973), the gravitational energy is locally radiated from the optically thin disc, and the plasma keeps its Keplerian rotation. However, *advection* has recently come to be thought of as an important process and results in a structure different from the standard model. The advection process physically means that

> the energy generated via viscous dissipation is restored as entropy of the accreting plasma rather than being radiated.

The advection effect can be very important since the radiation efficiency decreases under these circumstances (Section 6.3.4).

8.4.3 Jets near black holes

Jet-like phenomena, including relativistic jets (e.g., Birkinshaw, 1997), are observed on a wide range of scales in accretion disc systems. Active galactic nuclei show extremely energetic outflows extending even to scales beyond the outer edge of a galaxy in the form of strongly collimated radio jets. The luminosities of the radio jets give an appreciable fraction of the luminosity of the underlying central object. There is substantial evidence that **magnetic forces are involved in the driving mechanism** and that the magnetic fields also provide the collimation of relativistic flows (see also Section 15.1.3). So, numerucal simulations must incorporate relativistic MHD in a four-dimensional space-time (e.g., Nishikawa *et al.*, 1999; Koide *et al.*, 1999).

Rotating black holes are thought to be the prime-mover behind the activity detected in centers of galaxies. The gravitational field of rotating black holes is more complex than that of non-rotating ones. In addition to the ordinary gravitational force, $m\mathbf{g}$, the rotation generates the so-called *gravitomagnetic* force which is an analogy of the Lorentz force. In fact, the full weak-gravity (far from the hole) low-velocity (replacing the relativistic unified spacetime with an equivalent Galilean 'absolute-space-plus-universal-time') coordinate acceleration of uncharged particle (see Macdonald *et al.*, 1986 and references there; see also Chapter 4 in Novikov and Frolov, 1989)

$$\frac{d^2\mathbf{r}}{dt^2} = \mathbf{g} + \frac{d\mathbf{r}}{dt} \times \mathbf{H}_{gr} \tag{8.78}$$

looks like the Lorentz force with the electric field $\mathbf{E}$ replaced by $\mathbf{g}$, the magnetic field $\mathbf{B}$ replaced by the vector $\mathbf{H}_{gr} = \operatorname{curl} \mathbf{A}_{gr}$, and the electric charge e replaced by the particle mass m. These analogies lie behind the use of the words 'gravitoelectric' and 'gravitomagnetic' to describe the gravitational acceleration field $\mathbf{g}$ and to describe the 'shift function' $\mathbf{A}_{gr}$ and its derivatives (Problem 8.5).

The analogy with electromagnetism remains strong so long as all velocities are small compared with that of light and gravity is weak enough to be linear. Thus, far from the horizon, the gravitational acceleration

$$\mathbf{g} = -\frac{M}{r^2}\,\mathbf{e}_r \tag{8.79}$$

is the radial Newtonian acceleration and the gravitomagnetic field

$$\mathbf{H}_{gr} = 2\,\frac{\mathbf{J} - 3\,(\mathbf{J}\cdot\mathbf{e}_r)\,\mathbf{e}_r}{r^3} \tag{8.80}$$

is a dipole gravitomagnetic field with the role of dipole moment played by the hole's angular momentum

$$\mathbf{J} = \int (\mathbf{r} \times \rho_m \mathbf{v})\; dV\,. \tag{8.81}$$

A physical manifestation of the gravitomagnetic field (8.80) is the gravitomagnetic precession that is induced in gyroscopes far from the hole. The electromagnetic analogy suggests that not only should the gravitomagnetic field exert a torque on a gyroscope outside a black hole, it should also exert a force. **The gravitomagnetic force drives an accretion disc into the hole's equatorial plane** and holds it there indefinitely regardless of how the disc's angular momentum may change (Figure 8.3). Consequently, at radii where the bulk of the disc's gravitational energy is released and where the hole-disc interactions are strong,

> there is only one geometrically preferred direction along which a jet might emerge: the normal to the disc plane, which coincides with the rotation axis of the black hole.

In some cases the jet might be produced by winds off the disc, in other cases by electrodynamic acceleration of the disc, and in others by currents in the hole's magnetosphere (see Begelman, Blandford and Rees, 1984). However whatever the mechanism, the jet presumably is locked to the hole's rotation axis. **The black hole acts as a gyroscope to keep the jet aligned.** The fact that it is very difficult to torque a black hole accounts for the constancy of the observed jet directions over length scales as great as millions of light years and thus over time scales of millions of years or longer.

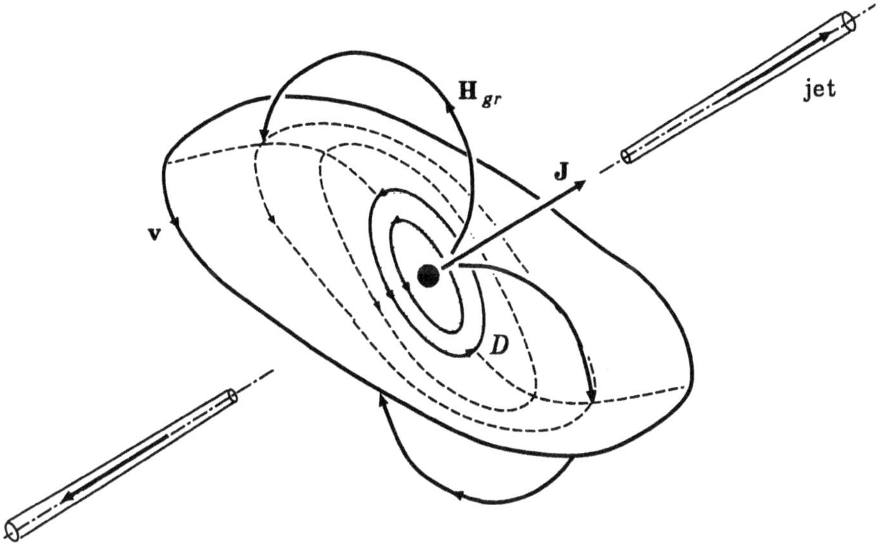

Figure 8.3: An accretion disc D around a rotating black hole is driven into the hole's equatorial plane at small radii by a combination of gravitomagnetic forces (action of the gravitomagnetic field $\mathbf{H}_{gr}$ on orbiting plasma) and viscous forces.

A black hole by itself is powerless to produce the observed jets. It does so only with the aid of surrounding plasma and magnetic fields. A supermassive hole in a galactic nucleus can acquire surrounding matter either by gravitationally pulling interstellar gas into its vicinity, or by tidally disrupting passing stars and smearing their matter out around itself. In either case the gas is likely to have so much angular momentum that, instead of being swallowed directly and radially into the hole, it forms an orbiting disc around the hole. The orientation of the disc at large radii is determined by the direction of the angular momentum of the recently acquired gas, see an external part of the accretion disc in Figure 8.3.

In the highly-conducting medium, the gravitomagnetic force couples with electromagnetic fields over Maxwell's equations. This effect has interesting consequences for the magnetic fields advected from the interstellar matter towards the black hole (e.g., Camenzind, 1990). It leads to a gravitomagnetic dynamo which amplifies any seed field near a rotating compact object. This process builds up the dipolar magnetic structures which may be behind the bipolar outflows seen as relativistic jets (for comparison with a non-relativistic process see Section 9.4).

Magnetic fields also influence the accretion towards the rotating black hole. For rapidly rotating holes, the accreation can carry negative angular momentum inwards, spinning down the black hole.

8.4.4 Flares in accretion disc coronae

Following the launch of several X-ray satellites, astrophysicists have tried to observe and analyze the violent variations of high energy flux from black hole candidates (e.g., Negoro *et al.*, 1995; see also review in Di Matteo *et al.*, 1999). So far, similar solar and astrophysical statistical studies have been done almost independently of each other. Ueno (1998) first compared X-ray light curves from the solar corona and from the accretion disc in Cyg X-1, a famous black hole candidate. He analyzed also the power spectral densities, the peak interval distributions (the interval of time between two consecutive flares), and the peak intensity distributions.

It has appeared that there are many relationships between flares in the solar corona and 'X-ray shots' in accretion discs. (Of course, there are many differences and unexplained features.) For example, the peak interval distribution of Cyg X-1 shows that the occurrence frequency of large X-ray shots is reduced. A second large shot does not occur soon after a previous large shot. This suggests the existence of energy-accumulation structures, such as magnetic fields in solar flares.

It is likely that accreation discs have a corona which interacts with a magnetic field generated inside a disc. Drawing on developments in solar physics, Galeev *et al.* (1979) suggested that the corona is confined and heated in strong magnetic loops which have buoyantly emerged from the disc. Buoyancy constitutes a mechanism able to channel a part of the energy released in the accretion process directly into the corona outside the disc.

> Magnetic reconnection of buoyant fields in the lower density surface regions may supply the energy source for a hot corona.

On the other hand, the coronal magnetic field can penetrate the disc and is stressed by its motions. The existence of a disc corona with a *strong* field (Section 8.3.3) raises the possibility of a wind flow similar to the solar wind. In principle, this would result in angular momentum transport away from the disc, which could have some influence on the inflow. Another feature related to the accretion disc corona is the possibility of a flare energy release similar to solar flares.

When a plasma in the disc corona is optically thin and has a dominant magnetic pressure, the circumstances are likely to be very similar to the solar corona. Therefore

> it is possible to imagine some similarity between the mechanisms of solar flares and X-ray shots in accretion discs.

Besides the effect of heating the the disc corona, reconnection is able to accelerate electrons and protons to relativistic energies (Lesch and Pohl, 1992; Bednarek and Protheroe, 1999). Some geometrical and physical properties of the flares in accretion disc coronae can be inferred almost directly from soft- and hard X-ray observations of Galactic black hole candidates (Beloborodov, 1999; Di Matteo *et al.*, 1999).

8.4.5 Relativistic jets from disc coronae

Subramanian *et al.* (1999) consider the possibility that the relativistic jets observed in many active galactic nuclei may be powered by the Fermi acceleration of protons in a tenuous corona above a two-temperature accretion disc (Section 6.3.4). The acceleration arises, in this scenario, as a consequence of the shearing motion of the magnetic field lines in the corona, that are anchored in the underlying Keplerian disc. The protons in the corona have a power-law distribution because the density there is too low for proton-proton collisions (formula (6.34)) to thermalize the energy supplied via Fermi acceleration.

The same mechanism also operates in the disc itself. However there the density is high enough for termalization to occur and consequently the disc protons have the Maxwellian distribution. Particle acceleration in the corona leads to the development of a pressure-driven wind that passes through a critical point and subsequently transforms into a relativistic jet at large distances from the black hole.

8.5 Practice: Problems and Answers

Problem 8.1. Estimate the magnetic diffusivity and the magnetic Reynolds number under typical conditions in the solar corona.

Answer. Let us take characteristic values of the parallel conductivity as they were estimated in Problem 7.7:

$$\sigma_{\parallel} = \sigma \sim 10^{16} - 10^{17}\ \mathrm{s}^{-1}\,.$$

Substituting these values in formula (8.24) we obtain

$$\nu_{\rm m} \approx 7.2 \times 10^{19} \, \frac{1}{\sigma} \sim 10^{3} - 10^{4} \ {\rm cm}^{2} \, {\rm s}^{-1} . \tag{8.82}$$

According to definition (8.36) the magnetic Reynolds number

$$\mathrm{Re}_{\rm m} = \frac{vL}{\nu_{\rm m}} \sim 10^{11} - 10^{12} , \tag{8.83}$$

if the characteristic values of length and velocity, $L \sim 10^{10}$ cm and $v \sim 10^{5}$ cm s^{-1}, are taken for the corona. Thus the ideal MHD approximation can be well used to consider, for example, magnetic field diffusion in coronal linear scales.

Problem 8.2. Evaluate the characteristic value of Alfvén speed in the solar corona above a large sunspot.

Answer. From definition (8.63) we find the following formula for Alfvén speed

$$V_{\rm A} \approx 2.18 \times 10^{11} \, \frac{B}{\sqrt{n}} \, , \ {\rm cm \, s}^{-1} . \tag{8.84}$$

In this formula we neglect a small contribution of the ions that are heavier than protons into the plasma density ρ.

Above a sunspot the field strength can be as high as $B \approx 3000$ G. Plasma density in the low corona $n \approx 2 \times 10^{8}\,{\rm cm}^{-3}$. For these values formula (8.84) gives unacceptably high values of the Alfvén speed: $V_{\rm A} \approx 5 \times 10^{10}$ cm s^{-1} $> c$. This means that

> in a strong magnetic field and low density plasma the Alfvén waves propagate with velocities approaching the light speed c.

So, formula (8.84) has to be corrected by a *relativistic* factor which takes this fact into account.

Alfvén (1950) pointed out that the 'magnetohydrodynamic waves' are just an extreme case of electromagnetic waves (Section 10.2.2 and Problem 10.3). Alfvén has shown that the transition between electromagnetic and Alfvén waves can be surveyed by the help of the following formula for the speed of propagation along the magnetic field:

$$V_{\rm A}^{rel} = \frac{B}{\sqrt{4\pi\rho}} \, \frac{1}{\sqrt{1 + B^{2}/4\pi\rho c^{2}}} \, , \tag{8.85}$$

which agrees with (8.63) when $B^2 \ll 4\pi\rho c^2$. Therefore the relativistic Alfvén wave speed is always smaller than the light speed:

$$V_{\rm A}^{rel} = \frac{c}{\sqrt{1 + 4\pi\rho c^2/B^2}} \leq c. \tag{8.86}$$

For values of the magnetic field and plasma density mentioned above, this formula gives $V_{\rm A}^{rel} \approx 2 \times 10^{10}\ {\rm cm\, s^{-1}} < c$.

Formula (8.86) shows that, in low desity cosmic plasmas, the Alfvén speed can easily approach the light speed c.

Problem 8.3. For the conditions in the low corona used in Problem 8.2 estimate the parameter γ^2.

Answer. Substitute $p_0 = 2n_0 k_{\rm B} T_0$ in definition (8.62):

$$\gamma^2 = \frac{n_0 k_{\rm B} T_0}{B_0^2/8\pi} \approx 3.47 \times 10^{-15}\, \frac{n_0 T_0}{B_0^2}\,. \tag{8.87}$$

Let us take as the characteristic values of temperature $T_0 \approx 2 \times 10^6$ K and magnetic field $B_0 \approx 3000$ G. For these values formula (8.87) gives the dimensionless parameter $\gamma^2 \sim 10^{-7}$. Hence, in the solar corona above sunspots the conditions (8.72) of a strong field can be satisfied well for a wide range of plasma parameters.

Problem 8.4. By using general formula (8.47) for the energy flux in ideal MHD, find the magnetic energy influx into a reconnecting current sheet.

Answer. Consider a current sheet as a neutral one (Figure 4.19). In this simplest approximation, near the sheet $\mathbf{B} \perp \mathbf{v}$. Therefore in formula (8.47) the scalar product $(\mathbf{B} \cdot \mathbf{v}) = 0$ and the energy flux density

$$\mathbf{G} = \rho \mathbf{v} \left(\frac{v^2}{2} + w \right) + \frac{B^2}{4\pi}\, \mathbf{v}\,. \tag{8.88}$$

If the approximation of a strong field is satisfied, the last term in (8.88) is dominating, and we find the magnetic energy flux density or the Poynting vector (cf. general definition (8.27)) directed into the current sheet

$$\mathbf{G}_{\rm P} = \frac{B^2}{4\pi}\, \mathbf{v}\,. \tag{8.89}$$

For a quarter of the current sheet assumed to be symmetrical and for a unit length along the current, the total flux of magnetic energy

$$\mathcal{E}_{mag}^{in} = \frac{B_0^2}{4\pi}\, v_0 b\,, \tag{8.90}$$

where b is half-width of the current sheet (Figure 4.19), B_0 is the field strength on the inflow sides of the sheet, v_0 is the inflow velocity.

Problem 8.5. Consider a weakly gravitating, slowly rotating body such as the Earth or the Sun, with all nonlinear gravitational effects neglected. Compute the gravitational force and gravitomagnetic force (Section 8.4.3) from the linearized Einstein equations (e.g., Landau and Lifshitz, *Classical Theory of Field*, 1971, Ch. 10). Show that, for a time-independent body, these equations are identical to the Maxwell-type equations (1.1)–(1.4):

$$\operatorname{curl}\mathbf{g} = 0\,, \quad \operatorname{div}\mathbf{g} = -4\pi G\rho_m\,, \tag{8.91}$$

$$\operatorname{curl}\mathbf{H}_{gr} = -16\pi G\rho_m\mathbf{v}\,, \quad \operatorname{div}\mathbf{H}_{gr} = 0\,. \tag{8.92}$$

Here the differences are minus signs due to gravity being attractive rather than repulsive, a factor 4 in the curl$\mathbf{H}_{gr}$ equation, the presence of the gravitational constant G, the replacement of charge density ρ^q by mass density ρ_m, and the replacement of current density $\mathbf{j}$ by the density of mass current $\rho_m\mathbf{v}$ with $\mathbf{v}$ the velocity of the mass.

Chapter 9

Cosmic Plasma Flows in a Strong Magnetic Field

A sufficiently strong magnetic field easily moves a comparatively rarified plasma in many non-stationary phenomena in space.

9.1 The general formulation of the problem

As was shown in Section 8.3.3, the set of MHD equations for an ideal medium in the approximation of strong field and cold plasma is characterized only by the small parameter $\varepsilon = v/V_{\rm A}$:

$$\varepsilon^2 \frac{d\mathbf{v}}{dt} = -\frac{1}{\rho}\,\mathbf{B} \times \operatorname{curl} \mathbf{B}\,, \tag{9.1}$$

$$\frac{\partial \mathbf{B}}{\partial t} = \operatorname{curl}\left(\mathbf{v} \times \mathbf{B}\right), \tag{9.2}$$

$$\frac{\partial \rho}{\partial t} + \operatorname{div} \rho \mathbf{v} = 0\,. \tag{9.3}$$

Let us try to find the solution to this set as a power series in the parameter ε^2, i.e. representing all the unknown quantities in the form

$$f(\mathbf{r},t) = f^{(0)}(\mathbf{r},t) + \varepsilon^2 f^{(1)}(\mathbf{r},t) + \dots\,. \tag{9.4}$$

Then we try to find the solution in three consequent steps.

(a) To zeroth order with respect to ε^2, the magnetic field is determined by the equation

$$\mathbf{B}^{(0)} \times \operatorname{curl} \mathbf{B}^{(0)} = 0\,. \tag{9.5}$$

This must be supplemented with a boundary condition, which generally depends on time:

$$\mathbf{B}^{(0)}(\mathbf{r},t)\,|_S = \mathbf{f}_1(\mathbf{r},t)\,. \tag{9.6}$$

Here S is the boundary of the region G, in the interior of which the force-free-field Equation (9.5) applies.

> The strong force-free magnetic field, changing in time according to the boundary condition (9.6), sets the plasma in motion.

(b) The kinematics of this motion is uniquely determined by two conditions. The first one follows from the equation of motion and signifies the orthogonality of acceleration to the magnetic field lines

$$\mathbf{B}^{(0)} \cdot \frac{d\mathbf{v}^{(0)}}{dt} = 0\,. \tag{9.7}$$

This equation is obtained by taking the scalar product of Equation (9.1) and the vector $\mathbf{B}^{(0)}$.

The second condition is a consequence of the freezing-in Equation (9.2)

$$\frac{\partial \mathbf{B}^{(0)}}{\partial t} = \operatorname{curl}\left(\mathbf{v}^{(0)} \times \mathbf{B}^{(0)}\right). \tag{9.8}$$

Equations (9.7) and (9.8) determine the velocity field $\mathbf{v}^{(0)}(\mathbf{r},t)$, if the initial condition inside the region G is given:

$$\mathbf{v}_{\parallel}^{(0)}(\mathbf{r},0)\,|_G = \mathbf{f}_2(\mathbf{r})\,. \tag{9.9}$$

Here $\mathbf{v}_{\parallel}^{(0)}$ is the velocity component along the field lines. The velocity component across the field lines is uniquely defined, once the field $\mathbf{B}^{(0)}(\mathbf{r},t)$ is known, by the freezing-in Equation (9.8) at any moment, including the initial one.

(c) Since we know the velocity field $\mathbf{v}^{(0)}(\mathbf{r},t)$, the continuity equation

$$\frac{\partial \rho^{(0)}}{\partial t} + \operatorname{div} \rho^{(0)}\mathbf{v}^{(0)} = 0 \tag{9.10}$$

allows us to find the plasma density distribution $\rho^{(0)}(\mathbf{r},t)$, if we know its initial distribution

$$\rho^{(0)}(\mathbf{r},0)\,|_G = f_3(\mathbf{r})\,. \tag{9.11}$$

Therefore Equations (9.5), (9.7) and (9.8), together with the continuity equation (9.10), completely determine the unknown zero-order quantities $\mathbf{B}^{(0)}(\mathbf{r},t)$, $\mathbf{v}^{(0)}(\mathbf{r},t)$ and $\rho^{(0)}(\mathbf{r},t)$, once the boundary condition (9.6) and the initial conditions (9.9) and (9.11) inside the region G are given.

At any moment of time, the field $\mathbf{B}^{(0)}(\mathbf{r},t)$ is found from Equation (9.5) and the boundary condition (9.6). Thereupon the velocity $\mathbf{v}^{(0)}(\mathbf{r},t)$ is determined from Equations (9.7) and (9.8) and the initial condition (9.9). Finally the continuity Equation (9.10) and the initial condition (9.11) give the plasma density distribution $\rho^{(0)}(\mathbf{r},t)$.

From here on we restrict our attention to the consideration of the zeroth order relative to the parameter ε^2, neglecting the magnetic field deviation from a force-free state. However, it can be seen that the consecutive application of the expansion (9.4) to the set of Equations (9.1)–(9.3) allows one to obtain a closed set of equations for determination of MHD quantities in any order relative to ε^2. An important point, however, is that, during the solution of the problem in the zeroth order relative to ε^2, regions can appear, where the gas pressure gradient cannot be ignored. Here effects proportional to the small parameter γ^2 must be taken into account (Section 8.3.3). This fact usually imposes a limitation on the applicability of the strong-field-cold-plasma approximation.

The question of the existence of general solutions to the MHD equations in this approximation will be considered in Section 9.3, using two-dimensional problems as an example.

9.2 The formalism of two-dimensional problems

While being relatively simple from the mathematical viewpoint, two-dimensional MHD problems allow one to gain some knowledge concerning the plasma flows with the frozen-in magnetic field. Moreover, the two-dimensional problems are sometimes a close approximation of the real three-dimensional flows and can be used to compare the theory with experiments and observations, both qualitatively and quantitatively.

There are two types of problems treating the plane plasma flows, i.e. the flows with the velocity field of the form

$$\mathbf{v} = \{\, v_x(x,y,t),\ v_y(x,y,t),\ 0\,\}\,. \qquad (9.12)$$

All the quantities are dependent on the variables x, y and t.

9.2.1 The first type of problems

The first type incorporates problems with a magnetic field which is everywhere parallel to the z axis of a Cartesian system of coordinates. The corresponding current is parallel to the (x, y) plane. Thus

$$\mathbf{B} = \{\, 0,\ 0,\ B\,(x, y, t)\,\}\,, \tag{9.13}$$

$$\mathbf{j} = \{\, j_x(x, y, t),\ j_y(x, y, t),\ 0\,\}\,. \tag{9.14}$$

As an example of a problem of the first type, consider the effect of a *longitudinal* magnetic field in reconnecting current sheets. Under real conditions, reconnection does not occur at the zeroth lines, but rather at the 'limiting lines' of the magnetic field or 'separators' (Section 16.5.2). The latter differ from the zeroth lines only in that the separators contain the longitudinal component of the field as shown in Figure 9.1.

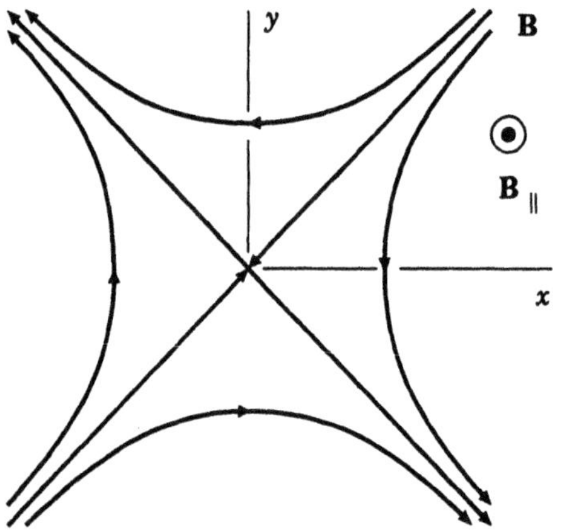

Figure 9.1: Structure of the magnetic field near a separator. A longitudinal field $\mathbf{B}_{\parallel}$ parallel to the z axis is superimposed on the two-dimensional hyperbolic field in the plane (x, y) (cf. Figure 4.16).

With the appearance of the longitudinal field, the force balance in the current sheet that is formed at the separator is changed. The field and plasma pressure outside the sheet must balance not only the gas pressure but also that of the longitudinal field inside the sheet (Figure 9.2)

$$\mathbf{B}_{\parallel} = \Big\{\, 0,\ 0,\ B_{\parallel}\,(x, y, t)\,\Big\}\,. \tag{9.15}$$

This effect is known in the so-called theta-pinch. In axially symmetric geometry, in cylindrical coordinates r, θ, z, an azimuthal current density j_θ crossed with an axial field B_z can support a radial pressure gradient.

If the longitudinal field accumulated in the current sheet during reconnection, the field pressure $B_{\parallel}^2/8\pi$ would considerably limit the sheet compression as well as the reconnection rate. However, the solution of the problem of the first type with respect to $\mathbf{B}_{\parallel}$ (see the discussion in Section 17.2.2 and references therein) shows that another effect is of importance in the real plasma with finite conductivity.

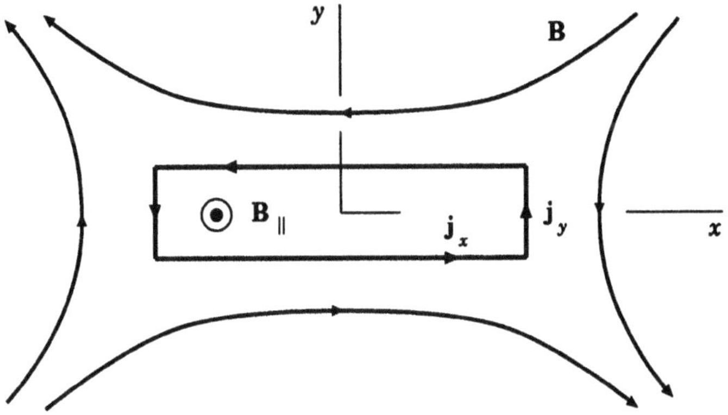

Figure 9.2: A model of a reconnecting current sheet with a longitudinal component of a magnetic field $\mathbf{B}_{\parallel}$.

The effect, in essence, is this: the **longitudinal field compression** in the current sheet produces a gradient of this field and a corresponding electric current circulating in the transversal (relative to the main current j_z in the sheet) plane (x, y). This current circulation is of the type (9.14); it is represented schematically in Figure 9.2. The circulating current plays just the same role as the j_θ-current in the theta-pinch, a one-dimensional equilibrium in a cylindric geometry with an axial field $B_z(r)$. **Ohmic dissipation of the circulating current** under conditions of finite conductivity leads to longitudinal field diffusion outwards from the sheet, thus limiting the longitudinal field accumulation in the reconnecting current sheet (Section 17.2.2).

9.2.2 The second type of problems

(a) Magnetic field and its vector potential

From this point on we shall be mainly interested in two-dimensional problems of the second type. They treat the plane plasma flows (9.12) associated with the plane magnetic field

$$\mathbf{B} = \{\, B_x(x, y, t),\ B_y(x, y, t),\ 0 \,\}\,. \tag{9.16}$$

The electric currents corresponding to this field are parallel to the z axis

$$\mathbf{j} = \{\,0,\,0,\,j\,(x,y,t)\,\}\,. \tag{9.17}$$

The vector-potential $\mathbf{A}$ of such a field has as its only non-zero component:

$$\mathbf{A} = \{\,0,\,0,\,A\,(x,y,t)\,\}\,.$$

The magnetic field $\mathbf{B}$ is defined by the z-component of the vector-potential:

$$\mathbf{B} = \left\{ \frac{\partial A}{\partial y}\,,\; -\frac{\partial A}{\partial x}\,,\; 0 \right\}. \tag{9.18}$$

The scalar function $A\,(x,y,t)$ is often termed the *vector potential*. This function is quite useful, owing to its properties.

Property 1. Substitute (9.18) in the differential equations describing the magnetic field lines

$$\frac{dx}{B_x} = \frac{dy}{B_y} = \frac{dz}{B_z}\,. \tag{9.19}$$

Equations (9.19) imply parallelism of the vector $d\mathbf{l} = \{dx,\,dy,\,dz\}$ to the vector $\mathbf{B} = \{B_x, B_y, B_z\}$. In the case under study $B_z = 0,\; dz = 0$, and

$$\frac{dx}{\partial A/\partial y} = -\frac{dy}{\partial A/\partial x}$$

or

$$\frac{\partial A}{\partial x}\,dx + \frac{\partial A}{\partial y}\,dy = 0\,.$$

On integrating the last, we come to the conclusion that the relation

$$\boxed{A\,(x,y,t) = \text{const} \quad \text{for} \quad t = \text{const}} \tag{9.20}$$

is the equation for a family of magnetic field lines in the plane $z = \text{const}$ at the moment t.

Property 2. Let L be some curve in the plane (x,y) and $d\mathbf{l}$ an arc element along the curve in Figure 9.3.

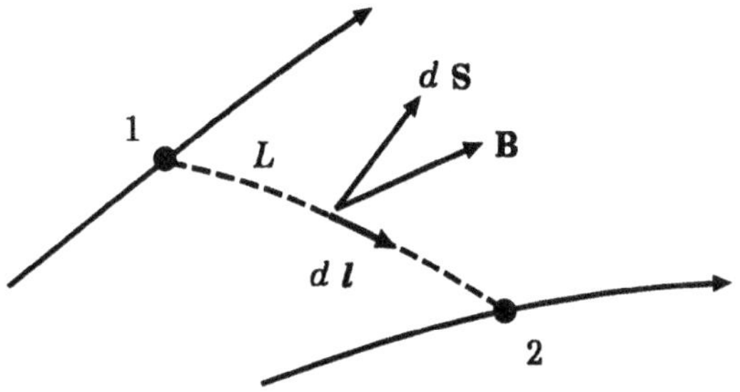

Figure 9.3: The curve L connects the points 1 and 2 situated in different field lines.

Calculate the magnetic flux $d\Phi$ through the arc element $d\mathbf{l}$. By definition,

$$d\Phi = \mathbf{B} \cdot d\mathbf{S} = \mathbf{B} \cdot (\mathbf{e}_z \times d\mathbf{l}) = \mathbf{B} \cdot \begin{vmatrix} \mathbf{e}_x & \mathbf{e}_y & \mathbf{e}_z \\ 0 & 0 & 1 \\ dx & dy & 0 \end{vmatrix} =$$

$$= \mathbf{B} \cdot \{ (-dy)\, \mathbf{e}_x + dx\, \mathbf{e}_y \} = -B_x\, dy + B_y\, dx\,. \tag{9.21}$$

On substituting definition (9.18) in formula (9.21) we find that

$$d\Phi = -\frac{\partial A}{\partial y}\, dy - \frac{\partial A}{\partial x}\, dx = -\, dA\,. \tag{9.22}$$

On integrating (9.22) along the curve L from point 1 to point 2 we obtain the magnetic flux

$$\Phi = A_2 - A_1\,. \tag{9.23}$$

Thus the fixed value of the vector potential A is not only the field line 'tag' determined by formula (9.20);

> the difference of values of the vector potential A on two field lines is equal to the magnetic flux between them.

From this, in particular, the following simple rule holds: we have to plot the field lines corresponding to equidistant values of A.

Property 3. Substitute definition (9.18) in the freezing-in Equation (9.2). We obtain the following general equation

$$\text{curl}\, \frac{d\mathbf{A}}{dt} = 0\,. \tag{9.24}$$

Disregarding a gradient of an arbitrary function, which can be eliminated by a gauge transformation, and considering the second type of problems, we have

$$\frac{dA}{dt} \equiv \frac{\partial A}{\partial t} + (\mathbf{v} \cdot \nabla) A = 0\,. \tag{9.25}$$

This equation means that the surfaces

$$A\,(x, y, t) = \text{const} \tag{9.26}$$

are *Lagrangian* surfaces, i.e. they move together with the plasma. According to (9.20) they are composed of the field lines, hence Equation (9.25) expresses the magnetic field freezing in plasma. Thus (formally it follows from (9.25) on passing to the Lagrangian variables) we have one of the integrals of motion

$$\boxed{A\,(x, y, t) = A\,(x_0, y_0, 0) \equiv A_0} \tag{9.27}$$

at an arbitrary t. Here x_0, y_0 are the coordinates of some 'fluid particle' at the initial moment of time; x, y are the coordinates of the same particle at a moment of time t or (by virtue of (9.27)) the coordinates of any other particle situated on the same field line A_0 at the moment t.

Property 4. Equation of motion (9.1) rewritten in terms of the vector potential $A(x, y, t)$ is of the form

$$\varepsilon^2 \, \frac{d\mathbf{v}}{dt} = -\frac{1}{\rho}\, \Delta A \, \nabla A \, . \tag{9.28}$$

In the zeroth order relative to ε^2, outside the zeroth points (where $\nabla A = 0$) and the magnetic field sources (where $\Delta A \neq 0$) we have:

$$\boxed{\Delta A = 0\,,} \tag{9.29}$$

i.e. the vector potential is a *harmonic* function of variables x and y. Hence, while considering the (x, y) plane as a complex plane $z = x + \mathrm{i}y$, it is convenient to relate an *analytic* function F to the vector potential A in the region under consideration:

$$F\,(z, t) = A\,(x, y, t) + \mathrm{i}\, A^{+}(x, y, t)\,. \tag{9.30}$$

Here $A^{+}(x, y, t)$ is a conjugate harmonic function connected with $A\,(x, y, t)$ by the Cauchy-Riemann condition

$$A^{+}(x, y, t) = \int \left(-\frac{\partial A}{\partial y}\, dx + \frac{\partial A}{\partial x}\, dy \right) + A^{+}(t) =$$

$$= - \int \mathbf{B} \cdot d\mathbf{l} + A^{+}(t)\,, \tag{9.31}$$

where $A^{+}(t)$ is a quantity independent of the coordinates x and y (see Lavrent'ev and Shabat, 1973, § 2).

The function $F(z,t)$ is termed the *complex potential*. The magnetic field vector, according to (9.18) and (9.30), is:

$$\mathbf{B} = B_x + \mathrm{i}\, B_y = -\,\mathrm{i} \left(\frac{dF}{dz} \right)^{*}, \tag{9.32}$$

the asterisk denoting the complex conjugation. After the introduction of the complex potential, we can widely apply the methods of the complex variable function theory, in particular the method of *conform mapping*, to determine the magnetic field in zeroth order in the small parameter ε^2 (e.g., Problem 9.4). This has been done, for example, to determine the structure of the magnetic field in solar coronal streamers (Somov and Syrovatskii, 1972b) and the field of the Earth's magnetosphere (Oberz, 1973).

(b) Motion of the plasma and its density

In the strong field approximation, the plasma motion kinematics due to changes in a potential field is uniquely determined by two conditions:

(i) the freezing-in condition (9.25) or its solution (9.27) and

(ii) the acceleration orthogonality with respect to the field lines

$$\frac{d\mathbf{v}^{(0)}}{dt} \times \nabla A^{(0)} = 0 \tag{9.33}$$

(cf. Equation (9.7)). A point to be noted is that Equation (9.33) is a result of eliminating the unknown $\Delta A^{(1)}$, which has a first order in ε^2, from two components of the vector equation

$$\frac{d\mathbf{v}^{(0)}}{dt} = -\,\frac{1}{\rho^{(0)}}\, \Delta A^{(1)}\, \nabla A^{(0)}\,. \tag{9.34}$$

Once the kinematic part of the problem is solved, the trajectories of fluid particles are known:

$$x = x\,(x_0\,,\, y_0\,, t)\,, \qquad y = y\,(x_0\,,\, y_0\,, t)\,. \tag{9.35}$$

In this case the continuity Equation (9.3) solution presents no problem. In fact, the fluid particle density change on moving along the found trajectory is determined by the continuity Equation (9.3), rewritten in the Lagrangian form, and is equal to

$$\frac{\rho\,(x,\,y,\,t)}{\rho_0\,(x_0,\,y_0)} = \frac{d\,U_0}{d\,U} = \frac{\mathcal{D}(x_0,\,y_0)}{\mathcal{D}(x,\,y)}\,. \tag{9.36}$$

Here dU_0 is the initial volume of a particle, dU is the volume of the same particle at a moment of time t;

$$\frac{\mathcal{D}(x_0,\,y_0)}{\mathcal{D}(x,\,y)} = \frac{\partial x_0}{\partial x}\frac{\partial y_0}{\partial y} - \frac{\partial x_0}{\partial y}\frac{\partial y_0}{\partial x} \tag{9.37}$$

is the Jacobian of the transformation that is inverse to the transformation (9.35) of coordinates at a fixed value of time t.

The two-dimentional equations of the strong-field-cold-plasma approximation (Somov and Syrovatskii, 1976a) in the problem of the second type are relatively simple but rather useful for applications to space plasmas. In particular, they enable us to study the fast plasma flows in the solar atmosphere (Syrovatskii and Somov, 1980) and to understand some aspects of the reconnection process (Chapters 16–22).

In spite of their numerous applications, the list of exact solutions to them is rather poor. Still, we can enrich it significantly,

relying on many astrophysical objects and some mathematical ideas.

Titov and Priest (1993) have shown that the equations of zeroth order can be reduced to a set of Cauchy-Riemann and ordinary differential equations, by using a conformal system of coordinates in which the positions of particles are fixed by magnitudes of two conjugate functions. These are the flux function and the potential of magnetic field. The set obtained has a special class of solutions. First, in such flows the conjugate potential (9.31) is frozen into the moving medium as well as the vector potential $A(x, y, t)$. Second, each flow is realized as a contiuous sequence of conformal mappings. A linear diffusion-like equation describes such flows. The equation was solved analytically for examples describing the *magnetic collapse* (cf. Section 16.4) in the neighbourhood of the X-point.

9.3 On the existence of continuous flows

Thus, in the strong-field-cold-plasma approximation, the MHD equations for a plane two-dimensional flow of ideally conducting plasma (for second-type problems) are reduced, in the zeroth order in the small parameter ε^2, to the following set of equations:

$$\Delta A = 0 , \tag{9.38}$$

$$\frac{d\mathbf{v}}{dt} \times \nabla A = 0 , \tag{9.39}$$

$$\frac{dA}{dt} = 0 , \tag{9.40}$$

$$\frac{\partial \rho}{\partial t} + \operatorname{div} \rho \mathbf{v} = 0 . \tag{9.41}$$

Seemingly, the solution of this set is completely defined inside some region G (Figure 9.4) on the plane (x, y), once the boundary condition is given

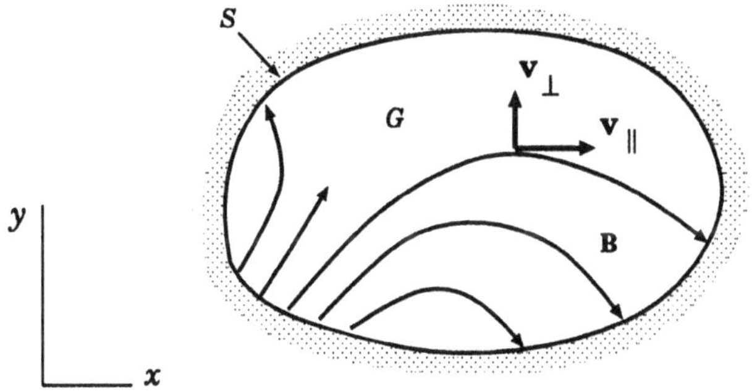

Figure 9.4: The boundary and initial conditions for second-type problems.

at the boundary S

$$A\,(x, y, t)\,|_{\,S} = f_1\,(x, y, t) \tag{9.42}$$

together with the initial conditions inside the region G

$$\mathbf{v}_{\parallel}\,(x, y, 0)\,|_{\,G} = \mathbf{f}_2\,(x, y) , \tag{9.43}$$

$$\rho\,(x, y, 0)\,|_{\,G} = f_3\,(x, y) . \tag{9.44}$$

Here $\mathbf{v}_{\parallel}$ is the velocity component along field lines. Once the potential $A\,(x, y, t)$ is known, the transversal velocity component is uniquely determined by the freezing-in Equation (9.40) and is equal, at any moment including the initial one, to

$$\mathbf{v}_{\perp}(x, y, t) = (\mathbf{v} \cdot \nabla A)\, \frac{\nabla A}{|\nabla A\,|^2} = - \frac{\partial A}{\partial t}\, \frac{\nabla A}{|\nabla A\,|^2} . \tag{9.45}$$

From Equation (9.38) and boundary condition (9.42) we find the vector potential $A\,(x, y, t)$ at any moment of time. Next, from Equations (9.39) and (9.40) and the initial condition (9.43), the velocity $\mathbf{v}\,(x, y, t)$ is determined; the density $\rho\,(x, y, t)$ is found from the continuity Equation (9.41) and the initial density distribution (9.44).

However, such a procedure is not always possible. This means that continuous solutions to the Equations (9.38)–(9.41) do not necessarily exist. Let the boundary and initial conditions be given. The vector potential $A\,(x, y, t)$ is uniquely determined by Equation (9.38) and the boundary condition (9.42). The latter can be chosen in such a way that the field $\mathbf{B}$ will contain zeroth points:

$$\mathbf{B} = \left\{ \frac{\partial A}{\partial y}\,,\; -\frac{\partial A}{\partial x}\,,\; 0 \right\} = 0\,. \tag{9.46}$$

Among them, there can exist ones in which the electric field is distinct from zero

$$\mathbf{E} = -\frac{1}{c}\,\frac{\partial \mathbf{A}}{\partial t} \neq 0\,. \tag{9.47}$$

Note that such points contradict the freezing-in Equation (9.40). We will call them the *peculiar* points.

> The freezing-in condition allows continuous deformation of the strong magnetic field and the corresponding continuous motion of plasma everywhere except at peculiar zeroth points,

i.e. the lines parallel to the z axis of the Cartesian system of coordinates, where the magnetic field is zero while the electric one is nonzero.

Note that simultaneous vanishing of both fields is quite unlikely. This is the reason why the peculiar points occur rather frequently. They will receive much attention in what follows because they represent the places where reconnecting current sheets are formed (Chapters 16–22). Here we only stress that

> if there is not a zeroth point inside the region G at the initial time, it does not mean that such a point will never appear there,

like in the simplest case of a magnetic dipole considered in the next section. An initial field can even be an homogeneous one (Parker, 1972). Following the continuous evolution of the boundary condition (9.42), a zeroth point may appear on the boundary S and, if the electric field in this point does not equal zero, it will create a magnetic field discontinuity which prevents a change of magnetic field topology in the approximation of an ideal plasma. This discontinuity is a *neutral* sheet of infinitesimal thickness (Chapter 16). In a plasma of finite conductivity, a *reconnecting* current sheet of finite thickness is formed in a peculiar zeroth point (Chapter 17).

The creation of a current sheet in the zeroth point which appears on the boundary S was used in the model of coronal streamers driven by the solar wind (Somov and Syrovatskii, 1972b). Just the same occurs in the model for interacting magnetic fluxes, compressed by a converging motion of magnetic footpoints in the photosphere (Low, 1987; Low and Wolfson, 1988).

Another case is an appearance of a couple of neutral points inside the region G. Anyway, and in all cases,

> the interaction of magnetic fluxes in the peculiar point changes the field topology and creates the reconnecting current sheet.

This kind of MHD discontinuous flows is of great importance for cosmic plasma physics. First, however, we treat a simple example of a plasma continuous flow in a strong magnetic field without the peculiar points. The next section is devoted to the consideration of this example, which may have interesting applications.

9.4 Flows in the field of a time-dependent dipole

9.4.1 Plane magnetic dipole fields

Two straight parallel currents, equal in magnitude but opposite in direction, engender the magnetic field which far enough from the currents can be described by a complex potential

$$F(z) = \frac{\mathrm{i}\,\mathbf{m}}{z}, \qquad \mathbf{m} = m\,e^{\mathrm{i}\psi} \tag{9.48}$$

and is called the plane *dipole* field. The quantity $m = 2Il/c$ has the meaning of the *dipole moment*, I is the current magnitude, l is the distance between the currents. Formula (9.48) corresponds to the plane dipole situated at the origin of coordinates in the plane (x, y) and directed at an angle of ψ to the x axis. The currents are parallel to the z axis of the Cartesian system of coordinates.

Consider the plasma flow caused by the change with time of the strong magnetic field of the plane dipole. Let $\psi = \pi/2$ and $m = m(t)$, $m(0) = m_0$.

(a) Let us find the first integral of motion. According to (9.30) and (9.48), the complex potential

$$F(z,t) = \frac{\mathrm{i}\, m(t)\, e^{\mathrm{i}\pi/2}}{x + \mathrm{i}\, y} = \frac{-m(t)\, x + \mathrm{i}\, m(t)\, y}{x^2 + y^2}\,. \tag{9.49}$$

So, according to (9.20), the field lines constitute a family of circles

$$A(x,y,t) = -\frac{m(t)\, x}{x^2 + y^2} = \text{const} \quad \text{for} \quad t = \text{const}\,. \tag{9.50}$$

They have centres on the axis x and the common point $x = 0$, $y = 0$ in Figure 9.5.

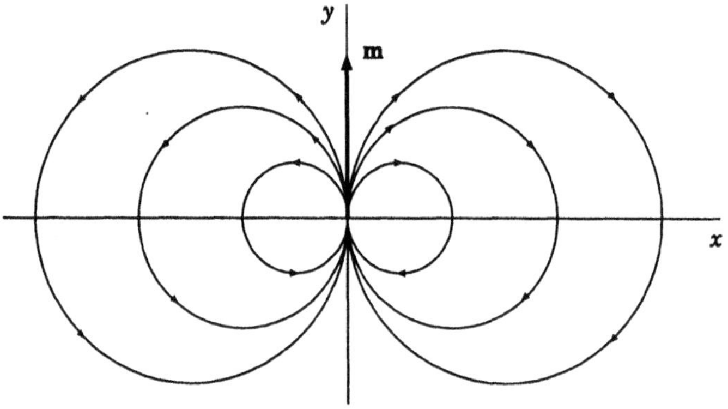

Figure 9.5: The field lines of a plane magnetic dipole.

Therefore the freezing-in condition (9.27) results in a first integral of motion

$$\frac{m\, x}{x^2 + y^2} = \frac{m_0\, x_0}{x_0^2 + y_0^2}\,. \tag{9.51}$$

Here x_0, y_0 are the coordinates of some fluid particle at the initial moment of time $t = 0$; Lagrangian variables x and y are the coordinates of the same particle at a moment t.

(b) The second integral is easily found in the limit of small changes of the dipole moment $m(t)$ and respectively *small* plasma displacements. Assuming the parameter $\delta = v_T/L$ to be small, Equation (8.74), which is *linear in velocity*, takes the place of (9.33). The integration over time (with zero initial values for the velocity) allows us to reduce Equation (8.74) to the form

$$\frac{\partial x}{\partial t} = K(x,y,t)\, \frac{\partial A}{\partial x}\,, \qquad \frac{\partial y}{\partial t} = K(x,y,t)\, \frac{\partial A}{\partial y}\,. \tag{9.52}$$

Here $K(x, y, t)$ is some function of coordinates and time. Eliminating it from two Equations (9.52), we arrive at

$$\frac{\partial y}{\partial x} = \frac{\partial A}{\partial y} \Big/ \frac{\partial A}{\partial x} \,. \tag{9.53}$$

Thus, in the approximation of small displacements, not only the acceleration but also the plasma displacements are normal to the field lines.

On substituting (9.50) in (9.53), we obtain an ordinary differential equation. Its integral

$$\frac{y}{x^2 + y^2} = \text{const}$$

describes **a family of circles, orthogonal to the field lines**, and presents fluid particle trajectories. In particular, the trajectory of a particle, situated at a point (x_0, y_0) at the initial moment of time $t = 0$, is an arc of the circle

$$\frac{y}{x^2 + y^2} = \frac{y_0}{x_0^2 + y_0^2} \tag{9.54}$$

from the point (x_0, y_0) to the point (x, y) on the field line (9.51).

Thus the integrals of motion (9.51) and (9.54) completely determine the plasma flow in terms of the Lagrangian coordinates

$$x = x\,(x_0, y_0, t)\,, \qquad y = y\,(x_0, y_0, t)\,. \tag{9.55}$$

This flow has a simple form: the particles are connected with the magnetic field lines and move together with them in a transversal direction. Such simple kinematics is a result of considering small plasma displacements (from the state having zero initial velocity) under the action of the force perpendicular to the field lines.

The plasma density change is defined by Equation (9.36). On calculating the Jacobian for the transformation implicitly given by formulae (9.51) and (9.54), we obtain (for the case of a homogeneous initial density distribution ρ_0) the formula

$$\frac{\rho\,(x, y, t)}{\rho_0} = \left(\frac{m}{m_0}\right) \frac{m_0^4}{(m^2 x^2 + m_0^2\, y^2)^4} \left\{ \left[m^2 x^4 + m_0^2\, y^4 + \right. \right.$$

$$\left. \left. + x^2 y^2 \left(3m^2 - m_0^2\right) \right]^2 - \left[2x^2 y^2 \left(m_0^2 - m^2\right) \right]^2 \right\} . \tag{9.56}$$

In particular, on the dipole axis ($x = 0$)

$$\boxed{\frac{\rho\,(0, y, t)}{\rho_0} = \frac{m}{m_0}\,,} \tag{9.57}$$

whereas in the 'equatorial plane' ($y = 0$)

$$\frac{\rho\,(x, 0, t)}{\rho_0} = \left(\frac{m_0}{m}\right)^3 . \tag{9.58}$$

> With increasing dipole moment m, the plasma density on the dipole axis grows proportionally to the moment,

whereas that at the equatorial plane falls in inverse proportion to the third power of the moment. The opposite process takes place as the moment decreases.

The result pertains to the case of small changes in the dipole moment and can demonstrate just the tendency of plasma behaviour in the strong magnetic field of a plane dipole. The exception is formula (9.57). It applies to any changes of the dipole moment. The reason is in the following. In the approximation of a strong field and cold plasma, the acceleration of plasma is perpendicular to the field lines and is zero at the dipole axis. Hence, if the plasma is motionless at the initial moment, arbitrary changes of the dipole moment do not cause a plasma motion on the dipole axis ($\mathbf{v} = 0$). Plasma displacements in the vicinity of the dipole axis always remain small ($\delta \ll 1$) and the solution obtained applies.

In the general case of arbitrarily large dipole moment changes,

> the inertial effects resulting in plasma flows along the magnetic field lines are of considerable importance

(Somov and Syrovatskii, 1972a). In this case, the solution of the problem requires the integration of Equation (9.33) or (9.34) together with the freezing-in Equation (9.25).

One can obtain exact analytical solutions for a linearly changing magnetic moment using the 'frozen-in coordinates' technique (Gorbachev and Kel'ner, 1988). These coordinates can be quite useful while solving nonstationary MHD problems. One introduces a set which is doubly Lagrangian: in the parameter s_1 along a stream line (along the velocity field $\mathbf{v}$) and in the parameter s_2 along a magnetic field line.

9.4.2 Axisymmetric dipole fields in cosmic plasma

Two-dimensional axisymmetric problems are better suited to astrophysical applications of the second-type problem considered. The MHD equations are written, using the approximation of a strong field and cold plasma, in spherical coordinates with due regard for axial symmetry. The role of the vector potential is fulfilled by the so-called *stream function*

$$\Phi\,(r,\theta,t) = r\sin\theta\, A_{\varphi}(r,\theta,t)\,. \tag{9.59}$$

Here A_{φ} is the only non-zero φ-component of the vector-potential $\mathbf{A}$.

In terms of the stream functions, the equations take the form

$$\frac{d\mathbf{v}}{dt} = \varepsilon^{-2} K(r,\theta,t)\,\nabla\Phi\,, \qquad \frac{d\,\Phi}{dt} = 0\,, \qquad \frac{d\rho}{dt} = -\rho\,\mathrm{div}\,\mathbf{v}\,, \tag{9.60}$$

where

$$K(r,\theta,t) = \frac{j_{\varphi}(r,\theta,t)}{\rho\, r\sin\theta} \tag{9.61}$$

(Somov and Syrovatskii, 1976a). The equations formally coincide with the corresponding Equations (9.28), (9.25) and (9.3) describing the plane flows in terms of the vector potential. Thus the condition (obtained in Section 9.3) for the existence of continuous solutions can be extended to the case of plasma flows in the axisymmetric poloidal field ($B_{\varphi} = 0$); namely, the continuous deformation of the strong axisymmetric poloidal field and the corresponding

> continuous (poloidal or meridional) plasma motions are possible everywhere except at the peculiar zeroth points of magnetic field,

i.e. the points where

$$B_r = \frac{1}{r^2\sin\theta}\,\frac{\partial\,\Phi}{\partial\theta} = 0\,, \qquad B_{\theta} = -\frac{1}{r\sin\theta}\,\frac{\partial\,\Phi}{\partial r} = 0\,, \tag{9.62}$$

$$E_{\varphi} = -\frac{1}{c}\,\frac{\partial}{\partial t}\left(\frac{\Phi\,(r,\theta,t)}{r\sin\theta}\right) \neq 0\,. \tag{9.63}$$

Thus there are grounds to assume that continuous flows do exist in the strong magnetic field of a dipole type, since the latter does not have such points. As a zeroth approximation in the small parameter ε^2, we may take, for example, the dipole field. In this case the stream function is of the form

$$\Phi^{(0)}\,(r,\theta,t) = m(t)\,\frac{\sin^2\theta}{r}\,, \tag{9.64}$$

where $m(t)$ is a time-varying moment.

Imagine a homogeneous magnetized ball of radius $R(t)$ with the frozen field $\mathbf{B}_{int}(t)$. The dipole moment of such a ball (a star or its envelope) is

$$m(t) = \frac{1}{2}\, B_{int}(t)\, R^3(t) = \frac{1}{2\pi} \left(B_0\, \pi R_0^2 \right) R(t) \, , \tag{9.65}$$

where B_0 and R_0 are the values of $B_{int}(t)$ and $R(t)$ at the initial moment of time $t = 0$. The second equality takes account of the magnetic field freezing-in as conservation of the flux $B_{int}(t)\, R^2(t)$ through the ball. Formula (9.65) shows that the dipole moment of the ball is thereby proportional to its radius $R(t)$.

The solution to the problem (Somov and Syrovatskii, 1972a) shows that as the dipole moment grows (when the ball expands)

> the magnetic field 'rakes the plasma up' to the dipole axis, compresses it and simultaneously accelerates it along the field lines.

A distinguishing characteristic of the solution is that the density at the axis grows in proportion to the dipole moment, just as in the two-dimensional plane case (formula (9.57)).

Envelopes of nova and supernova stars present a wide variety of different shapes. We can hardly find the ideally round envelopes, even among the ones of regular shape. It is more common to find either flattened or stretched envelopes. As a rule, their surface brightness is maximal at the ends of the main axes of an oval image. This phenomenon can sometimes be interpreted as a gaseous ring observed almost from an edge. However, if there is no luminous belt between the brightness maxima, which would be characteristic of the ring, then the remaining possibility is that single gaseous compressions – condensations – exist in the envelope.

At the early stages of the expansion during the explosion of a nova, the condensations reach such brightness that they give the impression that the nova 'bifurcates'. Consider one of the models in which a magnetic field plays a decisive role. Suppose that the star's magnetic field was a dipole one before the explosion. At the moment of the explosion a massive envelope with the frozen-in field separated from the star and began to expand. According to (9.65), the expansion results in the growth of the dipole moment. According to the solution of the problem considered above, the field will rake the interstellar plasma surrounding the envelope, as well as external layers of the envelope, up in the direction of the dipole axis.

The process of polar condensate formation can be conventionally divided into two stages (Somov and Syrovatskii, 1976a, Ch. 2). At the first one, the interstellar plasma is raked up by the magnetic field into the polar regions, a corresponding growth in density and pressure at the dipole axis taking place. At the second stage, the increased pressure hinders the growth of the density at the axis, thus stopping compression, but the plasma raking-up still continues. At the same time, the gas pressure gradient, arising ahead of the envelope, gives rise to the motion along the axis. As a result, by the time the magnetic force action stops, all the plasma is raked up into two compact condensates.

The plasma raking-up by the strong magnetic field seems to be capable of explaining some types of chromospheric ejections on the Sun (Somov and Syrovatskii, 1976a, Ch. 2, § 4).

Note, in closing, that if a magnetized ball compresses, plasma must flow from the poles to the equatorial plane, thus forming a **dense disc or ring**. This case is the old problem of cosmic electrodynamics concerning the compression of a gravitating plasma cloud with the frozen-in field. The process of magnetic raking-up of plasma into dense discs or rings can effectively work in the atmospheres of collapsing stars.

9.5 Practice: Problems and Answers

Problem 9.1. Consider the properties of the vector-potential **A** which is determined in terms of two scalar functions α and β:

$$\mathbf{A} = \alpha \nabla \beta + \psi \, . \tag{9.66}$$

Answer. Formula (9.66) permits **B** to be written as

$$\mathbf{B} = \nabla \alpha \times \nabla \beta \, , \tag{9.67}$$

where the last step follows from the fact that the curl of a gradient vanishes.

This representation of **B** provides another way to obtain information about the magnetic field in three-dimensional problems. According to (9.67)

$$\mathbf{B} \cdot \nabla \alpha = 0 \quad \text{and} \quad \mathbf{B} \cdot \nabla \beta = 0 \, . \tag{9.68}$$

Thus $\nabla \alpha$ and $\nabla \beta$ are perpendicular to the vector **B**, and functions α and β are constant along **B**. The surfaces $\alpha = \text{const}$ and $\beta = \text{const}$ are orthogonal to their gradients and targent to **B**. Hence a magnetic field line can be conveniently defined in terms of a pair of values: α and β.

A particular set of α and β labels a field line.

The functions α and β are referred to as Euler potentials or Clebsch variables. Depending on the problem to be examined, one form may have an advantage over another. The variables α and β, while in general not easily obtained, are available for some axisymmetric geometries. Another advantage of these variables appears in the study of field line motions in the context of MHD theory (Section 5.5 in Parks, 1991).

Problem 9.2. Evaluate the typical value of the dipole moment for a neutron star.

Answer. Typical neutron stars have $B \sim 10^{12}$ G. With $R \sim 10$ km, it follows from formula (9.65) that $m \sim 10^{30}\,\text{G cm}^3$. Some of neutron stars, related to the so-called 'Soft Gamma-ray Repeaters' (SGRs), are the spinning super-magnetized neutron stars created by supernova explosions. The rotation of such stars called *magnetars* is slowing down so rapidly that a superstrong field of the unprecedented strength, $B \sim 10^{15}$ G, could provide so fast braking. For a magnetar the dipole moment $m \sim 10^{33}\,\text{G cm}^3$.

Problem 9.3. Show that, prior to the onset of a solar flare, the magnetic energy density is of about three orders of magnitude greater than any of the other types. So the solar flares occur in a plasma environment well dominated by the magnetic field.

Hint. Take the coronal field of about 100 Gauss, and the coronal plasma velocity of order of 1 km s^{-1}.

Problem 9.4. By using the method of conform mapping, determine the shape of a magnetic cavity created by a plane dipole inside a perfectly conducting uniform plasma with a gas pressure p_0. Determine the magnetic field inside the cavity.

Answer. The conditions to be satisfied along the boundary S of the magnetic cavity G are equality of magnetic and gas pressure,

$$\left.\frac{B^2}{8\pi}\right|_S = p_0 = \text{const}\,, \tag{9.69}$$

and tangency of the magnetic field,

$$\mathbf{B}\cdot\mathbf{n}\big|_S = 0\,. \tag{9.70}$$

Condition (9.70) means that

$$\operatorname{Re} F(z) = A\,(x, y) = \text{const}\,, \tag{9.71}$$

where a complex potential $F(z)$ is an analytic function (9.30) within the region G in the complex plane z except at the point $z = 0$ of the dipole.

Let us assume that a conform transformation $w = w(z)$ maps the region G onto the circle $|w| \le 1$ in an auxiliary complex plane $w = u + iv$ so that the point $z = 0$ goes into the centre of the circle without rotation of the dipole as shown in Figure 9.6. The boundary $|w| = 1$ is a field line of the solution

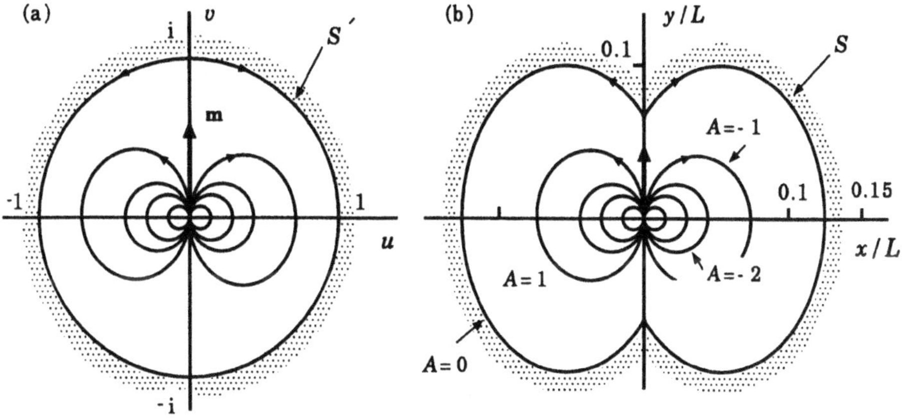

Figure 9.6: The field lines of a plane dipole $\mathbf{m}$ inside: (a) the unit circle in the plane w, (b) the cavity in a plasma of constant pressure.

in the plane w, which we easily construct:

$$F(w) = \left(w - \frac{1}{w} \right) . \tag{9.72}$$

Note that we have used only the boundary condition (9.70).

The other boundary condition (9.69) will allow us to find an unknown conform transformation $w = w(z)$. With account of definition (9.32) taken, condition (9.69) gives us the following relation

$$\left| \frac{dz}{dw} \right|^2 = \frac{1}{8\pi p_0} \left| \frac{dF}{dw} \right|^2 . \tag{9.73}$$

At the boundary $|w| = 1$, this condition reduces to an ordinary differential equation relative to the real part, $x = x(u)$, of an unknown function $z = z(w)$:

$$\left(\frac{dx}{du} \right)^2 = M^2 u^4 \,, \quad \text{where} \quad M^2 = \frac{1}{2\pi} \,. \tag{9.74}$$

By integrating this equation we find

$$x = \pm M \frac{u^3}{3} + c_1 = \pm \frac{M}{3} \cos^3 \varphi + c_1 \,, \tag{9.75}$$

here φ is an argument of the complex number w, and c_1 is a constant of integration.

Since we know the real part $x = x(\varphi)$ on the circle boundary, we find the complex function $z = z(w)$ in the entire region $|w| \leq 1$ (Oreshina and Somov, 1999b):

$$z(w) = \frac{M}{4} \left(w + \frac{w^3}{3} \right) . \tag{9.76}$$

The conform mapping (9.76) and the potential (9.72) determine the general solution of the problem, the complex potential:

$$F(z) = B_0 L^{2/3} \, \frac{K^4 - 3L^{2/3} K^2 + L^{4/3}}{K \left(K^2 - L^{2/3} \right)} \,. \tag{9.77}$$

Here $B_0 = p_0^{1/2}$ is the unit of magnetic field strength, the function

$$K(z) = \left(6\sqrt{2\pi} \cdot z + \sqrt{L^2 + 72\pi \cdot z^2} \right)^{1/3} , \tag{9.78}$$

and $L = m^{1/3} p_0^{-1/6}$ is the unit of length; it shows that, when the dipole moment m increases, the size of the magnetic cavity also increases. This is consistent with what we discussed in Section 9.4.

The field lines corresponding solution (9.77) are shown in Figure 9.6b. Therefore, in addition to the shape of the boundary (Cole and Huth, 1959), we have found an analytic solution for the magnetic field inside the static dipole cavity. This solution can be used in the zero-order approximation, described in Section 9.1, to analyse properties of plasma flows near collapsing or exploding astrophysical objects with strong magnetic fields.

Chapter 10

MHD Waves in Cosmic Plasma

Magnetohydrodynamic waves create turbulence, nonlinearly cascade in a wide range of wavenumbers, accelerate particles and produce a lot of effects in cosmic plasma.

10.1 The general dispersion equation in ideal MHD

Small disturbances in a conducting medium with a magnetic field propagate as waves, their properties being different from those of the usual sound waves in a gas or electromagnetic waves in a vacuum. First, the conducting medium with a magnetic field has a characteristic anisotropy: the wave propagation velocity depends upon the direction of propagation relative to the magnetic field. Second, as a result of the interplay of electromagnetic and hydrodynamic phenomena, the waves in MHD are generally neither longitudinal nor transversal.

The study of the behaviour of small-amplitude waves, apart from being important in itself, has a direct bearing on the analysis of large-amplitude waves, in particular shock waves and other discontinuous flows in MHD.

Initially we shall study the possible types of small waves, restricting ourselves to the *ideal* medium approximation, i.e. Equations (8.40). Suppose a medium in the initial stationary state is subjected to a small perturbation, so that velocity $\mathbf{v}_0$, magnetic field $\mathbf{B}_0$, density ρ_0, pressure p_0 and entropy s_0 acquire some small deviations $\mathbf{v}'$, $\mathbf{B}'$, ρ', p' and s':

$$\begin{aligned} &\mathbf{v} = \mathbf{v}_0 + \mathbf{v}' , \qquad \mathbf{B} = \mathbf{B}_0 + \mathbf{B}' , \\ &\rho = \rho_0 + \rho' , \qquad p = p_0 + p' , \qquad s = s_0 + s' . \end{aligned} \tag{10.1}$$

The initial state is assumed to be a uniform flow of an homogeneous medium in a constant magnetic field:

$$\begin{aligned} &\mathbf{v}_0 = \text{const}\,, \qquad \mathbf{B}_0 = \text{const}\,, \\ &\rho_0 = \text{const}\,, \qquad p_0 = \text{const}\,, \qquad s_0 = \text{const}\,. \end{aligned} \tag{10.2}$$

Needless to say, the latter simplification can be ignored, i.e. one may study waves in inhomogeneous media, the coefficients in linearized equations being dependent upon the coordinates. For the sake of simplicity we restrict our consideration to the case (10.2).

It is convenient to introduce the following designations:

$$\mathbf{u} = \frac{\mathbf{B}_0}{\sqrt{4\pi\rho_0}}\,, \qquad \mathbf{u}' = \frac{\mathbf{B}'}{\sqrt{4\pi\rho_0}}\,. \tag{10.3}$$

Linearize the initial set of MHD equations for an ideal medium. Substitute definitions (10.1)–(10.3) in the set of Equations (8.40), neglecting the products of small quantities. Hereafter the subscript '0' for undisturbed quantities will be omitted. We shall get the following set of *linear differential* equations for the primed quantities characterizing small perturbations:

$$\begin{aligned} &\partial \mathbf{u}' / \partial t + (\mathbf{v} \cdot \nabla)\, \mathbf{u}' = (\mathbf{u} \cdot \nabla)\, \mathbf{v}' - \mathbf{u}\, \text{div}\, \mathbf{v}'\,, \qquad \text{div}\, \mathbf{u}' = 0\,, \\ &\partial \mathbf{v}' / \partial t + (\mathbf{v} \cdot \nabla)\, \mathbf{v}' = -\,\rho^{-1}\, \nabla\, (p' + \rho\, \mathbf{u} \cdot \mathbf{u}') + (\mathbf{u} \cdot \nabla)\, \mathbf{u}'\,, \\ &\partial \rho' / \partial t + (\mathbf{v} \cdot \nabla)\, \rho' = -\,\rho\, \text{div}\, \mathbf{v}'\,, \\ &\partial s' / \partial t + (\mathbf{v} \cdot \nabla)\, s' = 0\,, \quad p' = (\partial p / \partial \rho)_s\, \rho' + (\partial p / \partial s)_\rho\, s'\,. \end{aligned} \tag{10.4}$$

The latter equation is the linearized equation of state. Rewrite it as follows:

$$p' = V_s^2 \rho' + b\, s'. \tag{10.5}$$

Here

$$V_s = (\partial p / \partial \rho)_s^{1/2} \tag{10.6}$$

is the velocity of *sound* in a medium without a magnetic field (Problem 10.1), the coefficient $b = (\partial p / \partial s)_\rho$.

By virtue of (10.2), the set of Equations (10.4) is that of linear differential equations with *constant coefficients*. That is why we may seek a solution in the form of a superposition of plane waves with a dependence on coordinates and time of the type

$$f'(\mathbf{r}, t) \sim \exp\left[\,\mathrm{i}\,(\mathbf{k} \cdot \mathbf{r} - \omega t)\,\right], \tag{10.7}$$

where ω is the wave frequency and $\mathbf{k}$ is the wave vector. An arbitrary disturbance can be expanded into such waves by means of a Fourier transform. As this takes place, the set of Equations (10.4) is reduced to the following set of *linear algebraic* equations:

$$\begin{aligned}
&(\omega - \mathbf{k}\cdot\mathbf{v})\,\mathbf{u}' + (\mathbf{k}\cdot\mathbf{u})\,\mathbf{v}' - \mathbf{u}\,(\mathbf{k}\cdot\mathbf{v}') = 0\,, \qquad \mathbf{k}\cdot\mathbf{u}' = 0\,,\\
&(\omega - \mathbf{k}\cdot\mathbf{v})\,\mathbf{v}' + (\mathbf{k}\cdot\mathbf{u})\,\mathbf{u}' - \rho^{-1}\,(p' + \rho\,\mathbf{u}\cdot\mathbf{u}')\,\mathbf{k} = 0\,,\\
&(\omega - \mathbf{k}\cdot\mathbf{v})\,\rho' - \rho\,(\mathbf{k}\cdot\mathbf{v}') = 0\,,\\
&(\omega - \mathbf{k}\cdot\mathbf{v})\,s' = 0\,, \qquad p' - V_s^{\,2}\,\rho' - b\,s' = 0\,.
\end{aligned} \tag{10.8}$$

The quantities $\mathbf{k}$ and ω appearing in this set are assumed to be known from the initial conditions. The unknown terms are the primed ones. With respect to these the set of Equations (10.8) is closed, linear and homogeneous (the right-hand sides equal zero). For this set to have nontrivial solutions, its determinant must be equal to zero.

The determinant can be conveniently calculated in a frame of reference with one of the axes along the wave vector $\mathbf{k}$. In addition, it is convenient to use the frequency

$$\omega_0 = \omega - \mathbf{k}\cdot\mathbf{v}\,, \tag{10.9}$$

i.e. the frequency in a frame of reference moving with the medium. Setting the determinant equal to zero, we get the following equation (Syrovatskii, 1957):

$$\omega_0^2 \left[\omega_0^2 - (\mathbf{k}\cdot\mathbf{u})^2\right] \times$$

$$\times \left[\omega_0^4 - k^2\left(V_s^{\,2} + u^2\right)\omega_0^2 + k^2 V_s^{\,2}\,(\mathbf{k}\cdot\mathbf{u})^2\right] = 0\,. \tag{10.10}$$

This equation is called the *dispersion* or *characteristic* equation. It defines four values of ω_0^2. Since they differ in absolute magnitude, **four different modes of waves** are defined, each of them having its own velocity of propagation with respect to the medium

$$\mathbf{V}_{\mathrm{ph}} = \frac{\omega_0}{\mathbf{k}}\,. \tag{10.11}$$

Clearly this is the *phase* velocity of the wave. It should be distinguished from the *group* velocity

$$\mathbf{V}_{\mathrm{gr}} = \frac{d\,\omega_0}{d\,\mathbf{k}}\,. \tag{10.12}$$

Let us consider the properties of the waves defined by the dispersion Equation (10.10) in greater detail.

10.2 Small-amplitude waves in ideal MHD

10.2.1 Entropy waves

The first root of the dispersion Equation (10.10)

$$\omega_0 = \omega - \mathbf{k} \cdot \mathbf{v} = 0 \tag{10.13}$$

corresponds to the small perturbation which is immobile with respect to the medium:

$$\mathbf{V}_{\text{ph}} = 0 \, . \tag{10.14}$$

If the medium is moving, the disturbance is carried with it.

Substituting (10.13) in (10.8), we obtain the following equations:

$$(\mathbf{k} \cdot \mathbf{u}) \, \mathbf{v}' - \mathbf{u} \, (\mathbf{k} \cdot \mathbf{v}') = 0 \, , \tag{10.15}$$

$$\mathbf{k} \cdot \mathbf{u}' = 0 \, , \tag{10.16}$$

$$(\mathbf{k} \cdot \mathbf{u}) \, \mathbf{u}' - \rho^{-1} \, (p' + \rho \, \mathbf{u} \cdot \mathbf{u}') \, \mathbf{k} = 0 \, , \tag{10.17}$$

$$\mathbf{k} \cdot \mathbf{v}' = 0 \, , \tag{10.18}$$

$$p' - V_s^{\,2} \rho' - b \, s' = 0 \, . \tag{10.19}$$

Make use of (10.18) in (10.15). Then take the scalar product of Equation (10.17) with the vector $\mathbf{k}$ and make allowance for (10.16). We write

$$(\mathbf{k} \cdot \mathbf{u}) \, \mathbf{v}' = 0 \, , \tag{10.20}$$

$$\mathbf{k} \cdot \mathbf{u}' = 0 \, , \tag{10.21}$$

$$p' + \rho \, \mathbf{u} \cdot \mathbf{u}' = 0 \, , \tag{10.22}$$

$$\mathbf{k} \cdot \mathbf{v}' = 0 \, , \tag{10.23}$$

$$p' - V_s^{\,2} \rho' - b \, s' = 0 \, . \tag{10.24}$$

Substitution of (10.22) in (10.17) gives us the following set of equations:

$$(\mathbf{k} \cdot \mathbf{u}) \, \mathbf{u}' = 0 \, , \qquad (\mathbf{k} \cdot \mathbf{u}) \, \mathbf{v}' = 0 \, , \tag{10.25}$$

$$p' + \rho \, \mathbf{u} \cdot \mathbf{u}' = 0 \, , \qquad p' - V_s^{\,2} \rho' - b \, s' = 0 \, . \tag{10.26}$$

Since generally $\mathbf{k} \cdot \mathbf{u} \neq 0$, the velocity, magnetic field and gas pressure are undisturbed in the wave under discussion:

$$\mathbf{v}' = 0 \, , \quad \mathbf{u}' = 0 \, , \quad p' = 0 \, . \tag{10.27}$$

The only disturbed quantities are the density and entropy related by the condition

$$\rho' = -\frac{b}{V_s^2}\, s' . \tag{10.28}$$

This is the reason why these disturbances are called the *entropy* waves. They are known in hydrodynamics (Problem 10.2). The meaning of an entropy wave is that regions containing hotter but more rarefied plasma can exist in a plasma flow.

The entropy waves are only arbitrarily termed *waves*, since their velocity of propagation with respect to the medium is zero. Nevertheless the entropy waves must be taken into account together with the real waves in such cases as the study of shock waves behaviour under small perturbations. Blochintsev (1945) has considered the passage of small perturbations through a shock in ordinary hydrodynamics. He came to the conclusion that the entropy wave must be taken into account in order to match the linearized solutions at the shock front (Problem 12.1).

10.2.2 Alfvén waves

The second root of the dispersion Equation (10.10),

$$\omega_0^2 = (\mathbf{k} \cdot \mathbf{u})^2 \quad \text{or} \quad \omega_0 = \pm\, \mathbf{k} \cdot \mathbf{u} , \tag{10.29}$$

corresponds to waves with the phase velocity

$$\boxed{V_{\rm A} = \pm \frac{B}{\sqrt{4\pi\rho}} \cos\theta .} \tag{10.30}$$

Here θ is the angle between the direction of wave propagation $\mathbf{k}/k$ and the ambient field vector $\mathbf{B}_0$ (Figure 10.1). In formula (10.30) the value $B = |\, \mathbf{B}_0 \,|$ and $\rho = \rho_0$. These are the *Alfvén* waves.

By substituting (10.29) in the algebraic Equations (10.8) we check that the thermodynamic characteristics of the medium remain unchanged

$$\rho' = 0 , \quad p' = 0 , \quad s' = 0 , \tag{10.31}$$

while the perturbations of the velocity and magnetic field are subject to the conditions

$$\mathbf{v}' = \mp\, \mathbf{u}' , \quad \mathbf{u} \cdot \mathbf{u}' = 0 , \quad \mathbf{k} \cdot \mathbf{u}' = 0 . \tag{10.32}$$

Thus the Alfvén waves are the displacements of plasma together with the magnetic field frozen into it. They are transversal with respect to both the field direction and the wave vector as shown in Figure 10.1.

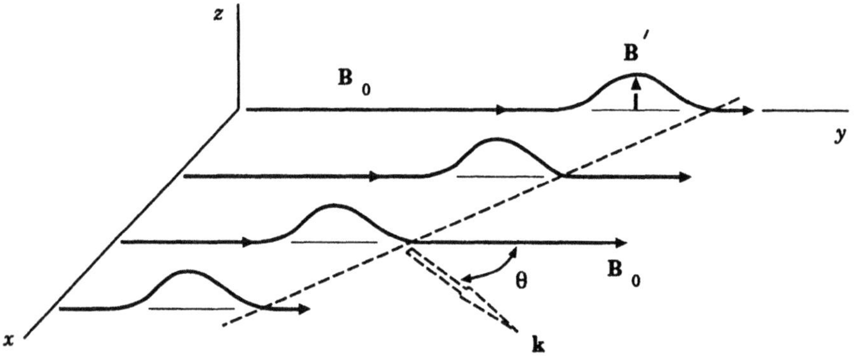

Figure 10.1: The transversal displacements of plasma and magnetic field in the Alfvén wave.

The Alfvén waves have no analogue in hydrodynamics. They are specific to MHD and were called the *magnetohydrodynamic* waves. This term emphasized that they do not change the density of a medium. The fact that the Alfvén waves are transversal signifies that

> a conducting plasma in a magnetic field has a characteristic elasticity resembling that of stretched strings under tension.

This is one of the characteristics of MHD. According to (10.32), the perturbed quantities are related by an energy equipartition:

$$\frac{1}{2}\rho\,(v')^2 = \frac{1}{8\pi}\,(B')^2 . \tag{10.33}$$

Note also that

> the energy of Alfvén waves, much like the energy of oscillations in a stretched string, propagates along the field lines only.

Unlike the phase velocity, the group velocity of the Alfvén waves (10.12)

$$\mathbf{V}_{\rm gr} = \frac{\mathbf{B}}{\sqrt{4\pi\rho}} \tag{10.34}$$

is directed strictly along the magnetic field; here $\mathbf{B} = \mathbf{B}_0$ of course.

In low density astrophysical plasmas with a strong field, like the solar corona, the Alfvén speed $V_{\rm A}$ can approach the light speed c (Problem 10.3). The discovery of Alfvén waves was a major stage in the development of cosmic plasma physics (Alfvén, 1950).

10.2.3 Magnetoacoustic waves

The dispersion Equation (10.10) has two other branches – two types of waves defined by a bi-square equation

$$\omega_0^4 - k^2 \left(u^2 + V_s^{\,2}\right) \omega_0^2 + k^2 \, V_s^{\,2} \, (\mathbf{k} \cdot \mathbf{u})^2 = 0 \, . \qquad (10.35)$$

Its solutions are two values of ω_0, which differ in absolute magnitude, corresponding to two different waves with the phase velocities V_+ and V_- which are equal to

$$V_\pm^2 = \frac{1}{2} \left[u^2 + V_s^{\,2} \pm \sqrt{(u^2 + V_s^{\,2})^2 - 4u^2 V_s^{\,2} \cos^2 \theta} \, \right] . \qquad (10.36)$$

These waves are called the *fast* (+) and the *slow* (−) *magnetoacoustic* waves, respectively (van de Hulst, 1951). The point is that the entropy of the medium, as follows from Equations (10.8) under condition (10.35), does not change in such waves

$$s' = 0 \, , \qquad (10.37)$$

as is also the case in an usual sound wave. Perturbations of the other quantities can be expressed in terms of the density perturbation

$$p' = V_s^{\,2} \rho' \, , \qquad (10.38)$$

$$\mathbf{v}' = - \frac{\omega_0}{\rho \, k^2} \left(\frac{k^2 (\mathbf{k} \cdot \mathbf{u}) \, \mathbf{u} - \omega_0^2 \, \mathbf{k}}{\omega_0^2 - (\mathbf{k} \cdot \mathbf{u})^2} \right) \rho' \, , \qquad (10.39)$$

$$\mathbf{u}' = \frac{\omega_0^2}{\rho \, k^2} \left(\frac{k^2 \, \mathbf{u} - (\mathbf{k} \cdot \mathbf{u}) \, \mathbf{k}}{\omega_0^2 - (\mathbf{k} \cdot \mathbf{u})^2} \right) \rho' \, . \qquad (10.40)$$

Formulae (10.39) and (10.40) show that the magnetoacoustic waves are neither longitudinal nor transversal. Perturbations of the velocity and magnetic field intensity, $\mathbf{v}'$ and $\mathbf{u}'$, as differentiated from the Alfvén wave, lie in the $(\mathbf{k}, \mathbf{B}_0)$ plane in Figure 10.1. They have components both in the direction of the wave propagation $\mathbf{k}/k$ and in the perpendicular direction. That is why the magnetoacoustic waves generally have a linearly polarized electric field $\mathbf{E}'$ normal to both $\mathbf{B}_0$ and $\mathbf{k}$.

Note that the perturbation of magnetic pressure $B^2/8\pi$ may be written in the form (see definition (10.3))

$$p_{\mathrm{m}}' = \rho \, \mathbf{u} \cdot \mathbf{u}' = \left(\frac{V_\pm^2}{V_s^{\,2}} - 1 \right) p' \, . \qquad (10.41)$$

Therefore for the fast wave, by virtue of that $V_+^2 > V_s^2$, the perturbation of magnetic pressure $p_{\rm m}'$ is of the same sign as that of gas pressure p'.

> The magnetic pressure and the gas pressure are added in the fast magnetoacoustic wave. The wave propagates faster, since the effective elasticity of the medium is greater.

A different situation arises with the slow magnetoacoustic wave. In this case $V_-^2 < V_s^2$ and $p_{\rm m}'$ is *opposite* in sign to p'. Magnetic and gas pressure deviations partially compensate each other. That is why such a slow wave propagates slowly.

10.2.4 The phase velocity diagram

The dependence of the wave velocities on the angle θ between the undisturbed field $\mathbf{B}_0$ and the wave vector $\mathbf{k}$ is clearly demonstrated in a polar diagram – the phase velocity diagram. In Figure 10.2, the radius-vector length from the origin of the coordinates to a curve is proportional to the corresponding phase velocity (10.11). The horizontal axis corresponds to the direction of the undisturbed magnetic field.

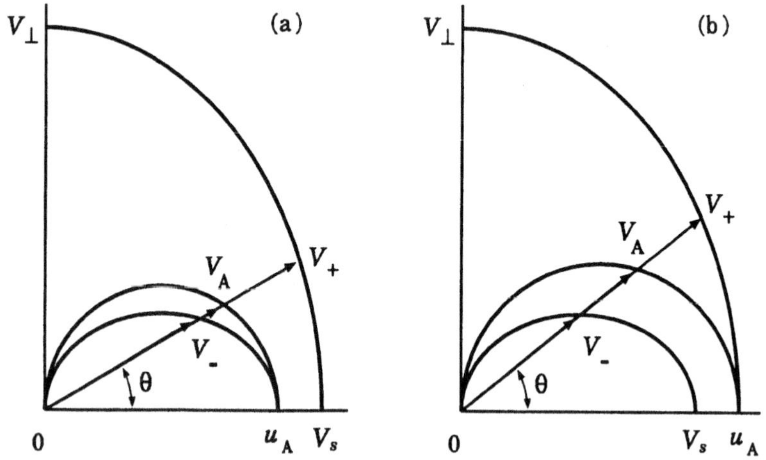

Figure 10.2: The phase velocities of MHD waves versus the angle θ for the two cases: (a) $u_{\rm A} < V_s$ and (b) $u_{\rm A} > V_s$.

As the angle $\theta \to 0$, the fast magnetoacoustic wave V_+ transforms to the usual sound one V_s if

$$V_s > V_{\rm A\parallel} = \frac{B}{\sqrt{4\pi\rho}} \equiv u_{\rm A} \qquad (10.42)$$

in Figure 10.2a or to the Alfvén wave if $V_s < u_A$ in Figure 10.2b.

For the angle $\theta \to \pi/2$, the propagation velocities of the Alfvén and slow waves approach zero. As this takes place, both waves convert to the weak tangential discontinuity in which disturbances of velocity and magnetic field are parallel to the front plane. As $\theta \to \pi/2$, the fast magnetoacoustic wave velocity tends to

$$V_{\perp} = \sqrt{V_{A\parallel}^{2} + V_{s}^{2}} = \sqrt{u_{A}^{2} + V_{s}^{2}}\,. \tag{10.43}$$

In the *strong* field limit ($V_{A\parallel}^{2} \gg V_{s}^{2}$) the diagram for the fast magnetoacoustic wave becomes practically isotropic as shown in Figure 10.3. Such a wave may

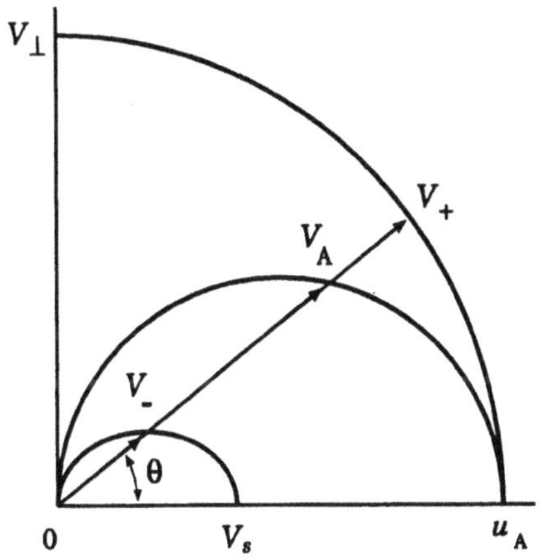

Figure 10.3: The phase velocity diagram for a plasma with a strong magnetic field.

be called the 'magnetic sound' wave since its phase velocity $V_+ \approx V_{A\parallel} \equiv u_A$ is almost independent of the angle θ.

Generally, the sound speed is the minimum velocity of disturbance propagation in ordinary hydrodynamics. By contrast, there is *no minimum velocity* in magnetohydrodynamics.

This general property is of fundamental importance for what follows in Chapters 11 and 12 – in study of the principal questions related to discontinuous flows of cosmic plasma. The first of these questions is what kinds of discontinuities can really exist?

MHD waves produce a lot of effects in cosmic plasma. LaRosa *et al.* (1996), Miller *et al.* (1996) consider the fast magnetoacoustic wave turbulence as a mechanism which accelerates electrons in solar flares (Section 5.2.3). Miller and Reames (1996) propose that the heavy ions observed in interplanetary space after impulsive flares result from stochastic acceleration by cascading Alfvén wave turbulence (Section 5.2.4).

10.3 Dissipative waves

10.3.1 Damping of Alfvén waves

We shall start by treating a plane Alfvén wave propagating along a uniform field $\mathbf{B}_0$; so the angle $\theta = 0$ in Figure 10.1. Perturbations of the magnetic field and the velocity are small and parallel to the z axis:

$$\mathbf{B}' = \{ 0, 0, b(t, y) \}, \quad \mathbf{v}' = \{ 0, 0, v(t, y) \}. \tag{10.44}$$

In general, the damping effects for such a wave are determined by the viscosity and by a finite conductivity. Let us consider, first, only the uniform finite conductivity σ. In this case we obtain the extended equation of the wave type with a dissipative term:

$$\frac{\partial^2 h}{\partial t^2} = u_{\rm A}^2 \frac{\partial^2 h}{\partial y^2} + \nu_{\rm m} \frac{\partial^3 h}{\partial^2 y \, \partial t} . \tag{10.45}$$

Here $u_{\rm A} = V_{\rm A\parallel}$ and $\nu_{\rm m}$ is the magnetic diffusivity (8.24). In the case of infinite conductivity Equation (10.45) is reduced to the wave equation and represents an Alfvén wave with velocity $u_{\rm A}$.

Suppose that the conductivity is finite. Suppose further that the small perturbations are functions of t and y only:

$$b(t, y) = b_0 \exp(\mathrm{i}\,\omega t + \alpha y), \quad v(t, y) = v_0 \exp(\mathrm{i}\,\omega t + \alpha y), \tag{10.46}$$

where ω, α, b_0, and v_0 are constants, all of which except ω may be complex numbers. Substituting (10.46) in (10.45) gives us the dispersion equation:

$$\omega^2 + \left(u_{\rm A}^2 + \mathrm{i}\,\nu_{\rm m}\,\omega \right) \alpha^2 = 0 \tag{10.47}$$

or

$$\alpha = \pm \mathrm{i} \frac{\omega}{u_{\rm A}} \left(1 + \mathrm{i} \frac{\nu_{\rm m}\,\omega}{u_{\rm A}^2} \right)^{-1/2} . \tag{10.48}$$

For small damping

$$\alpha = \pm \left(\mathrm{i} \frac{\omega}{u_{\rm A}} + \frac{\nu_{\rm m}\,\omega^2}{2 u_{\rm A}^3} \right) . \tag{10.49}$$

The distance y_0 in which the amplitude of the wave is reduced to $1/e$ is the inverse value of the real part of α. Thus we have

$$y_0 = \frac{2 u_{\rm A}^3}{\nu_{\rm m}\,\omega^2} = \frac{8 \pi \sigma u_{\rm A}^3}{\omega^2 c^2} = \frac{2 \sigma u_{\rm A}}{\pi c^2} \lambda^2 , \tag{10.50}$$

where $\lambda = 2\pi u_{\rm A}/\omega$ is the wave length. We see that **short wavelength waves suffer more damping** than do the long wavelength waves.

Since we treat the dissipative effects as small, the expression (10.50) is valid if $\lambda \ll y_0$. Thus we write

$$b\,(t, y) = b_0 \exp\left(-\frac{y}{y_0}\right) \exp\left[\mathrm{i}\,\omega\left(t - \frac{y}{u_{\rm A}}\right)\right], \tag{10.51}$$

$$v\,(t, y) = v_0 \exp\left(-\frac{y}{y_0}\right) \exp\left[\mathrm{i}\,\omega\left(t - \frac{y}{u_{\rm A}}\right)\right] \tag{10.52}$$

with

$$v_0 = u_{\rm A} \frac{b_0}{B_0}\left(1 - \mathrm{i}\,\frac{\nu_{\rm m}\,\omega}{2u_{\rm A}^2}\right). \tag{10.53}$$

The imaginary part indicates the phase shift of the velocity v in relation to the magnetic perturbation field b. Therefore we can write

$$v\,(t, y) = u_{\rm A} \frac{b_0}{B_0} \exp\left(-\frac{y}{y_0}\right) \exp\left\{\mathrm{i}\left[\omega\left(t - \frac{y}{u_{\rm A}}\right) - \varphi\right]\right\}, \tag{10.54}$$

where

$$\varphi = \frac{\nu_{\rm m}\,\omega}{2u_{\rm A}^2} = \frac{\omega\, c^2}{8\pi\sigma u_{\rm A}^2} = \frac{\omega\, c^2 \rho}{2\sigma B_0^2}\,. \tag{10.55}$$

So, the existence of Alfvén waves requires an external field B_0 enclosed between two limits.

> The magnetic field should be strong enough to make the damping effects small and yet weak enough to keep the Alfvén speed well below the velocity of light,

because otherwise the wave becomes an ordinary electromagnetic wave (Problem 8.2). In optical and radio frequencies it is not possible to satisfy both conditions. However, longer periods often observed in cosmic plasma leave a wide range between both limits so that Alfvén waves may exist.

One of favourable sites for excitation of MHD waves is the solar atmosphere. The chromosphere and corona are highly inhomogeneous media supporting a variety of filamentary structures in the form of arches and loops. The foot points of these structures are anchored in the poles of the photospheric magnetic fields. They undergo a continuous twisting and turning due to convective motions in the subphotospheric layers. This twisting and turning excite MHD waves. The waves then dissipate and heat the corona. Presumably this energy is enough to explain coronal heating, but the unambiguous detection of the MHD waves heating the corona is still awaited.

10.3.2 Slightly damped MHD waves

The damping effects due to a finite conductivity σ and due to a kinematic viscosity $\nu = \eta/\rho$ (Section 8.1.2) can be included in a general treatment of MHD waves of small amplitudes (van de Hulst, 1951). Well developed waves are the waves that travel at least a few wave lengths before they lose a considerable fraction of their energy if the two dimensionless parameters

$$p_\nu = \frac{\omega\,\nu}{c^2} \quad \text{and} \quad p_{\nu_{\rm m}} = \frac{\omega\,\nu_{\rm m}}{c^2}\,, \tag{10.56}$$

that characterize two dissipative processes, are much smaller than the two small dimensionless parameters

$$p_s = \frac{V_s^2}{c^2} \quad \text{and} \quad p_{\rm A} = \frac{u_{\rm A}^2}{c^2}\,, \tag{10.57}$$

that characterize the propagation speeds of undamped waves.

Let us postulate the form

$$X \equiv c^2 / V_{\rm ph}^2 = X_0\,(1 - {\rm i}\,q) \tag{10.58}$$

for a general solution of linearized equations of dissipative MHD. Here

$$X_0 = c^2 / V_{\rm ph,0}^2 \tag{10.59}$$

represents any solution for an undamped wave. We shall not review all special cases here but shall mention only one, the same case as in previous Section. For Alfvén wave we find the following solution

$$X = X_{\rm m} \equiv c^2/u_{\rm A}^2, \quad q = (p_\nu + p_{\nu_{\rm m}})\,X_{\rm m}\,. \tag{10.60}$$

This shows that, if dissipative effects are small,

> the relative importance of resistivity and viscosity as damping effects in Alfvén wave is independent of frequency ω.

The damping length, i.e., the distance l_d, in which the amplitude of a wave decreases by a factor $1/e$, and the damping time τ_d, in which this distance is covered by the wave, can be found:

$$l_d = \frac{1}{kq} = \frac{u_{\rm A}}{q\,\omega} = \frac{u_{\rm A}^3}{\omega^2\,(\nu + \nu_{\rm m})}\,, \tag{10.61}$$

$$\tau_d = \frac{l_d}{u_{\rm A}} = \frac{1}{q\,\omega} = \frac{u_{\rm A}^2}{\omega^2\,(\nu + \nu_{\rm m})}\,. \qquad (10.62)$$

So the high frequency waves have a short damping length and time.

The magnetoacoustic waves (Section 10.2.3), being compressional, have an additional contribution to their damping rate from compressibility of the plasma. If dissipative effects are not small, they result in additional waves propagating in a homogeneous medium (see Section 12.3).

10.4 Practice: Problems and Answers

Problem 10.1. Evaluate the sound speed in the solar corona.

Answer. For an ideal gas with constant specific heats c_p and c_v, the sound speed (10.6) is

$$V_s = \left(\gamma_g \,\frac{p}{\rho}\right)^{1/2}, \qquad (10.63)$$

where $\gamma_g = c_p/c_v$. Let us consider the coronal plasma as a 'monatomic gas' ($\gamma_g = 5/3$) of electrons and protons with $T_{\rm e} = T_{\rm p} = T \approx 2 \times 10^6$ K and $n_{\rm e} = n_{\rm e} = n$. So $p = 2nk_{\rm B}T$ and $\rho = nm_{\rm p}$. Hence

$$V_s = \left(\frac{10}{3}\,\frac{k_{\rm B}}{m_{\rm p}}\right)_s^{1/2} \sqrt{T} = 1.66 \times 10^4 \sqrt{T({\rm K})}\,,\ {\rm cm\ s^{-1}}\,. \qquad (10.64)$$

In the solar corona $V_s \approx 230$ km s^{-1}.

Problem 10.2. Consider entropy waves in hydrodynamics.

Answer. Let us take the linear algebraic Equations (10.25) and (10.26). In the absence of a magnetic field we put $\mathbf{u} = 0$ and $\mathbf{u}' = 0$. It follows from (10.25) that the perturbation of the velocity $\mathbf{v}'$ can be an arbitrary value except the gas pressure must be undisturbed. This follows from (10.26) and means that, instead of (10.27), we write

$$\mathbf{v}' \neq 0\,, \quad p' = 0\,. \qquad (10.65)$$

Perturbations of the density and entropy remain to be related by condition (10.28). So, the velocity perturbation is independent of the entropy perturbation and, according to (10.13) and (10.23), satisfies the equation

$$\mathbf{k} \cdot \mathbf{v}' = \frac{\omega}{v}\,v_x' + k_y v_y' = 0\,. \qquad (10.66)$$

This is in the reference frame in which $v = v_x$.

Note that for such velocity perturbation (see Landau and Lifshitz, *Fluid Mechanics*, 1959, Ch. 9):

$$\operatorname{curl} \mathbf{v}' \neq 0 \,. \tag{10.67}$$

That is why the wave is called the *entropy-vortex* wave.

> In the presence of a magnetic field in plasma, it is impossible to create a vortex without a perturbation of the magnetic field.

For this reason, in a MHD entropy wave, the only disturbed quantities are the entropy and the density (Equation (10.28)).

Problem 10.3. Show that the inclusion of the displacement current modifies the dispersion relation for the Alfvén waves (10.29) to the following equation

$$\omega_0^2 = \frac{(\mathbf{k} \cdot \mathbf{u})^2}{1 + u^2/c^2} \quad \text{or} \quad \omega_0 = \pm \frac{\mathbf{k} \cdot \mathbf{u}}{\sqrt{1 + u^2/c^2}} \,. \tag{10.68}$$

So the phase velocity of the relativistic Alfvén waves

$$V_{\mathrm{A}} = \pm \frac{B}{\sqrt{4\pi\rho}} \cos\theta \, \frac{1}{\sqrt{1 + B^2/4\pi\rho c^2}} \,, \tag{10.69}$$

which coinsides with the Alfvén formula (8.85).

Problem 10.4. Discuss the following situation. A star of the mass M moves along a uniform magnetic field $\mathbf{B}_0$ at a constant velocity $\mathbf{v}_0$ which exceeds the phase velocity of a fast magnetoacoustic wave (Dokuchaev, 1964).

Hint. The star emits magnetoacoustic waves by Chernkov radiation.

Chapter 11

Discontinuous Flows in a MHD Medium

The phenomena related to shock waves and other dicontinuous flows in cosmic plasma are so numerous that the study of MHD discontinuities on their own is of independent interest for space science.

11.1 Discontinuity surfaces in hydrodynamics

11.1.1 The origin of shocks in ordinary hydrodynamics

Recall the way shock waves are formed in ordinary hydrodynamic media without a magnetic field. Imagine a piston moving into a tube occupied by a gas. Let the piston velocity increase from zero by small jumps δv. As soon as the piston starts moving, it begins to rake the gas up and compress it. The front edge of the compression region thereby travels down the undisturbed gas inside the tube with the velocity of sound

$$V_s = \left(\frac{\partial p}{\partial \rho} \right)_s^{1/2} . \tag{11.1}$$

Each following impulse of compression $\delta \rho$ will propagate in a denser medium and hence with greater velocity. Actually, the derivative of the sound speed with respect to density

$$\frac{\partial V_s}{\partial \rho} = \frac{1}{2} \left(\frac{\partial^2 p}{\partial \rho^2} \right)_s \left(\frac{\partial p}{\partial \rho} \right)_s^{-1/2} \approx \sqrt{\gamma_g} \, (\gamma_g - 1) \, \rho^{(\gamma_g - 3)/2} > 0 \, ,$$

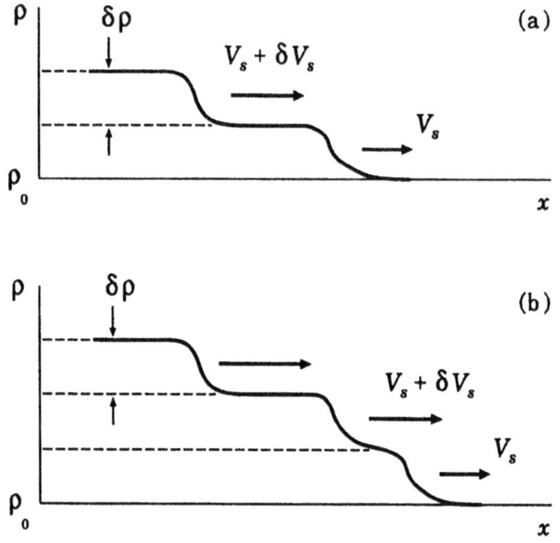

Figure 11.1: The behaviour of small perturbations in front of a piston.

since for all real substances $\gamma_g > 1$ in the adiabatic process $p \sim \rho^{\gamma_g}$. Therefore $\delta V_s > 0$.

As a consequence of this fact, successive compression impulses will catch up with each other as shown in Figure 11.1a. As a result, the compression region front steepens (Figure 11.1b). The gradients of the gas parameters become so large that the description of the gas as a hydrodynamic medium (Section 8.1) is no longer valid. The density, pressure and velocity of the gas change abruptly over a distance comparable to a particle's mean free path λ.

The physical processes inside such a jump, called a *shock wave*, are determined by the kinetic phenomena in the gas. As far as the hydrodynamic approximation is concerned, the surface at which the continuity of the hydrodynamic parameters of a medium is violated is some discontinuity surface – a discontinuous solution of the hydrodynamic equations. It stands to reason that some definite boundary conditions must hold at the discontinuity surface. What are they?

11.1.2 Boundary conditions and classification

Choose a frame of reference connected with a discontinuity surface. The frame is supposed to move with a constant velocity with respect to the medium. Generally, if the gas flow is non-stationary in the vicinity of the discontinuity, we could consider the discontinuity surface over a small period of time, so that the changes of velocity and other hydrodynamic quantities in time could be neglected.

In order to formulate the boundary conditions, consider an element of the discontinuity surface. Let the axis x be directed normally to it. **The flux of mass** through such a surface element must conserve:

$$\rho_1 \, v_{x1} = \rho_2 \, v_{x2} \, . \tag{11.2}$$

Here the indices 1 and 2 refer to the two sides of the discontinuity surface.

In this chapter, the difference in a quantity across the discontinuity surface will be designated by curly brackets, e.g.

$$\{ \rho \, v_x \} = \rho_1 \, v_{x1} - \rho_2 \, v_{x2} \, .$$

Then Equation (11.2) is rewritten as

$$\{ \rho \, v_x \} = 0 \, . \tag{11.3}$$

The energy flux must also be continuous at the discontinuity surface. For a hydrodynamic medium without a magnetic field (cf. (8.47)) we obtain the following condition for the energy flux conservation:

$$\left\{ \rho \, v_x \left(\frac{v^2}{2} + w \right) \right\} = 0 \, .$$

Here w is the specific enthalpy (7.24).

The momentum flux must be also continuous (cf. (8.48)):

$$\Pi_{\alpha\beta} = p \, \delta_{\alpha\beta} + \rho \, v_\alpha v_\beta \, , \qquad \alpha = x \, .$$

The continuity of the x-component of the momentum flux means that

$$\left\{ p + \rho \, v_x^2 \right\} = 0 \, ,$$

while the continuity of y- and z-components gives the two conditions

$$\{ \rho \, v_x v_y \} = 0 \, , \qquad \{ \rho \, v_x v_z \} = 0 \, .$$

Taking care of condition (11.3), rewrite the full set of boundary conditions at the discontinuity surface as follows:

$$\{ \rho \, v_x \} = 0 \, , \qquad \rho \, v_x \, \{ \mathbf{v}_\tau \} = 0 \, ,$$

$$\rho v_x \left\{ \frac{v^2}{2} + w \right\} = 0 , \quad \left\{ p + \rho v_x^2 \right\} = 0 . \tag{11.4}$$

Here the index τ identifies the tangential components of the velocity.

Obviously the set of Equations (11.4) falls into two *mutually exclusive* groups, depending on whether the matter flux across the discontinuity surface is zero or not. Consider these groups.

(a) If

$$v_x = 0$$

then the gas pressure is also continuous at the discontinuity surface,

$$\{ p \} = 0 , \tag{11.5}$$

while the tangential velocity component $\mathbf{v}_\tau$ as well as the density may experience an arbitrary jump:

$$\{ v_\tau \} \neq 0 , \quad \{ \rho \} \neq 0 , \quad \left\{ \frac{v^2}{2} + w \right\} \neq 0 .$$

Such discontinuities are called *tangential* (e.g., Landau and Lifshitz, *Fluid Mechanics*, 1959, § 84).

(b) By contrast, if

$$v_x \neq 0$$

then

$$\{ \rho v_x \} = 0 , \quad \{ v_\tau \} = 0 , \quad \left\{ p + \rho v_x^2 \right\} = 0 , \quad \left\{ \frac{v^2}{2} + w \right\} = 0 . \tag{11.6}$$

Discontinuities of this type are termed *shock waves*. Their properties are also well known in hydrodynamics (Landau and Lifshitz, *Fluid Mechanics*, 1959, Ch. 9).

Therefore the equations of ideal hydrodynamics in the conservation law form at the discontinuity surface allow just two *mutually exclusive* types of discontinuities to exist: the shock wave and the tangential discontinuity.

11.1.3 Dissipative processes and entropy

The equations of ideal hydrodynamics, as a specific case ($\mathbf{B} = 0$) of the ideal MHD Equations (8.41)–(8.46), do not take into account either viscosity or thermal conductivity. For this reason the ideal hydrodynamics equations describe three conservation laws: conservation of mass, momentum, and entropy. The last one is the specific form of the energy conservation law under assumption that the process under consideration is adiabatic. In Section 11.1.2 to obtain the boundary conditions at the discontinuity surface we used **conservation of mass, momentum, and energy**, but not entropy. The entropy increases across a shock (Problem 11.6).

The increase in entropy indicates that *irreversible* dissipative processes (which can be traced to the presence of viscosity and heat conduction in a medium) occur in the shock wave. The model which does not take into account these processes (Section 11.1.2) admits the existence of discontinuities but is not capable of describing the continuous transition from the initial to the final state. The ideal hydrodynamics cannot describe either the mechanism of shock compression or the structure of the very thin but finite layer where the plasma undergoes a transition from the initial to the final state.

> The entropy increase across the shock is entirely independent of the dissipative mechanism and is defined exclusively by the conservation laws of mass, momentum, and energy

(see Problem 11.6). Only the thickness of the discontinuity depends upon the rate of the irreversible heating of the plasma compressed by the shock. The following analogy in everyday life is interesting. A glass of hot water will invariably cool from a given temperature (the initial state) to a room temperature (the final state), independently of the mechanism of heat exchange with the surrounding air; the mechanism determines only the rate of cooling.

Recommended Reading: Zel'dovich and Raizer, *Physics of Shock Waves and High-Temperature Hydrodynamic Phenomena*, 1966, v. 1, Ch. 2.

11.2 Magnetohydrodynamic discontinuities

11.2.1 Boundary conditions at a discontinuity surface

Much like ordinary hydrodynamics, the equations of MHD for an ideal medium (Section 8.2) allow discontinuous solutions. De Hoffmann and Teller (1950) were the first to consider shock waves in MHD, based on the relativistic energy-momentum tensor for an ideal medium and the electromagnetic field.

Syrovatskii (1953) has given a more general formulation of the problem of the possible types of discontinuity surfaces in a conducting medium with a magnetic field. He has formulated a closed set of equations of ideal MHD and, using this, the *boundary conditions* at the discontinuity were written. We shall briefly reproduce the derivation of the boundary conditions.

We start from the equations of ideal MHD (8.41)–(8.46). Rewrite them (the equation of state is omitted) as follows:

$$\operatorname{div}\mathbf{B} = 0\,, \quad \frac{\partial \mathbf{B}}{\partial t} = \operatorname{curl}\,(\,\mathbf{v}\times\mathbf{B}\,)\,, \quad \frac{\partial \rho}{\partial t} = -\operatorname{div}\rho\mathbf{v}\,, \tag{11.7}$$

$$\frac{\partial}{\partial t}\left(\frac{\rho v^2}{2} + \rho\varepsilon + \frac{B^2}{8\pi}\right) = -\operatorname{div}\mathbf{G}\,, \quad \frac{\partial}{\partial t}\,(\rho\, v_\alpha) = -\,\frac{\partial}{\partial r_\beta}\,\Pi^*_{\alpha\beta}\,.$$

In a frame of reference moving with the discontinuity surface, all the conditions are stationary $(\partial/\partial t = 0)$. Hence,

$$\operatorname{div}\mathbf{B} = 0\,, \tag{11.8}$$

$$\operatorname{curl}\,(\,\mathbf{v}\times\mathbf{B}\,) = 0\,, \tag{11.9}$$

$$\operatorname{div}\rho\mathbf{v} = 0\,, \qquad \operatorname{div}\mathbf{G} = 0\,, \qquad \frac{\partial}{\partial r_\beta}\,\Pi^*_{\alpha\beta} = 0\,. \tag{11.10}$$

Four of these conditions have the divergent form and are therefore reduced in the integral form to the conservation of fluxes of vectors appearing at the divergence. Thus the following quantities must conserve at the discontinuity: the perpendicular (to the surface S) component of the magnetic field vector $B_{\rm n}$, the mass flux $\rho v_{\rm n}$, the energy flux $G_{\rm n}$, and the momentum flux $\Pi^*_{\alpha{\rm n}}$.

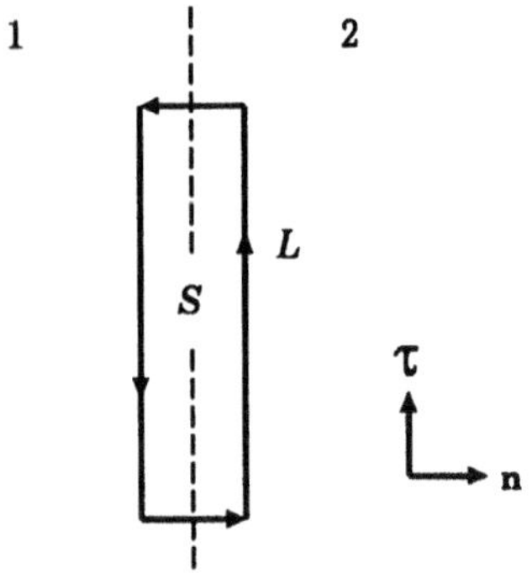

Figure 11.2: The contour L for the derivation of the boundary condition on electric field tangential component.

The exception is condition (11.9). It is written as the curl of $\mathbf{v}\times\mathbf{B}$. Integration of (11.9) over the area enclosed by the contour shown in Figure 11.2 gives, by virtue of the Stokes theorem,

$$\int_S \operatorname{curl}\,(\,\mathbf{v}\times\mathbf{B}\,)\cdot d\mathbf{S} = \oint_L (\,\mathbf{v}\times\mathbf{B}\,)\cdot d\mathbf{l} = 0\,.$$

Thus condition (11.9) demonstrates the continuity of the tangential component of the vector $(\mathbf{v} \times \mathbf{B})_\tau$, i.e. the electric field $\mathbf{E}_\tau$ in the discontinuity surface S.

As in the previous section, the jump of a quantity on crossing the discontinuity surface is designated by curly brackets. The full system of boundary conditions at the surface is written as follows:

$$\{ B_{\rm n} \} = 0 , \tag{11.11}$$

$$\{ (\mathbf{v} \times \mathbf{B})_\tau \} = 0 , \tag{11.12}$$

$$\{ \rho v_{\rm n} \} = 0 , \tag{11.13}$$

$$\{ G_{\rm n} \} = 0 , \tag{11.14}$$

$$\{ \Pi^*_{\alpha {\rm n}} \} = 0 . \tag{11.15}$$

The physical meaning of the boundary conditions obtained is obvious. The first two are the usual electrodynamic continuity conditions for the normal component of the magnetic field and the tangential component of the electric field. The last three equations represent the continuity of fluxes of mass, energy and momentum, respectively.

As distinct from that in ordinary hydrodynamics (see Equations (11.4)),

> the set of the MHD boundary conditions does not fall into mutually exclusive groups of equations.

What this means is that, with a few exceptions, any discontinuity, once accepted by these equations, can, generally speaking, transform to any other discontinuity under continuous change of the conditions of the motion (Syrovatskii, 1956).

Hence the classification of discontinuities in MHD is a matter of convention. Any classification is based on the external properties of the flow near the surface, such as the absence or presence of normal components of the velocity $v_{\rm n}$ and magnetic field $B_{\rm n}$, continuity or jump in density. The classification given below is due to Syrovatskii (1953). It is quite convenient for investigating MHD discontinuities.

Before turning our attention to the discussion of the classification mentioned above, let us rewrite the boundary conditions obtained, using (8.47) and (8.48) for the densities of the energy and momentum fluxes and substituting (11.11) in (11.12) and (11.13) in (11.15). We get

$$\{ B_{\rm n} \} = 0 , \tag{11.16}$$

$$\{ v_{\rm n} \mathbf{B}_\tau \} = B_{\rm n} \{ \mathbf{v}_\tau \} \, , \tag{11.17}$$

$$\{ \rho v_{\rm n} \} = 0 \, , \tag{11.18}$$

$$\left\{ \rho v_{\rm n} \left(\frac{v^2}{2} + w \right) + \frac{1}{4\pi} \left(B^2 v_{\rm n} - (\mathbf{v} \cdot \mathbf{B}) \, B_{\rm n} \right) \right\} = 0 \, , \tag{11.19}$$

$$\left\{ p + \rho v_{\rm n}^2 + \frac{B^2}{8\pi} \right\} = 0 \, , \tag{11.20}$$

$$\rho v_{\rm n} \{ \mathbf{v}_\tau \} = \frac{B_{\rm n}}{4\pi} \{ \mathbf{B}_\tau \} \, . \tag{11.21}$$

For later use, we write down the boundary conditions in the Cartesian frame of reference, the x axis being perpendicular to the discontinuity surface:

$$\{ B_x \} = 0 \, , \tag{11.22}$$

$$\{ v_x B_y - v_y B_x \} = 0 \, , \tag{11.23}$$

$$\{ v_x B_z - v_z B_x \} = 0 \, , \tag{11.24}$$

$$\{ \rho v_x \} = 0 \, , \tag{11.25}$$

$$\left\{ \rho v_x \left(\frac{v^2}{2} + w \right) + \frac{1}{4\pi} \left(B^2 v_x - (\mathbf{v} \cdot \mathbf{B}) \, B_x \right) \right\} = 0 \, , \tag{11.26}$$

$$\left\{ p + \rho v_x^2 + \frac{B^2}{8\pi} \right\} = 0 \, , \tag{11.27}$$

$$\left\{ \rho v_x v_y - \frac{1}{4\pi} B_x B_y \right\} = 0 \, , \tag{11.28}$$

$$\left\{ \rho v_x v_z - \frac{1}{4\pi} B_x B_z \right\} = 0 \, . \tag{11.29}$$

The set consists of eight boundary conditions. For $\mathbf{B} = 0$ it converts to the set of four Equations (11.4).

Consider the classification of discontinuity surfaces in MHD, which stems from the boundary conditions (11.16)–(11.21).

11.2.2 Discontinuities without plasma flows across them

Suppose the plasma flow through the discontinuity surface is absent

$$v_{\rm n} = 0 \,. \tag{11.30}$$

The discontinuity type depends on whether the magnetic field penetrates through the surface or not. Consider both possibilities.

(a) If the perpendicular component of the magnetic field

$$B_{\rm n} \neq 0 \,, \tag{11.31}$$

then the set of Equations (11.16)–(11.21) becomes

$$\{ B_{\rm n} \} = 0 \,, \quad B_{\rm n} \{ \mathbf{v}_\tau \} = 0 \,, \quad B_{\rm n} \{ \mathbf{B}_\tau \} = 0 \,,$$

$$\left\{ p + \frac{B^2}{8\pi} \right\} = 0 \,, \quad \{ \rho \} \neq 0 \,. \tag{11.32}$$

The velocity, magnetic field strength and (by virtue of the fourth equation) gas pressure are continuous at the surface. The density jump does not have to be zero; otherwise, all values change continuously.

The discontinuity type considered is called the *contact* discontinuity and constitutes just **a boundary between two media**, which moves together with them. It is schematically depicted in Figure 11.3a.

(b) On the other hand, if

$$B_{\rm n} = 0 \tag{11.33}$$

then the velocity and magnetic field are parallel to the discontinuity surface (plane $x = 0$). In this case all the boundary conditions (11.16)–(11.21) are satisfied identically, with the exception of one. The remaining equation is

$$\left\{ p + \frac{B^2}{8\pi} \right\} = 0 \,.$$

In other words, the velocity and magnetic field are parallel to the discontinuity surface and may experience arbitrary jumps in magnitude and direction, the only requirement being that the total pressure, that is the sum of the usual gas pressure and the magnetic one, remains continuous at the discontinuity surface:

$$p^* = p + \frac{B^2}{8\pi} \,. \tag{11.34}$$

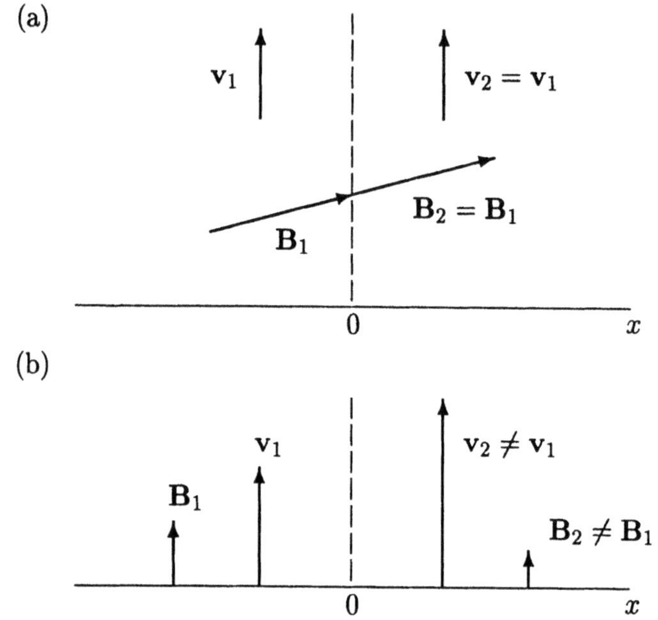

Figure 11.3: Discontinuity surfaces without a plasma flow across them: (a) contact discontinuity, (b) tangential discontinuity.

Such a discontinuity is called a *tangential* discontinuity (Figure 11.3b). As treated in MHD, it has a remarkable property. The tangential discontinuity in ordinary hydrodynamics is known to be unstable (Syrovatskii, 1954). The velocity jump engenders vortices, thus resulting in a turbulence near the discontinuity. Another situation occurs in MHD.

Syrovatskii (1953) has shown that the magnetic field exerts a stabilizing influence on the tangential discontinuity. In particular, if the density ρ_0 and magnetic field $\mathbf{B}_0$ are continuous, the only discontinuous quantity being the tangential velocity component, $\mathbf{v}_2 - \mathbf{v}_1 = \mathbf{v}_0 \neq 0$, then the condition for the tangential discontinuity stability is especially simple:

$$\boxed{\frac{B_0^2}{8\pi} \geq \frac{1}{4}\,\frac{\rho_0 v_0^2}{2}\,.} \tag{11.35}$$

To put it another way, such a discontinuity (Figure 11.4a) becomes stable with respect to small perturbations (of the general rather than a particular type) once the magnetic energy density reaches one quarter of the kinetic energy density.

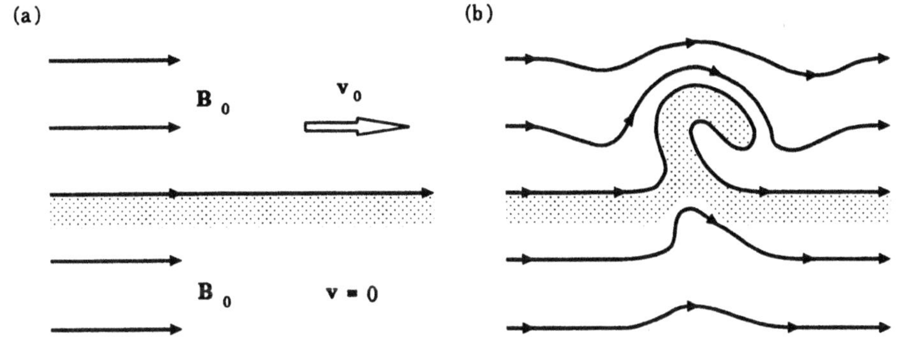

Figure 11.4: (a) The simplest type of tangential discontinuities. (b) Formation of a turbulent vortex gives rise to the magnetic field growth.

The general conclusion concerning the influence of the magnetic field on the stability of hydrodynamic motions of a conducting fluid is as follows:

> the magnetic field can only increase the stability of a given velocity distribution as compared to the stability of the same distribution in the absence of a magnetic field.

The point is that any flow instability and turbulence give rise, in view of the freezing-in of the field, to an increase of the magnetic energy (Figure 11.4b), which is always disadvantageous from the standpoint of the energetic principle of stability.

11.2.3 Perpendicular shock wave

Now let

$$v_{\rm n} \neq 0 \quad \text{and} \quad B_{\rm n} = 0\,, \tag{11.36}$$

i.e. a flow through the discontinuity surface is present whereas the magnetic field does not penetrate through the surface. Under these conditions, the following two statements result from Equations (11.16)–(11.21).

(a) From (11.21) the continuity of the tangential velocity component follows:

$$\{\, \mathbf{v}_\tau \} = 0\,. \tag{11.37}$$

This makes it possible to transform to such a frame of reference in which the tangential velocity component is absent on either side of the discontinuity: $\mathbf{v}_{\tau 1} = \mathbf{v}_{\tau 2} = 0$.

(b) The tangential electric field continuity (11.17) results in

$$\{ v_n \mathbf{B}_\tau \} = 0 . \tag{11.38}$$

If the frame of reference is rotated with respect to the x axis in such a way that $B_z = 0$ on one side of the surface, then the same is true on the other side (for clarity see (11.24)). Thus a frame of reference exists in which, in view of (a),

$$\mathbf{v} = (v_n, 0, 0) = (v, 0, 0)$$

and in addition, by virtue of (b),

$$\mathbf{B} = (0, B_\tau, 0) = (0, B, 0) .$$

In this frame of reference, the other boundary conditions take the form:

$$\{ \rho v \} = 0 , \tag{11.39}$$

$$\{ B / \rho \} = 0 , \tag{11.40}$$

$$\left\{ \rho v^2 + p + \frac{B^2}{8\pi} \right\} = 0 , \tag{11.41}$$

$$\left\{ \frac{v^2}{2} + w + \frac{B^2}{4\pi\rho} \right\} = 0 . \tag{11.42}$$

Such a discontinuity is called the *perpendicular* shock wave, since it constitutes the compression shock (see (11.6)) propagating perpendicular to the magnetic field as shown Figure 11.5. Condition (11.40) reflects the fact of the field 'freezing-in' into the plasma. The role of pressure in such a wave is played by the total pressure

$$p^* = p + \frac{B^2}{8\pi} , \tag{11.43}$$

whereas the role of the specific enthalpy is fulfilled by

$$w^* = w + \frac{B^2}{4\pi\rho} . \tag{11.44}$$

Therefore the role of the internal energy density is played by the total internal energy

$$\varepsilon^* = w^* - \frac{p^*}{\rho} = \varepsilon + \frac{B^2}{8\pi\rho} \tag{11.45}$$

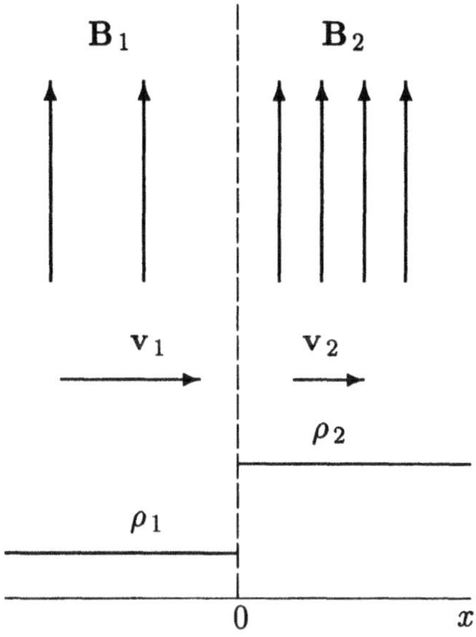

Figure 11.5: The character of the plasma motion and magnetic field compression ($B_2 > B_1$) in the perpendicular shock wave.

(cf. corresponding terms in Equations (8.41) and (8.42)).

For $\mathbf{B} = 0$, the perpendicular shock degenerates to the usual compression shock wave (Equations (11.6)).

For $\mathbf{B} \neq 0$, the propagation velocity of the perpendicular shock depends on the magnetic field strength.

A magnetic field decreases the compressibility of plasma while increasing its elasticity.

This is seen from (11.43) and the freezing-in condition (11.40). Accordingly, the magnetic field increases the shock wave propagation velocity.

If the intensity of a perpendicular shock is diminished, it converts to a fast magnetoacoustic wave propagating across the magnetic field ($\theta = \pi/2$ in Figure 10.2) with the speed (10.43), i.e.

$$V_{\perp} = \sqrt{V_s^2 + V_A^2}\,. \tag{11.46}$$

11.2.4 Oblique shock waves

The types of discontinuity surfaces treated above are the limiting cases of a more general discontinuity type for which

$$v_n \neq 0 \quad \text{and} \quad B_n \neq 0\,. \tag{11.47}$$

(a) The de Hoffmann-Teller frame of reference

In investigating the discontinuities (11.47), a frame of reference would be convenient in which $\mathbf{v}_1$ and $\mathbf{B}_1$ are parallel to each other. Such a frame does exist. It moves with respect to the laboratory one with the velocity

$$\mathbf{U} = \mathbf{v}_1 - \frac{v_{x1}}{B_{x1}}\,\mathbf{B}_1$$

parallel to the discontinuity surface. Actually, in this frame

$$\mathbf{v}_1\,(\mathbf{U}) = \mathbf{v}_1 - \mathbf{U} = \frac{v_{x1}}{B_{x1}}\,\mathbf{B}_1$$

and hence

$$\mathbf{v}_1 \times \mathbf{B}_1 = 0\,. \tag{11.48}$$

Then condition (11.17) in its coordinate form (11.23)–(11.24) can be used to obtain two equations valid to the right of the discontinuity, i.e. downstream of the shock:

$$v_{x2}B_{y2} - v_{y2}B_{x2} = 0\,, \qquad v_{x2}B_{z2} - v_{z2}B_{x2} = 0\,.$$

On rewriting these conditions as

$$\frac{v_{x2}}{v_{y2}} = \frac{B_{x2}}{B_{y2}} \quad \text{and} \quad \frac{v_{x2}}{v_{z2}} = \frac{B_{x2}}{B_{z2}}\,,$$

we ensure that the magnetic field is parallel to the velocity field (in the chosen reference frame) to the right of the discontinuity. In such frame of reference, called the de Hoffmann-Teller frame (de Hoffmann and Teller, 1950), the **electric field does not appear** according to (11.48).

This fact does not mean, of course, that the local cross-shock electric fields do not appear inside the shock transition layer, i.e. inside the discontinuity. The quasi-static electric and magnetic fields may determine the dynamics of particles in the shock front especially if Coulomb collisions play only a minor role. In collisionless shock waves, this dynamics depend on the particular mechanism of the energy redistribution among the perpendicular (with respect to the local magnetic field) and parallel degrees of freedom (see Section 11.4).

(b) Two types of shock waves

Thus, $\mathbf{v}$ is parallel to $\mathbf{B}$ on either side of the discontinuity. As a consequence, of the eight boundary conditions initially considered (see (11.22)–(11.29)), there remain six equations:

$$\{ B_x \} = 0 \,, \tag{11.49}$$

$$\{ \rho v_x \} = 0 \,, \tag{11.50}$$

$$\left\{ \frac{v^2}{2} + w \right\} = 0 \,, \tag{11.51}$$

$$\left\{ p + \rho v_x^2 + \frac{B^2}{8\pi} \right\} = 0 \,, \tag{11.52}$$

$$\left\{ \rho v_x v_y - \frac{B_x B_y}{4\pi} \right\} = 0 \,, \tag{11.53}$$

$$\left\{ \rho v_x v_z - \frac{B_x B_z}{4\pi} \right\} = 0 \,. \tag{11.54}$$

Let us take account of the parallelism of $\mathbf{v}$ and $\mathbf{B}$ in the chosen reference frame:

$$\mathbf{v}_1 = q_1 \mathbf{B}_1 \,, \qquad \mathbf{v}_2 = q_2 \mathbf{B}_2 \,, \tag{11.55}$$

where q_1 and q_2 are some proportionality coefficients. On substituting (11.55) in (11.49)–(11.54) we obtain the following three conditions from (11.50), (11.53), and (11.54):

$$\{ \rho q \} = 0 \,, \tag{11.56}$$

$$\left\{ \left(1 - \frac{1}{4\pi \rho q^2} \right) v_y \right\} = 0 \,, \tag{11.57}$$

$$\left\{ \left(1 - \frac{1}{4\pi \rho q^2} \right) v_z \right\} = 0 \,. \tag{11.58}$$

These equations admit two essentially different discontinuity types, depending on whether the density of the plasma is continuous or experiences a jump.

First we consider the discontinuity accompanied by a *density jump*:

$$\{ \rho \} \neq 0 \,. \tag{11.59}$$

Discontinuities of this type are called *oblique* shock waves.

Rotate the reference frame with respect to the x axis in such a way that

$$v_{z1} = 0 \, .$$

Then from (11.58) the following equation follows:

$$\left(1 - \frac{1}{4\pi\rho_2 \, q_2^2} \right) v_{z2} = 0 \, . \tag{11.60}$$

This suggests two possibilities: either

(**Case I**)

$$v_{z2} = 0 \, , \tag{11.61}$$

i.e. the motion is planar (the velocity and magnetic field are in the plane (x, y) on either side of the discontinuity), or

(**Case II**)

$$v_{z2} \neq 0 \quad \text{but} \quad q_2^2 = \frac{1}{4\pi\rho_2} \, . \tag{11.62}$$

Note that in the latter case

$$q_1^2 \neq \frac{1}{4\pi\rho_1} \tag{11.63}$$

since concurrently valid equations

$$q_2^2 = \frac{1}{4\pi\rho_2} \quad \text{and} \quad q_1^2 = \frac{1}{4\pi\rho_1}$$

would imply that

$$\rho_2 q_2 = \frac{1}{4\pi q_2} \quad \text{and} \quad \rho_1 q_1 = \frac{1}{4\pi q_1} \, ,$$

thus obviously contradicting (11.56) and (11.59). Therefore (11.63) must be valid.

Consider both cases indicated above.

(c) Fast and slow shock waves

Let us consider first the **Case I**. On the strength of (11.61), the boundary conditions (11.49)–(11.54) take the form

$$\{ B_x \} = 0 \, , \qquad \{ \rho v_x \} = 0 \, , \qquad \left\{ \frac{v^2}{2} + w \right\} = 0 \, ,$$

$$\left\{ p + \rho v_x^2 + \frac{B_y^2}{8\pi} \right\} = 0 \, , \quad \left\{ \rho v_x v_y - \frac{1}{4\pi} B_x B_y \right\} = 0 \, . \tag{11.64}$$

The compression oblique shock wave interacts with the magnetic field in an intricate way. The relationship between the parameters determining the state of a plasma before and after the wave passage is the topic of a large body of research (see reviews: Syrovatskii, 1957; Polovin, 1961; monographs: Anderson, 1963, Ch. 5; Priest, 1982, Ch. 5).

Boundary conditions (11.64) can be rewritten in such a way as to represent the Rankine-Hugoniot relation (see Problems 11.2 and 11.3 for an ordinary shock wave) for shocks in MHD (see Landau, Lifshitz and Pitaevskii, 1984, Ch. 8, § 72). Moreover, the Zemplen theorem on the **increase of density and pressure in a shock wave** can be proved in MHD (Iordanskii, 1958; Liubarskii and Polovin, 1958; Polovin and Liubarskii, 1958; see also Zank, 1991). The *fast* and the *slow* oblique shock waves are distinguished.

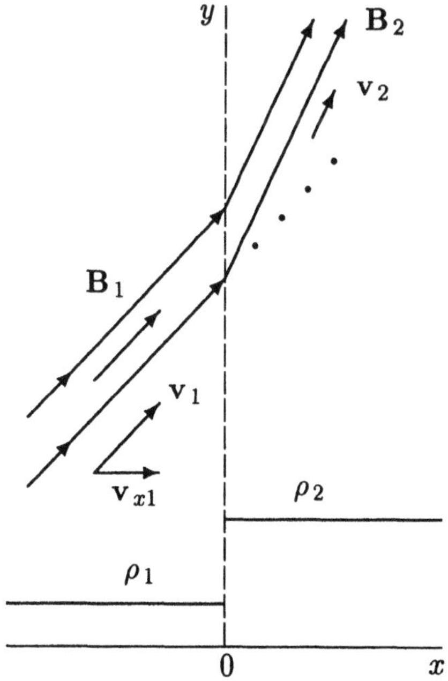

Figure 11.6: The magnetic field change ($B_2 > B_1$), velocity field and plasma density at the front of the fast shock wave.

In the fast shock wave, the magnetic field increases across the shock and is bent towards the shock front surface $x = 0$ (Figure 11.6). So the magnetic pressure increases as well as the gas pressure:

$$\delta p_{\rm m} > 0 \, , \quad \delta p > 0 \, . \tag{11.65}$$

In other words, and this seems to be a natural behaviour,

compression of the plasma in a fast shock wave is accompanied by compression of the magnetic field.

In the limiting case of small intensity, the fast shock converts to the fast magnetoacoustic wave (see (11.43)). The speed of the fast shock wave with respect to the medium equals v_{x1}. It is greater than or equal to the speed of the fast magnetoacoustic wave:

$$v_{x1} \geq V_{+} \,. \tag{11.66}$$

No small perturbation running in front of the shock can exist upstream of the fast shock wave.

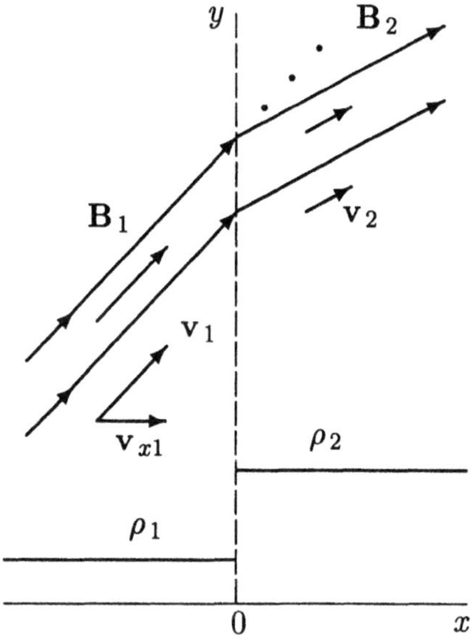

Figure 11.7: The magnetic field change ($B_2 < B_1$), velocity field and plasma density at the front of the slow shock wave.

In the slow shock wave, the magnetic field decreases across the shock and is bent towards the shock normal (Figure 11.7). Therefore

$$\delta p_{\rm m} < 0 \,, \quad \delta p > 0 \,. \tag{11.67}$$

Compression of the plasma is accompanied by a *decrease* of the magnetic field strength in the slow shock wave.

As the amplitude decreases, the slow shock wave will transform to the slow magnetoacoustic wave. The speed of the slow shock propagation is

$$V_{-} \leq v_{x1} \leq V_{\rm A} \,. \tag{11.68}$$

In the particular case

$$B_y = 0 \tag{11.69}$$

the set of boundary conditions (11.64) results in the set (11.4). This means that the oblique shock wave converts to the parallel (longitudinal) shock wave propagating along the magnetic field, mutual interaction being absent.

Note that the set of boundary conditions (11.49)–(11.54) formally admits four other types of discontinuous solutions (Section 12.4.2), apart from those indicated above. These are the so called *intermediate* or *transalfvénic* shock waves (e.g., Shercliff, 1965, Ch. 7).

> The peculiarity of these discontinuous solutions is that they have no counterpart among the small amplitude waves or simple waves.

This is the reason why the intermediate and transalfvénic shock waves are not included in the classification of discontinuities under consideration. What is more important is that the intermediate and transalfvénic shock waves are *non-evolutionary* (see Section 12.1).

The **Case II** shall be considered in the next Section.

11.2.5 Peculiar shock waves

We return to the consideration of the particular case (11.62) and (11.63):

$$v_{z2} \neq 0\,, \quad q_1^2 \neq \frac{1}{4\pi\rho_1}\,, \quad q_2^2 = \frac{1}{4\pi\rho_2}\,. \tag{11.70}$$

On the strength of (11.57) and (11.58), the following conditions must be satisfied at such a discontinuity:

$$\left(1 - \frac{1}{4\pi\rho_1 q_1^2}\right) v_{y1} = 0\,, \quad \left(1 - \frac{1}{4\pi\rho_1 q_1^2}\right) v_{z1} = 0\,.$$

Because the expression in the parentheses is not zero, we get

$$v_{y1} = v_{z1} = 0\,, \tag{11.71}$$

i.e. in front of such a discontinuity the tangential velocity component $\mathbf{v}_{\tau 1}$ is absent. The tangential field component $\mathbf{B}_{\tau 1}$ is also zero in front of the discontinuity, i.e. the motion follows the pattern seen in the parallel shock wave. However, arbitrary tangential components of the velocity and magnetic field are permissible downstream of the shock, the only condition being that

$$\mathbf{v}_2 = \frac{\mathbf{B}_2}{\sqrt{4\pi\rho_2}}\,. \tag{11.72}$$

Such a discontinuity is called the *switch-on* shock. The character of motion of this wave is shown in Figure 11.8.

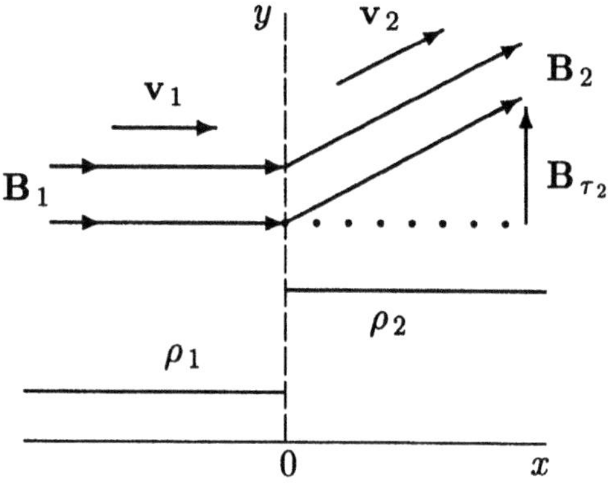

Figure 11.8: A switch-on wave: $\mathbf{B}_{\tau 1} = 0$, but $\mathbf{B}_{\tau 2} \neq 0$.

The switch-on shock exists in the interval

$$1 < \frac{v_{x1}^2}{V_{\mathrm{A}x1}^2} < \frac{4\, v_{x1}^2}{v_{x1}^2 + V_{\mathrm{s}1}^2}$$

(e.g., Liberman, 1978).

Assuming the tangential magnetic field component to be zero to the rear of the peculiar shock wave,

$$\mathbf{B}_{\tau 2} = 0\,, \tag{11.73}$$

the fluid velocity in front of the discontinuity is the Alfvén one:

$$\mathbf{v}_1 = \frac{\mathbf{B}_1}{\sqrt{4\pi\rho_1}}\,. \tag{11.74}$$

Such a peculiar shock wave is called the *switch-off* shock (Figure 11.9).

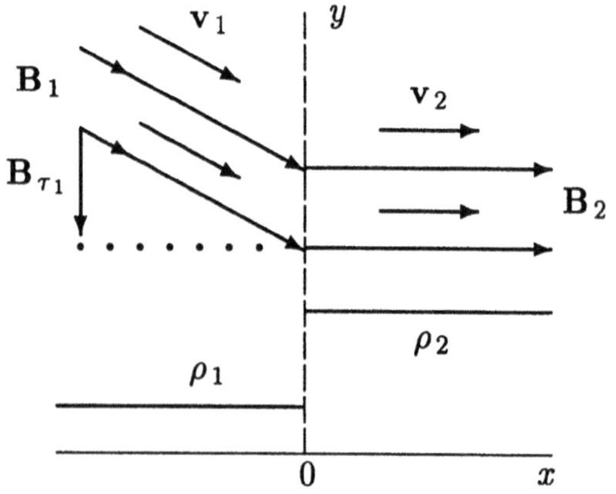

Figure 11.9: A switch-off wave: $\mathbf{B}_{\tau 2} = 0$, but $\mathbf{B}_{\tau 1} \neq 0$.

11.2.6 The Alfvén discontinuity

Returning to the general set of Equations (11.49)–(11.54), consider the discontinuity at which the density is constant:

$$\{\rho\} = 0\,. \tag{11.75}$$

On substituting this condition in (11.50), we see that the normal component of the velocity is continuous at the discontinuity:

$$\{v_x\} = 0\,.$$

Furthermore, in view of Equation (11.56), the quantity q does not change at the discontinuity:

$$\{q\} = 0\,.$$

Hence the quantity

$$\left(1 - \frac{1}{4\pi\rho\, q^2}\right)$$

is also continuous and may be factored out in Equations (11.57) and (11.58). Rewrite them as follows:

$$\left(1 - \frac{1}{4\pi\rho\, q^2}\right)\{\mathbf{v}_\tau\} = 0\,. \tag{11.76}$$

If the expression in the parentheses is not zero then the tangential velocity component is continuous and all other quantities are easily checked to be continuous solutions. So, to consider the discontinuous solutions, we put

$$q = \pm \frac{1}{\sqrt{4\pi\rho}}\,.$$

Thus the velocity vector is connected with the magnetic field strength through the relations

$$\mathbf{v}_1 = \pm \frac{\mathbf{B}_1}{\sqrt{4\pi\rho}}\,, \qquad \mathbf{v}_2 = \pm \frac{\mathbf{B}_2}{\sqrt{4\pi\rho}}\,. \tag{11.77}$$

The following relations also hold at the discontinuity surface

$$\{p\} = 0\,, \qquad \left\{\mathbf{B}_\tau^2\right\} = 0\,. \tag{11.78}$$

Therefore the normal components and the absolute values of the tangential components of the magnetic field and velocity as well as all thermodynamical parameters conserve at the discontinuity. For given values of $\mathbf{B}_1$ and $\mathbf{v}_1$,

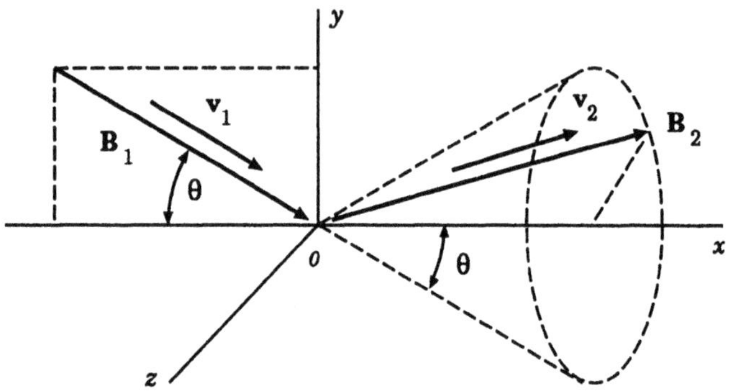

Figure 11.10: An Alfvén or rotational discontinuity.

possible values of $\mathbf{B}_2$ and $\mathbf{v}_2$ lie on a conical surface, the cone angle being equal to that between the normal to the discontinuity surface and the vector $\mathbf{B}_1$ (Figure 11.10). A discontinuity of this type is called *Alfvén* or *rotational*.

Its peculiarity is reflected in the second name. On passing the discontinuity surface, a medium can acquire a directionally arbitrary tangential momentum, so that the flow is not generally planar.

The speed of the discontinuity propagation relative to the plasma

$$v_{x1} = \mp \frac{B_{x1}}{\sqrt{4\pi\rho}} \, . \tag{11.79}$$

In the limiting case of small intensity, the *Alfvén* or *rotational* discontinuity converts to the Alfvén wave (see (10.29)).

11.3 Transitions between discontinuities

As was shown by Syrovatskii (1956), *continuous* transitions can occur between *discontinuous* MHD solutions of different types. This statement is easily verified on passing from the discontinuities (Section 11.2) to the limit of small-amplitude waves (Section 10.1). In this limit the fast and slow magnetoacoustic waves correspond to the oblique shocks, whereas the Alfvén wave corresponds to the Alfvén or rotational discontinuity.

Phase velocity diagrams for the small-amplitude waves are shown in Figure 10.2. Reasoning from it, the following scheme of continuous transitions between discontinuous solutions in ideal MHD can be suggested (Figure 11.11).

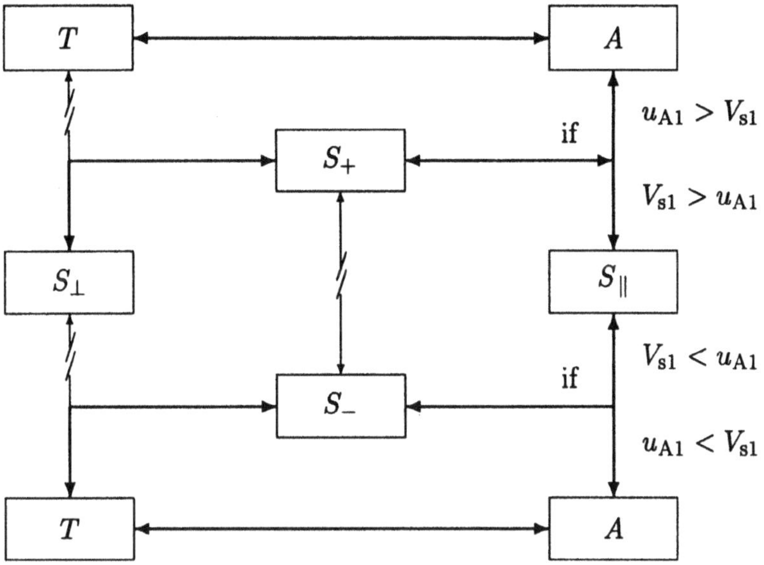

Figure 11.11: A scheme of the continuous transitions between discontinuous solutions in MHD, following from comparison of the properties of the discontinuities and small-amplitude waves on the phase velocity diagram.

Recall that θ is the angle between the wave vector $\mathbf{k}$ and the magnetic field direction $\mathbf{B}_0/B_0$, i.e. axis x in Figure 10.2. If $\theta \to \pi/2$ then the fast magnetoacoustic wave (V_+) converts to the perpendicular wave propagating across the field with the velocity $V_\perp$ (10.43). In the limit of large-amplitude waves this corresponds to the transition from the fast shock (S_+) to the perpendicular one ($S_\perp$).

As $\theta \to 0$, the fast magnetoacoustic wave (V_+) converts to the usual sound one (V_s) if $V_s > V_A$ or to the Alfvén wave (V_A) if $V_A > V_s$. Therefore the fast shock (S_+) must convert, when $\theta \to 0$, either to the longitudinal shock ($S_\parallel$) if $V_{s1} > V_{A1}$ or to the Alfvén discontinuity (A) if $V_{A1} > V_{s1}$.

In much the same way, we conclude, reasoning from Figure 10.2, that the slow shock (S_-) converts either to the longitudinal shock ($S_\parallel$) for $V_{s1} < V_{A1}$ or to the Alfvén discontinuity (A) for $V_{s1} > V_{A1}$. This transition takes place as $\theta \to 0$. For $\theta \to \pi/2$, both the slow shock wave (S_-) and Alfvén discontinuity (A) transform to the tangential discontinuity (T) as demonstrated by the fact that the corresponding phase velocities of the slow magnetoacoustic (V_-) and Alfvén (V_A) waves tend to zero for $\theta \to \pi/2$.

How are such transitions realized? They are effected through some discontinuities which may be called *transitional* since they conform to boundary conditions for both types of discontinuities and may be classified as either of the two. The existence of transitional discontinuities means that the discontinuity of one type can convert to the discontinuity of another type under a *continuous* change of parameters (Syrovatskii, 1956).

The absence of transitional discontinuities in MHD, manifested as the absence of transitions between small-amplitude waves in the phase velocity diagram (Figure 10.2), signifies the impossibility similar to that one in ordinary hydrodynamics because there exists a minimal velocity of shock propagation—the sound velocity V_s. That is why small perturbations in hydrodynamics cannot convert the shock wave (S) into the tangential discontinuity (T).

For the same reason the continuous transition between fast (S_+) and slow (S_-) shocks is impossible in MHD. This is shown in Figure 11.11 by the doubly crossed arrow. The fast shock (S_+) cannot continuously convert to the perpendicular one ($S_\perp$). These and other restrictions on continuous transitions between discontinuities in MHD will be explained in Chapter 12 from the viewpoint of evolutionarity conditions.

11.4 Shock waves in collisionless plasma

In ordinary collision-dominated gases or plasmas the density rise across a shock wave occurs in a distance of the order of a few collision mean free paths. The velocity distributions on both sides of the front are constrained by collisions to be Maxwellian and, if there is more than one kind of particles (for example, ions and electrons), the temperature of the various constituents of the plasma reach equality. Moreover, as we saw in Sections 11.1 and 11.2, the conditions (density, pressure, and flow velocity) on one side of the front are rigidly determined in terms of those on the other side by requirement that the flux of mass, momentum, and energy through the front be conserved. For weak shocks the front structure itself can be determined relatively simple, by taking into account collisional transfer coefficients representing viscosity, resistivity, and so on (Sirotina and Syrovatskii, 1960; Zel'dovich and Raizer, 1966; see also Section 12.4).

In a collisionless plasma the mechanisms by which the plasma state is changed by the passage of the shock front are more complex. Energy and momentum can be transferred from the plasma flow into electric and magnetic field oscillations for example by some **kinetic instabilities**. The energy of these collective motions must be taken into account when conservation laws are applied to relate the pre-shock state to the post-shock state. The ions and electrons are affected differently by instabilities. So, there is no reason for their temperatures to remain equal. Since kinetic instabilities are seldom isotropic, it is unlikely that the temperatures will remain isotropic. These anisotropies further change the jump conditions.

The change in state derives from the collective interactions between particles and electric and magnetic fields. In general these fields are of two types: they can be (a) **constant in time**, more exactly, quasi-static fields produced by charge separation, currents (e.g., Gedalin and Griv, 1999), or (b) **fluctuating in time**, produced by kinetic instabilities. The first situation is usually termed laminar, the second one turbulent. The fields often are turbulent. So, the scattering of particles by turbulence can play the role of dissipation in the collisionless shock structure. This turbulence can be either a small-scale one generated by plasma instabilities inside a laminar shock front, or a large-scale turbulence associated with the dominant mode of the shock interaction itself (see Tidman and Krall, 1971).

Since we are discussing the kinetic processes which occur on a time scale much shorter than the time scale of Coulomb collisions, we may efficiently use the Vlasov equation (2.47) or fluid-type descriptions derived from it to study the properties of shock waves in collisionless plasma.

11.5 Practice: Problems and Answers

Problem 11.1. Relate the flow variables ρ, v, and p at the surface of an ordinary shock wave (Section 11.1.2).

Answer. From formula (11.6) with $v_\tau = 0$ and $v_x = v$, we find

$$\rho_1 v_1 = \rho_2 v_2 \,, \tag{11.80}$$

$$p_1 + \rho_1 v_1^2 = p_2 + \rho_2 v_2^2 \,, \tag{11.81}$$

$$\frac{v_1^2}{2} + w_1 = \frac{v_2^2}{2} + w_2 \,. \tag{11.82}$$

Here

$$w = \varepsilon + \frac{p}{\rho} \tag{11.83}$$

is the specific enthalpy; the thermodynamic relationship for the specific internal energy $\varepsilon(p, \rho)$ is assumed to be known.

Problem 11.2. Assuming that the value of a parameter describing the strength of the shock in Problem 11.1 is known (for example, the relative velocity $\delta v = v_1 - v_2$ which playes the role of the 'piston' velocity), find the general relationships that follow from the conservation laws (11.80)–(11.82).

Answer. Instead of the density let us introduce the specific volume $U = 1/\rho$. From (11.80) we obtain

$$\frac{U_2}{U_1} = \frac{v_2}{v_1} . \tag{11.84}$$

Eliminating the velocities v_1 and v_2 from Equations (11.81) and (11.82), we find

$$v_1^2 = U_1^2 \, \frac{p_2 - p_1}{U_1 - U_2} , \tag{11.85}$$

$$v_2^2 = U_2^2 \, \frac{p_2 - p_1}{U_1 - U_2} . \tag{11.86}$$

The velocity of the compressed plasma with respect to the undisturbed one

$$\delta v = v_1 - v_2 = \left[(p_2 - p_1)(U_1 - U_2) \right]^{1/2} . \tag{11.87}$$

Substituting (11.85) and (11.86) in the energy equation (11.82), we obtain

$$\delta w = w_2 - w_1 = \frac{1}{2} \, (p_2 - p_1)(U_1 + U_2) . \tag{11.88}$$

This is the most general form of the Rankine-Hugoniot relation.

Problem 11.3. Consider the Rankine-Hugoniot relation for an ideal gas.

Answer. For an ideal gas with constant specific heats c_p and c_v, the specific enthalpy

$$w\,(p, U) = c_p \, T = \frac{\gamma_g}{\gamma_g - 1} \, pU , \tag{11.89}$$

where $\gamma_g = c_p / c_v$ is the specific heat ratio.

If we substitute (11.89) in (11.88), we obtain the Rankine-Hugoniot relation in the explicit form

$$\frac{p_2}{p_1} = \frac{(\gamma_g + 1)\, U_1 - (\gamma_g - 1)\, U_2}{(\gamma_g + 1)\, U_2 - (\gamma_g - 1)\, U_1} . \tag{11.90}$$

From here, the density ratio

$$r = \frac{\rho_2}{\rho_1} = \frac{U_1}{U_2} = \frac{(\gamma_g + 1)\, p_2 + (\gamma_g - 1)\, p_1}{(\gamma_g - 1)\, p_2 + (\gamma_g + 1)\, p_1} . \tag{11.91}$$

It is evident from (11.91) that **the density ratio** across a very strong shock, where the pressure p_2 behind the wave front is much higher than the initial pressure p_1, **does not increase infinitely** with increasing strength p_2/p_1, but approaches a certain finite value. This limiting density ratio is a fuction of the specific heat ratio γ_g only, and is equal to

$$r_\infty = \frac{\rho_2}{\rho_1} = \frac{\gamma_g + 1}{\gamma_g - 1} \, . \tag{11.92}$$

For a monatomic gas with $\gamma_g = 5/3$ the limiting compression ratio $r_\infty = 4$.

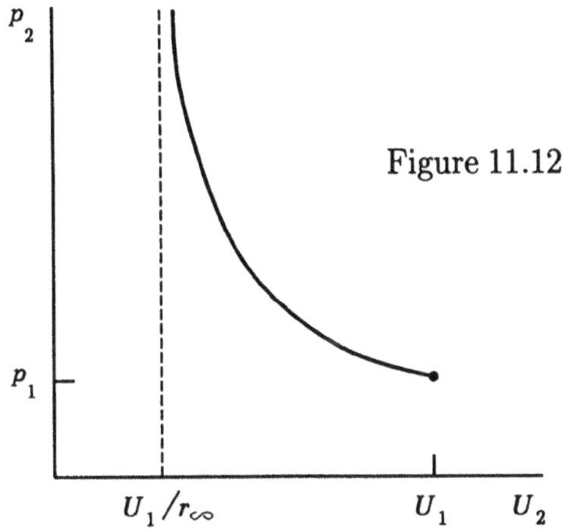

Figure 11.12: The Rankine-Hugoniot curve.

A curve on the diagram (p, U) passing through the initial state (p_1, U_1) according to (11.90) is called the Rankine-Hugoniot curve; it is shown in Figure 11.12.

Problem 11.4. What is the value of the limiting density ratio r in relativistic shock waves?

Answer. Note that Equation (11.80) is valid only for nonrelativistic flows. In relativistic shock waves, the Lorentz factor (4.6) for the upstream and downstream flows must be included, and we have (de Hoffmann and Teller, 1950):

$$\gamma_{L,1}\, \rho_1\, v_1 = \gamma_{L,2}\, \rho_2\, v_2 \, . \tag{11.93}$$

The density ratio

$$r = \frac{\rho_2}{\rho_1} = \frac{v_1}{v_2} \frac{\gamma_{L,1}}{\gamma_{L,2}} \, . \tag{11.94}$$

In highly relativistic shocks, the ratio v_1/v_2 remains finite, while the ratio $r \to \infty$. This is important for the problem of particle acceleration by shock waves (see Chapter 13).

Problem 11.5. Write the density ratio r as a function of the upstream Mach number.

Answer. Let us use the definition of the sound speed (11.1) in an ideal gas with constant specific heats

$$V_s = \left(\gamma_g \frac{p}{\rho} \right)^{1/2} = (\gamma_g \, pU)^{1/2} . \tag{11.95}$$

The upstream Mach number (to the second power)

$$M_1^2 = \frac{v_1^2}{V_{s1}^2} = \frac{U_1}{\gamma_g \, p_1} \, \frac{p_2 - p_1}{V_1 - V_2} . \tag{11.96}$$

Here the solution (11.85) has been taken into account.

Substituting (11.96) in (11.91) gives us the compression ratio as a function of the upstream Mach number

$$r = \frac{(\gamma_g + 1) \, M_1^2}{(\gamma_g - 1) \, M_1^2 + 2} . \tag{11.97}$$

When $M_1 \to \infty$, the density ratio $r \to (\gamma_g + 1)/(\gamma_g - 1)$ of course. This is the limiting case of a *strong* but nonrelativistic shock wave.

When $M_1 \to 1$, which is the limiting case of a *weak* shock wave, the density ratio $r \to 1$ too. By using formula (11.90), we see that the pressures on both sides of a weak shock wave are close to each other: $p_1 \approx p_2$ and $(p_2 - p_1)/p_1 \ll 1$. Thus a weak shock wave is practically the same as an acoustic compression wave.

For $M_1 < 1$ we could formally have an expansion shock wave with $r < 1$ and $p_2 < p_1$. However it can be shown (see the next Problem) that such a transition would involve a *decrease* of entropy rather than an increase. So such transitions are ruled out by the second law of thermodynamics.

Problem 11.6. Show that the entropy jump of a gas compressed by a shock increases with the strength of the shock wave but is entirely independent of the dissipative mechanism.

Answer. To within an arbitrary constant the entropy of an ideal gas with constant specific heats is given by formula (e.g., Landau and Lifshitz,

Statistical Physics, 1959, Ch. 4):

$$S = c_v \ln pU^{\gamma_g} \, . \tag{11.98}$$

The difference between the entropy on each side of the shock front, as derived from (11.91), is

$$S_2 - S_1 = c_v \ln \left\{ \left(\frac{p_2}{p_1} \right) \left[\frac{(\gamma_g - 1)(p_2/p_1) + (\gamma_g + 1)}{(\gamma_g + 1)(p_2/p_1) + (\gamma_g - 1)} \right]^{\gamma_g} \right\} . \tag{11.99}$$

In the limiting case of a weak wave ($p_2 \approx p_1$) the expression in braces is close to unity. Therefore $S_2 \approx S_1$ and $S_2 > S_1$ if $p_2 > p_1$. As the strength of the wave increases, that is, as the ratio p_2/p_1 increases beyond unity, the expression in braces increases monotonically and approaches infinity as $p_2/p_1 \to \infty$. Thus the entropy jump is positive and does increase with the strength of the shock wave.

> The increase in entropy indicates that irreversible dissipative processes occur in the shock front.

This can be traced to the presence of viscosity and heat conduction in the gas or plasma (see the discussion in Section 11.1.3).

Problem 11.7. Consider a collisionless gravitational system described by the gravitational analog of the Vlasov equation (Problem 2.5). Explain qualitatively why the Vlasov equation (2.80) does not predict the existence of a shock wave. In other words, unlike the case of gas or plasma, an evolution governed by the set of Equations (2.80)–(2.82) never leads to caustics or shocks.

Hint. By analogy with the discussion of the shock origin in ordinary hydrodynamics (Section 11.1.1), it is necessary to show that, given sufficiently smooth initial data, the distribution function will never diverge. So, the gravitational analog of the Vlasov equation manifests the so-called 'global existance' (e.g., Pfaffelmoser, 1992).

Problem 11.8. Discuss properties of the Petschek-type reconnecting region with the four slow MHD shocks shown in Figure 11.13 (Petschek, 1964).

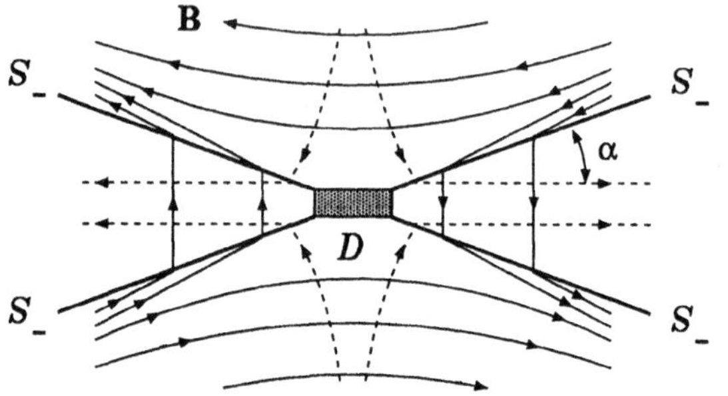

Figure 11.13: The Petschek-type reconnecting flow.

Chapter 12

Evolutionarity of MHD Discontinuities

A discontinuity cannot exist in a real plasma if small perturbations disintegrate it into other discontinuities or transform it to a more general nonsteady flow.

12.1 Conditions for evolutionarity

12.1.1 The physical meaning and definition

Of concern to us is the issue of the stability of MHD discontinuities with respect to their decomposition into more than one discontinuity. To answer this question small perturbations must be imposed on the discontinuity surface. If they do not instantaneously lead to large changes of the discontinuity, then the discontinuity is termed *evolutionary*.

Obviously the property of evolutionarity does not coincide with stability in the ordinary sense. The usual instability means exponential ($e^{\gamma t}$, $\gamma > 0$) growth of the disturbance, it remains small for some time ($t \leq \gamma^{-1}$). The discontinuity gradually evolves. By contrast,

> a disturbance **instantaneously** becomes large in a non-evolutionary discontinuity.

By way of illustration, the decomposition of a density jump $\rho(x)$ is shown in Figure 12.1. The disturbance $\delta\rho$ is not small, though it occupies an interval δx which is small for small t, when the two discontinuities have not become widely separated.

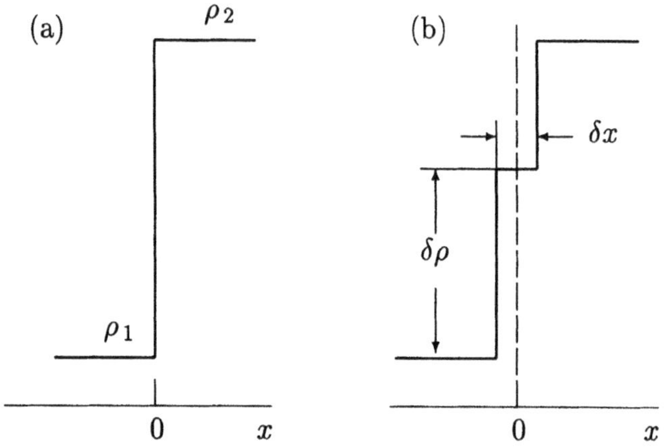

Figure 12.1: Disintegration of a density jump into two successive jumps.

The problem of disintegration of discontinuities has a long history. Kotchine (1926) first considered the disintegration of an arbitrary discontinuity into a set of other discontinuities and rarefaction waves in the frame of ordinary hydrodynamics. Bethe (1942) studied the disintegration of a shock wave. This effect is closely related to the problem of evolutionarity.

The mathematical idea of evolutionarity was expressed for the first time in the context of the study of discontinuities in ordinary hydrodynamics (e.g., Courant and Fridrichs, 1985; see also Gel'fand, 1959).

> With respect to evolutionary discontinuities, the usual problem of linear stability can be formulated,

i.e. we find solutions to the linearized equations giving rise to amplitudes, which grow or decay in time.

The evolutionarity criterion may be obtained by counting the number of equations supplied by linearized boundary conditions at the discontinuity surface, and the number of independent parameters determining an arbitrary, initially small disturbance of the discontinuity. If the numbers are equal, then the boundary conditions uniquely define further development – evolution – of the disturbance which remains small for small $t > 0$. Such a discontinuity is evolutionary.

By contrast, if the number of independent parameters characterizing the perturbation is greater or less than the number of equations, then the problem of a small perturbation of the discontinuity has an infinitely large number of solutions or no solutions at all. That indicates that the initial assumption of the smallness of the disturbance for small t is incorrect. The discontinuity is non-evolutionary.

Such a discontinuity cannot exist in a real medium as a stationary configuration, because a small perturbation leads to a finite variation of the initial flow. This variation is the disintegration of the discontinuity into other discontinuities, that move away from the place of their formation (Figure 12.1), or a transition to a more general nonsteady flow.

Let us count the number of equations which must be satisfied by an arbitrary small perturbation at the discontinuity. Let us take as the initial conditions the set of eight boundary conditions (11.22)–(11.29). It is to be linearized.

We will consider perturbations of the discontinuity, which generate plane waves propagating along the x axis. Then the quantity B_x remains constant on either side of the discontinuity, and condition (11.22) (both exact and linearized) is satisfied identically. Hence, on either side of the discontinuity, seven quantities are perturbed: three velocity components (v_x, v_y, v_z), two field components (B_y, B_z), density ρ and pressure p. Infinitesimally small perturbations of these quantities

$$\delta v_x \,,\; \delta v_y \,,\; \delta v_z \,,\; \delta B_y \,,\; \delta B_z \,,\; \delta \rho \,,\; \delta p \tag{12.1}$$

on either side of the discontinuity surface are characterized by the coordinate and time dependence

$$\delta f \,(t, x) \sim \exp \left[\, \mathrm{i} \, (kx - \omega t) \, \right] \tag{12.2}$$

typical of the plane wave.

If the number of small-amplitude waves leaving the discontinuity is equal to the number of independent boundary conditions, then the problem of small perturbations has only one solution and the discontinuity is evolutionary. This form of evolutionarity conditions has been obtained for the first time by Lax (1957). It has been applied to MHD shocks by Akhiezer, Lyubarskii, and Polovin (1959), Syrovatskii (1959), see for a review also Polovin (1961).

The small perturbations must obey the linearized boundary conditions, i.e. linear algebraic equations following from (11.23)–(11.29). In addition to these seven quantities, the velocity of the propagation of the discontinuity surface is disturbed. It acquires a small increment δu_x relative to the chosen frame of reference in which the undisturbed discontinuity is at rest.

12.1.2 Linearized boundary conditions

Let us write down the linearized boundary conditions in a reference frame rotated with respect to the x axis in such a way that the undisturbed values $B_z = 0$ and $v_z = 0$. Thus we restrict our consideration to those discontinuity surfaces in which the undisturbed fields $\mathbf{B}_1$, $\mathbf{B}_2$ and the velocities $\mathbf{v}_1$, $\mathbf{v}_2$ lie in the plane (x, y).

From the boundary conditions (11.22)–(11.29) we find a set of linear equations which falls into two groups describing different perturbations:

(a) Alfvén perturbations $(\delta v_z,\ \delta B_z)$

$$\left\{ \rho\, v_x\, \delta v_z - \frac{1}{4\pi}\, B_x\, \delta B_z \right\} = 0\,, \tag{12.3}$$

$$\{\, v_x\, \delta B_z - B_x\, \delta v_z \,\} = 0\,; \tag{12.4}$$

(b) magnetoacoustic and entropy perturbations $(\delta v_x,\ \delta v_y,\ \delta B_y,\ \delta \rho,\ \delta p)$

$$\{\, \rho\, (\delta v_x - \delta u_x) + v_x\, \delta \rho \,\} = 0\,, \tag{12.5}$$

$$\left\{ \rho\, v_x\, \delta v_y + v_y\, [\, \rho\ (\delta v_x - \delta u_x) + v_x\, \delta \rho \,] - \frac{1}{4\pi}\, B_x\, \delta B_y \right\} = 0\,, \tag{12.6}$$

$$\left\{ \delta p + v_x^2\, \delta \rho\ + 2 \rho\, v_x\, (\delta v_x - \delta u_x) + \frac{1}{4\pi}\, B_y\, \delta B_y \right\} = 0\,, \tag{12.7}$$

$$\{\, B_x\, \delta v_y - B_y\, (\delta v_x - \delta u_x) - v_x\, \delta B_y \,\} = 0\,, \tag{12.8}$$

$$\begin{aligned} &\{\, \rho\, v_x\ [\, v_x\, (\delta v_x - \delta u_x) + v_y\, \delta v_y + \delta w \,] + \\ &+ \left(\frac{v_x^2 + v_y^2}{2} + w \right) [\, \rho\ (\delta v_x - \delta u_x) + v_x\, \delta \rho \,] + \\ &+ \frac{B_y}{4\pi}\ [\, B_y\ (\delta v_x - \delta u_x) + v_x\, \delta B_y - B_x\, \delta v_y \,] + \\ &+ \frac{1}{4\pi}\, (v_x B_y - v_y B_x)\, \delta B_y \left. \right\} = 0\,. \end{aligned} \tag{12.9}$$

Condition (12.5) allows us to express the disturbance of the propagation velocity of the discontinuity surface δu_x in terms of perturbations of ρ and v_x:

$$\delta u_x\, \{\, \rho \,\} = \{\, \rho\, \delta v_x + v_x\, \delta \rho \,\}\ . \tag{12.10}$$

On substituting (12.10) in (12.6)–(12.9) there remain four independent equations in the second group of boundary conditions, since the disturbance of the velocity of the discontinuity surface δu_x can be eliminated from the set.

Therefore the MHD boundary conditions for perturbations of the discontinuity, which generate waves propagating perpendicular to the discontinuity surface, fall into two *isolated* groups. As this takes place,

> the conditions of evolutionarity (the number of waves leaving the MHD discontinuity is equal to the number of independent linearized boundary conditions) must hold not only for the variables in total but also for *each* isolated group

(Syrovatskii, 1959). The number of Alfvén waves leaving the discontinuity must be two, whereas there must be four magnetoacoustic and entropy waves. This make the evolutionary requirement more stringent.

Whether or not a discontinuity is evolutionary is clearly shown to be a *purely kinematic* problem. We have to count the number of small-amplitude waves leaving the discontinuity surface on either side: upstream and downstream of the discontinuity.

* * *

Concerning the boundary conditions the following comment should be made. As distinct from the unperturbed MHD equations, the perturbed ones are not stationary. Therefore the arguments used to derive Equations (11.16)–(11.21) from (11.7) are not always valid.

To derive boundary conditions at a disturbed discontinuity we have to transform to the reference frame connected with the surface. For example, for a perturbation (see Problem 12.2)

$$\xi_x(y,t) = \xi_0 \exp\left[\,\mathrm{i}\;(k_y\, y - \omega\, t)\right] ,$$

where ξ_x is a displacement of the surface, this is equivalent to the following substitution in the linearized MHD equations

$$\frac{\partial}{\partial t}\delta \to -\,\mathrm{i}\,\omega\left(\delta - \xi_0\frac{\partial}{\partial y}\right), \quad \frac{\partial}{\partial y}\delta \to \mathrm{i}\,k_y\left(\delta - \xi_0\frac{\partial}{\partial y}\right),$$

where $-\,\mathrm{i}\,\omega\,\xi_0 = \delta u_x$ is the amplitude of the time derivative of ξ. Consider, for example, the linearized continuity equation which after the integration over the discontinuity thickness takes the form

$$\mathrm{i}\int\limits_{-a}^{+a} (\omega - k_y v_y)\,\delta\rho\, dx - \mathrm{i}\,k_y \int\limits_{-a}^{+a} \rho\,\delta v_y\, dx =$$

$$= \{ v_x \, \delta\rho + \rho \, [\, \delta v_x + \mathrm{i} \, (\omega - k_y v_y) \, \xi_0 \,] \} \, . \tag{12.11}$$

If the integrals on the left-hand side of Equation (12.11) are equal to zero in the limit $a \to 0$ then, for $k_y = 0$, formula (12.11) transforms to (12.5). However this possibility is based on the supposition that $\delta\rho$ and δv_y inside the discontinuity do not increase in the limit $a \to 0$. We shall see below (in Chapter 19) that this supposition is not valid at least for more complicated, two-dimensional, configurations such as a reconnecting current sheet.

12.1.3 The number of small-amplitude waves

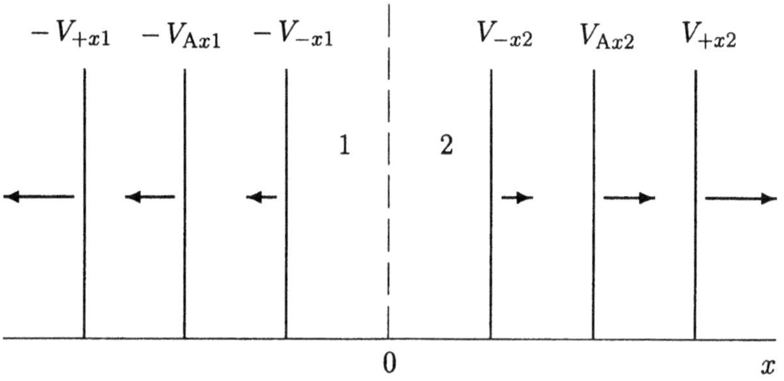

Figure 12.2: Six small-amplitude waves leaving an immovable discontinuity surface ($x = 0$) being perturbed.

If the discontinuity is immovable with respect to the plasma (no flow across the discontinuity), then *on either side* of the surface there exist three waves leaving it as shown in Figure 12.2:

$$-V_{+x1} \, , \; -V_{\mathrm{A}x1} \, , \; -V_{-x1} \, , \; V_{-x2} \, , \; V_{\mathrm{A}x2} \, , \; V_{+x2} \, . \tag{12.12}$$

Let the discontinuity move with a velocity v_{x1} relative to the plasma (Figure 12.3). The positive direction of the axis x is chosen to coincide with the direction of the plasma motion at the discontinuity surface. The index '1' refers to the region in front of the surface ($x < 0$) whereas the index '2' refers to the region behind the discontinuity ($x > 0$), i.e. downstream of the flow. Then there exist fourteen different phase velocities of propagation of small-amplitude waves:

$$v_{x1} \pm V_{+x1} \, , \; v_{x1} \pm V_{\mathrm{A}x1} \, , \; v_{x1} \pm V_{-x1} \, , \; v_{x1} \, ,$$
$$v_{x2} \pm V_{-x2} \, , \; v_{x2} \pm V_{\mathrm{A}x2} \, , \; v_{x2} \pm V_{+x2} \, , \; v_{x2} \, .$$

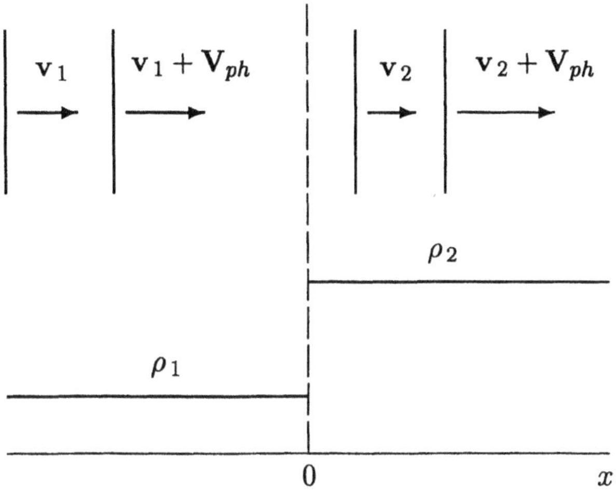

Figure 12.3: Small-amplitude waves in a plasma moving through the MHD discontinuity.

Waves leaving the discontinuity have negative phase velocities in the region 1 and positive phase velocities in the region 2.

In the region 1, four velocities, corresponding to the waves moving toward the discontinuity surface, can be immediately discarded:

$$v_{x1} + V_{+x1}\,, \quad v_{x1} + V_{\mathrm{A}x1}\,, \quad v_{x1} + V_{-x1}\,, \quad v_{x1}\,.$$

The remaining three waves $(7-4)$ can leave the discontinuity or propagate toward it, depending on the plasma flow velocity towards the discontinuity v_{x1}.

In the region 2, four waves always have positive phase velocities:

$$v_{x2} + V_{+x2}\,, \quad v_{x2} + V_{\mathrm{A}x2}\,, \quad v_{x2} + V_{-x2}\,, \quad v_{x2}\,. \tag{12.13}$$

These waves leave the discontinuity. Other waves will be converging or diverging, depending on relations between the quantities

$$v_{x2}\,, \quad V_{+x2}\,, \quad V_{\mathrm{A}x2}\,, \quad V_{-x2}\,.$$

Let

$$0 < v_{x1} < V_{-x1}\,. \tag{12.14}$$

Then there are three waves leaving the discontinuity in the region 1:

$$v_{x1} - V_{-x1}\,, \; v_{x1} - V_{\mathrm{A}x1}\,, \; v_{x1} - V_{-x1}\,.$$

If

$$0 < v_{x2} < V_{-x2} \,, \tag{12.15}$$

then four waves (12.13) propagate downstream of the discontinuity since the waves

$$v_{x2} - V_{-x2} \,, \;\; v_{x2} - V_{\mathrm{A}x2} \,, \;\; v_{x2} - V_{+x2}$$

converge to the discontinuity.

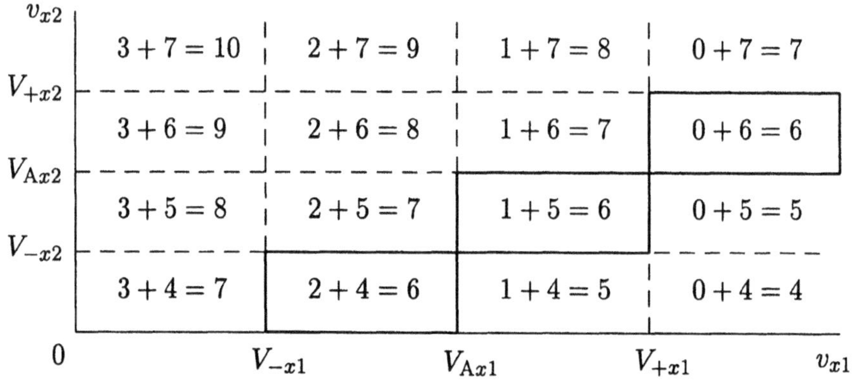

Figure 12.4: The number of small-amplitude waves leaving a discontinuity surface.

We shall write down the number of diverging waves to the left (in front of) and to the right (behind) the discontinuity as their sum (e.g. $3 + 4 = 7$ in the case considered) in the corresponding rectangle in the plane (v_{x1}, v_{x2}) presented in Figure 12.4. This rectangle is the lower left one. In the rectangle situated to the right of this one, two rather than three waves are diverging in the region 1:

$$v_{x1} - V_{\mathrm{A}x1} \,, \;\; v_{x1} - V_{+x1} \,.$$

The wave $v_{x1} - V_{-x1}$ is carried by the flow to the discontinuity since

$$\boxed{V_{-x1} < v_{x1} < V_{\mathrm{A}x1} \,.} \tag{12.16}$$

Thus we write $2 + 4 = 6$ in this rectangle. The whole table is filled up in a similar manner.

12.1.4 Domains of evolutionarity

If one considers the total number of boundary conditions (six), without allowance being made for their falling into two groups, then just three rectangles in Figure 12.4 should be inspected for possible evolutionarity. The boundaries of these rectangles are shown by solid lines.

However, as indicated above, the equality of the total number of independent boundary conditions to the number of diverging waves is insufficient for the existence and uniqueness of the solutions in the class of small perturbations (Syrovatskii, 1959). Take into account that

> the linearized boundary conditions fall into two groups, and hence the number of Alfvén waves must equal two and that of diverging magnetoacoustic and entropy waves must equal four.

Then one of the three rectangles becomes the point A in Figure 12.5.

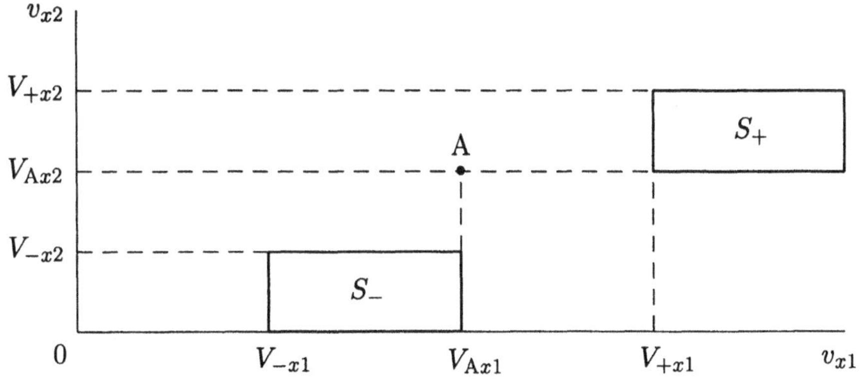

Figure 12.5: The evolutionarity domains for the fast (S_+) and slow (S_-) shocks and the Alfvén discontinuity.

The figure shows that there exist two domains of evolutionarity of shock waves:

(a) fast shock waves (S_+) for which

$$v_{x1} > V_{+x1}\,, \qquad V_{Ax2} < v_{x2} < V_{+x2}\,, \tag{12.17}$$

(b) slow shock waves (S_-) for which

$$V_{-x1} < v_{x1} < V_{Ax1}\,, \qquad v_{x2} < V_{-x2}\,. \tag{12.18}$$

Recall that our treatment of the Alfvén discontinuity was not quite satisfactory. It was treated as a flow in the plane (x, y). Generally this is not the case (Figure 11.10). The result of the above analysis is also not quite satisfactory: the evolutionarity of the Alfvén discontinuity, as well as the switch-on and switch-off shocks, is more complicated. While investigating the evolutionarity of these discontinuities, dissipative effects must be allowed for (Section 12.3).

Although *dissipative* waves quickly damp as they propagate away from the discontinuity surface, they play an important role in the system of small-amplitude waves leaving the discontinuity. Thus only one solution exists for the switch-off shock, i.e. it is evolutionary. By contrast,

> the switch-on shock wave, as well as the Alfvén or rotational discontinuity, are non-evolutionary

in the linear approximation.

Roikhvarger and Syrovatskii (1974) have shown that attention to dissipation in the dispersion equation for magnetoacoustic and entropy waves leads to the appearance of dissipative waves and, as a consequence, to the non-evolutionarity of tangential and contact discontinuities. While being quite natural from the physical standpoint, this effect has not yet gained a formal mathematical proof.

Recall that in an ideal medium the disintegration of a discontinuity is instantaneous in the sense that the secondary discontinuities become separated in the beginning of the disintegration process (Figure 12.1). In a dissipative medium the spatial profiles of the MHD discontinuities are continuous. Nevertheless, the principal result remains the same. The steady flow is rearranged toward a nonsteady state, and after a large enough period of time the disintegration manifests itself (Section 12.4).

12.2 Consequences of evolutionarity conditions

12.2.1 The order of wave propagation

Some interesting inferences concerning the order of shock propagation result from the evolutionarity conditions (12.17) and (12.18).

If a shock wave follows another one of the same type (fast or slow), the back shock will catch up with the front one (Akhiezer, Lyubarskii, and Polovin, 1959). Consider, as an example, two slow shock waves, S_{-}^{A} and S_{-}^{B}, propagating in the direction of the x axis as shown in Figure 12.6.

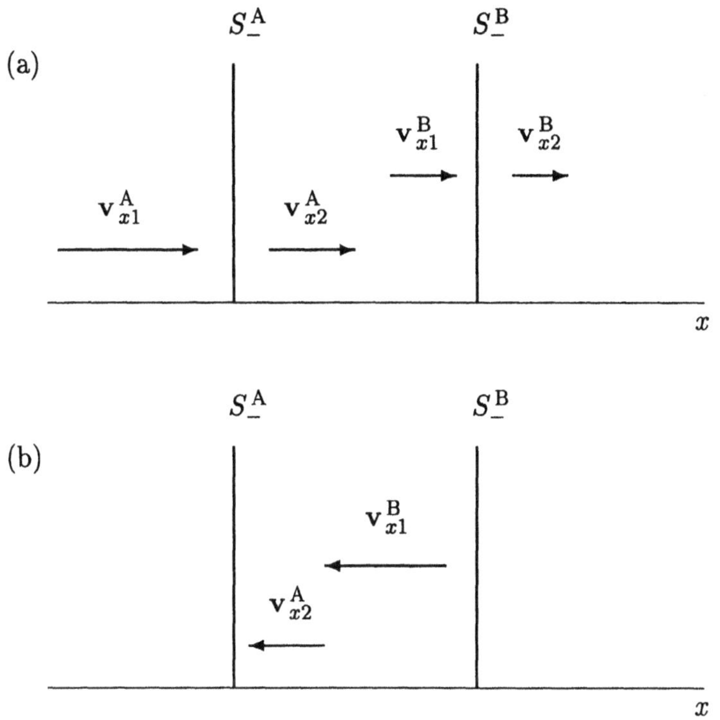

Figure 12.6: Plasma flow velocities relative to: (a) shock wave fronts, (b) the plasma between the shock waves.

In a reference frame connected with the front of the first shock $S_-^{\rm A}$, we get, by virtue of the evolutionary condition (12.18),

$$V_{-x1}^{\rm A} < v_{x1}^{\rm A} < V_{{\rm A}x1}^{\rm A}\,, \quad v_{x2}^{\rm A} < V_{-x2}^{\rm A}\,. \tag{12.19}$$

In a reference frame connected with the front of the second shock $S_-^{\rm B}$, analogous conditions hold:

$$V_{-x1}^{\rm B} < v_{x1}^{\rm B} < V_{{\rm A}x1}^{\rm B}\,, \quad v_{x2}^{\rm B} < V_{-x2}^{\rm B}\,. \tag{12.20}$$

Since the velocities of slow magnetoacoustic waves of small amplitude $V_{-x2}^{\rm A}$ and $V_{-x1}^{\rm B}$ refer to the same region (between the shocks), they are equal

$$V_{-x2}^{\rm A} = V_{-x1}^{\rm B}\,. \tag{12.21}$$

Substituting (12.21) in the second part of (12.19) and in the first part of (12.20) gives the inequality

$$v_{x2}^{\rm A} < v_{x1}^{\rm B}\,. \tag{12.22}$$

Hence, relative to the plasma between the shocks (Figure 12.6b), the shock $S_-^{\rm B}$ catches up with the shock $S_-^{\rm A}$, which was to be proved.

As for different types of waves, the following inferences can be drawn: the Alfvén discontinuity will catch up with the slow shock, whereas the fast shock will catch up with all possible types of discontinuities. If shock waves are generated by a single source (for example, a flare in the solar atmosphere), then no more than three shocks can move in the same direction: the fast shock is followed by the Alfvén discontinuity, the slow shock being to the rear of the Alfvén discontinuity.

12.2.2 Continuous transitions between discontinuities

Reasoning from the polar diagram for phase velocities of small-amplitude waves, in Section 11.3 we have treated the possibility of continuous transitions between different types of discontinuous solutions in MHD. However, the evolutionarity conditions have not been taken into account. They are known to impose limitations on possible continuous transitions between the discontinuities under changes of external parameters (magnetic field, flow velocity, etc.).

Continuous transition is impossible between the fast and slow shock waves. This stems from the fact that the evolutionarity domains for fast (S_+) and slow (S_-) shocks have no common points (Figure 12.5). Similarly, the lines of phase velocities V_+ and V_- in polar diagrams (Figures 10.2 and 10.3) are out of contact. That was the basis for banning transitions between the fast and slow shocks in Figure 11.11.

The fast shock (S_+) cannot continuously convert to the tangential discontinuity (T) since that would go against the evolutionarity condition $v_{x1} > V_{{\rm A}x1}$. The same ban stems from the consideration of the phase velocity diagram (Section 11.3). The perpendicular shock ($S_\perp$) is the limiting case of the fast shock. That is why the continuous transition of the perpendicular shock to the tangential discontinuity is forbidden, as shown in Figure 11.11.

As was indicated in the previous section, the issue of evolutionarity of the Alfvén discontinuity has no satisfactory solution in the framework of ideal MHD. The established viewpoint is that the continuous transition of shock waves (S_- and S_+) to the Alfvén discontinuity (A) is impossible, as is predicted by the phase velocity diagram with $\theta \to 0$ (Polovin, 1961). Transitions between the Alfvén (A) and tangential (T) discontinuities, between the tangential discontinuity and the slow shock (S_-), between the tangential and contact (C) discontinuities are assumed to be possible. These discontinuities convert to the tangential discontinuity in the limiting case $B_x \to 0$ (see Polovin, 1961; Akhiezer *et al.*, 1975).

In Chapter 19 we shall consider the consequences of the evolutionarity conditions for reconnecting current sheets as a MHD discontinuity.

12.3 Dissipative effects in evolutionarity

The effect of dissipation on the peculiar shocks was taken into account by Roikhvarger and Syrovatskii (1974). In this case the dispersion relation of the Alfvén waves has the form:

$$k^2 V_{\mathrm{A}x}^2 - \left(\omega - kv_x - \mathrm{i}k^2 \nu_{\mathrm{m}}\right)\left(\omega - kv_x - \mathrm{i}k^2 \nu\right) = 0 . \tag{12.23}$$

Here $\mathbf{k}$ is directed along the x axis, ν_{m} is the magnetic diffusivity, and $\nu = \eta/\rho$ is the kinematic viscosity. After expansion of the solutions of this equation in powers of a small ω (the conditions under which ω is small will be discussed below) the expression for k reads as follows:
(a) for $v_x = V_{\mathrm{A}x}$

$$k^d = \pm \sqrt{\frac{\omega}{\nu_{\mathrm{m}} + \nu}}\,(1 - \mathrm{i}) , \tag{12.24}$$

$$k^{\mathrm{A}} = \frac{\omega}{2v_x} - \mathrm{i}\,\frac{(\nu_{\mathrm{m}} + \nu)\,\omega^2}{16\,v_x^3} , \tag{12.25}$$

$$k^* = -\frac{\omega\,(\nu_{\mathrm{m}}^2 + \nu^2)}{v_x\,(\nu_{\mathrm{m}} + \nu)^2} + \mathrm{i}\,\frac{v_x\,(\nu_{\mathrm{m}} + \nu)}{\nu_{\mathrm{m}}\,\nu} ; \tag{12.26}$$

(b) for $v_x \neq V_{\mathrm{A}x}$

$$k^{\mathrm{A}} = \frac{\omega}{v_x \pm V_{\mathrm{A}x}} - \mathrm{i}\,\frac{(\nu_{\mathrm{m}} + \nu)\,\omega^2}{2\,(v_x \pm V_{\mathrm{A}x})^3} , \tag{12.27}$$

$$k^* = -\frac{\omega\,[\,(\nu_{\mathrm{m}} - \nu)^2\, v_x \pm (\nu_{\mathrm{m}} + \nu)\, K\,]}{4\,V_{\mathrm{A}x}^2\,\nu_{\mathrm{m}}\nu + v_x^2\,(\nu_{\mathrm{m}} - \nu)^2 \pm v_x\,(\nu_{\mathrm{m}} + \nu)\,K} + {}$$
$$+\,\mathrm{i}\,\frac{v_x\,(\nu_{\mathrm{m}} + \nu) \pm K}{2\,\nu_{\mathrm{m}}\,\nu} , \tag{12.28}$$

where

$$K = \left[\,v_x^2\,(\nu_{\mathrm{m}} - \nu)^2 + 4\,V_{\mathrm{A}x}^2\,\nu_{\mathrm{m}}\,\nu\,\right]^2 .$$

Thus

the dissipative effects result in additional small-amplitude waves propagating in a homogeneous MHD medium.

The width of an MHD shock (at least of small amplitude) is proportional, in order of magnitude, to the dissipative transport coefficients and inversally proportional to the shock intensity (Sirotina and Syrovatskii, 1960). The intensity is determined by the difference $v_x - V_{Ax}$ on the side of the discontinuity on which it is not zero. Since the switch-off shock, as a slow one, has a finite intensity, and the switch-on shock exists in the interval (see Section 11.2.5)

$$1 < \frac{v_{x1}^2}{V_{Ax1}^2} < \frac{4v_{x1}^2}{v_{x1}^2 + V_{s1}^2},$$

the width of the peculiar shock can be estimated as

$$l \sim \frac{\nu_{\rm m} + \nu}{\mid v_x - V_{Ax} \mid} \tag{12.29}$$

(Roikhvarger and Syrovatskii, 1974). It is just this distance within which the perturbations k^* from (12.26) and (12.28) damp considerably.

Therefore outside the shock front these waves are absent, and their amplitudes do not enter into the boundary conditions which relate perturbations outside the shock front.

The situation is different for the remaining perturbations, in particular, for the *purely dissipative* waves k^d from (12.24). For small enough ω their wave numbers are much larger than the thickness l of the shock. This is true under the condition

$$\omega \ll \frac{(v_x - V_{Ax})^2}{\nu_{\rm m} + \nu}, \tag{12.30}$$

which coinsides with that used to derive (12.24)–(12.28). Since the characteristic length scale of such perturbations is much larger than the shock thickness l, their amplitudes satisfy the boundary conditions at the discontinuity surface (12.3) and (12.4) obtained for an ideal medium.

The classification of dissipative perturbations on incoming and outgoing waves should be made according to the sign of the imaginary part of the wave vector, because in a stable medium such waves damp in the direction of the propagation (Section 10.3). Consequently, there are two outgoing perturbations leaving the peculiar shock, one of them being the dissipative wave. Much like the case of non-peculiar shocks, both waves propagate downstream away from the (fast) switch-on shock, while there is one outgoing wave on each side of the (slow) switch-off shock (Figure 12.7).

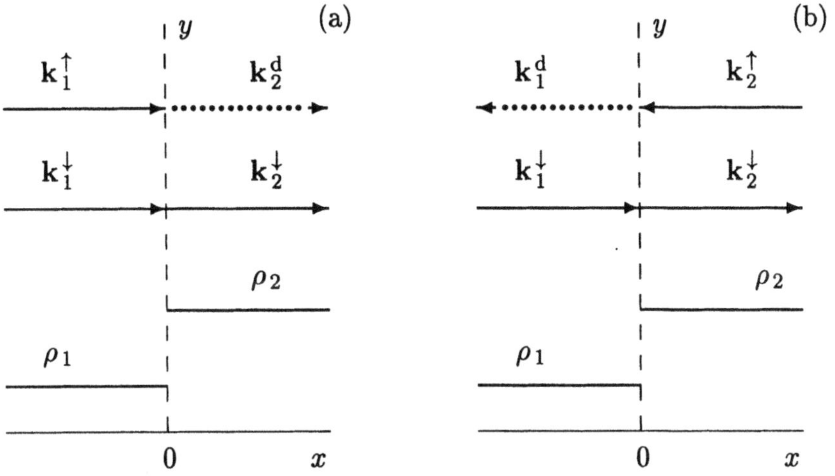

Figure 12.7: The direction of the wave propagation in the case of a switch-on shock (a) and a switch-off shock (b).

With the precision adopted when deriving (12.24)–(12.28), the perturbations δv_z and δB_z in the dissipative wave k^* from (12.24) are related by the formula

$$\delta v_z^{\,d} = \left(1 \pm \frac{\nu_{\rm m} - \nu}{v_x} \sqrt{\frac{{\rm i}\,\omega}{2\,(\nu_{\rm m} + \nu)}} \right) \frac{\delta B_z^{\,d}}{\sqrt{4\pi\rho}} \,. \tag{12.31}$$

From here and, (12.3) and (12.4), it follows that if an Alfvén wave is incident onto the switch-off shock from upstream or downstream then the amplitude of the dissipative wave equals respectively

$$\delta B_{z1}^{\,d} = -\frac{2\,v_{x1}}{\nu_{\rm m} - \nu} \sqrt{\frac{2\,(\nu_{\rm m} + \nu)}{{\rm i}\,\omega}} \;\delta B_{z1}^{\,\downarrow} \,, \tag{12.32}$$

or

$$\delta B_{z1}^{\,d} = -\frac{2\,v_{x1}}{\nu_{\rm m} - \nu} \sqrt{\frac{2\,(\nu_{\rm m} + \nu)}{{\rm i}\,\omega}} \;\delta B_{z2}^{\,\uparrow} \,. \tag{12.33}$$

The amplitude $\delta B_{z2}^{\,\downarrow}$ of the travelling (non-dissipative) wave equals zero in the first case and $-\delta B_{z2}^{\,\uparrow}$ in the second case. Thus only one solution exists for the switch-off shock. Consequently, the switch-off shock is evolutionary.

On the contrary, the switch-on shock is non-evolutionary. Indeed, Equations (12.3) and (12.4), with regard for the relation at the switch-on shock

$$v_{x1}\, v_{x2} = V_{{\rm A}x1}^{\,2} \quad \text{and} \quad \frac{\rho_2}{\rho_1} = \frac{v_{x1}^{\,2}}{V_{{\rm A}x1}^{\,2}} \,, \tag{12.34}$$

can be rewritten as

$$v_{x1}\left(\delta v_{z2}-\frac{\delta B_{z2}}{\sqrt{4\pi\rho}}\right)=v_{x1}\,\delta v_{z1}-V_{\mathrm{A}x1}\,\frac{\delta B_{z1}}{\sqrt{4\pi\rho}}\,, \tag{12.35}$$

$$V_{\mathrm{A}x1}\left(\delta v_{z2}-\frac{\delta B_{z2}}{\sqrt{4\pi\rho}}\right)=V_{\mathrm{A}x1}\,\delta v_{z1}-v_{x1}\,\frac{\delta B_{z1}}{\sqrt{4\pi\rho}}\,. \tag{12.36}$$

The set of Equations (12.35) and (12.36) is incompatible with a non-zero amplitude of the incident wave, i.e. when δv_{z1} and δB_{z1} are not equal to zero. Note that if the incident wave is absent, this set has an infinite number of solutions. Hence the switch-on shock is non-evolutionary.

Finally it should be mentioned that the additional dissipative waves appear only for $v_x = V_{\mathrm{A}x}$. This means that

> dissipative effects do not alter the evolutionarity conditions for non-peculiar (fast and slow) shock waves.

At the same time the Alfvén discontinuity becomes non-evolutionary with respect to dissipative Alfvén waves. This is consistent with the fact that in the presence of dissipation it cannot have a stationary thickness and smooths out with time (e.g., Landau, Lifshitz, and Pitaevskii, 1984).

It was also pointed out by Roikhvarger and Syrovatskii (1974) that the inclusion of dissipation into the dispersion relation for magnetoacoustic and entropy waves results in the appearence of dissipative waves, and, as a consequence, in non-evolutionarity of tangential, contact, and *weak* discontinuities (discontinuities of the derivatives of the MHD properties).

12.4 Discontinuity structure and evolutionarity

12.4.1 Perpendicular shock waves

It is natural to assume that

> the stationary problem of the structure of an evolutionary discontinuity has a unique solution, while for the non-evolutionary one this problem does not have a solution.

To illustrate this assumption let us obtain the structure of the perpendicular shock. With this aim the one-dimensional dissipative MHD equations should be integrated over x. After that the conservation laws of mass, momentum, and energy, and Maxwell equations take the form (e.g., Polovin and Demutskii, 1990):

$$\rho v = J, \tag{12.37}$$

$$Jv + p + \frac{B^2}{8\pi} - \mu \frac{dv}{dx} = S, \tag{12.38}$$

$$J \left[\frac{v^2}{2} + \frac{p}{\rho(\gamma_g - 1)} \right] + pv + \frac{vB^2}{4\pi} - \mu v \frac{dv}{dz} - \frac{\nu_m}{4\pi} B \frac{dB}{dx} = Q, \tag{12.39}$$

$$vB - \nu_m \frac{dB}{dx} = cE, \tag{12.40}$$

where the thermal conductivity of the medium is assumed to be zero. Here J, S, and Q are constants of integration, γ_g is the adiabatic index, $\mu = (4/3)\eta + \zeta$, and ζ is a bulk viscosity (the indexes x and y at the quantities v_x and B_y are omitted).

From (12.37)–(12.40) we obtain the set of ordinary differential equations which describes the structure of the perpendicular shock:

$$\mu \frac{dv}{dx} = f(v, B), \tag{12.41}$$

$$\nu_m \frac{dB}{dx} = g(v, B), \tag{12.42}$$

where

$$f(v, B) = \frac{\gamma_g + 1}{2} Jv - \gamma_g \left(S - \frac{B^2}{2\pi} \right) + \frac{\gamma_g - 1}{v} \left(Q - \frac{cEB}{4\pi} \right), \tag{12.43}$$

$$g(v, B) = vB - cE. \tag{12.44}$$

The curves $f(v, B) = 0$ and $g(v, B) = 0$ on the plane (v, B) are shown schematically in Figure 12.8. At the points 1 and 2 of intersection of these curves the derivatives dv/dx and dB/dx equal zero simultaneously. The

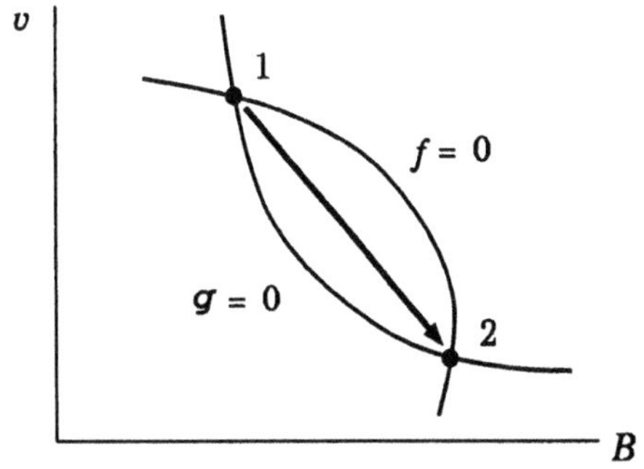

Figure 12.8: The structure of the perpendicular shock (bold arrow) connecting the states 1 and 2.

points (B_1, v_1) and (B_2, v_2) correspond to the states ahead of the shock ($x \to -\infty$) and behind the shock ($x \to +\infty$). These are stationary points of the set of differential equations (12.41) and (12.42). The structure of the shock

$$v = v\,(x)\,, \qquad B = B\,(x) \tag{12.45}$$

is a solution to the set (12.41), (12.42) which leaves the initial point 1 and enters into the final point 2.

To consider the behaviour of the integral curves in the vicinity of the stationary points 1 and 2 (Figure 12.8) the quantities J, S, and Q should be expressed in termes of the MHD properties v_i and B_i ahead of the shock ($i = 1$) and behind the shock ($i = 2$). Then, by virtue of the fact that the derivatives $\mathrm{d}v/\mathrm{d}x$ and $\mathrm{d}B/\mathrm{d}x$ tend to zero for $x \to \pm\infty$, Equations (12.37)–(12.39) yield

$$J = \rho_i\, v_i\,, \tag{12.46}$$

$$S = J v_i + p_i + \frac{B_i^2}{8\pi}\,, \tag{12.47}$$

$$Q = J\left(\frac{v_i^2}{2} + \frac{\gamma_g}{\gamma_g - 1}\,\frac{p_i}{\rho_i}\right) + \frac{v_i B_i^2}{4\pi}\,, \tag{12.48}$$

where $i = 1, 2$.

Let us now represent the quantities B and v in the form

$$B = B_i + \delta B_i\,, \qquad v = v_i + \delta v_i\,, \tag{12.49}$$

with δ being a small perturbation. Substituting this together with (12.46)–(12.48) in (12.43) and (12.44), and expanding the result in powers of δB_i and δv_i, we find to the first order

$$\mu\,\frac{\mathrm{d}\delta v_i}{\mathrm{d}x} = \frac{\rho_i}{v_i}\left(v_i^2 - V_{si}^2\right)\delta v_i + \frac{B_i}{4\pi}\,\delta B_i\,, \tag{12.50}$$

$$\nu_{\mathrm{m}}\,\frac{\mathrm{d}\delta B_i}{\mathrm{d}x} = B_i\,\delta v_i + v_i\,\delta B_i\,. \tag{12.51}$$

As is known (e.g., Fedoryuk, 1985), a stationary point $\delta v_i = 0$, $\delta B_i = 0$ of the set of autonomous differential equations

$$\frac{\mathrm{d}\delta v_i}{\mathrm{d}x} = a_{11}\,\delta v_i + a_{12}\,\delta B_i\,, \tag{12.52}$$

$$\frac{\mathrm{d}\delta B_i}{\mathrm{d}x} = a_{21}\,\delta v_i + a_{22}\,\delta B_i \tag{12.53}$$

is a saddle if the roots of characteristic equation

$$(a_{11} - \lambda)(a_{22} - \lambda) - a_{12} a_{21} = 0 \tag{12.54}$$

are real numbers and have opposite signs, i.e. if

$$(a_{11} - a_{22})^2 + 4a_{12} a_{21} > 0, \quad a_{11} a_{22} - a_{12} a_{21} < 0. \tag{12.55}$$

In this case only two integral curves enter the stationary point $\delta v_i = 0$, $\delta B_i = 0$ from the opposite directions (Figure 12.9a). And in the orthogonal way only two curves leave the stationary point.

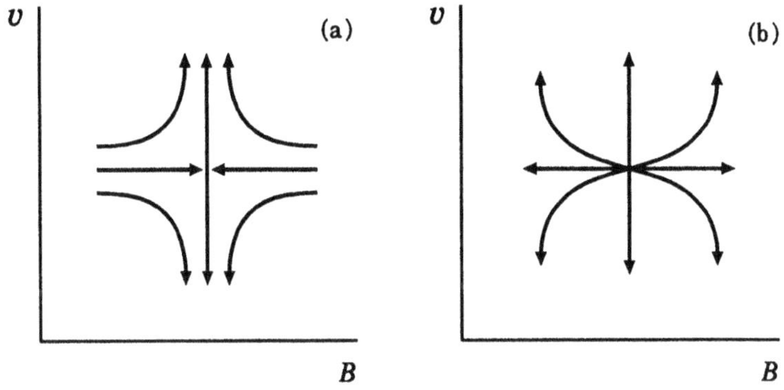

Figure 12.9: Stationary points of the set of autonomous differential equations. (a) Saddle. (b) Unstable node.

In the case when the roots of characteristic Equation (12.54) are real numbers and have the same sign, i.e. if

$$(a_{11} - a_{22})^2 + 4a_{12} a_{21} > 0, \quad a_{11} a_{22} - a_{12} a_{21} > 0, \tag{12.56}$$

then the stationary point is a node. If in addition

$$a_{11} + a_{22} > 0 \tag{12.57}$$

then the node is unstable, and all the integral curves leave the stationary point (Figure 12.9b).

In a perpendicular MHD shock

$$a_{11} a_{22} - a_{12} a_{21} = \frac{\rho_i \,(v_i^2 - V_{\perp i}^2)}{\mu \nu_{\rm m}}, \tag{12.58}$$

as follows from Equations (12.50) and (12.51). Here

$$V_{\perp} = \sqrt{V_{\mathrm{A}\,\|}^{2} + V_{s}^{2}} = \sqrt{u_{\mathrm{A}}^{2} + V_{s}^{2}}\,. \tag{12.59}$$

(Section 10.2.4). So the second inequality (12.56) is always valid. As for the quantity $a_{11} + a_{22}$, it equals

$$a_{11} + a_{22} = \frac{\rho_i \left(v_i^2 - V_{si}^2\right)}{\mu\, v_i} + \frac{v_i}{\nu_m}\,. \tag{12.60}$$

It follows from (12.58) and (12.60) that in the case of the perpendicular shock the stationary points of (12.50), (12.51) can be only of two types: either a saddle or an unstable node (recall that v_i is assumed to be positive).

Consider at first the case when

$$v_1 > V_{\perp 1}\,, \qquad v_2 < V_{\perp 2}\,. \tag{12.61}$$

Then point 2 is a saddle, while point 1 is an unstable node. The only integral curve enters into point 2 in Figure 12.8 from the side of larger values of v. If the quantities v and B vary along this curve in the opposite direction, i.e. upstream of the shock, then they will inevitably reach the values (v_1, B_1), i.e. point 1, because all integral curves leave point 1 (unstable node in the case under consideration). This curve describes a unique structure of the perpendicular shock. The inequalities (12.61) coincide with the conditions of evolutionarity of the perpendicular shock (see (12.17)), because $V_+ = V_\perp$ for perpendicular propagation. Therefore

> the conditions that the perpendicular shock wave has the unique structure coincide with the conditions of its evolutionarity.

Consider now the structure of a non-evolutionary perpendicular shock. If

$$v_2 > V_{\perp 2} \tag{12.62}$$

then point 2 is an unstable node. Neither integral curve enters this point, i.e. the problem of structure of the shock does not have a solution.

If

$$v_1 < V_{\perp 1}\,, \qquad v_2 < V_{\perp 2}\,, \tag{12.63}$$

then both stationary points 1 and 2 are saddles. In this case one of two integral curves, leaving point 1, may coinside with one of two curves entering point 2. However, this takes place only for the definite exclusive values of the parameters ahead of the shock front. An infinitesimal perturbation of the state upstream of the shock destroys its structure. In other words, the integral curve cannot connect the states 1 and 2 in a general case.

12.4.2 Discontinuities with penetrating magnetic field

Let us turn to the discontinuity type for which

$$v_x \neq 0 \quad \text{and} \quad B_x \neq 0 \tag{12.64}$$

(Sections 11.2.4 and 11.2.5). Consider at first the discontinuity accompanied by a density jump:

$$\{\rho\} \neq 0 . \tag{12.65}$$

(oblique shock waves). In this case the boundary conditions (11.64) can be rewritten in such a way as to represent the Rankine-Hugoniot relation for shock waves in MHD. Germain (1960) and Shercliff (1965) have shown that the boundary conditions allow four states

$$\begin{aligned} \text{I} &: \quad v_x > V_+ \,, \\ \text{II} &: \quad V_+ > v_x > V_{Ax} \,, \\ \text{III} &: \quad V_{Ax} > v_x > V_- \,, \\ \text{IV} &: \quad V_- > v_x \,. \end{aligned} \tag{12.66}$$

The states are arranged in order of increasing entropy. The second law of thermodynamics requires that a shock transition is possible only from a lower state of entropy to an upper one. There are thus six transitions shown in Figure 12.10.

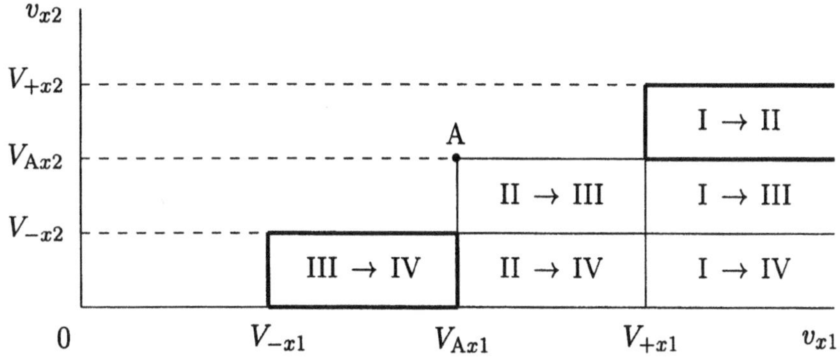

Figure 12.10: Transitions with increasing entropy. Evolutionarity domains (bold rectangles) for the fast (I → II) and slow (III → IV) shock waves.

The evolutionarity of an oblique shock is related to its structure in the following way (Germain, 1960; Kulikovskii and Lyubimov, 1961; Anderson, 1963). **The evolutionary fast and slow shocks always have a**

unique structure. The shock transition II → III has a unique structure only for the definite relationship between the dissipative transport coefficients. If these coefficients fall into the certain intervals, the I → III and II → IV shocks may have a unique structure, while the I → IV transition may be connected by an infinite number of integral curves.

Besides, as shown by Liberman (1978) with the help of the method discussed in Section 12.4.1, the switch-on shock, which is not evolutionary with respect to dissipative waves, has a unique structure. The possible reason is that the peculiarity of the switch-on and switch-off shocks is related to the absence of B_τ on one side of the discontinuity surface. The small asymmetry, that is assumed when studying the stationary points, removes the degeneration, and thus makes the shock evolutionary.

12.5 Practice: Problems and Answers

Problem 12.1. Show that an ordinary shock wave (Section 11.1.2) is evalutionary.

Answer. From (11.6) it follows that there exist three boundary conditions at the surface of a shock wave in ordinary hydrodynamics

$$\{ \rho v_x \} = 0 , \quad \left\{ p + \rho v_x^2 \right\} = 0 , \quad \left\{ \frac{v^2}{2} + w \right\} = 0 . \tag{12.67}$$

The boundary condition

$$\{ v_\tau \} = 0 \tag{12.68}$$

makes it possible to transform to such a frame of reference in which the tangential velocity component is absent on either side of the discontinuity: $\mathbf{v}_{\tau 1} = \mathbf{v}_{\tau 2} = 0$. So we obtain three linearized conditions for small perturbations. Since the disturbance of the velocity of the shock front surface δu_x can be eliminated from the set of boundary conditions, there remain two independent equations in the set.

Let us count the number of outgoing small-amplitude waves. There are no such waves upstream the shock because of the condition

$$v_{x1} > V_{s1} = 0 , \tag{12.69}$$

where V_{s1} is the upstream sound velocity. At the downstream side of the shock there are two waves: the sound wave propagating with velocity $v_{x2} + V_{s2}$ and the entropy-vortex wave (Problem 10.2) propagating with velocity v_{x2} as

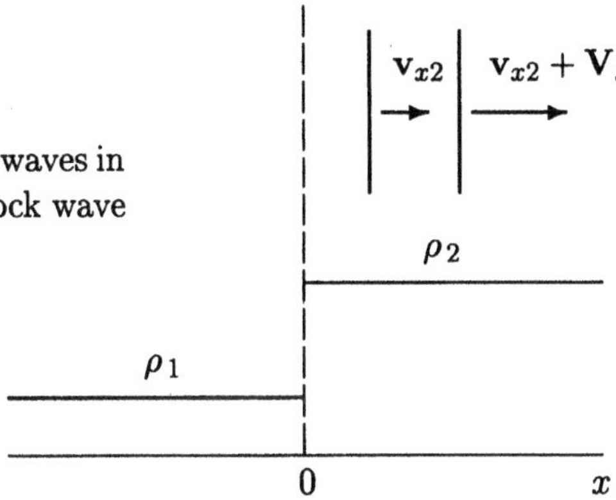

Figure 12.11: Small-amplitude waves in a plasma moving through a shock wave without a magnetic field.

shown in Figure 12.11. Therefore the number of waves leaving the shock is equal to the number of independent linearized boundary conditions. Q.E.D.

Problem 12.2. Since an ordinary shock wave is evolutionary, formulate and consider the problem of its stability in the ordinary sense of small perturbations.

Answer. Suppose that the surface of a shock is perturbed in the following way:

$$\xi = \xi_0 \exp\left[\, \mathrm{i}\, (k_y\, y - \omega\, t)\right], \tag{12.70}$$

where ξ is a displacement of the surface. The shock front thus becomes corrugated. The corrugation causes a perturbation of the flow. An arbitrary hydrodynamic perturbation is represented as a sum of the entropy-vortex wave and the sound wave. Since the flow is stationary and homogeneous in the y direction, all perturbations have the same frequency ω and tangential component of the wave vector k_y.

Since the flow velocity ahead of the shock $v_1 > V_{s1}$, only the downstream flow is perturbed. The usual condition of compatibility of the linear equation set is that the determinant of the coefficients at unknown quantities is zero, which yields the dispersion equation

$$\frac{\omega v_2}{v_1}\left(k_y^2 + \frac{\omega^2}{v_2^2}\right) - \left(\frac{\omega^2}{v_1 v_2} + k_x^2\right)(\omega - k_y v_2)\left[1 + J^2\left(\frac{\partial U_2}{\partial p_2}\right)_{\mathrm{RH}}\right] = 0\,. \tag{12.71}$$

Here $U = 1/\rho$ is a specific volume, $J = \rho_1 v_1 = \rho_2 v_2$. The subscript RH means that the derivative is taken along the Rankine–Hugoniot curve.

The shock front as a discontinuity is unstable if

$$\operatorname{Im}\omega > 0\,, \quad \operatorname{Im} k_x > 0\,. \tag{12.72}$$

The second condition (12.72) means that the perturbation is excited by the shock itself, but not by some external source. As shown by D'yakov (1954), Equation (12.71) has solution which satisfies the condition (12.72), when

$$J^2 \left(\frac{\partial U_2}{\partial p_2} \right)_{\rm RH} < -1 \tag{12.73}$$

or

$$J^2 \left(\frac{\partial U_2}{\partial p_2} \right)_{\rm RH} > 1 + 2\,\frac{v_2}{V_{s2}}\,. \tag{12.74}$$

If the parameters of the flow fall into the interval (12.73) or (12.74) then the small perturbation of the shock grows exponentially with time. This is the so-called *corrugational* instability of shock waves in ordinary hydrodynamics.

Along with this there is a possibility that Equation (12.71) has solutions with real ω and k_x which correspond to non-damping waves outgoing from the discontinuity (D'yakov, 1954). In this case

> the shock spontaneously radiates sound and entropy-vortex waves, with the energy being supplied from the whole moving medium.

Apparently this instability is the reason of the flow inhomogeneities observed, for example, in laboratory experiments when a strong shock propagates in a gas (for more detail see review by Markovskii and Somov, 1996b).

Problem 12.3. Show that an ordinary tangential discontinuity (Section 11.1.2) is non-evalutionary.

Answer. From (11.5) it follows that there exists only one boundary condition at the surface of a tangential discontinuity in ordinary hydrodynamics. However, two sound waves can propagate from the discontinuity at its both sides. Therefore the number of small-amplitude waves is greater than the number of linearized boundary equations. Q.E.D.

Chapter 13

Particle Acceleration by Shock Waves

Sir Charles Darwin (1949) presumably thought that shock waves are responsible for accelerating cosmic rays. Nowadays shocks are widely recognized as a key to understanding particle acceleration in a variety of astrophysical environments.

13.1 Two basic mechanisms

Cosmic plasmas, being tenuous, differ from laboratory plasmas in many ways (see Introduction); one of them is the following. In most environments where accelerated paricles are observed, typical sound speeds are considerably less than easily obtainable bulk flow velocities, and shock waves are expected to develop. In fact, shocks are associated with most energetic particle populations seen in space.

In the heliosphere, collisionless shocks are directly observable with spacecrafts and they have been well studied. In every case where direct observations have been made, shocks are seen to accelerate particles, often to power-law distributions. Investigations of heliospheric shocks, along with a great deal of theoretical work, also show that collective field-particle interactions control the shock dissipation and structure. The physics of shock dissipation and particle acceleration seem to be intimately related.

In this Chapter, we introduce only the most important aspects of the shock acceleration theory including two fundamental mechanisms of particle acceleration by a shock wave. Analytical models and numerical simulations (see reviews by Jones and Ellison, 1991; Blandford, 1994) illustrate the possible high efficiency of *diffusive* and *drift* accelerations to high energies.

13.2 Shock diffusive acceleration

13.2.1 The canonical model of diffusive mechanism

Axford *et al*. (1977) and Krymskii (1977) considered the idealized problem of the particle acceleration by a shock wave of plane geometry propagating in a medium containing small-scale inhomogeneities of a magnetic field which scatter fast particles. The origin of these scatterers will be discussed later on. This may be, for example, the case of parallel or nearly parallel MHD shocks. In shocks of this kind (see case (11.69)) the avarage magnetic field plays essentially no role since it is homogeneous, while fluctuations in the avarage field play a secondary role producing particle scattering. Assuming this, we consider a shock wave as an ordinary hydrodynamic shock with scatterers.

If the medium is homogeneous, and the propagation of the shock is stationary, then the front of the shock separates the two half-spaces: $x < 0$ and $x > 0$, and the velocity of the medium is given by the following formula:

$$v(x) = \begin{cases} v_1 & \text{for} \quad x < 0\,, \\ v_2 = r^{-1} v_1 & \text{for} \quad x > 0\,. \end{cases} \tag{13.1}$$

Here

$$r = \frac{\rho_2}{\rho_1} = \frac{v_1}{v_2} \tag{13.2}$$

is the compression ratio. It follows from formula (11.92) that, in a very strong (but nonrelativistic) shock wave, the ratio

$$r \to r_\infty = \frac{\gamma_g + 1}{\gamma_g - 1}$$

and

$$v_2 = \frac{v_1}{r_\infty} = \frac{\gamma_g - 1}{\gamma_g + 1}\, v_1\,. \tag{13.3}$$

The adiabatic index γ_g is considered constant on both sides of the shock front $x = 0$.

Following Axford *et al*. (1977) and Krymskii (1977), let us assume that the distribution function in space and the scalar momentum of the accelerated particles, $f(\mathbf{r}, p)$, is isotropic to first order (see discussion and generalization in Gieseler, *et al.*, 1999; Ruffolo, 1999). This means that $f(\mathbf{r}, p)$ is the same in all reference frames to first order in the small parameter v/v_p, where v_p and p are the individual particle velocity and momentum measured in the local plasma frame.

As long as scattering is strong enough to insure the isotropy assumption, the kinetic equation (2.13) describing the transport of particles with $v_p \gg v$ in space and velocity can be written in the form of a *diffusion-convection* equation (see Krymskii (1977) and references therein):

$$\frac{\partial f}{\partial t} = \nabla_{\mathbf{r}} \left(D \, \nabla_{\mathbf{r}} f \right) - \nabla_{\mathbf{r}} \left(f \mathbf{v} \right) + \frac{1}{3} \frac{\partial \left(f p \right)}{\partial p} \operatorname{div} \mathbf{v} \,. \tag{13.4}$$

Here $D = D(\mathbf{r}, p)$ is the coefficient of diffusion of fast particles.

For our problem under consideration, with one-dimensional geometry, we have in the *stationary* case

$$\frac{\partial}{\partial x} \left[v f(x,p) - D(x,p) \, \frac{\partial f(x,p)}{\partial x} \right] = \frac{1}{3} \frac{\partial v}{\partial x} \frac{\partial}{\partial p} \left[p f(x,p) \right] . \tag{13.5}$$

Let us integrate Equation (13.5) over x from $x = -\infty$ to $x = +\infty$. By employing the boundary conditions

$$f(x = -\infty, p) = f_1(p) \quad \text{and} \quad f(x = +\infty, p) = f_2(p) \,, \tag{13.6}$$

where $f_2(p)$ is an unknown spectrum of accelerated particles, we obtain the following differential equation in p

$$v_2 f_2(p) - v_1 f_1(p) - 0 + 0 = \frac{1}{3} \left(v_2 - v_1 \right) \frac{d}{dp} \left[p f_2(p) \right] . \tag{13.7}$$

Using the definition of the compression ratio (13.2), we obtain an ordinary differential equation for the downstream distribution function $f_2(p)$ in the form

$$p \, \frac{d}{dp} \, f_2(p) + \frac{r+2}{r-1} \, f_2(p) = \frac{3r}{r-1} \, f_1(p) \,; \tag{13.8}$$

recall that $r > 1$.

The general solution of this equation is

$$f_2(p) = \frac{3}{r-1} \, p^{-\gamma_p} \int\limits_{p_0}^{p} f_1(p') \, (p')^{-\gamma_p} \, dp' + c_1 p^{-\gamma_p} \,. \tag{13.9}$$

Here

$$\boxed{\gamma_p = \frac{r+2}{r-1}} \tag{13.10}$$

plays the role of the *spectral index* of the accelerated particles, c_1 is an arbitrary constant of integration which multiplies the homogenious term, the distribution function $f_1(p)$ is the far upstream spectrum of ambient particles that are accelerated by the shock, and p_0 is large enough so that the assumption $v_p \gg v$ holds.

So the solution of the diffusion-convection equation does show that a shock, propagating through a region in which fast particles are diffusing, produces a superthermal population of particles with the power-law momentum distribution

$$f_2(p) \sim p^{-\gamma_p} . \tag{13.11}$$

The property which gave the diffusive acceleration process a wide appeal is the fact that, with the simplest assumptions made above,

> the spectral index (13.10) of the accelerated particles depends only on the compression ratio r of the shock wave.

Most astrophysical shocks, since they are strong, have compression ratios constrained to a rather narrow range of values near $r_\infty = 4$ assuming $\gamma_g = 5/3$. For a shock with Mach number M (see Problem 11.5) greater than 3 say, as we see in Figure 13.1, the compression ratio $3 < r < 4$ and the spectral index $2 < \gamma_p < 2.5$.

A spectral index of $\gamma_p \approx 2$ is characteristic of energy particle spectra observed in a wide range of astrophysical environments (see Jones and Ellison, 1991; Blandford, 1994). For example, $\gamma_p \approx 2$ closely fits the inferred source spectrum of **galactic cosmic rays** for high energies below approximately 10^{15} eV (e.g., Gombosi, 1999).

In the **solar wind** the shock-associated low-energy-proton events seem to be well studied. The most intensive of them have a power-low energy spectrum, suggesting that protons are accelerated by the diffusive-shock acceleration mechanism (e.g., Rodriguez-Pacheco *et al.*, 1998). Nevertheless, the correlation between the spectral exponent γ with the solar wind velocity compression ratio is found to be linear. This result differs from that presented above. The discrepancy of the spectral-exponent dependence on the shock-wave parameters could lie on the event selection criterion or on the account of nonlinear effects (Section 13.2.3) or on another mechanism of acceleration.

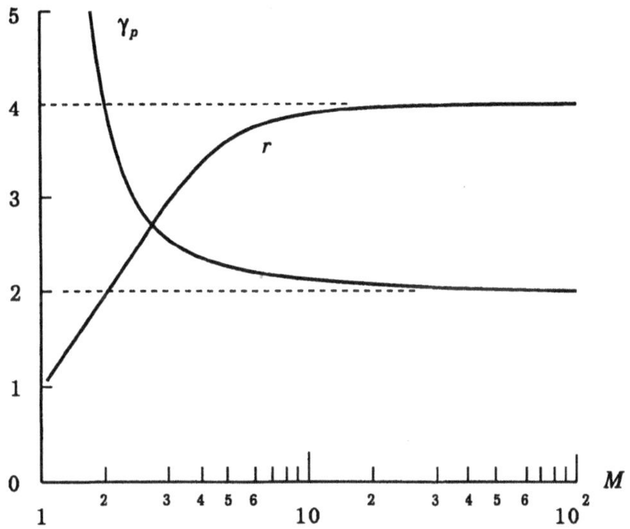

Figure 13.1: The compression ratio r and spectral index γ_p versus the Mach number M.

13.2.2 Some properties of diffusive mechanism

As we have seen above, the spectral index γ_p of energetic particles produced by diffusive shock acceleration does not depend on the diffusion coefficient D. However the diffusion coefficient D, together with the characteristic flow velocity $v \sim v_1$, determines the overall length scale of the acceleration region

$$l_D \sim D(p)/v \tag{13.12}$$

and acceleration time

$$t_D \sim D(p)/v^2 \, . \tag{13.13}$$

Particle energies are derived just from the relative motion (the converging flow with velocity $v_1 - v_2$) between scatterers (waves) on either side of a shock front.

This is a main advantage of the diffusive mechanism. Its disadvantage in applying it to some astrophysical phenomena consists of the lack of actual knowledge about the assumed scattering waves. However, as we saw above, diffusion determines only the length scale (13.12) and characteristic time (13.13) of the acceleration process.

In this context, let us recall once more (Section 11.1.3) the following analogy from everyday life. A glass of hot water with a temperature T_1 will invariably cool to a given room temperature T_2, independently of the mechanism of heat exchange with the surrounding medium, while the mechanism determines only the time of cooling.

* * *

In the presence of a magnetic field in cosmic plasma, diffusive shock acceleration requires that the particles are able to traverse the shock front in both directions either along the magnetic field or by scattering across the field, in order that they may couple to the shock compression by pitch-angle scattering both upstream and downstream of the shock. At quasi-parallel shocks this condition on particle mobility is easily met. For sufficiently fast shocks, downstream shock-heated particles can be kinematically able to return to the shock along the downstream magnetic field to initiate the process of diffusive shock acceleration. At quasi-perpendicular shocks (Section 13.3.2), however, this condition is stringent. Although the diffusive mechanism is rapid since particles are confined closer to the shock front, there is a **high threshold speed**, significantly exceeding v_1, in order that diffusive shock acceleration can occur (e.g., Webb *et al.*, 1995).

13.2.3 Nonlinear effects in diffusive acceleration

The test particle (i.e., linear) model demonstrated above is straightforward and yields the most important result, namely that the power law (13.11) with the spectral index (13.10) which is relatively insensitive to the ambient conditions is the natural product of the diffusive acceleration mechanism in shock waves. The equally important question of the actual efficiency of the process can only be adequately addressed to a fully *nonlinear* (and therefore complex) theory.

Using observations of the Earth bow shock and interplanetary observations, numerical modeling of different shocks (see Ellison et al., 1996, and references therein) shows that the inherent efficiency of shock acceleration implies that

> hydrodynamic feedback effects between the accelerated particles and the shock structure are very important

and therefore essential to any complete description of the process. This has turned out to be a formidable task because of the wide range of spatial and energy scales that must be self-consistently included in numerical simulations.

On the one hand, the plasma microprocesses of the shock dissipation control injection from the thermal population. On the other hand, the highest energy particles (extending to $10^{14} - 10^{15}$ eV in the case of galactic cosmic rays) with extremely long diffusion lengths (13.12) are dynamically significant in strong shock waves and feed back on the shock structure. Ranges of interacting scales of many orders of magnitudes must be described self-consistently.

13.3 Shock drift acceleration

The principal process whereby a particle gains energy upon crossing a shock wave with a magnetic field may be the so-called shock drift acceleration (Hudson, 1965; Alekseyev and Kropotkin, 1970; see also Toptyghin, 1980).

The drift mechanism, in contrast to the diffusive one, neglects any shock-front associated turbulence. So many not-well-justified assumptions concerning the physics of scatterers have not to be made in applying the drift acceleration model to an astrophysical phenomenon.

If the fast particle Larmor radius (Section 4.1.2)

$$r_{\mathrm{L}} = \frac{c p_{\perp}}{eB} \gg l_f \, , \tag{13.14}$$

where l_f is the front thickness, we can replace the shock by a simple discontinuity and can approximate the particle motion as scatter-free on both sides of the shock. Let us begin by considering an interaction of individual particles with such a discontinuity. We shall consider very fast particles:

$$v_p \gg v_1 > v_2 \, . \tag{13.15}$$

These assumptions are basic for further considerations that we start from the simplest case – a perpendicular shock (Section 11.2.3).

13.3.1 Perpendicular shock waves

As shown in Figure 11.5, the magnetic fields $\mathbf{B}_1$ and $\mathbf{B}_2$ are parallel to the shock front $x = 0$; and plasma moves perpendicularly to the front. According to (11.38), there exists an identical electric field on both sides of the shock:

$$\mathbf{E} = -\frac{1}{c}\, \mathbf{v}_1 \times \mathbf{B}_1 = -\frac{1}{c}\, \mathbf{v}_2 \times \mathbf{B}_2 \, . \tag{13.16}$$

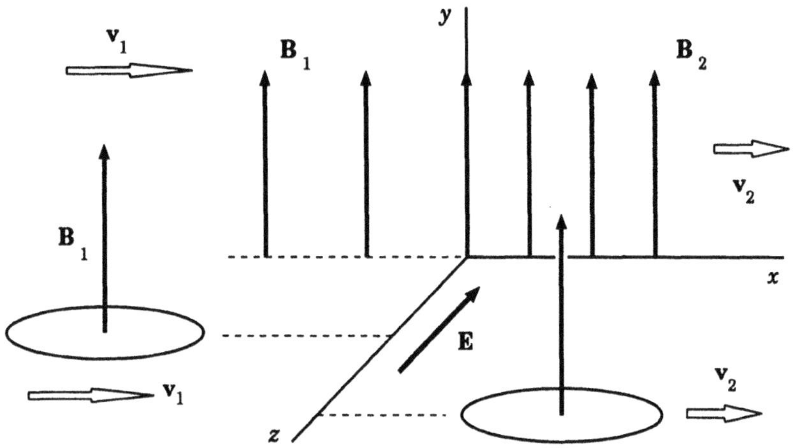

Figure 13.2: The Larmor ring moves together with the plasma and the magnetic field across the perpendicular shock front.

The fast particles rotate on the magnetic field lines and move together with the field lines with the plasma speed across the front as shown in Figure 13.2.

Nothing will happen before the Larmor ring touches the front; a particle simply drifts to the front. For what follows it is important that the particle will make many rotations (Figure 13.3) during the motion of the Larmor ring across the front because of the condition (13.15). Because of the difference between the Larmor radius ahead of and behind the front, a drift parallel to the front will appear, accompanying the drift across the front.

During each rotation, the electric field $\mathbf{E}$ accelerates a particle on the upstream side ($x < 0$) of the shock and decelerates it on the downstream side ($x > 0$). However, the work of the field $\mathbf{E}$ on a larger circle exceeds the work on the smaller circle:

$$\delta A_1 = + eE \times AC > - \delta A_2 = eE \times BC \, , \tag{13.17}$$

since the length AC is larger than the length AB. Therefore, during each rotation, the particle is slightly accelerated. How much energy does the particle take during the motion of its Larmor ring across the shock front?

Since we consider the shock as a discontinuity, the adiabatic approximation is formally not suitable. However it appears that the *transversal* invariant (Section 4.3) conserves:

$$\frac{p_{\perp}^2}{B} = \text{const} \tag{13.18}$$

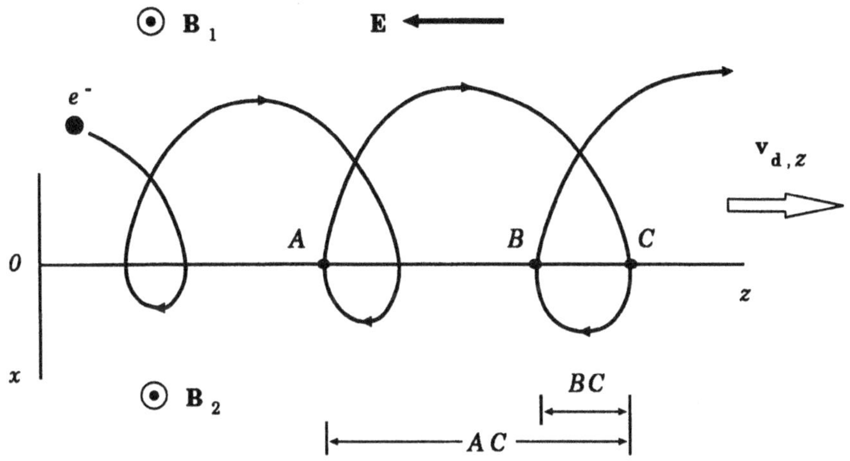

Figure 13.3: The trajectory of a negatively charged particle (an electron) multiply crossing the perpendicular shock front.

(Hudson, 1965; Alekseyev and Kropotkin, 1970). From (13.18) it follows that

$$p_{\perp 2}^2 = p_{\perp 1}^2 \times \frac{B_2}{B_1} .$$

Therefore the transversal kinetic energy of a nonrelativistic particle

$$\frac{\mathcal{E}_{kin\perp 2}}{\mathcal{E}_{kin\perp 1}} = \frac{p_{\perp 2}^2}{p_{\perp 1}^2} \propto \frac{B_2}{B_1} = r . \qquad (13.19)$$

An increase of transversal energy (13.19) is relatively small when the Larmor ring of a particle crosses the front only once. Multiple interactions of a particle with the shock is a necessary condition for a considerable increase of energy.

Drift acceleration typically involves several shock crossings and results from a net displacement δz of an ion (electron) guiding center parallel (anti-parallel) to the convection electric field **E**. The energy gain is proportional to this displacement, which in general depends upon the plasma and shock parameters, the particle species and velocity, and the intensity of possible electromagnetic fluctuations in the vicinity of the shock as well as within the shock front itself.

It is popular to discuss the displacement δz as the consequence of a gradient drift (see formula (5.14) in Jones and Ellison, 1991). Such a treatment is not reasonable, at least when we consider the shock as a discontinuity; so, formally $\nabla B \to \infty$. A wondeful thing is that the adiabatic approximation is not applicable for such a situation but the first adiabatic invariant (13.18) conserves.

13.3.2 Quasi-perpendicular shock waves

(a) Classical model

The basic aspects of drift acceleration of fast particles by an almost perpendicular shock wave, as a discontinuity, emerge from a simple model which allows us to derive analytical expressions for the reflection and transmission coefficients, the energy and the angular distributions (e.g., Toptyghin, 1980). By definition, in a quasi-perpendicular shock, the angle ψ_1 (Figure 13.4) between the shock normal **n** and the upstream magnetic field vector $\mathbf{B}_1$ is greater than about 80°. Hence the field lines form small angles α_1 and α_2 with the shock

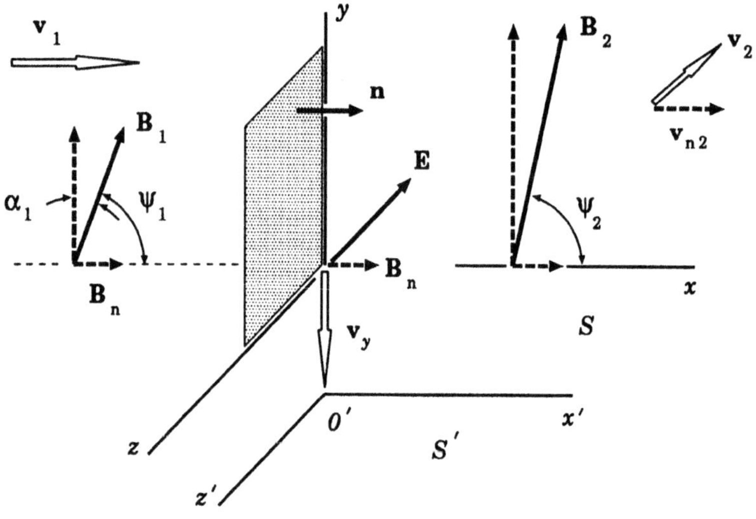

Figure 13.4: A quasi-perpendicular shock wave in the frame of reference S where $\mathbf{v}_1 \parallel \mathbf{n}$.

front plane $x = 0$. Under this condition, as well as for the perpendicular shock case considered above, **the first adiabatic invariant is conserved** (Hudson, 1965; Alekseyev and Kropotkin, 1970). This enables analytical calculations of the energy increase on the front of a quasi-perpendicular shock as well as the reflection and transmission of fast particles (e.g., Sarris and Van Allen, 1974).

Since the particles conserve the first adiabatic invariant (Section 4.3.2), all particles with pitch angles

$$\theta > \theta_0 \tag{13.20}$$

will be reflected. To find the critical pitch angle θ_0, consider two frames of reference: S and S'.

In the frame of reference S, where the shock front is in the plane (y, z) and the shock normal n is along the x axis, there is an electric field

$$\mathbf{E} = -\frac{1}{c}\,\mathbf{v}_1 \times \mathbf{B}_1 = -\frac{1}{c}\,\mathbf{v}_2 \times \mathbf{B}_2 \,. \tag{13.21}$$

In the frame of reference S', where $\mathbf{B}_1 \parallel \mathbf{v}_1$ and $\mathbf{B}_2 \parallel \mathbf{v}_2$ (Section 11.2.4), there is no electric field. The system S' moves along the y axis (perpendicular to the vector $\mathbf{E}$) with velocity

$$\mathbf{v}_y = c\,\frac{\mathbf{E} \times \mathbf{B}_n}{B_n^2}\,, \tag{13.22}$$

where $\mathbf{B}_n$ is the normal component of the magnetic field. Since $\mathbf{E}' = 0$, there is no change in the energy of a fast particle after reflection from the front: $\delta\mathcal{E}' = 0$.

We shall assume that B_n is very small but $v_y < c$. Using the relativistic Lorentz transformation for the energy-momentum 4D-vector with condition $\delta\mathcal{E}' = 0$, we obtain the relative energy increment of the reflected fast particles (see Problem 13.1):

$$\frac{\delta\mathcal{E}_{kin}}{\mathcal{E}_{kin}} \approx \frac{4v_1^2}{v_p^2}\left[\frac{v_p\cos\theta}{v_1} + \mathrm{tg}\,\psi_1\right]\mathrm{tg}\,\psi_1 \,. \tag{13.23}$$

Here $\mathcal{E}_{kin} = mv_p^2/2$ is the kinetic energy of a particle in the shock wave frame of reference S, v_p is the particle velocity in the same frame, and θ is the pitch angle also in the frame S. The connection between θ' and θ is given by

$$\cos\theta' = \frac{v_p\cos\theta + v_1\mathrm{tg}\,\psi_1}{\left[v_p^2 + (v_1\mathrm{tg}\,\psi_1)^2 + 2v_p\,(v_1\mathrm{tg}\,\psi_1)\,\cos\theta\right]^{1/2}}\,. \tag{13.24}$$

In the S' frame of reference, where the electric field is zero, the first adiabatic invariant can be written as (see definition (4.54)):

$$\frac{\sin^2\theta'}{B} = \mathrm{const}\,. \tag{13.25}$$

So the critical pitch angle θ_0' satisfies equation

$$\sin^2 \theta_0' = \frac{B_1}{B_2} \, . \tag{13.26}$$

This allows us to calculate the critical pitch angle θ_0 in the shock-front frame S. For example, if a non-relativistic proton has an initial energy $\mathcal{E}_{kin}$ = 0.3 MeV and if a shock wave has an upstream velocity v_1 = 150 km/s, the ratio $B_1/B_2 = 1/3$, and the angle $\psi_1 = 88°$ and 89°, then we find, correspondingly, $\theta_0 = 55°$ and 77°. As the angle ψ_1 increases toward 90°, most of the particles are really transmitted into the downstream side. At $\psi_1 = 90°$, which is the perpendicular shock case, there are no reflecting particles.

Formula (13.23) shows that

> the relative increment of kinetic energy of a fast particle increases when the angle ψ_1 increases toward 90°.

The model under consideration predicts **high field-aligned anisotropies** for a large ψ_1 because of conservation of first adiabatic invariant and the large energy gains.

It is widely believed that the slow **thermal particles** inside the shock front can also be considered as adiabatic, at least, in thick collisionless shocks: the electron magnetic moment is conserved throughout the shock and $v_\perp^2/B$ = const (Feldman *et al.*, 1982). In very thin collisionless shock (with a large cross-shock potential) the adiabaticity may break down, so that electrons become demagnetized. It means that the magnetic moment is no longer conserved, and a more substantial part of the energy may be transferred into the perpendicular degree of freedom (Balikhin *et al.*, 1993; Gedalin and Griv, 1999).

(b) Some astrophysical applications

Observations of interplanetary shocks (e.g., Sarris and Van Allen, 1974; Balogh and Erdös, 1991) show that the intensive acceleration of protons occurs when the upstream magnetic field is almost parallel to the shock front. Energetic particles entering the shock front stay with it, crossing it many times and being accelerated by the electric field of the front. After the direction of the interplanetary magnetic field changes again away from the parallel to the front, the intensive acceleration ceases.

Owing to interplanetary magnetic field fluctuations the upstream field vector $\mathbf{B}_1$, if it is found to be parallel to the shock front, stays as such for only a short time (a few minutes, in general). This time is enough for the low-energy protons ($\mathcal{E}_{kin,p} < 1$ MeV) to be accelerated to about 2–3 times their original energy but not enough for the high-energy protons ($\mathcal{E}_{kin,p} < 10$ MeV) to be noticeably affected by the shock wave.

Single scatter-free shock drift interactions at quasi-perpendicular shocks can accelerate particles to at most a few times the shock compression ratio. Weak scattering during single drift interactions can increase this upper limit for a small fraction of an incident particle distribution, but the energy spectra will be still rather steep. One anticipates large energy gains and flatter spectra that extend to high energies if some particles can return to the shock for many drift interactions.

This is suggestive of the classical case of a **collapsing magnetic trap** (Section 4.3.2), and is the basis of the model of proton trapping and acceleration due to multiple drift interactions along magnetic loops that convect through a planar quasi-perpendicular shock (Gisler and Lemons, 1990; Balogh and Erdös, 1991). Figure 13.5 represents a quasi-perpendicular shock, with

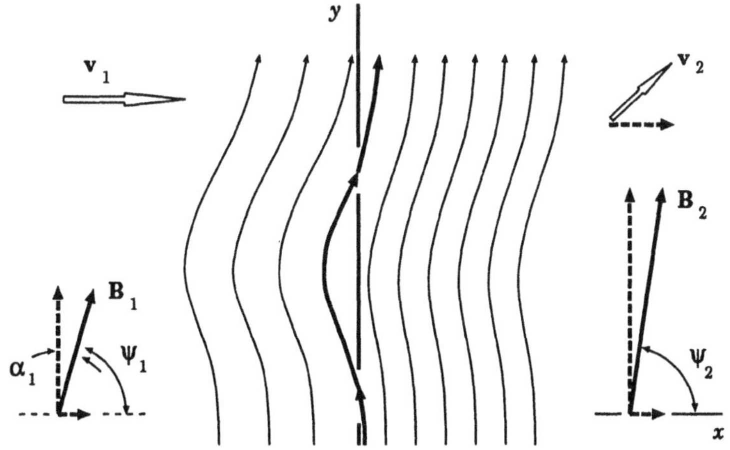

Figure 13.5: A magnetic trap on the upstream side of a quasi-perpendicular shock wave.

a small perturbation of the magnetic field superimposed on the unshocked homogeneous field $\mathbf{B}_1$. The heavy line displays a particular field line which intersects the shock front plane $x = 0$ two times, forming a magnetic loop in the upstream region.

Upstream particles bounce back and forth along a loop and gain parallel energy at each reflection until they fall within the loss cone and transmit downstream. A detailed numerical study has shown that the collapse of the trap by the convection of the loop field lines through the shock is accompanied by a considerable increase of the accelerated proton flux, which may be responsible for the 'shock spike' events observed near fast mode interplanetary shock waves (Erdös and Balogh, 1994).

13.3.3 Oblique shock waves

If values of the angle α between the magnetic field and the shock front plane are arbitrary, then the phase-averaged coefficients of reflection and transmission are complicated and can be found, in principle, by numerical calculations. When

$$\frac{v_1}{v_p} \leq \alpha_1 \leq \frac{\pi}{2} \tag{13.27}$$

and the pitch angle θ is arbitrary, the order of magnitude of the energy increase

$$\delta \mathcal{E} \approx \frac{p\, v_1}{\alpha_1} \ll \mathcal{E} - mc^2 \tag{13.28}$$

is small in comparison with the initial kinetic energy. In a general case, the increase of particle energy is small when the Larmor ring of a particle crosses the front once. Multiple interactions of a particle with the shock front is the necessary condition for a considerable increase of energy.

One possibility for multiple interactions of a particle with the shock is a strong supersonic turbulence. More exactly, it is assumed that in a sufficiently large region of space there exists an ensemble of MHD shocks which interact successively with the particles. The investigation of particle acceleration by a random shock wave ensemble is of certain interest in astrophysical applications but the conditions of such an acceleration mechanism are not totally clear yet.

Another possibity is the propagation of one shock in a turbulent medium. It is important to emphasize, however, that the particle acceleration near the shock front in a turbulent medium, i.e. the diffusive mechanism (Section 13.2) will take place in the absence of a regular electric field. No terms should be added to the basic diffusion-convection equation (13.4) to take account of the shock drift mechanism in an oblique shock. This process is already included in the energy change which is proportional to the div $\mathbf{v}$ term. This, of course, assumes that there is sufficient scattering and that other assumptions used in deriving the diffusion-convection equations are also valid. That is not trivial.

The interesting possibility discussed in Section 13.3.2 is a combination of a magnetic trap with an oblique shock wave. In the next Section, this idea is applied to the particle acceleration problem in solar flares.

13.4 The collapsing trap effect in solar flares

Magnetic reconnection plays a key role in the dynamics of flares, it serves as a highly efficient engine to convert magnetic energy into thermal and kinetic energies of plasma flows and accelerated particles (Section 16.5). The collisionless reconnection theory – more exactly, the model of a high-temperature turbulent-current sheet (Section 17.3) – under the coronal conditions derived from the *Yohkoh* data on the site and mechanism of magnetic energy transformation into kinetic and thermal energies of 'superhot' ($T_e \geq 30$–40 MK) plasma and accelerated particles – shows that the high-temperature turbulent-current sheet (HTTCS) can be considered as the source of flare energy and, at least, the first-step mechanism in a two-step acceleration of electrons and ions to high energies (Somov and Kosugi, 1997).

13.4.1 Fast plasma outflows and shocks

Fast outflows of superhot collisionless plasma create complicated dynamics in an external (relative to the HTTCS) region; this dynamics should be a topic of special research. It is clear, however, that the interaction of the fast flow of superhot plasma with an external plasma and magnetic field strongly depends on the initial and boundary conditions, especially on the relative position of the outflow source (the HTTCS) and the magnetic 'obstacle' – the region of the strong external field. Near the boundary of this region the energy density of the fast outflow becomes equal to the energy density of the external magnetic field which tries to stop the flow. In Figure 13.6 the magnetic obstacle is shown as a shadowed loop above two sunspots N and S in the photosphere Ph.

Not exactly the same but something similar was observed by the Soft X-ray Telescope (SXT) on the *Yohkoh* during the long-duration event (LDE), the limb flare in 1999 January 20. Images from the SXT show the formation of a large arcade of magnetic loops, similar to many arcades observed by the SXT, as well as high-speed flows in the region immediately above the flare loops (McKenzie and Hudson, 1999). Downward-traveling dark voids appear in the X-ray images. They presumably represent the cross-section of flux tubes; their downward motion would be interpretable as shrinkage of the

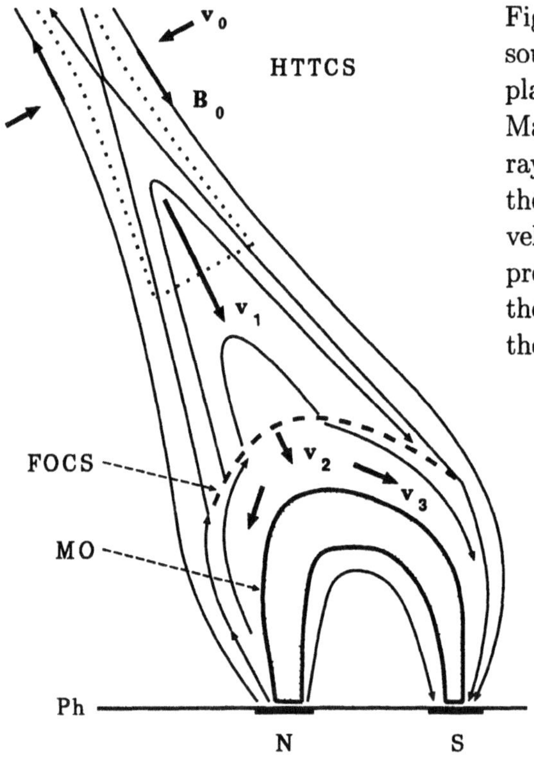

Figure 13.6: A HTTCS as the source of flare energy; superhot plasma outflow with velocity $\mathbf{v}_1$. Magnetic obstacle is the soft X-ray loop shown by shadow. $\mathbf{v}_2$ is the postshock velocity, $\mathbf{v}_3$ is the velocity of expansion of the compressed and heated plasma along the field lines toward the feet of the loop.

field lines due to magnetic tension. Some of the voids slow down and stop as they approach the top of the arcade.

Let us assume that the distance l_1 between the source of a fast outflow (an edge of the HTTCS) and the stagnation point O at the obstacle is not too large (Figure 13.7). This means that the outflow becomes *wider* but does not relax in the coronal plasma before reaching the obstacle. Moreover, if the flow velocity still exceeds the local fast magnetoacoustic wave velocity, a fast MHD shock appears ahead the obstacle, which is similar to the terrestrial bow shock ahead the magnetosphere.

By analogy with the ordinary hydrodynamics of supersonic flows, we assume that the shock front reproduces the shape of the obstacle smoothly and on a larger scale (Figure 13.7), more exactly, the shape of the upper part of the obstacle facing the incoming flow. This is true if the incoming flow is uniform or quasi-uniform. Generally, the incoming flow may significantly differ from a quasi-uniform one. In this case, the shock may have a more complicated shape. This is, however, not of crucial importance for the effect of the collapsing magnetic trap discussed below. For simplicity, in Figure 13.7, we

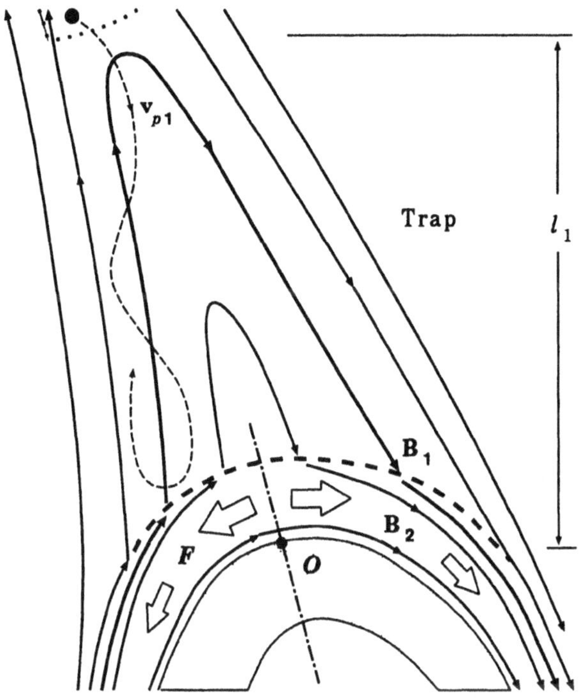

Figure 13.7: A magnetic trap between the HTTCS and the shock front; accelerated particles move with velocity $\mathbf{v}_{p1}$ along the field lines. Big arrows F show heat fluxes, directed along the field lines.

assume that all the field lines ejected by the HTTCS penetrate through the shock. This means that all superhot plasma and all particles pre-accelerated by the HTTCS, being frozen into the reconnected field lines, interact with the shock.

For what follows the most important point is that with respect to the particles pre-accelerated and to superhot particles energized by the HTTCS, the shock should be considered as a *fast oblique collisionless* shock (FOCS).

13.4.2 Particle acceleration in collapsing trap

Being frozen into superhot plasma, the reconnected field lines move out of the HTTCS and form magnetic loops at the height l_1 above the magnetic obstacle. The top of each loop moves with a high velocity $v_1 \approx 1400 - 2000$ km s^{-1}. The local fast magnetoacoustic wave speed ≈ 1000 km s^{-1}. So, a fast shock may appear between the HTTCS and the obstacle. Let us assume that both feet of a loop penetrate through the shock front ahead the obstacle.

Depending on the velocity and pitch-angle, some of the particles pre-accelerated by the HTTCS may pass directly through the magnetic field jump related to the shock. Others may either be simply reflected by the shock or interact with it.

For the particles reflected by the shock the magnetic loop represents a trap whose length decreases from $\approx 2l_1$ to zero (*collapses*) with the velocity $\approx 2v_1$. Therefore the lifetime of each magnetic field line – of each collapsing trap – is equal to

$$t_1 \approx l_1/v_1 \approx 10 \text{ s}, \tag{13.29}$$

if $l_1 \approx 10^4$ km and $v_1 \approx 10^3$ km s^{-1} are taken as the characteristic values for the length and velocity.

During the trap lifetime t_1 the reflected fast particles move between two magnetic corks – the reflecting points where the field line crosses the shock front. Since these corks move to each other with the high velocity $2v_1$, the particles trapped inside the trap are 'heated' quickly by the first-order Fermi mechanism.

For the electrons pre-accelerated by the HTTCS we estimate the characteristic value of the velocity as $v_{e,1} \approx 10^{10}$ cm s^{-1}. Hence the characteristic time between two subsequent reflections of a particle can be estimated as

$$\tau_1 \approx 2l_1/v_{e,1} \approx 0.2 \text{ s}. \tag{13.30}$$

Since $\tau_1 \ll t_1$, the conditions of the periodic longitudinal motions change ***adiabatically*** slowly (Section 4.3.1). Then the ***longitudinal*** adiabatic invariant is conserved (Section 4.3.2):

$$I = \oint p_{\|}\, dl \approx p_{\|}(t) \cdot 4\,l(t) = \text{const}. \tag{13.31}$$

Here $p_{\|} = p \cos\theta$ is the particle longitudinal momentum, θ is its pitch angle. From (13.31) it follows that

$$p_{\|}(t) = p_{\|}(0)\, \frac{l_1}{l(t)} \approx p_{\|}(0)\, \frac{1}{1-(t/t_1)}\,. \tag{13.32}$$

> When the magnetic trap collapses, the longitudinal momentum grows ***infinitely*** within the ***finite*** lifetime t_1.

Neglecting an unknown change of the transversal momentum, we see that the particle kinetic energy increases within the time scale t_1:

$$\mathcal{E}_{kin} = \frac{p^2}{2m} = \mathcal{E}_{kin}(0)\, \frac{1}{[1-(t/t_1)]^2}\,. \tag{13.33}$$

That is why we assume that just the trap lifetime t_1 can be responsible for the observed few-second delay in the higher energies of the hard X-ray (HXR) and gamma-ray emission (Bai et al., 1983).

The main objection usually raised against Fermi acceleration is that the Fermi mechanism is 'neither efficient nor selective'. A magnetic mirror reflects particles on a non-selective basis: thermal particles may be reflected as well as supra-thermal ones. Hence most of the primary energy – the kinetic energy of the fast flow of superhot plasma – goes into bulk heating of the plasma rather than the selective acceleration of only a small minority of the fast particles. This 'disadvantage' appears to be the main *advantage* of the Fermi mechanism when applied to solar flares in the frame of the collapsing trap model (Somov and Kosugi, 1997).

First, the collapsing trap heats and compresses the superhot plasma. Thus it becomes more visible in HXR emission. Second, the same Fermi mechanism can lift some electrons from a quasi-thermal distribution and accelerate them to higher energies; even better, it can further accelerate the electrons pre-accelerated by the HTTCS. The trap of the accelerated electrons can be seen as the non-thermal component of the coronal HXR source in flares. Third,

> being non-selective, the collapsing magnetic trap can accelerate not only electrons but also protons and other ions to high energies.

This is a big problem for many other acceleration mechanisms.

Superhot plasma trapped inside the collapsing loops certainly also contribute to the HXR and radio emission above the soft X-ray (SXR) emitting loop. The total coronal HXR emission consists of two parts: non-thermal and quasi-thermal. The model predicts, however, a significant difference between them. Being more collisional, the superhot plasma is less confined inside the trap. For this reason the non-thermal emission dominates at higher energies and occupies a more compact 'vertical' (Figure 13.8) HXR source in comparison with more extended 'horizontal' distribution of a quasi-thermal emission at lower energies. This seems to be consistent with the *Yohkoh* results (e.g., Tsuneta *et al.*, 1997).

Electron acceleration in the collapsing trap seems to be consistent with the results of the wavelet analysis of the solar flare HXR (Aschwanden *et al.*, 1998). This analysis yields a dynamic decomposion of the power at different timescales τ. The lifetime t_1 may correspond to the dominant peak time τ_{peak} detected in the wavelet scalegrams. The collapsing trap scenario is also consistent with the observed correlations, because the acceleration time is proportional to the spatial size of the collapsing trap ($\tau_{\text{min}} \sim l_1$).

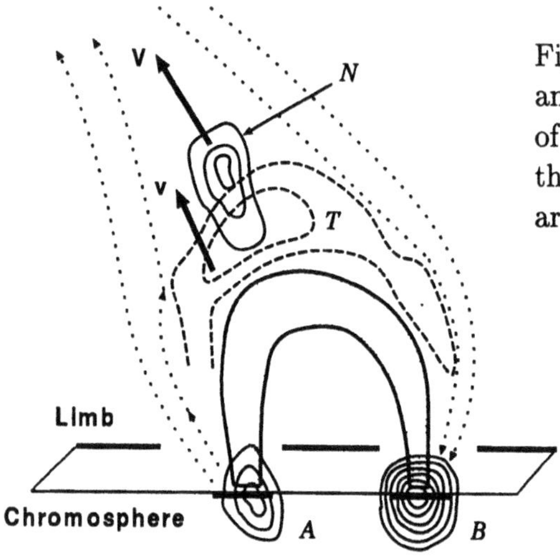

Figure 13.8: The non-thermal (N) and quasi-thermal (T) components of the coronal HXR emission and their apparent motion. A and B are the chromospheric footpoints.

13.4.3 The upward motion of coronal HXR source

Further development required for the trap model is a quantitative consideration of the upward motion of the coronal hard and soft X-ray sources predicted by the model (Somov *et al.*, 1999). It is clear that the superhot plasma heated and compressed inside the trap will unavoidably relax in the downstream flow behind the shock. This relaxation is strongly influenced by thermal conductive cooling, hydrodynamic expansion as well as by radiative energy losses. The dynamics of relaxation may not be simple and will depend on the initial and boundary conditions.

The behaviour of the magnetic field behind the shock seems to be more determined – the incoming field lines simply accumulate between the obstacle and the shock. Hence the shock must move upward together with the HXR source in the upstream side (Figure 13.8) and the SXR source in the downstream side.

In the adiabatic approximation, the postshock pressure reach extremely high values. As a result, the shock is accelerated to speeds of order 1000 km/s. This value exceeds by two orders of magnitude the upward speed of the coronal HXR source observed in flares, which usually does not exceed 10–20 km/s.

Postshock energy losses considerably change shock parameters. Bogachev *et al.*, (1998) have considered three mechanisms of energy losses from the shock-compressed superhot plasma: anomalous heat conduction, hydrodynamic expansion, and radiation. According to estimates, timescales of the first two processes do not exceed a few seconds, whereas radiative losses are much slower and can be initially neglected.

A fast removal of heat from the postshock superhot plasma and its expansion lead to a considerable decrease of the temperature and, as a consequence, of the gas pressure. As a result, the shock speed v_2 noticeably decreases. For large flow speeds v_1, the shock speed v_2 is proportional to the Alfvén speed upstream, i.e. directly proportional to the field B_1, frozen into the plasma, and inversely proportional to the square root of electron number density n_1. In particular, if we adopt $n_1 \approx 2 \times 10^9$ cm^{-3} and $B_1 \approx 0.5$ G, then the shock is moving at a speed of order 10 km/s, which coincides with the observed upward speed. Of course, this combination of n_1 and B_1 is not unique; we give it here just as the most plausible one on the basis of the *Yohkoh* observations.

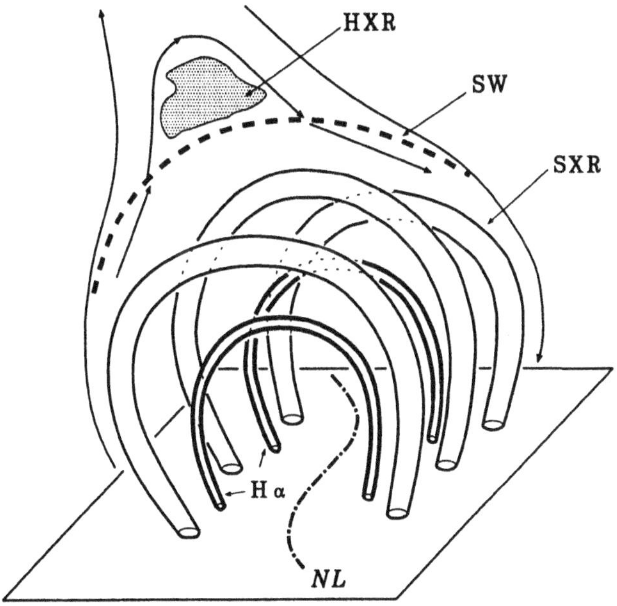

Figure 13.9: The two level structure of the SXR and Hα loops in the corona, created as a result of an instability of the magnetic obstacle.

However, if we assume higher densities of the flow, we have to assume stronger fields frozen into superhot plasma. This is acceptable. On the other hand, the shock speed only very weakly depends on the temperature and on the upstream speed. For this reason, a considerable uncertainty in these quantities (especially in the latter one) practically does not affect the results. Moreover, taking into account that the magnetic obstacle is not ideal (Somov *et al.*, 1999) and hence some of plasma with the frozen-in field can 'filter through' it (Figure 13.9) with speeds $v_4 \approx v_2$, allows us to obtain better agreement of the upward shock speed v_2 with observations for stronger magnetic fields in the corona above the shock.

To conclude, a fast MHD or collisionless shock wave with heat-conduction cooling of the postshock plasma may play an important role in the dynamics of a coronal source of HXR during a solar flare. The upward speed of the shock is determined by two processes: accumulation of magnetic flux behind the shock and 'filtering' of cold dense filaments (toghether with the frozen-in field) through the magnetic obstacle. This scenario agrees well with the observed hierarchy of hot (SXR) and cool (Hα) loops. For a more detailed comparison of the observed distributions of temperature and emission measure of the source, a more accurate model of the phenomenon is required: the model must take into account the actual structure of interaction of the super-Alfvén flow of superhot magnetized plasma with a magnetic obstacle – the region with a strong coronal field, created by sunspots.

13.5 Practice: Problems and Answers

Problem 13.1. Derive formula (13.23) in Section 13.3.2.

Answer. According to the geometry shown in Figure 13.4, the frame of reference S' moves with respect to the shock wave frame of reference S with velocity (13.22):

$$\mathbf{v}_y = -\frac{cE}{B_n}\,\mathbf{e}_y\,. \tag{13.34}$$

In the frame S' there is no electric field; therefore there is no change in the energy of a particle reflecting at the shock front, that is $\delta \mathcal{E}\,' = 0$, where $\mathcal{E}\,'$ is the energy of the particle in S'. Let us transform this condition back to S by using the Lorentz transformation of the energy-momentum 4D-vector (see Landau and Lifshitz, *Classical Theory of Field*, 1971, Ch. 2, § 9):

$$p_x = p_x', \quad p_y = \gamma_{\rm L}\left(p_y' + \frac{v_y}{c^2}\,\mathcal{E}\,'\right), \quad p_z = p_z',$$

$$\mathcal{E} = \gamma_{\rm L}\left(\mathcal{E}\,' + v_y\, p_y'\right). \tag{13.35}$$

Since $\delta \mathcal{E}\,' = 0$, it follows from (13.35) that

$$\delta \mathcal{E} = \gamma_{\rm L} v_y\, \delta p_y'. \tag{13.36}$$

The change in the y component of momentum of the reflected particle in the frame of reference S' is

$$\delta\,\mathbf{p}'_y = -2\,\mathbf{p}'_y = 2\gamma_{\mathrm{L}}\left(\mathbf{p}_y - \frac{v_y}{c^2}\,\mathcal{E}\,\mathbf{e}_y\right). \tag{13.37}$$

Note that vectors $\mathbf{p}_y$ and $\mathbf{v}_y$ point in opposite directions. Substituting (13.37) into (13.36) gives us

$$\delta\mathcal{E} = \frac{2v_y}{1 - v_y^2/c^2}\left[p_y + \frac{v_y}{c^2}\left(\mathcal{E}_{kin} + mc^2\right)\right], \tag{13.38}$$

where $\mathcal{E}_{kin} = mv_p^2/2$ is kinetic energy of a particle. Assuming $\mathcal{E}_{kin} \ll mc^2$ and using (13.34), we obtain

$$\delta\mathcal{E} = \delta\mathcal{E}_{kin} = \frac{2E}{B_n^2 - E^2}\left(\frac{v_{p,y}}{c}\,B_n + E\right)mc^2, \tag{13.39}$$

where $v_{p,y}$ is the y component of the particle velocity.

According to (13.21) the electric field

$$E = \frac{1}{c}\,v_1 B_{y1}\,, \tag{13.40}$$

where B_{y1} is the y component of the vector $\mathbf{B}_1$. So we rewrite formula (13.39) as follows

$$\delta\mathcal{E}_{kin} = 2mv_1^2\,\frac{(v_{p,y}/v_1)(B_n/B_{y1}) + 1}{(B_n/B_{y1})^2 - (v_1/c)^2}\,. \tag{13.41}$$

The condition $v_y < c$ can equivalently be written as

$$\frac{B_n}{B_{y1}} > \frac{v_1}{c} \quad \text{or} \quad \operatorname{tg}\Psi_1 < \frac{v_1}{c}\,. \tag{13.42}$$

If we further assume that

$$\frac{B_n}{B_{y1}} \gg \frac{v_1}{c}\,, \tag{13.43}$$

we obtain from (13.41) the following formula

$$\delta\mathcal{E}_{kin} = 2mv_1^2\left(\frac{v_p\cos\theta}{v_1}\operatorname{tg}\Psi_1 + \operatorname{tg}^2\Psi_1\right), \tag{13.44}$$

where θ is the pitch angle in the shock-front frame of reference S. Dividing (13.44) by $\mathcal{E}_{kin}$, we obtain formula (13.23).

Chapter 14

Cosmic Plasma Equilibrium in Magnetic Field

The concept of equilibrium is fundamental to any discussion of the energy contained in an astrophysical object or phenomenon. The MHD non-equilibrium is often related to the onset of dynamic phenomena in cosmic plasma.

14.1 The virial theorem in MHD

14.1.1 A brief pre-history

An integral equality relating different kinds of energy (kinetic, thermal, gravitational, etc.) of some region with a volume V and a surface S, is commonly referred to as the *virial theorem*. It has been proved for mechanical systems for the first time by Clausius (1870). The derivation of the virial theorem for a mechanical system executing a motion in some finite region of space, velocities also being finite, can be found, for example, in Landau and Lifshitz (1960, *Mechanics*, Ch. 2, § 10). Its relativistic form is presented in Landau and Lifshitz (1971, *Classical Theory of Field*, Ch. 4, § 34).

The generalization of the virial theorem to include the magnetic energy in the context of MHD was achieved by Chandrasekhar and Fermi (1953) when addressing the question of the gravitational stability of infinitely conductive masses of cosmic dimensions in the presense of a magnetic field. Although "most students of physics will recognize the name of the virial theorem, few can state it correctly and even fewer appreciate its power" (Collins, 1978).

14.1.2 Deduction of the scalar virial theorem

The virial theorem is deduced from the momentum conservation law (see the MHD Equation (8.42) or Equation (8.50)) rather than the energy conservation law. We have

$$\rho \, \frac{dv_\alpha}{dt} \equiv \rho \left(\frac{\partial v_\alpha}{\partial t} + v_\beta \, \frac{\partial v_\alpha}{\partial r_\beta} \right) = - \, \frac{\partial p}{\partial r_\alpha} - \frac{\partial T_{\alpha\beta}}{\partial r_\beta} - \rho \, \frac{\partial \phi}{\partial r_\alpha} \, . \tag{14.1}$$

Here

$$T_{\alpha\beta} = \frac{1}{4\pi} \left(\frac{B^2}{2} \, \delta_{\alpha\beta} - B_\alpha B_\beta \right) \tag{14.2}$$

is the Maxwellian stress tensor. The gravitational potential is

$$\phi = - \int \frac{G \, \rho \, (\mathbf{r}')}{|\, \mathbf{r} - \mathbf{r}' \,|} \, d^3 \mathbf{r}' , \tag{14.3}$$

where G is the gravitational constant (Appendix 3), $d^3\mathbf{r}' = dx' \, dy' \, dz'$.

Multiply the plasma motion Equation (14.1) by r_α and integrate it over the volume V. We observe in passing that multiplication of (14.1) by r_γ rather than r_α would result, on integrating, in the *tensor* virial theorem and not in the *scalar* one (Chandrasekhar, 1981; see also Strittmatter, 1966; Choudhuri, 1988).

First integrate the left-hand side of Equation (14.1) multiplied by r_α. We get

$$\int \rho \, r_\alpha \, \frac{dv_\alpha}{dt} \, dV = \int r_\alpha \, \frac{d^2 r_\alpha}{dt^2} \, \rho \, dV = \int r_\alpha \, \frac{d^2 r_\alpha}{dt^2} \, dm \, . \tag{14.4}$$

Here we have passed from the integration over volume to integration over mass: $dm = \rho \, dV$. Rearrange formula (14.4) as follows

$$r_\alpha \, \frac{d^2 r_\alpha}{dt^2} = \frac{d}{dt} \left(r_\alpha \, \frac{dr_\alpha}{dt} \right) - \left(\frac{dr_\alpha}{dt} \right)^2 =$$

$$= \frac{d}{dt} \left(\frac{1}{2} \, \frac{dr_\alpha^2}{dt} \right) - \left(\frac{dr_\alpha}{dt} \right)^2 = \frac{1}{2} \, \frac{d^2}{dt^2} \, r_\alpha^2 - v_\alpha^2 \, .$$

On substituting this into (14.4), we obtain

$$\int \rho \, r_\alpha \, \frac{dv_\alpha}{dt} \, dV = \frac{1}{2} \, \frac{d^2}{dt^2} \int r^2 \, dm - \int v^2 \, dm = \frac{1}{2} \, \frac{d^2 I}{dt^2} - 2T. \tag{14.5}$$

Here

$$I = \int r^2 \, dm \tag{14.6}$$

is the *moment of inertia* in the reference frame related to the mass center of the system under discussion,

$$T = \int \frac{v^2}{2} \, dm \tag{14.7}$$

is its *kinetic energy* or (to be more specific) the kinetic energy of macroscopic motions inside the system.

Multiply the first term on the right-hand side of Equation (14.1) by r_α and integrate it over volume:

$$-\int_V r_\alpha \frac{\partial p}{\partial r_\alpha} \, dV = -\oint_S p \, r_\alpha \, dS_\alpha + 3 \int_V p \, dV, \tag{14.8}$$

since

$$\frac{\partial}{\partial r_\alpha} (p r_\alpha) = r_\alpha \frac{\partial p}{\partial r_\alpha} + p \frac{\partial r_\alpha}{\partial r_\alpha} = r_\alpha \frac{\partial p}{\partial r_\alpha} + 3p \, .$$

The Gauss theorem was used to integrate the divergence over the volume in formula (14.8).

If U_{th} is the *thermal energy* of the plasma, γ_g is the ratio of specific heats at constant pressure and at constant volume, then

$$\int_V p \, dV = (\gamma_g - 1) \, U_{th} \, . \tag{14.9}$$

Therefore

$$-\int_V r_\alpha \frac{\partial p}{\partial r_\alpha} \, dV = -\oint_S p \, (\mathbf{r} \cdot d\mathbf{S}) + 3 \, (\gamma_g - 1) \, U_{th} \, . \tag{14.10}$$

Similarly we calculate the integral

$$-\int_V r_\alpha \frac{\partial T_{\alpha\beta}}{\partial r_\beta} \, dV = -\int_S T_{\alpha\beta} \, r_\alpha \, dS_\beta + \int_V T_{\alpha\beta} \, \delta_{\alpha\beta} \, dV \tag{14.11}$$

since

$$\frac{\partial}{\partial r_\beta} (r_\alpha T_{\alpha\beta}) = r_\alpha \frac{\partial T_{\alpha\beta}}{\partial r_\beta} + T_{\alpha\beta} \, \delta_{\alpha\beta} \, .$$

On rearranging, we find from (14.11) and (14.2)

$$-\int_V r_\alpha \frac{\partial T_{\alpha\beta}}{\partial r_\beta}\, dV = \mathcal{M} - \int_S \left[\frac{B^2}{8\pi}\,(\mathbf{r}\cdot d\mathbf{S}) - \frac{1}{4\pi}\,(\mathbf{B}\cdot\mathbf{r})\,(\mathbf{B}\cdot d\mathbf{S}) \right], \tag{14.12}$$

where

$$\mathcal{M} = \int_V \frac{B^2}{8\pi}\, dV \tag{14.13}$$

is the *magnetic energy* of the system.

The third term on the right-hand side of Equation (14.1) gives

$$-\int_V r_\alpha \frac{\partial \phi}{\partial r_\alpha}\, \rho\, dV = \int_V \rho\, r_\alpha \frac{\partial}{\partial r_\alpha} \int_{V'} \frac{G\,\rho(\mathbf{r}')}{|\mathbf{r}-\mathbf{r}'|}\, dV'\, dV =$$

$$= G \int_V \int_{V'} \rho\,\rho'\, r_\alpha \frac{\partial}{\partial r_\alpha} \frac{1}{\sqrt{\left(r_\beta - r'_\beta\right)^2}}\, dV\, dV'. \tag{14.14}$$

Rewrite the expression as follows. Let the distance $R = \sqrt{\left(r_\beta - r'_\beta\right)^2}$. Then

$$r_\alpha \frac{\partial}{\partial r_\alpha} \frac{1}{R} = \frac{1}{2} \left(r_\alpha \frac{\partial}{\partial r_\alpha} \frac{1}{R} + r'_\alpha \frac{\partial}{\partial r'_\alpha} \frac{1}{R} \right) = -\frac{1}{R}$$

and

$$-\int_V r_\alpha \frac{\partial \phi}{\partial r_\alpha}\, \rho\, dV = \Omega\,, \tag{14.15}$$

where

$$\Omega = -\frac{G}{2} \int_V \int_{V'} \frac{\rho\,\rho'}{R}\, dV\, dV', \tag{14.16}$$

is the *gravitational energy* of the system. Obviously, the energy is negative.

Combining (14.5), (14.10), (14.12), and (14.15) into a single equation, we finally obtain

$$\frac{1}{2}\frac{d^2 I}{dt^2} = 2T + 3\,(\gamma_g - 1)\, U_{th} + \mathcal{M} + \Omega - \oint_S p\,(\mathbf{r}\cdot d\mathbf{S}) -$$

$$- \oint_S \left[\frac{B^2}{8\pi}\,(\mathbf{r}\cdot d\mathbf{S}) - \frac{1}{4\pi}\,(\mathbf{B}\cdot\mathbf{r})\,(\mathbf{B}\cdot d\mathbf{S}) \right]. \tag{14.17}$$

Formula (14.17) is called the virial theorem. It has repeatedly been used in astrophysics when 'discussing the question of the stability' of equilibrium systems of various types. More exactly, this integral force balance relation is nothing more than a **necessary condition for equilibrium**. So it may be well used as a *non-existance theorem* for the equilibrium problem to find circumstances when non-equilibrium may occur.

14.1.3 Some astrophysical applications

By way of illustration, consider some consequences of the virial theorem for the case of a *steady* system, i.e. when

$$\frac{d^2 I}{dt^2} = 0 . \tag{14.18}$$

Moreover, let the kinetic energy of macroscopic motions be equal to zero

$$T = 0 , \tag{14.19}$$

i.e. the system is in *static* equilibrium. Both assumptions must be justified carefully, if they are applied to actual cosmic plasmas.

Suppose also that the system is finite and the surface S, over which the integration in (14.10) and (14.12) is performed, can be moved sufficiently far away (formally speaking, to infinity), so that

$$\oint_S p\,(\mathbf{r} \cdot d\mathbf{S}) = 0 \tag{14.20}$$

and

$$\oint_S \left[\frac{B^2}{8\pi}\,(\mathbf{r} \cdot d\mathbf{S}) - \frac{1}{4\pi}\,(\mathbf{B} \cdot \mathbf{r})\,(\mathbf{B} \cdot d\mathbf{S}) \right] = 0 . \tag{14.21}$$

Then from the virial theorem (14.17) it follows that

$$3\,(\gamma_g - 1)\; U_{th} + \mathcal{M} + \Omega = 0 . \tag{14.22}$$

Introduce the 'total' (without what has been neglected) energy of the system

$$\mathcal{E} = U_{th} + \mathcal{M} + \Omega . \tag{14.23}$$

Eliminating the thermal energy U_{th} from Equations (14.22) and (14.23), the total energy is expressed as follows

$$\boxed{\mathcal{E} = -\frac{(3\gamma_g - 4)}{3\,(\gamma_g - 1)}\,(\,|\,\Omega\,| - \mathcal{M}\,)\,.} \tag{14.24}$$

In a sense, the equilibrium is stable if $\mathcal{E} < 0$, i.e.

$$\frac{(3\gamma_g - 4)}{3\,(\gamma_g - 1)}\,(\,|\,\Omega\,| - \mathcal{M}\,) > 0\,, \tag{14.25}$$

which is equivalent, once $\gamma_g > 4/3$, to

$$|\,\Omega\,| > \mathcal{M}\,. \tag{14.26}$$

It is self-evident that inequality (14.26) is just a *necessary* condition for the dynamical *global* stability of a system. The condition is by no means sufficient. It can be used to show a non-existence of equilibrium of the system.

Consider **two particular cases of astrophysical interest**.

(a) If $\mathcal{M} = 0$ then the system can be stable only for $\gamma_g > 4/3$. This condition is easy to understand. The pressure inside the system under adiabatic compression ($p \sim \rho^{\gamma_g}$) must grow faster than the gravitational pressure $p_g \sim \rho^{4/3}$. It is in this case that the system, for instance a star, can be sufficiently resilient to resist the gravitational collapse. That is why a star consisting of a monatomic gas (with $\gamma_g = 5/3$) can be dynamically stable.

(b) Let $\mathcal{M} > 0$. Generally, the necessary condition for stability (14.25) can be, in principle, violated. What this means is that the field diminishes the stability of a star. Given a sufficiently strong field, gravitational attraction forces cannot balance the magnetic repulsion of the constituents of the system. However, such a situation is difficult to conceive.

In actuality, **gravitational compression** cannot result in $\mathcal{M} > |\,\Omega\,|$ since, given the freezing-in condition and isotropic compression, $p_{\rm mag} \sim \rho^{4/3}$ in common with $p_g \sim \rho^{4/3}$. It is also impossible to obtain $\mathcal{M} > |\,\Omega\,|$ by dint of magnetic field amplification owing to **differential rotation**, since $|\,\Omega\,| > 2T$ in a gravitationally bound system. On the other hand, the energy of a magnetic field generated by differential rotation must remain less than the kinetic energy T of the rotation motion, i.e. $\mathcal{M} < T$. Hence $\mathcal{M} < |\,\Omega\,|$.

At most, the condition $\mathcal{M} \sim |\,\Omega\,|$ can be realized. This situation is probably realized in stars of the cold giant type with a large radius. Perhaps such stars are at the limit of stability, which reveals itself as non-steady behaviour.

Condition (14.26) allows us to evaluate the upper limit of the mean intensity of a magnetic field inside a star or other equilibrium configuration. Substitute the gravitational energy of a uniform ball,

$$\Omega = -\frac{3}{5}\frac{GM^2}{R}\,, \tag{14.27}$$

in (14.26). The result is (Syrovatskii, 1957, p. 282)

$$B < B_{cr} = 2 \times 10^8 \left(\frac{M}{M_\odot}\right)\left(\frac{R}{R_\odot}\right)^{-2} . \tag{14.28}$$

For the Sun, magnetic field B must be less than 2×10^8 Gauss. For the most magnetic stars of the spectral class A, which are observed to have fields $\sim 10^4$ Gauss, the condition $B < 3 \times 10^7$ Gauss must hold. Hence these magnetic stars called the Ap stars, because they possess some peculiar properties, are very far from the stability limit. As is seen from the Syrovatskii condition (14.28), the cold giants with large radii could be closer to such a limit.

Given a uniform field inside a star, on approaching the limit established by (14.28), the form of the star increasingly deviates from a sphere:

> the magnetic field resists gravitational compression of a collapsing star in the direction perpendicular to the field, whereas the plasma may freely flow along the field lines.

As a result, the equilibrium configuration is represented by a rotation ellipsoid compressed in the field direction. The virial theorem can be written (e.g., Nakano, 1998) for an axisymmetric oblate magnetic cloud of mass M and semimajor axes $a_\perp$ and $a_\parallel$, respectively, embedded in a medium of pressure p_s. This is typical for the problem of star formation in magnetic clouds.

The action of a magnetic field is quite analogous to rotation (e.g., Strittmatter, 1966). Furthermore, both the **strong field and fast rotation are typical of pulsars**, especially of the magnetars (see Problem 9.2). So both these factors determine the real flattening of a neutron star. The flattening can be calculated using the tensor virial theorem. Note, however, that for a neutron star with $M \sim M_\odot$ and $R \sim 10$ km the critical magnetic field (14.28) is still unprecedentelly high: $B_{cr} \sim 10^{18}$ G.

The virial theorem is sometimes applied in solar physics, for example, while studying active regions (Section 14.5). It allows us to evaluate the energy of equilibrium electric currents and show that the energy can be large enough to explain the flaring activity (Litvinenko and Somov, 1991a).

14.2 Force-free fields and Shafranov's theorem

14.2.1 The simplest examples of force-free configurations

A particular case of equilibrium configurations of a cosmic plasma in a magnetic field is the *force-free field*, i.e. the field which does not require external forces. As was noted in Section 8.3, force-free fields naturally occur when the magnetic force dominates all the others, and hence the magnetic field must balance itself

$$\mathbf{B} \times \operatorname{curl} \mathbf{B} = 0 \,. \tag{14.29}$$

Consider several examples of such equilibrium configurations.

(a) The Syrovatskii force-free field

Let the magnetic field vector be situated in the plane parallel to the plane (x, y), but depend only on z

$$\mathbf{B} = \{ \, B_x(z), \; B_y(z), \; 0 \, \} \,. \tag{14.30}$$

Substitute (14.30) in Equation (14.29):

$$\operatorname{curl} \mathbf{B} = \left\{ -\frac{\partial B_y}{\partial z} \,, \; \frac{\partial B_x}{\partial z} \,, \; 0 \right\} , \tag{14.31}$$

$$\mathbf{B} \times \operatorname{curl} \mathbf{B} = \left\{ 0, \; 0, \; B_x \frac{\partial B_x}{\partial z} + B_y \frac{\partial B_y}{\partial z} \right\} = 0 \,. \tag{14.32}$$

The resulting equation is

$$\frac{\partial}{\partial z} \left(B_x^2 + B_y^2 \right) = 0 \,, \tag{14.33}$$

with the solution

$$B^2 = B_x^2 + B_y^2 = \text{const} \,. \tag{14.34}$$

This is the simplest example of a force-free field. The magnitude of the field vector is independent of z. A one-dimensional force-free field of this type may be considered to be a *local* approximation of an arbitrary force-free field in a region of the magnetic 'shear' in the solar atmosphere. As a particular example, suitable for formal analysis, one may adopt the force-free field of the type

$$\mathbf{B} = \{ B_0 \cos kz \, , \, B_0 \sin kz \, , \, 0 \} \tag{14.35}$$

(Bobrova and Syrovatskii, 1979). The field lines, and hence the electric current, lie in the plane (x, y). The direction of the lines rotates with increasing z.

(b) The Lundquist force-free field

The magnetic field of a direct current flowing along the z axis tends to compress the plasma to the axis, owing to the tension of the field lines (see Section 8.3.3). By contrast,

> a bundle of parallel field lines tends to expand by the action of the magnetic pressure gradient.

Given the superposition of these fields for a certain relationship between them, the total magnetic force can be zero. Field lines for such a force-free field have the shape of spirals shown in Figure 14.1.

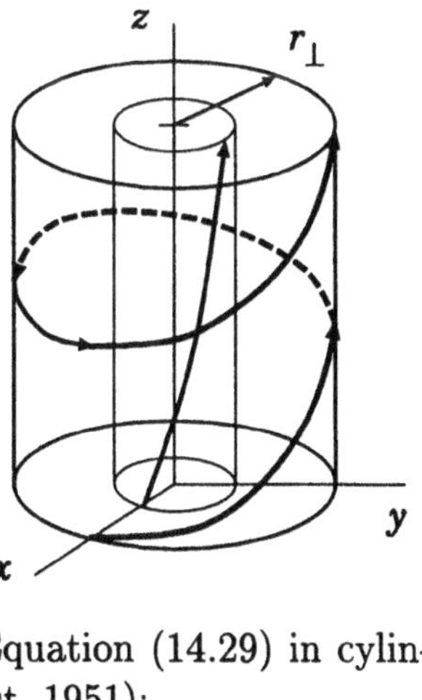

Figure 14.1: A helical magnetic field in the form of a spiral of constant slope on a cylindrical surface $r_\perp$ = const.

The corresponding *axially symmetric* solution to Equation (14.29) in cylindrical coordinates $r_\perp$, ϕ, z is of the form (Lundquist, 1951):

$$B_z = A \, J_0 \, (\alpha r_\perp) \, , \quad B_\phi = A \, J_1 \, (\alpha r_\perp) \, , \quad B_r = 0 \, . \tag{14.36}$$

Here J_0 and J_1 are the Bessel functions, A and α are constants.

A distinguishing feature of the field is that $B^2 \sim r_\perp^{-1}$ for large $r_\perp$ since Bessel functions $J_n \sim r_\perp^{-1/2}$ as $r_\perp \to \infty$ $(n = 0, \, 1)$. The magnetic energy

$$\mathcal{M} = \int \frac{B^2}{8\pi} \, dV \sim r_\perp^{-1} r_\perp^2 \sim r_\perp \tag{14.37}$$

diverges for large $r_\perp$. Such a **divergence of magnetic energy** is known to be typical of force-free fields and will be explained below.

14.2.2 The energy of a force-free field

Let us retain only magnetic terms in the virial theorem; we have

$$\mathcal{M} - \oint_S \left[\frac{B^2}{8\pi} (\mathbf{r} \cdot d\mathbf{S}) - \frac{1}{4\pi} (\mathbf{B} \cdot \mathbf{r}) (\mathbf{B} \cdot d\mathbf{S}) \right] = 0 . \tag{14.38}$$

Provided the electric currents occupy a *finite* region, the value of the magnetic field is proportional to r^{-3} (or higher degrees of r^{-1}). Once the surface of integration S is expanded to infinity, the surface integral tends to zero. Equality (14.38) becomes impossible.

Therefore any finite magnetic field cannot contain itself. There must be *external* forces to balance the outwardly directed pressure due to the *total magnetic* energy $\mathcal{M}$.

The same statement may be formulated as follows. **The force-free field cannot be created in the whole space**. This is the so-called *Shafranov theorem* (Shafranov, 1966). While stresses may be eliminated in a given region V, they cannot be canceled everywhere. In general

> a force-free configuration requires the forces needed to balance the outward pressure of the magnetic field to be reduced in magnitude by spreading them out over the bounding surface S.

In this way, the virial theorem sets limits on the space volume V that can be force-free.

The Shafranov theorem is the counterpart of the known Irnshow theorem (e.g., Sivukhin, 1977, Ch. 1, § 9) concerning the equilibrium configuration of a system of electric charges. Such a configuration also can be stable only in the case that some external forces, other than the electric ones, act in the system.

In fact Shafranov (1966) has proved a stronger statement than the above theorem on the force-free field. He has taken into account not only the terms corresponding to the magnetic force in (14.17) but the gas pressure as well:

$$\int_V \left(3p + \frac{B^2}{8\pi} \right) dV =$$

$$= \oint_S \left[\left(p + \frac{B^2}{8\pi} \right) (\mathbf{r} \cdot d\mathbf{S}) - \frac{1}{4\pi} (\mathbf{B} \cdot \mathbf{r}) (\mathbf{B} \cdot d\mathbf{S}) \right] . \qquad (14.39)$$

If the plasma occupies some finite volume, V, the pressure outside of this volume being zero, and if electric currents occupy a finite region, then the surface integral tends to zero, once the surface of integration, S, is expanded to infinity. On the other hand, the expression under the integral sign on the left-hand side is always positive. Hence the integral is positive. Thus the equality (14.39) turns out to be impossible. Therefore

> any finite equilibrium configuration of a plasma with a magnetic field can exist only in the presence of external forces which, *apart from the gas pressure*, serve to fix the electric currents.

In a laboratory, fixed current conductors must be present. In this case the right-hand side of (14.39) is reduced to the integral over the surface of the conductors.

Under astrophysical conditions, the role of the external force is frequently played by the gravitational force or by an external magnetic field having its sources outside the volume under investigation. However, these sources must be kept and driven by non-magnetic forces.

A typical example of such a situation is the magnetic field of an active region on the Sun. This is the sum of the proper field created by currents flowing inside the active region, and the external field with the sources situated (and fixed) below the photosphere (Litvinenko and Somov, 1991a). In this case the formula for the magnetic energy of the equilibrium system contains a term due to the interaction of internal currents (in particular current sheets in the regions of reconnection) with the external magnetic field.

14.3 Properties of equilibrium configurations

14.3.1 Magnetic surfaces

Consider the case of *magnetostatic* equilibrium

$$-\nabla p + \frac{1}{4\pi} \, \mathrm{curl}\, \mathbf{B} \times \mathbf{B} - \rho \nabla \phi = 0 . \qquad (14.40)$$

The gravitational force is supposed to be negligible

$$\rho \nabla \phi = 0 , \qquad (14.41)$$

On dropping the third term in Equation (14.40) and taking the scalar product with vector **B** we obtain

$$\boxed{\mathbf{B} \cdot \nabla p = 0\,,} \tag{14.42}$$

i.e. magnetic field lines in an equilibrium configuration are situated on the surface $p = \text{const}$. Therefore

> in order to contain a plasma by the magnetic field, the field lines are forbidden to leave the volume occupied by the plasma.

There is a common viewpoint that, by virtue of the condition

$$\operatorname{div} \mathbf{B} = 0\,, \tag{14.43}$$

field lines may either close or go to infinity. However, the other variant is possible, when a field line fills up an entire surface – *magnetic surface*.

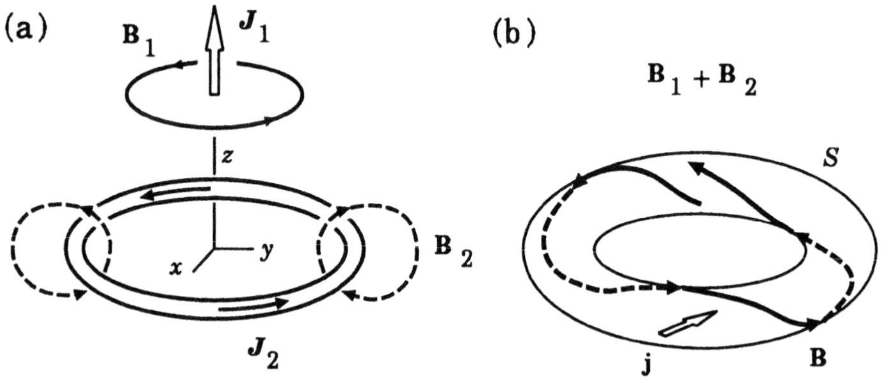

Figure 14.2: (a) A line current J_1 and a ring current J_2. (b) The field lines of the total field $\mathbf{B}_1 + \mathbf{B}_2$ form a toroidal surface S.

Consider the field of two electric currents – a line current J_1 flowing along the vertical z axis (Figure 14.2a) and a plane current ring J_2 (e.g., Tamm, 1989, Ch. 4, § 53). If there were only the current J_1, the field lines of this current $\mathbf{B}_1$ would constitute circumferences centred at the z axis. The field lines $\mathbf{B}_2$ of the ring current J_2 lie in meridional planes. The total field $\mathbf{B} = \mathbf{B}_1 + \mathbf{B}_2$ forms a helical line on a toroidal surface S. The course of this spiral depends on the ratio B_1/B_2. Once this is a rational number, the spiral will close. However, in general it does not close but continuously fills up the entire toroidal surface S (Figure 14.2b).

By virtue of condition (14.42), the plasma pressure at such a surface (called the magnetic one) is constant. Such a magnetic field can serve as a trap for the plasma. This fact constitutes the basis for constructing laboratory devices for plasma containment in stellarators, suggested by Spitzer.

Take the scalar product of Equation (14.40) with

$$\mathbf{j} = \frac{c}{4\pi} \, \mathrm{curl} \, \mathbf{B} \, . \tag{14.44}$$

The result is

$$\boxed{\mathbf{j} \cdot \nabla p = 0} \tag{14.45}$$

which, in combination with (14.42), signifies that, in an equilibrium configuration, the electric current flows on magnetic surfaces (Figure 14.2b).

14.3.2 The specific volume of a magnetic tube

Consider two closed magnetic surfaces: $p = \mathrm{const}$ and $p + dp = \mathrm{const}$. Construct a system of noncrossing partitions between them (Figure 14.3). Let $d\mathbf{l}_1$ be the line element directed normally to the surface $p = \mathrm{const}$:

$$d\mathbf{l}_1 = \frac{\nabla p}{|\nabla p|^2} \, dp \, . \tag{14.46}$$

The vectors $d\mathbf{l}_2$ and $d\mathbf{l}_3$ are directed along the two independent contours l_2 and l_3 which may be drawn on a toroidal surface: for example the curve l_2 is directed along a large circle of the toroid while l_3 lies along the small one. The surface element of this partition is

$$d\mathbf{S}_3 = d\mathbf{l}_1 \times d\mathbf{l}_2 \, . \tag{14.47}$$

The total current dJ_3 flowing through the partition situated on the contour l_2 is

$$dJ_3 = \oint_{l_2} \mathbf{j} \cdot (d\mathbf{l}_1 \times d\mathbf{l}_2) \, . \tag{14.48}$$

According to Equation (14.45), the total current flowing through the system of noncrossing partitions between the two magnetic surfaces is constant. In other words, dJ_3 is independent of the choice of the integration contour. We are concerned with the physical consequences of this fact.

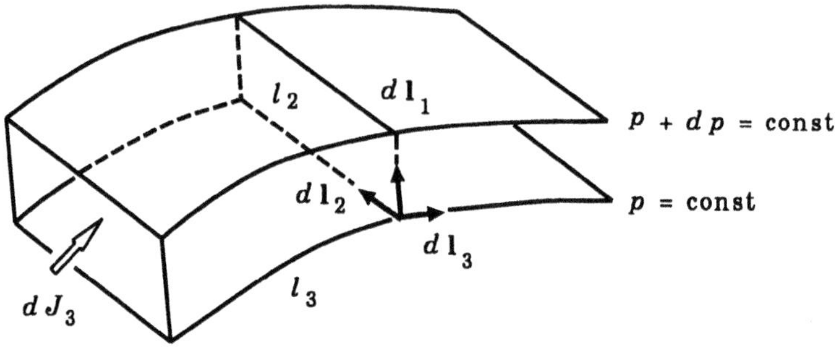

Figure 14.3: The calculation of the electric current between two magnetic surfaces.

In order to find the expression for the current density $\mathbf{j}$ in an equilibrium configuration, take the vector product of Equation (14.40) with the magnetic field $\mathbf{B}$. The result is

$$\mathbf{B} \times \nabla p = \frac{1}{c} \, \mathbf{B} \times (\mathbf{j} \times \mathbf{B}) \, ,$$

which, on applying the formula for a double vector product to the right-hand side, becomes

$$c\mathbf{B} \times \nabla p = \mathbf{j} \, B^2 - \mathbf{B} \, (\mathbf{j} \times \mathbf{B}) \, .$$

Thus we have

$$\mathbf{j} = c \, \frac{\mathbf{B} \times \nabla p}{B^2} + f \, \mathbf{B} \, , \tag{14.49}$$

where $f = f(\mathbf{r})$ is an arbitrary function. If need be, it can be found from the condition $\operatorname{div} \mathbf{j} = 0$.

Substitute (14.49) in the integral (14.48):

$$\mathbf{j} \cdot (d\mathbf{l}_1 \times d\mathbf{l}_2) = \frac{c}{B^2} \, (\mathbf{B} \times \nabla p) \cdot (d\mathbf{l}_1 \times d\mathbf{l}_2) + f \, \mathbf{B} \cdot (d\mathbf{l}_1 \times d\mathbf{l}_2) \, . \tag{14.50}$$

Rearrange the first item, based on the well-known Lagrange identity in vector analysis:

$$(\mathbf{a} \times \mathbf{b}) \cdot (\mathbf{c} \times \mathbf{d}) = (\mathbf{a} \cdot \mathbf{c}) \, (\mathbf{b} \cdot \mathbf{d}) - (\mathbf{b} \cdot \mathbf{c}) \, (\mathbf{a} \cdot \mathbf{d}) \, .$$

We get

$$(\mathbf{B} \times \nabla p) \cdot (d\mathbf{l}_1 \times d\mathbf{l}_2) = (\mathbf{B} \cdot d\mathbf{l}_1) \, (\nabla p \cdot d\mathbf{l}_2) - (\mathbf{B} \cdot d\mathbf{l}_2) \, (\nabla p \cdot d\mathbf{l}_1) \, .$$

By virtue of (14.42) and (14.46),

$$(\mathbf{B} \cdot d\mathbf{l}_1) = (\mathbf{B} \cdot \nabla p) \, \frac{dp}{|\nabla p|^2} = 0 \, , \quad (\nabla p \cdot d\mathbf{l}_1) = dp \, .$$

Hence

$$(\mathbf{B} \times \nabla p) \cdot (d\mathbf{l}_1 \times d\mathbf{l}_2) = -(\mathbf{B} \cdot d\mathbf{l}_2)\, dp\,. \tag{14.51}$$

Substitute (14.51) in (14.50):

$$\mathbf{j} \cdot (d\mathbf{l}_1 \times d\mathbf{l}_2) = -c\,\frac{dp}{B^2}\,(\mathbf{B} \cdot d\mathbf{l}_2) + f(\mathbf{r})\,\mathbf{B} \cdot (d\mathbf{l}_1 \times d\mathbf{l}_2)\,.$$

Thus the expression (14.48) for current dJ_3 assumes the form

$$dJ_3 = -c\,dp \oint\limits_{l_2} \frac{\mathbf{B} \cdot d\mathbf{l}_2}{B^2} + \oint\limits_{l_2} f(\mathbf{r})\,\mathbf{B} \cdot (d\mathbf{l}_1 \times d\mathbf{l}_2)\,. \tag{14.52}$$

Provided the contour l_2 coincides with a closed field line,

$$d\mathbf{l}_2 = \frac{\mathbf{B}}{B}\, dl$$

and the second term on the right of (14.52) vanishes.

Once a field line closes on making one circuit of the toroid, the expression

$$dJ_{\rm n} = -c\,dp \oint \frac{dl}{B} \tag{14.53}$$

defines the total current flowing between neighbouring magnetic surfaces normal (the subscript n) to the field line. Since the magnitude of this current is independent of the choice of contour, for each field line on a magnetic surface the integral

$$\boxed{U = \oint \frac{dl}{B}} \tag{14.54}$$

is constant. The condition of constancy of U can be generalized to include the surface with unclosed field lines (Shafranov, 1966). Thus (Kadomtsev, 1966),

> under the condition of magnetostatic equilibrium, the magnetic surface consists of the field lines with the same value of U.

Introduce the notion of the *specific volume* of a magnetic tube (Rosenbluth and Longmire, 1957) as the ratio of its geometric volume dV to the magnetic flux $d\Phi$ through the tube. If $dS_{\rm n}$ is the cross-sectional surface of the tube, its geometric volume is

$$dV = \oint dS_{\rm n}\, dl$$

whereas the magnetic flux

$$d\,\Phi = B\,dS_{\rm n}\,.$$

On the basis of the magnetic flux constancy inside the tube of field lines, i.e. $d\,\Phi = \text{const}$, we deduce that

$$\frac{d\,V}{d\,\Phi} = \oint \frac{dS_{\rm n}}{B\,dS_{\rm n}}\,dl = \oint \frac{dl}{B} = U. \tag{14.55}$$

The stability of an equilibrium configuration can be judged by the condition (14.54). This property will be discussed in the next Section.

14.3.3 The flute or convective instability

Much like any gas with a finite temperature, the plasma in a magnetic field tends to expand. However, given a high conductivity, it cannot move independently of the magnetic field. The plasma moves together with the field in such a way that it travels to a region of the field characterized by a greater specific volume.

In order for an equilibrium configuration to be stable with respect to a given perturbation type – deformation of a tube of magnetic field lines – the following condition is necessary (Rosenbluth and Longmire, 1957):

$$\delta U = \delta \oint \frac{dl}{B} < 0\,. \tag{14.56}$$

To put it another way,

> the magnetostatic equilibrium is stable once the given type of deformation does not facilitate the plasma spreading,

i.e. increasing its specific volume.

As an example, consider the plasma in the magnetic field of a linear current J:

$$B_{\varphi} = \frac{2J}{cr}\,, \tag{14.57}$$

here r, z, φ are cylindrical coordinates. In such a field there exists an equilibrium plasma configuration in the form of an infinite hollow cylinder C as shown in Figure 14.4a.

Calculate the specific volume for such a configuration. The geometric volume of the tube of field lines is

$$d\,V = 2\pi r\,dr\,dz\,,$$

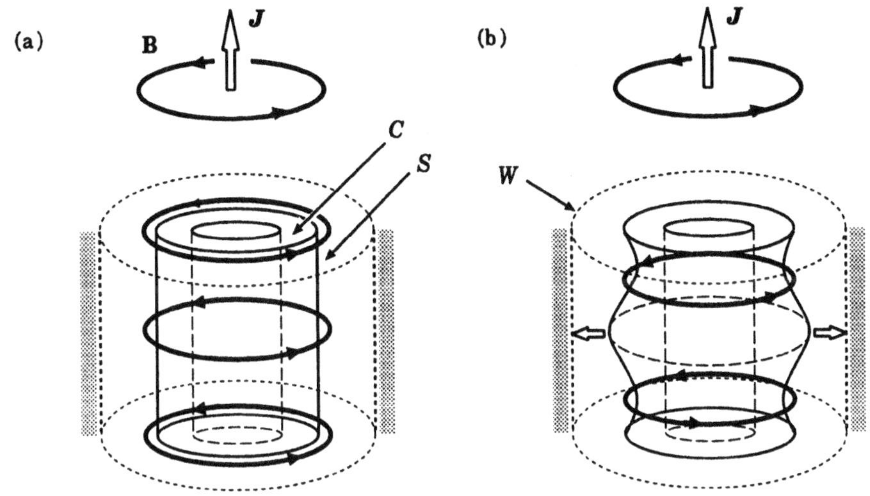

Figure 14.4: (a) An equilibrium plasma configuration. (b) Unstable perturbations of the outer boundary.

whereas the magnetic flux

$$d\,\Phi = B_\varphi\, dr\, dz = \frac{2J}{cr}\, dr\, dz\,.$$

Hence the specific volume

$$U = \frac{dV}{d\,\Phi} = \frac{\pi c}{J}\, r^2. \tag{14.58}$$

It is seen from (14.58) that the specific volume grows with the radius. In particular, for small perturbations δr of the external surface S of the plasma cylinder C

$$\delta U = \frac{2\pi c}{J}\, r\, \delta r > 0$$

once $\delta r > 0$. It is sufficient to have a small perturbation of the external boundary of the plasma to obtain ring flutes which will rapidly grow towards the wall W of the chamber as shown in Figure 14.4b.

The instability connected with the specific volume increase towards the walls of the chamber is termed the *flute* or *convective* instability.

Return to the toroid example (Figure 14.2b). The boundary of the toroidal plasma configuration is a magnetic surface. It is also unstable with respect to, amongst other things, convective perturbations (Kadomtsev, 1960).

The concept of the specific volume of a magnetic flux tube is also useful to demonstrate the creation of reconnecting current sheets as a discontinuity of a magnetic field in cosmic plasma (Chapter 22).

* * *

The problems of plasma **equilibrium and stability** are of great value for astrophysics as a whole (Chandrasekhar, 1981; Zel'dovich and Novikov, 1971), and especially for solar physics (e.g., Parker, 1979; Priest, 1982).

> The Sun seems to maintain stability of solar prominences and coronal loops with great ease

(e.g., Tandberg-Hanssen, 1995; Acton, 1996) in contrast to the immense difficulty of containing plasmas in a laboratory. So, sometimes we need to explain how an equilibrium can remain stable for a long time. This is, for example, the case of reconnecting current sheets in the solar atmosphere and the geomagnetic tail (Chapter 20). At other times, we want to understand

> why magnetic structures on the Sun suddenly become unstable and produce dynamic events

of great beauty such as eruptive prominences and solar flares, coronal transients, and coronal mass ejections (CMEs).

The methods employed to investigate the stability of an equilibrium MHD system are natural generalizations of those for studying a particle in one-dimensional motion. One approach is to seek normal mode solutions (as we did it in Chapter 10). An alternative approach for tackling stability is to consider the change in potential energy due to a displacement from equilibrium.

Recommended Reading: Morozov and Solov'ev (1966a), Kadomtsev (1966), Shu (1992).

14.4 Archimedean forces in MHD

14.4.1 A general formulation of the problem

Return to the equation of magnetostatic equilibrium (14.40). Rewrite it as follows (see Somov, 1994b):

$$\nabla p = \rho \mathbf{g} + \mathbf{f} , \tag{14.59}$$

where

$$\mathbf{f} = \frac{1}{c}\,\mathbf{j}\times\mathbf{B} \tag{14.60}$$

is the Lorentz force, $\mathbf{g} = -g\,\mathbf{e}_z$ is the gravity acceleration.

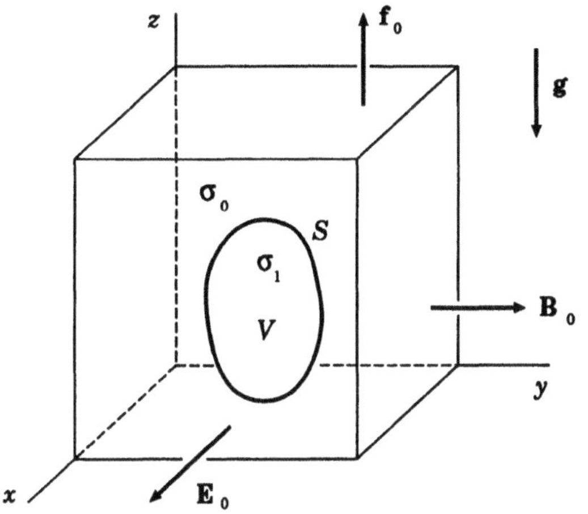

Figure 14.5: Formulation of the problem concerning the Archimedean force in magnetohydrodynamics.

Consider an incompressible conducting fluid situated in a uniform magnetic field $\mathbf{B}_0$ and electric field $\mathbf{E}_0$ (Figure 14.5). Provided the current $\mathbf{j}_0$ flowing in the fluid is *uniform*, the Lorentz force created is uniform as well:

$$\mathbf{f}_0 = \frac{1}{c}\,\mathbf{j}_0\times\mathbf{B}_0\,. \tag{14.61}$$

By virtue of Equation (14.59), **the Lorentz force makes the fluid heavier or lighter**. In both cases the uniform volume force is potential and, much like the gravity force, will be balanced by an additional pressure gradient appearing in the fluid. As will be shown later, that allows the creation of a regulated *expulsion* force (Figure 14.5) analogous to the Archimedean force in ordinary hydrodynamics.

A body plunged into the fluid is acted upon by the force

$$\mathbf{F} = \int_V (\rho_1\,\mathbf{g} + \mathbf{f}_1)\,dV + \oint_S p_0\,\mathbf{n}\,dS. \tag{14.62}$$

Here ρ_1 is the density of the submerged body, which is generally not equal to that of the fluid ρ_0;

$$\mathbf{f}_1 = \frac{1}{c}\,\mathbf{j}_1\times\mathbf{B}_0 \tag{14.63}$$

is the volume Lorentz force, $\mathbf{j}_1$ is the current inside the body, $\mathbf{n}$ is the inward normal to the surface S, and p_0 is the pressure on the body from the fluid, resulted from (14.59):

$$\nabla p_0 = \rho_0 \, \mathbf{g} + \mathbf{f}_0 \, . \tag{14.64}$$

14.4.2 An oversimplified consideration of the effect

If the current $\mathbf{j}_0$ was uniform, the right-hand side of Equation (14.64) would be a uniform force, and formula (14.62) could be rewritten as

$$\mathbf{F} = \int\limits_V (\rho_1 \mathbf{g} + \mathbf{f}_1) \, dV - \int\limits_V \nabla p_0 \, dV \tag{14.65}$$

or

$$\boxed{\mathbf{F} = \int\limits_V (\rho_1 - \rho_0) \, \mathbf{g} \, dV + \frac{1}{c} \int\limits_V (\mathbf{j}_1 - \mathbf{j}_0) \times \mathbf{B}_0 \, dV.} \tag{14.66}$$

The first term in (14.66) corresponds to the usual Archimedean force in hydrodynamics. It equals zero once $\rho_1 = \rho_0$. When $\rho_1 > \rho_0$, the direction of this force coinsides with the gravitational acceleration $\mathbf{g}$. The second term describes the *magnetic expulsion* force. It vanishes once $\mathbf{j}_1 = \mathbf{j}_0$, i.e. $\sigma_1 = \sigma_0$.

The second term in formula (14.66) shows that the magnetic expulsion force, different from the known Parker's magnetic buoyancy force (Ch. 8 in Parker, 1979) by its origin, appears provided $\sigma_1 \neq \sigma_0$. This fact has been used to construct, for example, MHD devices for the separation of mechanical mixtures. In what follows we shall call the second term in (14.66) the magnetic *σ-dependent* force:

$$\mathbf{F}_\sigma = \frac{1}{c} \int\limits_V (\sigma_1 - \sigma_0) \, \mathbf{E}_0 \times \mathbf{B}_0 \, dV. \tag{14.67}$$

Note, however, that the simplest formula (14.66) is of purely illustrative value, since **the electric field and current density are not uniform in the presence of a body** with conductivity σ_1 which is not equal to that of the fluid σ_0 (Figure 14.6). In this case, the appearing *σ-dependent* force is generally not potential. Hence it cannot be balanced by potential forces. That is the reason why

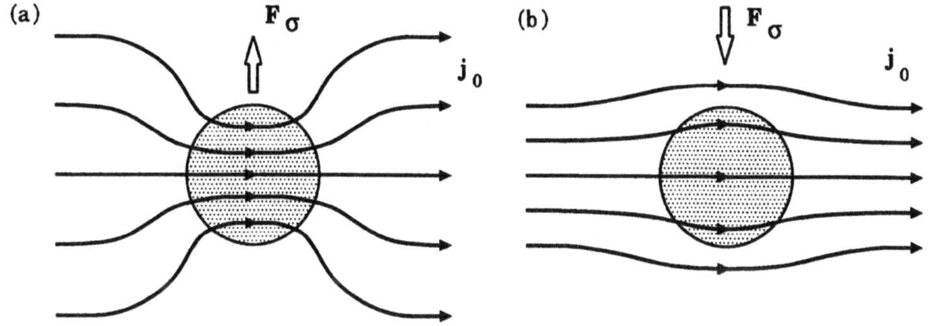

Figure 14.6: Opposite orientation of the σ-dependent force in two opposite cases: (a) $\sigma_1 > \sigma_0$ and (b) $\sigma_1 < \sigma_0$. Appearence of a non-uniform distribution of electric current is shown.

vortex flows of the conducting fluid must be generated by the magnetic σ-dependent force.

The general analysis of the corresponding MHD problem is contained in the paper by Andres, Polak, and Syrovatskii (1963). The stationary solutions for a ball and a cylinder were obtained by Syrovatskii and Chesalin (1963) for the specific case when both the magnetic and usual Reynolds numbers are small; similar stationary solutions for a cylinder see also in Marty and Alemany (1983), Gerbeth *et al.* (1990). The character of vortex flows and the forces acting on submerged bodies will be analyzed in Sections 15.3 and 15.4.

14.5 MHD equilibrium in the solar atmosphere

The magnetic configuration in an active region in the solar atmosphere is, in general, very complex and modelling of dynamical processes in these regions requires a high degree of idealization. First, as regards the most powerful and fascinating of these processes, the two-ribbon flare, the typical preflare magnetic field distribution seems to conform to a certain standard picture: a magnetic arcade including a more or less pronounced plage filament, prominence. Second, instead of dynamics, models deal with a static or steady-state equilibrium in order to understand the causes of a flare or another transient activity in the solar atmosphere as a result of some instability or lack of equilibrium.

So it is assumed that initially the configuration of prominence and overlying arcade is in equilibrium but later the eruption takes place. **Either the**

equilibrium has become unstable or the equilibrium has been lost. One limiting possibility is that the magnetic field around the prominence evolves into an unstable or non-equilibrium configuration and then drives the overlying magnetic arcade. However observations imply that this is unlikely. An alternative is that the overlying arcade evolves until it is no longer in stable equilibrium and then its eruption stimulates the prominence to erupt by removing stabilising field lines. Presumably this is the case of a coronal loop transient and coronal mass ejection (CME).

The idealized models used in theoretical and numerical studies of this problem usually consider two-dimensional force-free arcade configurations with foot points anchored in the photosphere which are energized, for example, by photospheric shear flows in the direction along the arcade (see Biskamp and Welter, 1989, for reviews). Some other models take into account the gas pressure gradient and the gravitational force (e.g., Webb, 1986).

However it is important to investigate more general circumstances when equilibrium and non-equilibrium may occur. The electromagnetic expulsion force – a MHD analogue of the usual Archimedean force – plays an important part in the dynamics of coronal plasma with a non-uniform distribution of temperature and, hence, electric conductivity. More exactly, the condensation mode of the radiatively-driven thermal instability in an active region may result in the formation of cold dense loops or filaments surrounded by hot rarified plasma (see Somov, 1992; Somov *et al.*, 1999).

The effect results from the **great difference of electric conductivities** outside and inside the filaments. The force can generate fast vortex flows inside and in the vicinity of the filaments as well as initiate the non-equilibrium responsible for transient activities: flares, CMEs etc. The virial theorem confirms this possibility and clarifies the role of pre-flare reconnecting current sheets in MHD equilibrium and non-equilibrium of an active region. Correct use of the virial theorem confirms the applicability of reconnection in current sheets for explaining the energetics of flares (Litvinenko and Somov, 1991a) and other non-steady phenomena in the solar atmosphere.

14.6 Practice: Problems and Answers

Problem 14.1. Show that, apart from the trivial case of a potential field, fields for which

$$\text{curl}\, \mathbf{B} = \alpha \mathbf{B} \tag{14.68}$$

will be force-free. In the most general case, α will be spatially dependent.

Answer. Just substitute formula (14.68) in Equation (14.29).

Problem 14.2. Show that the force-free fields with $\alpha = \text{const}$ represent the state of *minimal* magnetic energy in a closed system (Woltjer, 1958).

Hint. First, assume perfect conductivity and rewrite the freezing-in equation (8.44) by using $\mathbf{B} = \text{curl}\,\mathbf{A}$ as follows

$$\frac{\partial\,\mathbf{A}}{\partial t} = \mathbf{v} \times (\nabla \times \mathbf{A})\,. \tag{14.69}$$

Here $\mathbf{A}$ is the vector potential. Using Equation (14.69), show that

$$\boxed{\mathcal{H} = \int\limits_V \mathbf{A} \cdot (\nabla \times \mathbf{A})\, dV = \text{const}} \tag{14.70}$$

for all $\mathbf{A}$ which are constant on the boundary S of the region V. The integral $\mathcal{H}$ is called the *global magnetic helicity* of the closed system under consideration (Section 21.1).

Second, examine the stationary values of the magnetic energy

$$\mathcal{M} = \int\limits_V \frac{B^2}{8\pi}\, dV = \int\limits_V \frac{1}{8\pi}\, (\text{curl}\,\mathbf{A})^2\, dV. \tag{14.71}$$

Introduce a Lagrangian multiplier $\alpha/8\pi$ and obtain the following condition for stationary values

$$\delta \int\limits_V \Big[(\text{curl}\,\mathbf{A})^2 - \alpha\,\mathbf{A} \cdot \text{curl}\,\mathbf{A} \Big]\, dV = 0\,. \tag{14.72}$$

Performing the variation, the Equation (14.68) follows with $\alpha = \text{const}$. Such fields are called *linear* force-free fields.

Problem 14.3. The highly-conductive plasma in the solar corona can support an electric field $\mathbf{E}_{\parallel}$ if $E_{\parallel} \ll E_{\text{Dr}}$ where E_{Dr} is the Dreicer field (6.65). In the corona $E_{\text{Dr}} \approx 7 \times 10^{-6}\,\text{V cm}^{-1}$ (Problem 6.4). Evaluate the characteristic values of the magnetic field B and the velocity v of plasma motions in the corona which allow us to consider an equilibrium of moving plasma in the corona as a force-free one.

Answer. Let us evaluate an electric field as the electric field related to a motion of magnetic field lines in the corona

$$E_{\parallel} \approx E \approx \frac{1}{c}\, vB \approx 10^{-8}\, v\,(\mathrm{cm\ s^{-1}})\, B\,(\mathrm{G})\,, \ \ \mathrm{V\ cm^{-1}}\,. \tag{14.73}$$

From the condition that this field must be much smaller than the Dreicer field we find that

$$v\,(\mathrm{cm\ s^{-1}})\, B\,(\mathrm{G}) \ll 10^{8}\, E_{\mathrm{Dr}} \approx 7 \times 10^{2}\,. \tag{14.74}$$

So, with the magnetic field in the corona $B \sim 100$ G, the plasma motion velocity must be very small: $v \ll 10$ cm s^{-1}. Hence the solar corona hardly can remain force-free with ordinary collisional conductivity because of the motion of magnetic field lines. The electric runaway effects (Section 6.4.2) can become important even at very slow motions of the field lines in the corona. The *minimum current corona* (see Sections 16.5.4 and 16.5.6) seems to be a more realistic approximation everywhere except strongly twisted flux tubes.

Chapter 15

Stationary Flows in a Magnetic Field

There exist two different sorts of stationary flows depending on whether or not a medium can be considered as ideal or non-ideal.

15.1 Ideal plasma flows

Stationary motions of an ideal conducting medium in a magnetic field are subject to the following set of MHD equations (cf. (8.40)):

$$(\mathbf{v} \cdot \nabla)\, \mathbf{v} = -\frac{1}{\rho} \nabla \left(p + \frac{B^2}{8\pi} \right) + \frac{1}{4\pi\rho} \, (\mathbf{B} \cdot \nabla)\, \mathbf{B} \,, \tag{15.1}$$

$$\operatorname{curl} (\mathbf{v} \times \mathbf{B}) = 0 \,, \tag{15.2}$$

$$\operatorname{div} \rho \mathbf{v} = 0 \,, \tag{15.3}$$

$$\operatorname{div} \mathbf{B} = 0 \,, \tag{15.4}$$

$$(\mathbf{v} \cdot \nabla)\, s = 0 \,, \tag{15.5}$$

$$p = p\,(\rho, s) \,. \tag{15.6}$$

The induction Equation (15.2) is satisfied identically, provided the motion of the medium occurs along the magnetic field lines, i.e.

$$\boxed{\mathbf{v} \parallel \mathbf{B} \,.} \tag{15.7}$$

15.1.1 Incompressible medium

In the case of an incompressible fluid (ρ = const) Equations (15.1)–(15.6) have the general solution (Syrovatskii, 1956, 1957):

$$\mathbf{v} = \pm \frac{\mathbf{B}}{\sqrt{4\pi\rho}}\,, \tag{15.8}$$

$$\nabla\left(p + \frac{B^2}{8\pi}\right) = 0\,. \tag{15.9}$$

Here **B** is an arbitrary magnetic field: the form of the field lines is unimportant, once condition (15.4) holds. A conducting fluid flows parallel or anti-parallel to the magnetic field. We shall learn more about such equilibrium flows later on.

It follows from (15.8) that

$$\frac{\rho v^2}{2} = \frac{B^2}{8\pi}\,, \tag{15.10}$$

while Equation (15.9) gives

$$p + \frac{B^2}{8\pi} = \text{const}\,. \tag{15.11}$$

For the considered class of plasma motions along the field lines, the equipartition of energy between that of the magnetic field and the kinetic energy of the medium takes place, whereas the sum of the gas pressure and the magnetic pressure is everywhere constant.

The existence of the indicated solution means that

> an arbitrary magnetic field and an ideal incompressible medium in motion are in equilibrium, provided the motion of the medium occurs with the Alfvén speed along magnetic field lines.

Stationary flows of this type can be continuous in the whole space as well as discontinuous at some surfaces. For example, the solution (15.10) and (15.11) can be realized as a stream or non-relativistic jet of an arbitrary form, flowing in an immovable medium without a magnetic field.

Note that the tangential discontinuity at the boundary of such a jet is *stable*, since, by virtue of (15.10), the condition (11.35) by Syrovatskii is valid:

$$\frac{B^2}{8\pi} > \frac{1}{4}\,\frac{\rho v^2}{2}\,. \tag{15.12}$$

Such stable stationary jets of an incompressible fluid can close in *rings* and *loops* of an arbitrary type.

15.1.2 Compressible medium

In a compressible plasma ($\rho \neq$ const) the solution (15.8) is still possible, once the density of the plasma does not change along the field lines:

$$\mathbf{B} \cdot \nabla \rho = 0\,. \tag{15.13}$$

Obviously, this condition is necessary, but not sufficient. On substituting the solution (15.8) in Equation (15.3), we get

$$\operatorname{div} \rho \mathbf{v} = \pm \frac{1}{\sqrt{4\pi}} \left[\frac{1}{\sqrt{\rho}} \operatorname{div} \mathbf{B} - \frac{1}{2} \rho^{-3/2} \, \mathbf{B} \cdot \nabla \rho \right] = 0$$

by virtue of (15.4) and (15.13). Thus the condition (15.13) is enough for Equation (15.3) to be satisfied identically. However, to ensure the fulfilment of condition (15.9), we must require constancy of the gas and the magnetic pressure or the absolute value of the magnetic field intensity. The latter means that each magnetic flux tube must have a *constant cross-section*. Hence, by virtue of (15.8), the flow velocity along the tube will be constant as well.

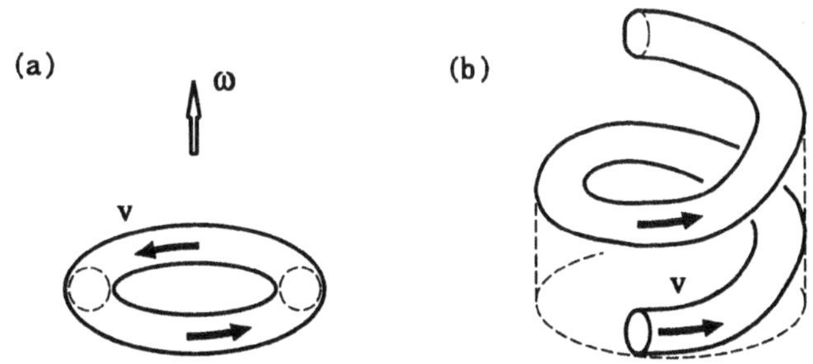

Figure 15.1: Rotational (a) and helical (b) stationary flows of a compressible plasma.

Therefore stationary flows corresponding to the solutions (15.8) and (15.9), which are **flows with a constant velocity in magnetic tubes of a constant cross-section**, are possible in a compressible medium. An example of such a flow is the plasma rotation in a ring tube (Figure 15.1a). We can envisage spiral motions of the plasma, belonging to the same type of stationary solutions in MHD (Figure 15.1b). This may be the case of an astrophysical jet when plasma presumably moves along a spiral trajectory.

15.1.3 Astrophysical collimated streams (jets)

Powerful extragalactic radio sources comprise two extended regions containing magnetic field and synchrotron-emitting relativistic electrons, each linked by a jet to a central compact radio source located in the nucleus of the associated active galaxy (e.g., Begelman, Blandford, and Rees, 1984). These jets are **collimated streams of plasma** that emerge from the nucleus in opposite directions, along which flow mass, momentum, energy, and magnetic flux. The oscillations of jets about their mean directions are observed. The origin of the jet is crucial to understanding all active nuclei (Section 8.4).

The microquasars recently discovered in our Galaxy offer a unique opportunity for a deep insight into the physical processes in relativistic jets observed in different source populations (e.g., Mirabel and Rodriguez, 1998; Atoyan and Aharonian, 1999). Microquasars are stellar-mass black holes in our Galaxy that mimic, on a small scale, many of the phenomena seen in quasars. Their discovery opens the way to study the connection between the accretion of plasma onto the black holes and the origin of the relativistic jets observed in remote quasars (Section 8.4).

In spite of the vast differences in luminosity and the sizes of microquasars in our Galaxy and those in active galaxies both phenomena are believed to be powered by gravitational energy released during the accretion of plasmas onto black holes. Since the accreting plasmas have non-zero angular momentum, they form accretion discs orbiting around black holes. If the accreting plasmas have non-zero poloidal magnetic field, the magnetic flux accumulates in the inner region of the disc to form a global poloidal field penetrating the disc. Such poloidal fields could also be generated by dynamo action inside the accretion disc.

In either case, poloidal fields are twisted by the rotating disc toward the azimuthal direction. Moreover this process extracts angular momentum from the disc, enabling efficient accretion of disc plasmas onto black holes. In addition, magnetic twist generated during this process accelerates plasmas in the surface layer of the disc toward the polar direction by the Lorentz force to form bi-directional relativistic jets which are also collimated by the magnetic force (e.g., Lovelace, 1976).

15.1.4 MHD waves of arbitrary amplitude

Let us return to the case of an incompressible medium. Consider a steady flow of the type (15.8) and (15.9) in the field shown in Figure 15.2.

Figure 15.2: A MHD wave of arbitrary amplitude.

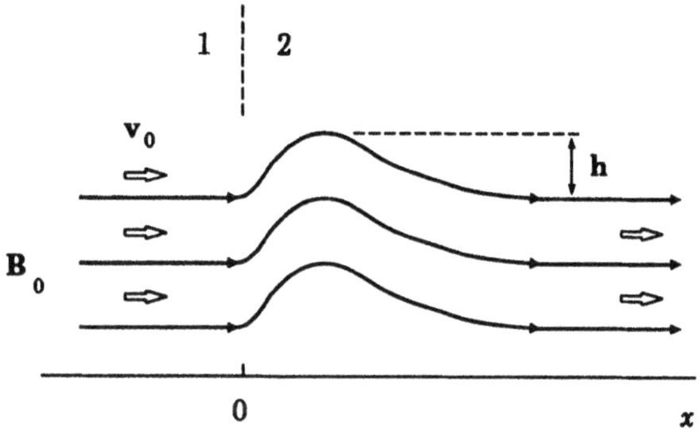

In the region 1, transformed to the frame of reference, the wave front of an arbitrary amplitude $\mathbf{h} = \mathbf{h}(x)$ runs against the immovable plasma in the uniform magnetic field $\mathbf{B}_0$, the front velocity being the Alfvén one:

$$\mathbf{v}_0 = \frac{\mathbf{B}_0}{\sqrt{4\pi\rho}} \,. \tag{15.14}$$

On the strength of condition (15.11), in such a wave

$$p + \frac{(\mathbf{B}_0 + \mathbf{h})^2}{8\pi} = p_0 + \frac{B_0^2}{8\pi} = \text{const}\,, \tag{15.15}$$

i.e. the gas pressure is balanced everywhere by the magnetic pressure.

The non-compensated part of the Lorentz force $(\mathbf{B}\cdot\nabla)\,\mathbf{B}/4\pi$, which is magnetic tensions along the field lines, provides the wave motion (cf. Section 10.2.2). In this sense, the MHD waves are analogous to elastic waves in a string. MHD waves of an arbitrary amplitude were found for the first time by Alfvén (1950) as non-stationary solutions of the MHD equations for an incompressible medium (see also Alfvén, 1981).

The Alfvén or rotational discontinuity considered in Section 11.2 is a particular case of the solutions (15.8) and (15.9), corresponding to a discontinuous velocity profile.

Behaviour of Alfvén waves in the isotropic and anisotropic cosmic plasmas can be essentially different (see Section 5.3).

15.1.5 Differential rotation and isorotation

Consider another exact solution to the stationary equations of ideal MHD. Suppose an equilibrium configuration (for example, a star) rigidly rotates about the symmetry axis of the cylindrically symmetric ($\partial/\partial\varphi = 0$) magnetic field. The angular velocity $\boldsymbol{\omega}$ is a constant vector. Then

$$\mathbf{v} = \mathbf{r} \times \boldsymbol{\omega} = \{0,\, 0,\, v_\varphi\}\,, \tag{15.16}$$

where

$$v_\varphi = \omega r .$$

The induction Equation (15.2) is satisfied identically in this case.

Now relax the assumption that $\boldsymbol{\omega}$ is a constant. Consider the case of the so-called *differential rotation*. Let the vector $\boldsymbol{\omega}$ be everywhere parallel to the z axis, i.e. the symmetry axis of the field $\mathbf{B}$, but the quantity $|\,\boldsymbol{\omega}\,| = \omega$ be dependent on the coordinates r and z, where r is the cylindrical radius:

$$\omega = \omega\,(r, z) .$$

Hence

$$v_\varphi = \omega\,(r, z)\, r . \tag{15.17}$$

Substitution of (15.17) in the induction Equation (15.2), with allowance being made for $\partial/\partial\varphi = 0$ and (15.4), gives

$$\operatorname{curl}\,(\mathbf{v} \times \mathbf{B}) = \mathbf{e}_\varphi\, r\, (\mathbf{B} \cdot \nabla\omega) = 0 .$$

Therefore

$$\boxed{\mathbf{B} \cdot \nabla\omega = 0 ,} \tag{15.18}$$

i.e. the magnetic field lines are situated at $\omega =$ const surfaces. When treated in astrophysics, this case is called *isorotation*.

As a consequence of cylindrical symmetry, the $\omega =$ const surfaces are those of rotation, hence **isorotation does not change the magnetic field**.

On the other hand, if the condition for isorotation (15.18) is not valid, **differential rotation twists the field lines** as shown in Figure 15.3, creating a *toroidal* field B_φ. The magnetic field is amplified.

Rigid rotation and isorotation are widely discussed, when applied to stellar physics, because rotation is an inherent property of the majority of the stars having strong magnetic fields (e.g., Schrijver and Zwaan, 1999). What is the actual motion of the plasma in the interior of stars?

Suppose there is no tangential stress at the surface of a star. The rigid rotation must be gradually established owing to viscosity in the star. However, the observed motion of the Sun, as a well studied example, is by no means rigid: **the equator rotates faster than the poles**. This effect cannot be explained by surface rotation. Deep layers of the Sun and fast-rotating solar-type stars participate in complex motions: differential rotation, convection, and meridional circulation (e.g., Rüdiger and von Rekowski, 1998).

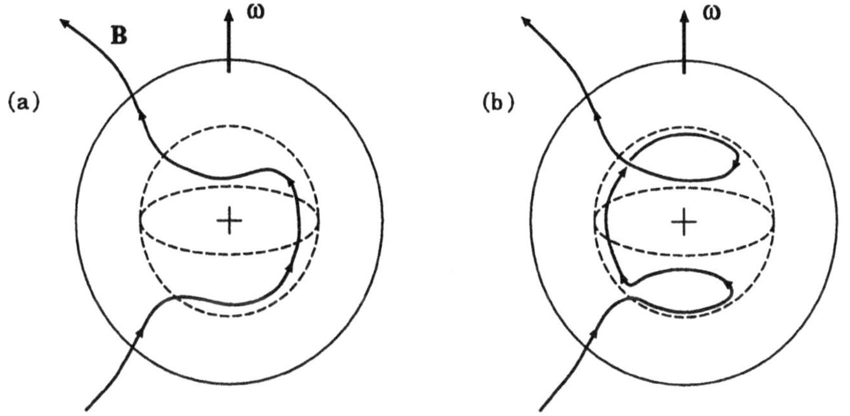

Figure 15.3: Differential rotation creates the toroidal (B_φ) component of a magnetic field inside a star.

Such motions ensure mixing of deep solar layers down to the solar core. The circumstantial evidence for this comes from observations of the solar neutrino flux as well as helioseismological data. The latter show, in particular, that **the solar core rotates faster than the surface**. Recent results of the SOHO helioseismology enable us to know the structure of the solar internal differential rotation (e.g., Schou *et al.*, 1998).

Recommended Reading: Elsasser (1956), Parker (1979), Moreau (1990).

15.2 Flows at small magnetic Reynolds numbers

While investigating MHD flows in a laboratory, the finite conductivity being significant, one has to account for the magnetic field dissipation. Furthermore, one has to take account of the fact that the freezing-in condition breaks down owing to the smallness of the magnetic Reynolds number (8.36):

$$\mathrm{Re_m} = \frac{vL}{\nu_\mathrm{m}} . \qquad (15.19)$$

The analogous situation takes place, for example, in deep layers of the solar atmosphere near the temperature minimum. The conductivity is small here, since the number of neutral atoms is relatively large (e.g., Hénoux and Somov, 1987, 1991, 1997).

Stationary flows are also possible in the case of finite conductivity. However, they differ greatly from the ideal medium flows considered in the previous section. The difference manifests itself in the fact that, given dissipative processes, steady flows can be realized only under the action of some constant external force, a pressure gradient, for instance. A second difference is that **the plasma of finite conductivity can flow across the field lines.**

15.2.1 Stationary flows inside a duct

We shall examine a flow which has been well studied for reasons of practical importance. Consider the steady flow of a viscous conducting fluid along a duct with a transversal magnetic field. Let the x axis of the Cartesian system

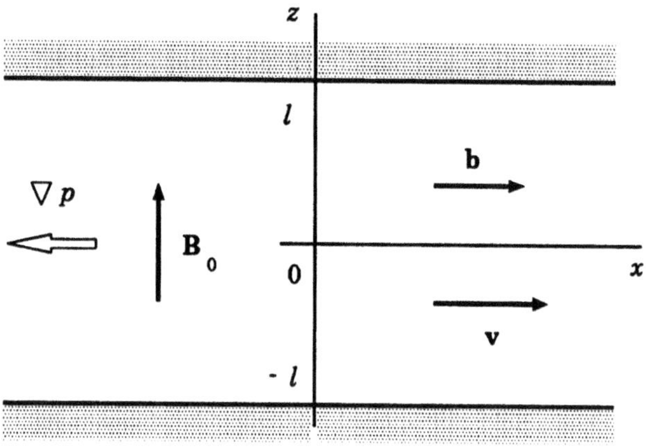

Figure 15.4: Formulation of the problem on the finite conductivity plasma flow in a duct.

(Figure 15.4) be chosen in the flow direction, the external uniform field $\mathbf{B}_0$ coinciding with the z axis:

$$\mathbf{v} = \{ v(z), 0, 0 \} , \quad \mathbf{B}_0 = \{ 0, 0, B_0 \} . \tag{15.20}$$

Let the width of the duct be $2l$.

We start from the set of Equations (8.18)–(8.23) for a steady flow of an incompressible medium:

$$\rho = \text{const} . \tag{15.21}$$

Consider two equations:

$$\text{curl} \, (\mathbf{v} \times \mathbf{B}) + \nu_{\text{m}} \Delta \mathbf{B} = 0 , \tag{15.22}$$

$$(\mathbf{v} \cdot \nabla) \mathbf{v} = - \frac{\nabla p}{\rho} - \frac{\mathbf{B} \times \text{curl} \, \mathbf{B}}{4 \pi \rho} + \nu \Delta \mathbf{v} . \tag{15.23}$$

The pressure gradient $\partial p/\partial x$ along the x axis, which is independent of x, is assumed to be the cause of the motion. Supposing the flow to be relatively slow, neglect the term on the left-hand side of Equation (15.23).

Let $b = b(z)$ be the magnetic field component along the velocity. In the coordinate form, Equations (15.22) and (15.23) are reduced to the following three equations:

$$B_0 \frac{\partial v}{\partial z} + \nu_{\rm m} \frac{\partial^2 b}{\partial z^2} = 0 , \tag{15.24}$$

$$\rho \nu \frac{\partial^2 v}{\partial z^2} + \frac{B_0}{4\pi} \frac{\partial b}{\partial z} - \frac{\partial p}{\partial x} = 0 , \tag{15.25}$$

$$\frac{\partial}{\partial z} \left(p + \frac{b^2}{8\pi} \right) = 0 . \tag{15.26}$$

Differentiating Equation (15.26) with respect to x gives

$$\frac{\partial^2 p}{\partial x \, \partial z} = 0 . \tag{15.27}$$

Differentiating (15.25) with respect to z, with care taken of (15.27), gives

$$\rho \nu \frac{\partial^3 v}{\partial z^3} + \frac{B_0}{4\pi} \frac{\partial^2 b}{\partial z^2} = 0 . \tag{15.28}$$

Eliminate $\partial^2 b/\partial z^2$ between Equations (15.24) and (15.28). The result is

$$\frac{d^3 v}{dz^3} - \frac{B_0^2}{4\pi \rho \nu \nu_{\rm m}} \frac{dv}{dz} = 0 . \tag{15.29}$$

This equation is completed by the boundary conditions on the duct walls

$$v(l) = v(-l) = 0 . \tag{15.30}$$

The corresponding solution is of the form

$$v(z) = v_0 \frac{\cosh \mathrm{Ha} - \cosh (\mathrm{Ha}\, z/l)}{\cosh \mathrm{Ha} - 1} . \tag{15.31}$$

Here $v_0 = v(0)$ is the flow velocity at the centre of the duct, the dimensionless parameter characterizing the flow is

$$\boxed{\mathrm{Ha} = \frac{l B_0}{\sqrt{4\pi \rho \nu \nu_{\rm m}}} .} \tag{15.32}$$

It is called the Hartmann number, the flow (15.31) being the Hartmann flow. As Ha $\rightarrow 0$, formula (15.31) converts to the usual parabolic velocity profile which is typical of viscous flows in a duct without a magnetic field:

$$v(z) = v_0 \left(1 - \frac{z^2}{l^2} \right) . \tag{15.33}$$

The influence of a transversal magnetic field shows itself as the appearance of **an additional drag to the plasma flow** and the change of the velocity profile which becomes *flatter* in the central part of the duct (Figure 15.5). In

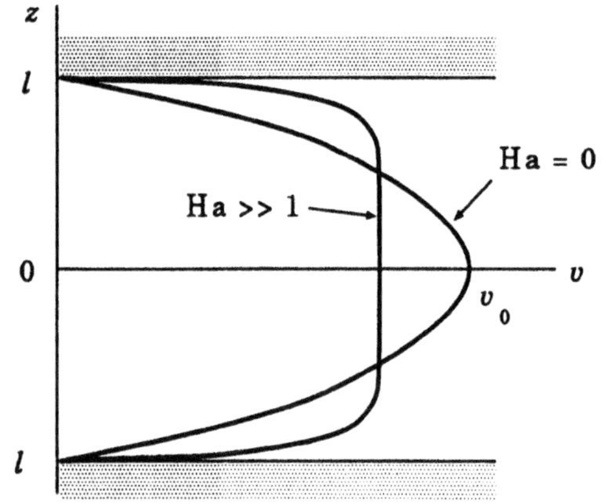

Figure 15.5: Usual parabolic (Ha = 0) and Hartmann profiles of the viscous flow velocity in a duct with a transverse magnetic field.

the limit Ha $\rightarrow \infty$, the Hartmann formula (15.31) gives

$$v(z) = v_0 \left\{ 1 - \exp \left[-\text{Ha} \left(1 - \frac{z}{l} \right) \right] \right\} . \tag{15.34}$$

Such a velocity profile is flat, $v(z) \approx v_0$, the exception being a thin layer near the walls, the *boundary layer* of the thickness l/Ha.

15.2.2 The MHD generator or pump

What factors determine the value of velocity v_0 at the center of the duct? Calculate the electric current density in the duct

$$j_y = \frac{c}{4\pi} \frac{\partial b}{\partial z} = \frac{c}{4\pi} \left(\frac{4\pi}{B_0} \frac{\partial p}{\partial x} - \rho \nu \frac{4\pi}{B_0} \frac{\partial^2 v}{\partial z^2} \right) =$$

$$= \frac{c}{B_0} \left(\frac{\partial p}{\partial x} - \rho \nu \frac{\partial^2 v}{\partial z^2} \right) . \tag{15.35}$$

Here the use is made of formula (15.25) to find the derivative $\partial b/\partial z$. Substitute in (15.35) an expression for velocity of the type (15.31), i.e.

$$v(z) = A\left(\cosh \mathrm{Ha} - \cosh \frac{\mathrm{Ha}\, z}{l}\right). \tag{15.36}$$

We get the following equation

$$\frac{j_y B_0}{c} = \frac{\partial p}{\partial x} - \rho\nu A \left(\frac{\mathrm{Ha}}{l}\right)^2 \cosh \frac{\mathrm{Ha}\, z}{l}. \tag{15.37}$$

Integrate (15.37) over z from $-l$ to $+l$. The result is

$$\frac{IB_0}{c} = 2l\,\frac{\partial p}{\partial x} - A\, 2\rho\nu \left(\frac{\mathrm{Ha}}{l}\right) \sinh \mathrm{Ha}, \tag{15.38}$$

where

$$I = \int_{-l}^{l} j_y \, dz \tag{15.39}$$

is the total current per unit length of the duct.

Finally it follows from Equation (15.38) that the sought-after coefficient in formula (15.36) is

$$\boxed{A = \frac{\partial p/\partial x - (1/2lc)\, IB_0}{(\rho\nu/l^2)\, \mathrm{Ha} \sinh \mathrm{Ha}}\,.} \tag{15.40}$$

Thus

> the velocity of the plasma flow in the duct is proportional to the gas pressure gradient and the magnetic Lorentz force.

This is why two different operational regimes are possible for the duct.

If the flow in the duct is realized under the action of an external pressure gradient, the duct operates as the MHD generator shown in Figure 15.6. The same principle explains the action of flowmeters which are important, for example, in controling the flow of the metallic heat conductor in reactors.

The second operating mode of the duct occurs when an external electromagnetic force (instead of a passive load R in Figure 15.6) creates the electric current I between the walls of the duct. Interaction of the current with the external magnetic field $\mathbf{B}_0$ gives rise to the Lorentz force that makes the plasma move along the duct, i.e. in the direction of the x axis. Hence the duct operates as the MHD pump, and this is also used in some technical applications.

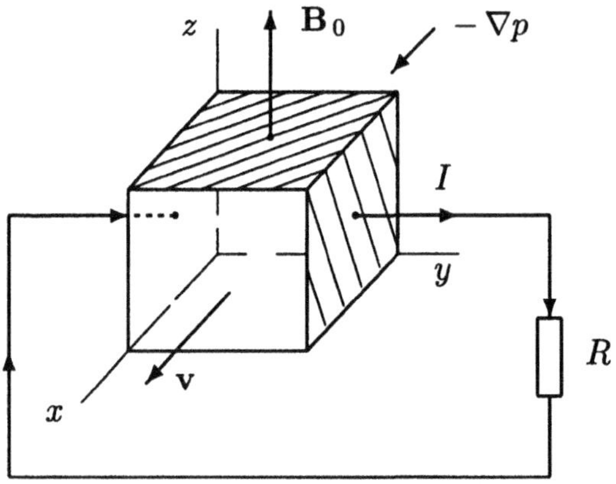

Figure 15.6: Utilization of the MHD duct as the generator of the current I; R is an external load.

15.2.3 Weakly-ionized plasma in space

Under astrophysical conditions, both operating modes of the MHD duct can be realized, once the plasma conductivity is small due, for instance, to its low temperature. In the solar atmosphere, in the minimum temperature region, neutral atoms move in the directions of convective flows and collide with ions, thus setting them in motion. At the same time, electrons remain 'frozen' in the magnetic field. Obviously, this effect (termed the 'photospheric dynamo') can generate electric currents and amplify the magnetic field in the photosphere and the chromosphere (Section 21.3).

A violent outflow of high-velocity weakly-ionized plasma is one of the first manifestations of the formation of a new stars (Bachiller, 1996; Bontemps *et al.*, 1996). Such outflows emerge bipolarly from the young object and involve amounts of energy similar to those involved in accretion processes. The youngest proto-stellar low-mass objects known to date (the class 0 protostars) present a particularly efficient outflow activity, indicating that outflow and infall motions happen simultaneously and are closely linked since the very first stages of the star formation processes.

The idea of a new star forming from relatively simple hydrodynamic infall of weakly-ionized plasma is giving place to a picture in which magnetic fields play a crucial role and stars are born through the formation of complex engines of accreation/ejection. It seems inevitable that future theories of star formation will have to take into account, together with the structure of the protostar and its surrounding accretion disc, the processes related to **multi-fluid hydrodynamics of weakly-ionized plasma**. These are the effects similar to the photospheric dynamo (Section 21.3) and magnetic reconnection in weakly-ionized plasma (Section 21.2).

Recommended Reading: Sutton and Sherman (1965), Ramos and Winowich (1986).

15.3 The σ-dependent force and vortex flows

15.3.1 Simplifications and problem formulation

As was shown in Section 14.4, a body plunged into a conducting fluid with magnetic and electric fields is acted upon by an *expulsion* force which is the counterpart of the Archimedean force in ordinary hydrodynamics or, more exactly, by the magnetic ***σ*-*dependent*** force. As this takes place, the electric field $\mathbf{E}$ and current density $\mathbf{j}$ are generally non-uniform, and the volume **Lorentz force inside the fluid is non-potential.** The force generates vortex flows of the fluid in the vicinity of the body. It is of interest to determine the character of these flows and the forces acting on the submerged body.

(a) Consider the stationary problem for an incompressible fluid having uniform constant viscosity ν and magnetic diffusivity $\nu_{\rm m}$ (Syrovatskii and Chesalin, 1963; Marty and Alemany, 1983; Gerbeth *et al.*, 1990). Let, at first, **both the usual and magnetic Reynolds numbers be small**:

$$\mathrm{Re} = \frac{vL}{\nu} \ll 1\,, \tag{15.41}$$

$$\mathrm{Re}_{\rm m} = \frac{vL}{\nu_{\rm m}} \ll 1\,. \tag{15.42}$$

The freezing-in condition (8.37) can be rewritten in the form

$$\Delta \mathbf{B} + \mathrm{Re}_{\rm m}\, \mathrm{curl}\,(\, \mathbf{v} \times \mathbf{B}\,) = 0\,, \tag{15.43}$$

where, in view of (15.42), $\mathrm{Re}_{\rm m}$ is a small parameter. In a zeroth approximation in this parameter, the magnetic field is potential:

$$\Delta \mathbf{B} = 0\,.$$

Moreover, the magnetic field will be assumed to be uniform, in accordance with the formulation of the problem discussed in Section 14.4. Strictly speaking, the assumption of **a uniform magnetic field** implies the inequality

$$B \gg \frac{4\pi}{c}\, Lj\,. \tag{15.44}$$

Its applicability will be discussed later on, in connection with the simplified form of Ohm's law to be used while solving the problem.

(b) Assuming the stationary flows occurring in the fluid to be slow, the inertial force (proportional to v^2) will be ignored in the equation of motion (15.23) as compared to the other forces: **pressure gradient, Lorentz force, viscous force**. The term describing the gravity force will be dropped, since its effect has already been studied in Section 14.4. Finally, on multiplying the equation

$$0 = -\frac{\nabla p}{\rho} - \frac{\mathbf{B} \times \operatorname{curl} \mathbf{B}}{4\pi\rho} + \nu\, \Delta \mathbf{v}$$

by the fluid density $\rho = \rho_0$, it is rewritten in the form

$$\eta\, \Delta \mathbf{v} = \nabla p - \mathbf{f}\,. \tag{15.45}$$

Here $\eta = \rho_0\, \nu$ is the *dynamic* viscosity coefficient, and

$$\mathbf{f} = \frac{1}{c}\, \mathbf{j} \times \mathbf{B}_0 \tag{15.46}$$

is the Lorentz force in the same approximation.

Recall that, in view of the assumed incompressibility of the fluid, the velocity field obeys the equation

$$\operatorname{div} \mathbf{v} = 0\,, \tag{15.47}$$

which formally follows from the continuity Equation (15.3).

(c) The electric field $\mathbf{E}$ is assumed to be uniform at infinity

$$\mathbf{E} \to \mathbf{E}_0\,, \quad r \to \infty\,. \tag{15.48}$$

Then, given the conductivities of the fluid σ_0 and of the submerged body σ_1, we can find the current $\mathbf{j}$ in the whole space using the following conditions:

$$\operatorname{div} \mathbf{j} = 0\,, \tag{15.49}$$

$$\mathbf{j} = \sigma \mathbf{E}\,, \tag{15.50}$$

$$\operatorname{curl} \mathbf{E} = 0\,. \tag{15.51}$$

The current $(\sigma/c)\, \mathbf{v} \times \mathbf{B}$ has been ignored in Ohm's law (15.50). This may be done, once the velocity of engendered vortex flows is much less than the drift velocity, i.e. once the inequality

$$v \ll v_{\mathrm{d}} = c\, \frac{E}{B} \tag{15.52}$$

holds. Note that substituting (15.44) in (15.52) results in the inequality

$$\frac{vL}{(c^2/4\pi\sigma)} \ll 1\,, \tag{15.53}$$

which coincides with the initial assumption (15.42).

15.3.2 The solution for a spherical ball

Let us solve the problem for a ball of radius a. Choose the Cartesian frame of reference, in which the direction of the x axis is parallel to $\mathbf{E}_0$, and the origin of coordinates coincides with the center of the ball as shown in Figure 15.7.

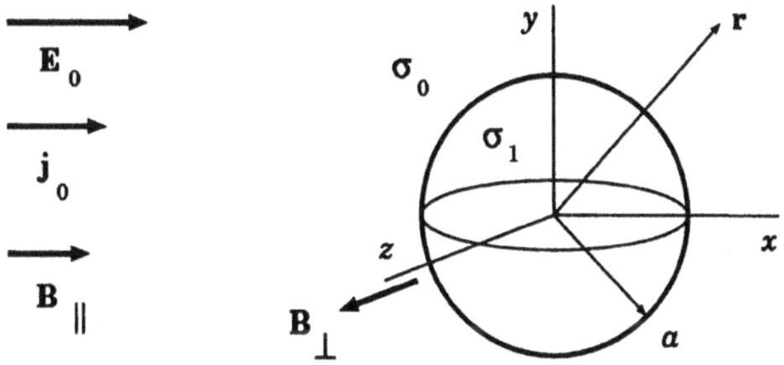

Figure 15.7: An uniform conducting ball of radius a, submerged in a conducting fluid with electric and magnetic fields.

By virtue of Ohm's law (15.50), the electric current at infinity

$$\mathbf{j}_0 = \sigma_0\,\mathbf{E}_0 \tag{15.54}$$

is also parallel to the x axis.

It follows from Equation (15.51) that the current can be represented in the form

$$\mathbf{j} = \nabla\psi\,. \tag{15.55}$$

Here a scalar function ψ, in view of Equation (15.49), satisfies the Laplace equation

$$\Delta\psi = 0\,. \tag{15.56}$$

Let us try to find the solution to the problem in the form of uniform and dipole components:

$$\psi = \mathbf{j}_0 \cdot \mathbf{r} + c_0\,\mathbf{j}_0 \cdot \nabla\frac{1}{r}\,, \quad r \geq a\,, \tag{15.57}$$

and

$$\psi = \mathbf{j}_1 \cdot \mathbf{r}, \qquad r < a. \tag{15.58}$$

Here c_0 is an unknown constant, $\mathbf{j}_1 = \{j_1, 0, 0\}$ is an unknown current density inside the ball. Both unknowns are to be found from the matching conditions at the surface of the ball:

$$\{j_r\} = 0 \quad \text{and} \quad \{\mathbf{E}_\tau\} = 0.$$

Obviously, these conditions can be rewritten as follows

$$\frac{\mathbf{j} \cdot \mathbf{r}}{r} = \frac{\mathbf{j}_1 \cdot \mathbf{r}}{r} \quad \text{at} \quad r = a, \tag{15.59}$$

and

$$\frac{\mathbf{j}_\tau}{\sigma_0} = \frac{\mathbf{j}_{\tau 1}}{\sigma_1} \quad \text{at} \quad r = a. \tag{15.60}$$

On substituting (15.57) and (15.58) in (15.59) and (15.60), the constants c_0 and j_1 are found. The result is

$$\psi = \left[1 + \beta \left(\frac{a}{r} \right)^3 \right] \mathbf{j}_0 \cdot \mathbf{r} \quad \text{for} \quad r \geq a, \tag{15.61}$$

and

$$\psi = (1 - 2\beta)\, \mathbf{j}_0 \cdot \mathbf{r} \quad \text{for} \quad r < a. \tag{15.62}$$

Here the constant

$$\beta = \frac{\sigma_0 - \sigma_1}{2\sigma_0 + \sigma_1}. \tag{15.63}$$

Specifically, inside the ball

$$\mathbf{j}_1 = (1 - 2\beta)\, \mathbf{j}_0, \tag{15.64}$$

and $\mathbf{j}_1 = \mathbf{j}_0$, once $\sigma_1 = \sigma_0$.

15.3.3 Forces and flows near a spherical ball

Knowing the current in the whole space, the Lorentz force (15.46) can be found

$$\mathbf{f} = \frac{1}{c} \nabla\psi \times \mathbf{B}_0 = \operatorname{curl} \frac{\psi \mathbf{B}_0}{c}. \tag{15.65}$$

In the case at hand,

the volume Lorentz force has a rotational character and hence generates vortex flows in the conducting fluid.

Operate with curl curl on Equation (15.45). Using the known vector identity

$$\operatorname{curl}\operatorname{curl}\mathbf{a} = \nabla(\nabla\mathbf{a}) - \Delta\,\mathbf{a}$$

and taking account of relations (15.49)–(15.51), a biharmonic equation for the velocity field is obtained

$$\Delta\Delta\,\mathbf{v} = 0\,. \tag{15.66}$$

Operating with divergence on (15.45) and taking account of (15.49)–(15.51), we get

$$\Delta p = 0\,. \tag{15.67}$$

Equations (15.66) and (15.67) are to be solved together with Equations (15.45) and (15.47). For bodies with spherical or cylindrical symmetry, it is convenient to make use of the identity

$$\mathbf{r}\cdot\Delta\mathbf{q} = \Delta\,(\mathbf{q}\cdot\mathbf{r})\,, \tag{15.68}$$

where $\mathbf{q}$ is any vector satisfying the condition $\operatorname{div}\mathbf{q} = 0$. Then from Equation (15.66) subject to the condition (15.47) we find

$$\Delta\Delta\,(v_r\,r) = 0\,. \tag{15.69}$$

The boundary conditions are taken to be

$$v\,|_S = 0\,, \qquad v\,|_\infty = 0\,. \tag{15.70}$$

Here S is the surface of the submerged body which is assumed to be a ball of radius a (cf. Figure 15.7). At its surface $r = a = \text{const}$; Equation (15.47) and the first of conditions (15.70) gives

$$\left.\frac{\partial v_r}{\partial r}\right|_S = 0\,. \tag{15.71}$$

The solution of Equation (15.69), satisfying the boundary condition (15.71) and the second of conditions (15.70), is clearly seen to be

$$v_r \equiv 0\,. \tag{15.72}$$

Thus

in the case of a spherical ball, the flow lines of a conducting incompressible fluid are situated at $r = \text{const}$ surfaces.

Next an equation for the pressure is found using Equation (15.45) and taking into account that, by virtue of (15.68),

$$\mathbf{r} \cdot \Delta \mathbf{v} = \Delta \left(v_r r \right) = 0 \, .$$

The resulting equation is

$$\frac{\partial p}{\partial r} = f_r \, . \tag{15.73}$$

The function f_r occuring on the right-hand side is the radial component of the above mentioned Lorentz force (15.65).

Once the pressure has been found by integrating Equations (15.73) and (15.67), the velocity is determined from Equation (15.45) with the known right-hand side.

Choose the Cartesian frame of reference in which

$$\mathbf{B}_0 = \{ B_{0x}, \; 0, \; B_{0z} \} \, ,$$

$B_{0x} = B_{\parallel}$ and $B_{0z} = B_{\perp}$ being the magnetic field components parallel and perpendicular to $\mathbf{j}_0$, respectively (see Figure 15.7). The current in the conducting fluid (cf. formula (15.61)) is

$$\mathbf{j} = \nabla \psi \, , \qquad \psi = j_0 \, x + j_0 \, \frac{\beta \, a^3 x}{r^3} \, , \tag{15.74}$$

the current inside the ball being defined by formula (15.64). The pressure in the fluid

$$p = \frac{1}{c} \, j_0 B_{\perp} \, y \left(\frac{\beta a^3}{2 r^3} - 1 \right) + \text{const} \, . \tag{15.75}$$

It is convenient to rewrite the velocity distribution in spherical coordinates

$$\mathbf{v} = \{ v_r, \; v_\theta, \; v_\varphi \} \tag{15.76}$$

(cf. Syrovatskii and Chesalin, 1963):

$$\begin{aligned}
v_r &= 0 \, , \\
v_\theta &= \frac{\beta j_0 a^2}{4 c \eta} \, \frac{a}{r} \left(1 - \frac{a^2}{r^2} \right) \left(- B_{\perp} \cos\theta \, \sin\varphi + B_{\parallel} \sin\theta \, \sin 2\varphi \right) , \\
v_\varphi &= \frac{\beta j_0 a^2}{4 c \eta} \, \frac{a}{r} \left(1 - \frac{a^2}{r^2} \right) \left(B_{\perp} \cos 2\theta \, \cos\varphi + B_{\parallel} \sin 2\theta \, \cos^2 \varphi \right) .
\end{aligned}$$

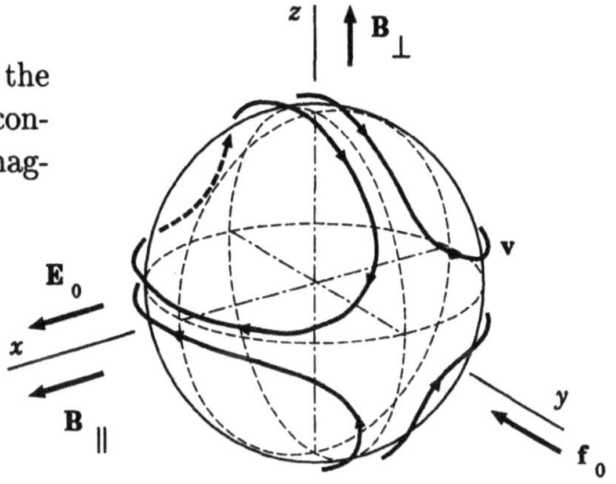

Figure 15.8: Vortex flows near the conducting ball submerged in a conducting fluid with electric and magnetic fields.

This velocity field pattern is shown in Figure 15.8.

The force acting on the body is defined to be (cf. formula (14.62))

$$\mathbf{F} = \frac{1}{c} \int\limits_V \mathbf{j} \times \mathbf{B}_0 \, dV + \oint\limits_S p \, \mathbf{n} \, dS - \oint\limits_S \sigma'_{\mathrm{n}} \, dS , \tag{15.77}$$

where $\mathbf{n}$ is the inward normal to the sphere;

$$\sigma'_{\mathrm{n}} = (\sigma'_{\alpha\beta} n_\beta)_{\mathrm{n}} , \tag{15.78}$$

$\sigma'_{\alpha\beta}$ being the viscous stress tensor, see definition (8.28).

On substituting the velocity distribution (15.76) in the viscous force formula (15.78) and integrating (15.77) over the surface S of the ball,

> the sum of the viscous forces is concluded to be zero. The moment of the viscous forces acting on the ball is also zero.

The remaining force determined by (15.77) is directed along the y axis and is equal to

$$F = \frac{4\pi a^3}{3} \frac{j_0 B_\perp}{c} \left\{ -(1-2\beta) + \left(1 - \frac{\beta}{2}\right) \right\} . \tag{15.79}$$

The constant β is defined by formula (15.63):

$$\beta = \frac{\sigma_0 - \sigma_1}{2\sigma_0 + \sigma_1} .$$

The first term in the curly brackets corresponds to the force $\mathbf{j}_1 \times \mathbf{B}_0 / c$ which immediately acts on the current $\mathbf{j}_1$ inside the ball. Note that

$$1 - 2\beta = \frac{3\sigma_1}{2\sigma_0 + \sigma_1} > 0 ,$$

in agreement with the direction of the vector product $\mathbf{j}_1 \times \mathbf{B}_0$ or $\mathbf{j}_0 \times \mathbf{B}_0$ (Figure 15.7). Moreover, provided $\sigma_1 = 0$, the term $(1 - 2\beta) = 0$ as it should be the case for a non-conducting ball, since there is no current inside it.

The second term in the curly brackets of formula (15.79) expresses the sum of the forces of the pressure on the surface of the ball. The coefficient

$$1 - \frac{\beta}{2} = \frac{3\,(\sigma_0 + \sigma_1)}{2\,(2\sigma_0 + \sigma_1)} > 0\,,$$

signifying that

> the actual σ-dependent force is always somewhat less than the force owing to the interaction of the current $\mathbf{j}_1$ and the magnetic field $\mathbf{B}_0$. Moreover, the total force can be opposite in sign.

In the particular case $\sigma_1 = 0$, when the current $j_1 = 0$

$$1 - \frac{\beta}{2} = \frac{3}{4}\,.$$

Hence $F > 0$. The non-conducting ball is expelled in the direction opposite to that of the vector product $\mathbf{j}_0 \times \mathbf{B}_0$ (Figure 15.9). The above properties of

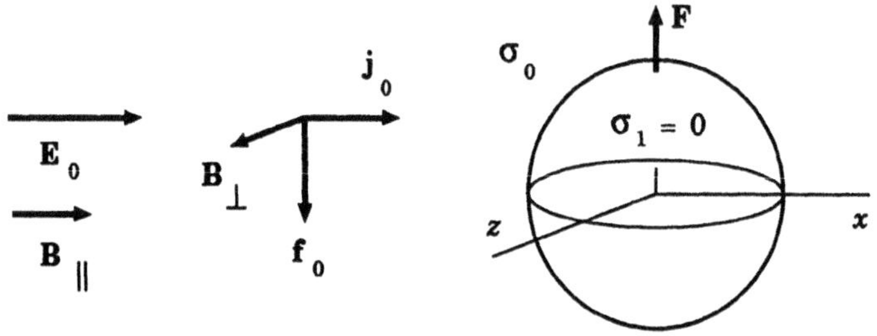

Figure 15.9: The expulsion force **F** acting on the non-conducting ball submerged in a conducting fluid with electric and magnetic fields.

the magnetic σ-dependent force are used in technical MHD. They constitute the principle of action for magnetic separators which are intended for dividing mechanical mixtures having different conductivities.

Having the physical sense of the two terms determining the magnetic σ-dependent force (15.79), let us combine them in the following descriptive formula:

$$\boxed{\mathbf{F} = -\,\mathbf{f}_0\,V \times \frac{3}{2}\beta\,.} \tag{15.80}$$

Here $V = 4\pi a^3/3$ is the volume of the ball, $\mathbf{f}_0 = \mathbf{j}_0 \times \mathbf{B}_0/c$ is the Lorentz force in the conducting fluid with uniform magnetic $\mathbf{B}_0$ and electric $\mathbf{E}_0$ fields (cf. (14.61)), the coefficient β being determined by formula (15.63).

15.4 Large magnetic Reynolds numbers

In the previous section we have considered the solution to the MHD problem concerning the magnetic σ-dependent force in the limit of small (usual and magnetic) Reynolds numbers. Leenov and Kolin (1954) seem to have been the first to obtain similar solutions in connection with the problem of electromagnetophoresis.

As a rule the opposite limiting case is applicable for astrophysical use. In this case, the problem of the magnetic σ-dependent force is difficult and can hardly be solved completely, especially given

$$\mathrm{Re} \ll 1\,, \qquad \mathrm{Re}_{\mathrm{m}} \gg 1\,. \tag{15.81}$$

A situation of this kind occurs, for example, in solar prominences (Section 15.4.2). In what follows we will show (Litvinenko and Somov, 1994a; Somov, 1994) that an expression for the magnetic σ-dependent force can be found for large magnetic Reynolds numbers, without rigorous calculations of the characteristics of the plasma flow near a body.

15.4.1 The general formula for the σ-dependent force

The equations of stationary MHD for flows of an incompressible fluid with density ρ_0 and dynamic viscosity $\eta = \rho_0 \nu$ are of the form:

$$\rho\,(\mathbf{v}\cdot\nabla)\,\mathbf{v} = -\nabla p + \frac{1}{c}\,\mathbf{j}\times\mathbf{B} + \eta\,\Delta\mathbf{v}\,,$$

$$\mathrm{curl}\,(\mathbf{v}\times\mathbf{B}) + \nu_{\mathrm{m}}\,\Delta\mathbf{B} = 0\,, \tag{15.82}$$

$$\mathrm{curl}\,\mathbf{B} = \frac{4\pi}{c}\,\mathbf{j}\,, \qquad \mathrm{div}\,\mathbf{v} = 0\,, \qquad \mathrm{div}\,\mathbf{B} = 0\,.$$

Let us find the σ-dependent force density $\mathbf{f}$ on the basis of *similarity* considerations. The given set of equations implies that five quantities are the determining parameters of the problem: ν, ν_{m}, a, ρ_0, and $\mathbf{f}_0$. By way of example, velocity v_0 depends on these parameters. Hence v_0 rather than

ρ_0 may be treated as a determining parameter. The standard procedure of *dimensional analysis*, described by Bridgman (1931), gives us the formula

$$\mathbf{f} = -\,\mathbf{f}_0\,\Phi\,(\mathrm{Re}, \mathrm{Re_m})\,. \tag{15.83}$$

In the limit $\mathrm{Re_m} = 0$ it reproduces (in a slightly different notation) the result presented in the theoretical part of the paper by Andres *et al.* (1963). Experimental data, which are stated in the same paper for $\mathrm{Re} < 10^2$, allow one to conclude that, with an accuracy which is completely sufficient for astrophysical applications,

$$\Phi\,(\mathrm{Re}, \mathrm{Re_m}) \approx \Phi_1(\mathrm{Re_m})\,, \tag{15.84}$$

where $\Phi_1(0) \approx 1$.

Generally, the behaviour of the magnetic field lines near the body for $\mathrm{Re_m} \neq 0$ can become nonregular and intricate, as a consequence of the electric current redistribution and vortex flow generation. For example, if $\mathrm{Re_m} < 1$, then the value of the nonregular field component $\delta B \approx \mathrm{Re_m}\, B_0$. The effective magnitude of the field and the magnetic σ-dependent force decrease as compared to the case $\mathrm{Re_m} = 0$.

The form of the decreasing function Φ_1 for $\mathrm{Re_m} \gg 1$ can be determined as follows. Far from the body, at infinity, the electromagnetic energy flux is equal to

$$\mathbf{G}_0 = \frac{c}{4\pi}\,\mathbf{E}_0 \times \mathbf{B}_0\,. \tag{15.85}$$

In close proximity to the body, the magnitude of the Poynting vector must diminish once the disordered behaviour of lines of force is assumed. The difference $(G_0 - G)$ is equal to the power of engendered vortex flows, hence generally we get

$$f a^3\, v_0 \leq G_0\, a^2. \tag{15.86}$$

The equality (15.86) is achieved in the limit $\mathrm{Re_m} \to \infty$. Here the characteristic velocity v_0 is determined from the equation of motion in the set (15.82):

$$v_0 = f a^2/\eta \qquad \text{for} \quad \mathrm{Re} \ll 1\,, \tag{15.87}$$

$$v_0 = (f a/\rho_0)^{1/2} \quad \text{for} \quad \mathrm{Re} \gg 1\,. \tag{15.88}$$

When $\mathrm{Re_m} \to \infty$, relations (15.84)–(15.88) allow us to obtain the sought-after function appearing in formula (15.83):

$$\Phi\,(\mathrm{Re}, \mathrm{Re_m}) = \begin{cases} 1 & \text{for} \quad \mathrm{Re_m} < 1\,, \\ \mathrm{Re_m^{-1}} & \text{for} \quad \mathrm{Re_m} > 1\,. \end{cases} \tag{15.89}$$

The case $\mathrm{Re_m} < 1$ was treated by Leenov and Kolin (1954).

Strictly speaking, we could take also into account the dependence of the function Φ on the usual Reynolds number Re. We could obtain

$$\Phi\,(\mathrm{Re}, \mathrm{Re_m}) = \frac{1}{\mathrm{Re_m}}\,\Phi_2\,(\mathrm{Re})\,, \tag{15.90}$$

where the function $\Phi_2\,(\mathrm{Re})$ is practically constant.

Note that formula (15.90) can be interpreted as a manifestation of an *incomplete self-similarity* of the function Φ relative to the similarity parameter $\mathrm{Re_m}$ (Barenblatt, 1979). The point is that, from the viewpoint of a 'naive' analysis, the function Φ does not depend on a dimensionless parameter whose magnitude is much greater (or less) than unity. This statement is true only if there exists a final non-zero limit of the function Φ as the parameter at hand tends to infinity (or zero). However, in general this is not the case, as is clearly demonstrated by (15.90). In fact, $\Phi \to 0$ when $\mathrm{Re_m} \to \infty$. At the same time the function Φ is a power-law one in $\mathrm{Re_m}$; that allows us to write down an expression for the force density $\mathbf{f}$ in a self-similar form. Naturally, as this takes place, the exact form of dimensionless combinations cannot be determined from the formal dimensional analysis alone.

Therefore an order-of magnitude expression is obtained for the density of the magnetic σ-dependent force acting on a body submerged into a conducting liquid or plasma (Litvinenko and Somov, 1994a; Somov, 1994):

$$\mathbf{f} = -\frac{c}{4\pi v_0\, a}\,\mathbf{E}_0 \times \mathbf{B}_0\,. \tag{15.91}$$

The expression (15.91) is valid in the limit of large magnetic Reynolds numbers. For a body with a non-zero conductivity σ_1, the electric current flowing inside the body must be taken care of in formula (15.91). The corresponding treatment was presented in Section 15.3.

The physical sense of formula (15.91) is obvious. Comparison of (15.91) with formula (14.61) for the σ-dependent force, which then holds a uniform current flow in the medium, shows that for $\mathrm{Re_m} \to \infty$ ($\sigma \to \infty$) the plasma in the vicinity of the body possesses, as it were, an *effective* conductivity

$$\boxed{\sigma_{\mathrm{ef}} \approx \frac{c^2}{v_0 a}\,.} \tag{15.92}$$

This finite conductivity of a plasma is a result of the electromagnetic energy losses to generation of *macroscopic* vortex flows.

This mechanism of conductivity of a plasma is different from the usual microscopic one, in which energy losses result from Coulomb collisions of current-carrying electrons with thermal electrons and ions of the plasma. It is no accident that an expression for conductivity, which is equivalent to (15.92), has emerged in quite another problem – while calculating the electrical resistivity of necks in Z-pinches appearing in a highly conductive plasma (Chernov and Yan'kov, 1982).

Note in this connection that the σ-dependent force, as well as the characteristic velocity of the plasma flow, depends in a *non-linear* way on the quantity $E_0 B_0$. Using (15.88), (15.88) and (15.91), we see that

$$f \sim \begin{cases} (E_0 B_0)^{1/2}\,, & \mathrm{Re} \ll 1\,, \\ (E_0 B_0)^{2/3}\,, & \mathrm{Re} \gg 1\,. \end{cases} \tag{15.93}$$

Litvinenko and Somov (1994a) have supposed that

the magnetic σ-dependent force may play an important part in the dynamics of a cosmic plasma with a non-uniform distribution of temperature and, hence, electric conductivity.

It is this force that can generate large-scale vortex flows of plasma in space. This possibility is illustrated in the next Section.

15.4.2 The σ-dependent force in solar prominences

The solar corona is a natural 'astrophysical laboratory' where formula (15.91), which is applicable at large magnetic Reynolds numbers, can be tested. Recall several of its characteristics: low density $\rho_0 \approx 10^{-16}\,\mathrm{g\,cm^{-3}}$, high temperature $T_0 \approx 10^6\,\mathrm{K}$, dynamic viscosity $\eta \approx 1\,\mathrm{g\,cm^{-1}\,s^{-1}}$, magnetic field $B_0 \approx 10 - 100\,\mathrm{G}$, electric field $E_0 \approx 10^{-5}$ CGSE units.

On the other hand, according to observational data (Tandberg-Hanssen, 1995), prominences consist of numerous fine threads – *cold dense* formations having a transversal scale $a \approx 10^7\,\mathrm{cm}$ and temperature $T_1 \approx 10^4\,\mathrm{K}$. Hence the ratio

$$\sigma_1 / \sigma_0 \approx 10^{-3} \ll 1\,,$$

as applied to prominences in the corona. In the vicinity of the threads, as well as near a prominence as a whole, rather fast plasma flows are actually observed.

According to the model under discussion, these flows can be generated by the vortex component of the magnetic σ-dependent force. For Re $\ll$ 1, their maximum velocity, as follows from relations (15.88) and (15.91), is determined by the expression

$$v_0 \approx \left(\frac{cE_0 B_0 a}{4\pi\eta} \right)^{1/2} \approx 10 - 30 \text{ km s}^{-1}, \tag{15.94}$$

that, generally speaking, corresponds to the characteristic values of observed velocities. However, the spatial resolution of modern optical, UV and soft X-ray observations is smaller than is necessary for the model to be confirmed or refuted. Consider another possibility.

The symmetric distribution of velocities on the line-of-sight projection (i.e., in the direction towards the observer) is a distinguishing feature of the model since it predicts the presence of **a large number of vortex flows of plasma** inside the prominence. Such a distribution can be observed as a symmetric broadening of spectral lines, which it will be necessary to study if one wishes to study the effect quantitatevly. A similar observational effect can be related to the **existence of reconnecting current sheets** in the same region (Antonucci and Somov, 1992; Antonucci *et al.*, 1996).

The gravity force acting on the prominences is supposed to be balanced by the σ-dependent expulsion. The equilibrium condition makes it possible to evaluate the characteristic value of the plasma density inside the fine threads forming the prominence

$$(\rho_1 - \rho_0)\, g_\odot \approx f\,. \tag{15.95}$$

Here the specific gravity of the Sun $g_\odot \approx 3 \times 10^4 \text{ cm s}^{-2}$. Formulae (15.91), (15.92), and (15.95) result in

$$\rho_1 \approx \left(\frac{cE_0 B_0\, \eta}{4\pi\, g_\odot^2 a^3} \right)^{1/2} \approx 3 \times 10^{-13} \text{ g cm}^{-3}, \tag{15.96}$$

in accordance with observational data.

Even faster motions with characteristic velocities $10^2 - 10^3 \text{ km s}^{-1}$ in so-called *eruptive* prominences are probably a consequence of the fact that the coronal fields $\mathbf{E}_0$ and $\mathbf{B}_0$ can change (in magnitude or direction) during the course of evolution. As this takes place, the equilibrium described by equation (15.95) can be violated.

Observations with high spectral resolution in UV and soft X-ray ranges are necessary to study the effect of the magnetic force stimulated by the presence of plasma regions with considerably different conductivity in the solar atmosphere.

15.5 Practice: Problems and Answers

Problem 15.1. Discuss a possible behavior of electrically conducting spheres in an insulating bounded liquid placed in a vertical traveling magnetic field.

Hint. The spheres move in response to the induced electromagnetic forces, the motion being influenced by gravity, viscous drag, vessel boundary reaction, and collisions. The range of possible behaviors, stable, unstable, and chaotic, is very wide. The term "electromagnetic billiards" seems appropriate to describe this phenomenon (Bolcato *et al.*, 1993).

Chapter 16

Magnetic Reconnection in Current Sheets

Reconnection in cosmic plasma serves as a highly efficient engine to convert magnetic energy into thermal and kinetic energies of plasma flows and accelerated particles.

16.1 Small perturbations near a neutral line

16.1.1 Historical comments

The notion of *reconnection* of magnetic field lines – magnetic reconnection – came into existence in the context of the interpretation of solar flare observations. The review of early works in the field is contained, for example in the eminent paper by Sweet (1969). From the viewpoint of reconnection, **points and lines where the magnetic field is zero are peculiarities**. This special feature, which is of a topological nature, has already been mentioned in Section 4.4 (see Figure 4.16).

Giovanelli (1947) pointed out that a *highly concentrated* electric current appears readily at an X-type zeroth (or neutral) point in a highly conducting plasma. This is really true and important. Dungey (1958) seems to have been the first to put forward the idea that

> unusual *electrodynamic* properties of a plasma emerge in the vicinity of a neutral (or zeroth) point of type X.

Since there was no clear view of the physical essence of reconnection, the notion has been accepted uncritically. It was assumed, for instance, that the mere existence of a zeroth point inevitably leads to spontaneous compression of a magneto-plasma configuration and rapid dissipation of the magnetic field, i.e. a flare (Dungey, 1958; Severnyi, 1963).

However, as was shown by Syrovatskii (1963), given magnetostatic equilibrium near a zeroth point, the plasma is stable with respect to spontaneous compression. The situation changes once **the plasma near the zeroth point is subject to an outside action due to an electric field** as shown in Figure 4.18 or due to a MHD wave which is created, for instance, by changes of the magnetic field sources at the photosphere (Figure 4.15).

This action gives rise to an original *cumulative effect* (Syrovatskii, 1966a). We attempted to understand this fundamental property at the qualitative level in Section 4.4. Let us illustrate it by the example of the behaviour of *small* MHD perturbations near the zeroth line. Bearing the solar flare case in mind, we consider the reconnection process in the approximation of a strong magnetic field at first.

16.1.2 Reconnection of strong magnetic fields

Let two equal currents I flow parallel to the axis z on lines $x = 0$, $y = \pm l$ (see Figure 4.17). The magnetic field of these currents is expressed with the aid of the vector-potential $\mathbf{A}_0$ having only the z component (Section 9.2):

$$\mathbf{A}_0 = \{ 0, 0, A_0 (x, y) \} ,$$

where

$$A_0 (x, y) = \frac{I}{c} \left\{ \ln \left[x^2 + (y - l)^2 \right] + \ln \left[x^2 + (y + l)^2 \right] \right\} . \tag{16.1}$$

Near the zeroth line situated on the z axis, formula (16.1) may be expanded in a Teylor series, the square terms of the expansion being sufficient for our purposes:

$$A_0 (x, y) = \frac{2I}{c} (x^2 - y^2)$$

or

$$A_0 (x, y) = \frac{h_0}{2} (x^2 - y^2) . \tag{16.2}$$

Here $h_0 = 4I/c$ is the magnetic field gradient in the vicinity of the zeroth line. The gradient of the field is an important characteristic of a reconnection region. In fact,

$$\mathbf{B}_0 = \text{curl}\, \mathbf{A}_0 = \left\{ \frac{\partial A_0}{\partial y} , - \frac{\partial A_0}{\partial x} , 0 \right\} = \{ - h_0 y , - h_0 x , 0 \} . \tag{16.3}$$

The field lines of the hyperbolic field (16.3) are shown in Figure 4.16.

Let us assume the field $\mathbf{B}_0$ to be sufficiently strong (Section 8.3.3), so that the Alfvén speed V_{A} should be much greater than that of sound V_{s} everywhere, the exception being a small region near the zeroth line. On the strength of formula (16.3),

$$V_{\text{A}}^2 = \frac{h_0^2 \, r^2}{4\pi \rho_0} \, ,$$

where $r = (x^2 + y^2)^{1/2}$ is the radius in the cylindrical frame of reference, i.e. in the plane (x, y). Hence the condition

$$V_{\text{A}}^2 \gg V_{\text{s}}^2$$

can be rewritten in the form:

$$r \gg r_{\text{s}} \, . \tag{16.4}$$

Here

$$r_{\text{s}} = \left(\frac{4\pi \, n_0 k_{\text{B}} T_0}{h_0^2} \right)^{1/2} , \tag{16.5}$$

n_0 and T_0 being the number density and temperature of the plasma at the initial stage of the process, k_{B} is Boltzmann's constant.

Let $l = 1$ in formula (16.1). Then the assumed condition (16.4), together with the condition for applicability of the approximate expression (16.2) for the potential $\mathbf{A}_0$, means that the domain of admissible values is

$$r_{\text{s}} \ll r \ll 1 \, . \tag{16.6}$$

16.1.3 A linearized problem in ideal MHD

Of concern to us are *small perturbations* in the region (16.6) relative to the initial equilibrium state

$$\mathbf{v}_0 = 0 \, , \qquad \rho_0 = \text{const} \, , \qquad p_0 = \text{const} \, , \qquad \Delta A_0 = 0 \, .$$

Consider plane flows of a plasma with a frozen magnetic field in the plane (x, y):

$$\mathbf{v} = \{ \, v_x(x, y, t), \; v_y(x, y, t), \; 0 \, \} \, , \qquad \mathbf{B} = \mathbf{B}_0 + \mathbf{b} \, ,$$

the small perturbation of magnetic field being

$$\mathbf{b} = \{ \, b_x(x, y, t), \; b_y(x, y, t), \; 0 \, \} \, .$$

Thus, from the mathematical standpoint (Section 9.2), the problem at hand belongs to the two-dimensional problems of the *second* type.

For small perturbations $\mathbf{v}$, p, ρ, and A (instead of $\mathbf{b}$), the linearized equations of ideal MHD can be written in the form

$$\frac{\partial A}{\partial t} = -\,\mathbf{v} \cdot \nabla A_0 \,,$$

$$\frac{\partial \mathbf{v}}{\partial t} = -\,\frac{\nabla p}{\rho_0} - \frac{1}{4\pi\rho_0}\,\nabla A_0\,\Delta A \,, \tag{16.7}$$

$$\frac{\partial \rho}{\partial t} = -\,\rho_0 \operatorname{div} \mathbf{v} \,.$$

The gas pressure gradient in the region (16.6) can be ignored. If we did not ignore the term ∇p, the set of Equations (16.7), on differentiating with respect to t, could be transformed to give us

$$\underline{\frac{\partial^2 A}{\partial t^2} = \frac{(\nabla A_0)^2}{4\pi\rho_0}}\,\Delta A + \frac{V_s^{\,2}}{\rho_0}\,\nabla A_0 \cdot \nabla \rho \,,$$

$$\frac{\partial^2 \mathbf{v}}{\partial t^2} = \frac{\nabla A_0}{4\pi\rho_0}\,\Delta\,(\mathbf{v} \cdot \nabla A_0) + V_s^{\,2}\,\nabla \operatorname{div} \mathbf{v} \,, \tag{16.8}$$

$$\frac{\partial^2 \rho}{\partial t^2} = \frac{1}{4\pi}\,\nabla A_0 \cdot \nabla\,\Delta A + V_s^{\,2}\,\Delta \rho \,.$$

So, perturbations in the region (16.6) are seen (see the underlined terms in the first equation) to propagate with the *local* Alfvén velocity V_A:

$$V_{A0}^{\,2} = V_{A0}^{\,2}\,(r) = \frac{\left(\nabla A_0\,(r)\right)^2}{4\pi\rho_0}\,, \tag{16.9}$$

the result being accurate to small corrections of the order of $V_s^{\,2}/V_{A0}^{\,2}$. This is the case of plasma with a strong magnetic field; see the isotropic wave V_+ in Figure 10.3.

The displacement of the plasma under the action of the perturbation, $\boldsymbol{\xi}$, is convenient to introduce instead of the velocity perturbation $\mathbf{v}$:

$$\mathbf{v} = \frac{\partial \boldsymbol{\xi}}{\partial t} \,. \tag{16.10}$$

Dropping the terms depending on the pressure gradient, the initial set of Equations (16.8) is recast as follows (Syrovatskii, 1966b):

$$\frac{\partial^2 A}{\partial t^2} = V_{A0}^{\,2}\,(r)\,\Delta A \,, \tag{16.11}$$

$$\frac{\partial^2 \boldsymbol{\xi}}{\partial t^2} = \frac{V_{\text{A0}}^2(r)}{\sqrt{4\pi\rho_0}} \, \Delta\,(\boldsymbol{\xi} \cdot \nabla A_0)\,, \tag{16.12}$$

$$\rho = -\,\rho_0 \,\mathrm{div}\,\boldsymbol{\xi}\,, \tag{16.13}$$

$$A = -\,(\boldsymbol{\xi} \cdot \nabla)\,A_0\,. \tag{16.14}$$

Rewrite Equation (16.11) in the cylindrical frame of reference

$$\frac{\partial^2 A}{\partial t^2} = \frac{h_0^2}{4\pi\rho_0} \left[r\,\frac{\partial}{\partial r}\left(r\,\frac{\partial A}{\partial r} \right) + \frac{\partial^2 A}{\partial \varphi^2} \right].$$

On substituting $x = \ln r$, this equation is reduced to the usual wave equation in the variables (x, φ)

$$\frac{\partial^2 A}{\partial t^2} = V_{\text{a}}^2 \left(\frac{\partial^2 A}{\partial x^2} + \frac{\partial^2 A}{\partial \varphi^2} \right), \tag{16.15}$$

where $V_{\text{a}} = h_0/\sqrt{4\pi\rho_0}$ is a constant playing the role of the wave velocity.

16.1.4 Converging waves and the cumulative effect

Consider an initial perturbation of the potential, which is independent of the cylindrical frame angle φ:

$$A\,(r, \varphi, 0) = \Phi\,(r)\,,$$

where $\Phi\,(r)$ is an arbitrary function of r. In this case the general solution of Equation (16.15) is

$$A\,(r, t) = \Phi\,(\ln r + V_{\text{a}}\, t)\,. \tag{16.16}$$

The sign +, which we have chosen, by $V_{\text{a}}\, t$ corresponds to the *converging* cylindrical wave, its velocity being

$$V(r) = \frac{dr}{dt} = -\,r\,V_{\text{a}} = -V_{\text{A0}}(r)\,,$$

i.e. the wave propagates with the Alfvén velocity (see definition (16.9)). The following properties of the wave are of interest.

(a) The magnetic field intensity in such a wave is

$$B_r = \frac{1}{r}\,\frac{\partial A}{\partial \varphi} = 0\,, \qquad B_\varphi = -\,\frac{\partial A}{\partial r} = -\,\frac{\Phi}{r}\,.$$

As the wave approaches the zeroth line, the field intensity grows

$$B\,(r) = B\,(R) \times \frac{R}{r}\,.$$

Here $B\,(R)$ is the field intensity in the wave when its front is at a distance R from the zeroth line.

(b) The magnetic field gradient increases as well

$$\frac{\partial B}{\partial r}(r) = \frac{\partial B}{\partial r}(R) \times \left(\frac{R}{r}\right)^2 .$$

Thus

> as the cylindrical wave converges to zero it gives rise to a *cumulative* effect in regard to the magnetic field and its gradient.

(c) The character of the plasma displacement $\boldsymbol{\xi}$ in such a wave can be judged from the motion Equation (16.12). It contains the scalar product $\boldsymbol{\xi} \cdot \nabla A_0$. Hence the displacements directed along the field lines are absent in the wave under consideration. The perpendicular displacements

$$\boldsymbol{\xi} = -\frac{A}{(\nabla A_0)^2}\,\nabla A_0\,, \tag{16.17}$$

whence, in view of (16.16), it follows that $|\,\boldsymbol{\xi}\,| \sim r^{-1}$. So

> the quantity of the displacement also grows, as the wave approaches the zeroth line of the magnetic field.

(d) As for the change in plasma density, we find from Equation (16.13), using formulae (16.17) and (16.16), that

$$\rho = -\,\rho_0\,\mathrm{div}\,\boldsymbol{\xi} \sim \frac{1}{r^2}\,\cos 2\varphi\,. \tag{16.18}$$

The plasma density increases in a pair of opposite quadrants while decreasing in the other pair (Figure 16.1). The first pair of quadrants ($-\pi/4 \le \varphi \le \pi/4$ and $3\pi/4 \le \varphi \le 5\pi/4$) corresponds to the regions where the plasma flows are convergent. In the second pair ($\pi/4 < \varphi < 3\pi/4$ and $5\pi/4 < \varphi < 7\pi/4$) of quadrants, the trajectories of the fluid particles diverge, resulting in a decrease of the plasma density.

Therefore, even in a linear approximation,

> small perturbations grow in the vicinity of the magnetic field zeroth line. As this takes place, regions appear in which the field and its gradients increase, whereas the plasma density decreases.

The so-called linear-reconnection theory takes into account the dissipative processes in the linear approximation (see Sections 21.2.1 and 21.2.3).

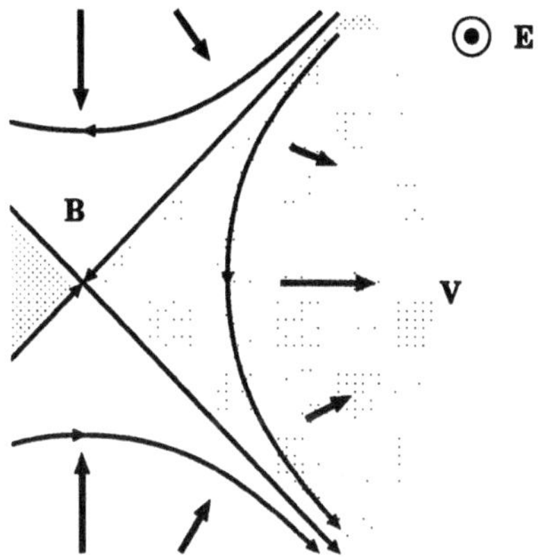

Figure 16.1: Plasma flows and the density change in small perpurbations in the vicinity of a hyperbolic zeroth point.

16.2 Large perturbations near the neutral line

Let us relax the assumption concerning the smallness of the perturbations in the vicinity of a zeroth line. Then, instead of linearized MHD equations, we shall deal with the exact set of two-dimensional equations in the approximation of the strong field and the cold plasma, taken in a zeroth order with respect to the small parameter $\varepsilon^2 = v^2/V_{\rm A}^2$, i.e. Equations (9.38)–(9.41):

$$\Delta A = 0\,, \tag{16.19}$$

$$\frac{d\mathbf{v}}{dt} \times \nabla A = 0\,, \tag{16.20}$$

$$\frac{dA}{dt} = 0\,, \tag{16.21}$$

$$\frac{\partial \rho}{\partial t} + \operatorname{div} \rho \mathbf{v} = 0\,. \tag{16.22}$$

Here it is implied that the region in the vicinity of the zeroth line is to be restricted by the condition (16.4).

16.2.1 Magnetic field line deformations

As was shown in Section 9.2, Equations (16.21) and (16.22) are integrated on passing to Lagrangian coordinates

$$\mathbf{r}\,(\mathbf{r}_0, t) = \mathbf{r}_0 + \boldsymbol{\xi}\,(\mathbf{r}_0, t) \tag{16.23}$$

(cf. definition (9.35)). Here $\mathbf{r}_0$ is the coordinate of a fluid particle before displacement, i.e. at the initial moment, $\mathbf{r}$ is its coordinate at a moment of time t, $\boldsymbol{\xi}(\mathbf{r}_0, t)$ is the ***displacement vector*** (cf. definition (16.10)). Rewrite Equation (16.23) as the inverse transformation

$$\mathbf{r}_0(\mathbf{r}, t) = \mathbf{r} - \boldsymbol{\xi}(\mathbf{r}, t) .$$

Then the continuity Equation (16.22) can be written in its Lagrangian form:

$$\rho(\mathbf{r}, t) = \rho_0(\mathbf{r} - \boldsymbol{\xi}(\mathbf{r}, t)) \, \frac{\mathcal{D}(\mathbf{r} - \boldsymbol{\xi}(\mathbf{r}, t))}{\mathcal{D}(\mathbf{r})} , \qquad (16.24)$$

where $\mathcal{D}(\mathbf{r}_0)/\mathcal{D}(\mathbf{r})$ is the Jacobian transformation from $\mathbf{r}_0$ coordinates to $\mathbf{r}$ coordinates (see definition (9.36)).

The integral of the freezing-in Equation (16.21) (cf. (9.27))

$$A(\mathbf{r}, t) = A_0(\mathbf{r} - \boldsymbol{\xi}(\mathbf{r}, t)) , \qquad (16.25)$$

where $A_0(\mathbf{r}_0)$ is an initial value of the vector-potential.

Had the displacement $\boldsymbol{\xi}(\mathbf{r}, t)$ been known, formulae (16.25) and (16.24) would have allowed us to uniquely determine the field line deformation and plasma density change in the vicinity of the zeroth line, given the displacement of the currents I. However, to find $\boldsymbol{\xi}(\mathbf{r}, t)$ generally, we must simultaneously solve Equations (16.19) and (16.20), i.e. the set of equations

$$\Delta A = 0 , \qquad (16.26)$$

$$\frac{\partial^2 \boldsymbol{\xi}}{\partial t^2} \times \nabla A = 0 . \qquad (16.27)$$

As a rule, to integrate Equation (16.27), we must have recourse to numerical methods (Somov and Syrovatskii, 1976a). Let us try to circumvent the difficulty.

Suppose the displacement of the currents occurs sufficiently fast as compared with the speed of sound but sufficiently slow as compared with the Alfvén speed. With these assumptions, the boundary conditions of the problem (see (9.42)) change slowly in comparison with the speed of fast magnetoacoustic waves, which allows us to **consider the field as being in equilibrium at each stage** of the process (see Equation (16.26)).

The latter assumption actually means that the total displacement $\boldsymbol{\xi}$ can be held to be a sum of successive small perturbations $\delta\boldsymbol{\xi}$ of the type (16.17), each of them transferring the system to a close equilibrial state. Since the small displacement $\delta\boldsymbol{\xi}$ is directed across the magnetic field lines, the total displacement $\boldsymbol{\xi}$ is also orthogonal to the picture of field lines. To put it another way, the lines of the plasma flow constitute a family of curves orthogonal to the magnetic field lines, i.e. the family of hyperbolae

$$x\,y = x_0\,y_0\,. \tag{16.28}$$

A numerical solution of the problem (Somov and Syrovatskii, 1976a) shows that such a flow is actually realized for comparably small t or sufficiently far from the zeroth line.

Let us make use of the freezing-in Equation (16.25) to find another equation relating the coordinates of a fluid particle (x, y) with their initial values (x_0, y_0). In view of the formula (16.1) for the initial vector-potential $A_0\,(x, y)$, the magnetic field potential of displaced currents is

$$A\,(x,y) = \frac{h_0}{4}\left\{\ln\left[x^2 + (y-1+\delta)^2\right] + \ln\left[x^2 + (y+1-\delta)^2\right]\right\}.$$

Relative to formula (16.1), $I/c = h_0/4$, $l = 1$, and $\delta l = \delta$.

Near the zeroth line, with the accuracy of the terms of order δ, we find

$$A\,(x,y) = \frac{h_0}{2}\left(x^2 - y^2 - 2\delta\right). \tag{16.29}$$

Substitution of (16.29) in (16.25) gives

$$y^2 - x^2 + 2\delta = y_0^2 - x_0^2\,. \tag{16.30}$$

Equations (16.28) and (16.30) allow us to express the initial coordinates of a fluid particle (x_0, y_0) in terms of its coordinates (x, y) at the moment of time t (Syrovatskii, 1966a):

$$x_0^2 = \frac{1}{2}\left\{\left[\left(x^2 - y^2 - 2\delta\right)^2 + 4x^2y^2\right]^{1/2} + \left(x^2 - y^2 - 2\delta\right)\right\},$$

$$y_0^2 = \frac{1}{2}\left\{\left[\left(x^2 - y^2 - 2\delta\right)^2 + 4x^2y^2\right]^{1/2} - \left(x^2 - y^2 - 2\delta\right)\right\}. \tag{16.31}$$

The displacements determined by these expressions are such that the field lines which crossed the y axis at points 0, $\sqrt{\delta}$, $\sqrt{2\delta}$, would take the place of the field lines which crossed the x axis at points $\sqrt{2\delta}$, $\sqrt{\delta}$, 0, respectively (see Figure 16.2 in the region $r \gg r_{\rm s}$).

The plasma displacements and frozen-in field line deformations obtained pertain only to the region $r \gg r_{\rm s}$. The approximation of a strong field and

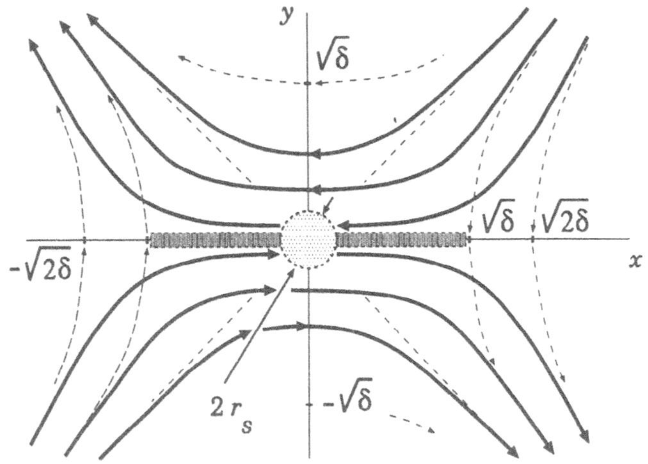

Figure 16.2: The deformation of the magnetic field lines in the neighbourhood of a zeroth line.

a cold plasma is inapplicable outside this region, i.e. $r \leq r_s$. It must also be considered that a region of *strong plasma compression* can arise in the course of the displacement. The conditions for applicability of the strong-field-cold-plasma approximation can be broken down in such regions, thus making it necessary to solve a more general problem. In particular, field deformations can be distinctly different here, owing to *strong electric currents* flowing in these regions. They will be discussed in the next Section.

The main effect demonstrated above is the deformation of the field lines which is schematically shown as two long dashed areas along the x axis. Here

> a current sheet formation is confirmed by the presence of oppositely directed magnetic field lines

near the origin of the coordinates in Figure 16.2. The current inside the current sheet is parallel to the z axis, i.e. parallel to the electric field $\mathbf{E}$ related to the magnetic field line motion (cf. Figure 4.18). However, at the edges of the sheet, the currents are sometimes opposite in direction (the so-called reverse currents) to the one inside the main current sheet which is formed at the zeroth line as shown above.

16.2.2 Plasma density variations

Let us find the density distribution (16.24) by calculating the Jacobian of the reverse transformation of the Lagrangian variables, with the aid of the formulae (16.31). Assuming an homogeneous initial distribution of plasma, we get

$$\frac{\rho(x,y)}{\rho_0} = \frac{x^2+y^2}{\left[(x^2+y^2)^2 + 4\delta\,(y^2-x^2) + 4\delta^2 \right]^{1/2}} \, . \tag{16.32}$$

The formula obtained shows that in the region

$$x^2 < y^2 + \delta \tag{16.33}$$

the displacement of the currents leads to plasma rarefaction. As this takes place, the largest rarefaction occurs for small r ($r^2 \ll \delta$):

$$\frac{\rho(x,y)}{\rho_0} \sim \frac{r^2}{2\delta} \, . \tag{16.34}$$

By contrast, in the region $x^2 > y^2 + \delta$ the plasma is compressed, its density tending to infinity at the points (Figure 16.3):

$$y = 0 \, , \qquad x = \pm\sqrt{2\delta} \, . \tag{16.35}$$

The approximation of a strong field and a cold plasma is inapplicable in the vicinity of these points, and the actual deformation of the field lines can differ significantly from that found above.

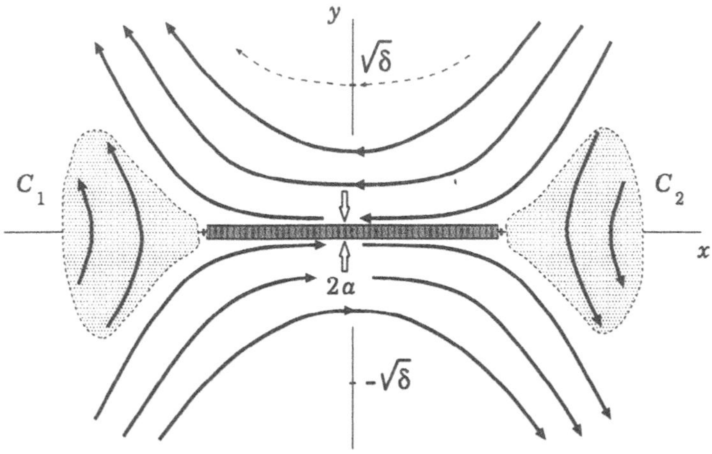

Figure 16.3: The plasma distribution near a forming current sheet. $2a$ is the thickness of the current sheet.

Figure 16.3 illustrates a characteristic distribution of plasma near a reconnecting current sheet ($-\sqrt{\delta} \leq x \leq \sqrt{\delta}$). The regions of strong plasma compression near the points (16.35) are shown by the shadowed regions C_1 and C_2 outside of the current sheet.

16.3 The dynamic dissipation of a magnetic field

16.3.1 Conditions of appearance

In the region between the points (16.35), where the plasma density formally tends to infinity, the character of the displacements can be determined by using the freezing-in condition for the magnetic field lines and taking into account that, as was mentioned in the previous section, plasma spread along the field lines during the rapid displacement of the currents may be neglected. Under these assumptions, the magnetic field deformation is of the form shown in Figures 16.2 and 16.3.

It is important for the following discussion that the whole magnetic flux which crossed the axis y in the region $0 < y < \sqrt{2\delta}$, namely

$$\Phi = A_0\,(0, \sqrt{2\delta}) - A_0\,(0, 0) = h_0\,\delta\,, \tag{16.36}$$

is now confined to the strip $y \leq r_{\rm s}$. The thickness of this strip $r_{\rm s} \approx a$ in Figure 16.3. The field lines of this flux 'spread' along the x axis in the negative direction. The same flux of field lines, but oppositely directed, is situated along the x axis in the lower half-plane.

Therefore, in the region $|\,x\,| \leq \sqrt{\delta}$, $|\,y\,| \leq r_s$, the field lines of opposite directions are compressed to form a thin *reconnecting current sheet* (RCS). The region of the magnetic field compression is shown in Figures 16.2 and 16.3 as the long dashed area along the x axis. The magnetic field gradient in this region is evaluated as

$$h \approx \frac{B}{r_{\rm s}} \approx \frac{\Phi}{r_{\rm s}^2} \approx \frac{h_0}{r_{\rm s}^2}\,\delta\,. \tag{16.37}$$

The field gradient h in the region of the magnetic compression is $\delta / r_{\rm s}^2$ times its initial value h_0. In other words,

> the magnetic field gradient inside the current sheet is proportional to the value of the external currents displacement δ,

with the proportionality coefficient, by virtue of definition (16.5), being larger, the smaller is the gas pressure as compared with the magnetic one in the reconnecting plasma.

At the same time, according to (16.34) the plasma density in the region $r^2 < \delta$ decreases by a factor of $r^2/2\delta$. This conclusion applies for $r \gg r_s$ and is of a qualitative character. Nonetheless, it is of fundamental importance that we can make an order-of-magnitude evaluation of the ratio of the field gradient to the plasma concentration in the region of the magnetic compression ($r \approx r_s$)

$$\frac{h}{n} \approx \frac{h_0}{n_0} \frac{\delta^2}{r_s^4} \,. \tag{16.38}$$

Recall that in the MHD approximation (Section 8.1) we usually neglect the displacement current $(1/c)\, \partial \mathbf{E}/\partial t$ as compared with the conductive one

$$\mathbf{j} = ne\, \mathbf{u} \,.$$

Here e is the charge on a particle, $\mathbf{u}$ is the current velocity, i.e. the velocity of current carriers. Subject to the condition (8.8), we may use the 'truncated' Maxwell equation

$$\operatorname{curl} \mathbf{B} = \frac{4\pi}{c} \mathbf{j} \,, \tag{16.39}$$

whence, on setting $|\operatorname{curl} \mathbf{B}| \approx h$, the following estimate is obtained

$$\frac{h}{n} \approx 4\pi e \left(\frac{u}{c} \right) .$$

Since the particle velocity u cannot exceed the speed of light c, the current density is limited by the value $j = nec$ and the ratio

$$\frac{h}{n} < 4\pi e \,. \tag{16.40}$$

On the other hand, from (16.38) this ratio is determined by the value of the displacement δ and by the parameters r_s and h_0/n_0. Once the condition (16.40) breaks down, by virtue of (16.38), i.e.

$$\frac{h_0}{n_0} \frac{\delta^2}{r_s^4} \geq 4\pi e \,, \tag{16.41}$$

the displacement current $(1/c)\, \partial \mathbf{E}/\partial t$ must be accounted for in Equation (16.39). It means that, under condition (16.41), a strong electric field of an inductive nature arises in the region where magnetic fluxes interact.

A quantitative description of the physical processes in the region involved is difficult and is the subject of the theory of reconnection in current sheets. The qualitative effects are as follows.

16.3.2 The physical meaning of dynamic dissipation

The appearance of the inductive electric field, independent of the plasma motion, signifies the violation of the freezing-in condition. Thus the motion of the field lines relative to the plasma, which is necessary for their reconnection in the region of interaction of the magnetic fluxes, is allowed. The important aspect of the situation under discussion is that these processes are *independent* of Joule dissipation and can take place in a collisionless plasma. This is the reason why this phenomenon may be termed *dynamic dissipation* (Syrovatskii, 1966) or, in fact, *collisionless reconnection* (Section 16.4.3).

An essential peculiarity of the dynamic dissipation of a magnetic field is that the inductive electric field is directed along the main current $\mathbf{j}$ in the reconnection region. Hence the field does positive work on charged particles, thus increasing their energy. It is this process that provides the transformation of the magnetic energy into the kinetic one, i.e. dynamic dissipation.

As opposed to Joule dissipation, there is no direct proportionality of the current density $\mathbf{j}$ to the electric field intensity $\mathbf{E}$ in the case of dynamic dissipation. Given the condition (16.41),

the current density is saturated at the value $j \approx nec$, the field energy going to increase the total energy of the particles,

$$\mathcal{E} = \frac{mc^2}{\sqrt{1 - v^2/c^2}}, \tag{16.42}$$

i.e. the acceleration by the electric field. Thus, under the conditions considered, the field energy converts directly to that of the accelerated particles.

Acceleration occurs along zeroth lines (parallel to the z axis) which are formed in the current sheet region. Recall that the particle motion along neutral lines (or surfaces, see Section 4.4) is stable: the magnetic field returns deviating particles to the neutral line (or surface), as is clear from immediate consideration of the Lorentz force $(e/c)\,\mathbf{v} \times \mathbf{B}$. More realistic analysis of the acceleration problem will be given in Chapter 18.

The condition (16.41) is, in fact, extreme. This implies the regular acceleration of particles to relativistic energies. In fact, acceleration may take place under much more modest conditions, when the dynamic dissipation of a magnetic field is, in essence, related to the known phenomenon of the electric runaway of particles (primarily electrons; see Section 6.4.2). The condition which in this case replaces the extreme condition (16.41) was derived by Syrovatskii (1966).

Needless to say, relativistic energies are not always reached in the acceleration process. Some instabilities are, as a rule, excited in the plasma-beam system in the acceleration region. As this takes place, particle scattering and acceleration with the created wave turbulence must be accounted for. However, it is important that the general inference as to **the possibility of particle acceleration by an electric field in the magnetic reconnection region** (i.e. dynamic dissipation of the magnetic field) remains valid, in particular, when applied to the solar flare problem (see Section 16.5, Chapters 17 and 18).

16.4 Nonstationary analytical models of the RCS

16.4.1 Self-similar 2D MHD solutions

In connection with the 2D problem of the equilibrium state of a plasma near the X-type zeroth point of magnetic field, Chapman and Kendall (1963) had obtained the exact particular solution of the ideal MHD equations for an *incompressible* fluid. This *self-similar* analytical solution has a perfectly defined character. A fixed mass of a plasma near the zeroth point receives energy from the outside in the form of an electromagnetic-field energy flux. Finally, a cumulative effect is developed and arbitrarily large energy densities are attained. The solution demonstrates the tendency to form a current sheet near the zeroth point.

Imshennik and Syrovatskii (1967) had found a self-similar solution for an ideal *compressible* fluid. Let us also start from the set of the ideal MHD Equations (8.50)–(8.55). Consider the 2D problem of the second type (see Section 9.2.2). Substitute definition (9.18) of the vector potential **A** in the first three equations, we have the following set:

$$\rho \frac{d\mathbf{v}}{dt} = -\nabla p - \frac{1}{4\pi} \Delta A \, \nabla A \,, \tag{16.43}$$

$$\operatorname{curl} \frac{d\mathbf{A}}{dt} = 0 \,, \tag{16.44}$$

$$\frac{d\rho}{dt} + \rho \operatorname{div} \mathbf{v} = 0 \,. \tag{16.45}$$

We assume that the pressure p is a function of the density ρ only. This condition is satisfied by any politropic equation of state. Moreover, as it was shown by Imshennik and Syrovatskii, for the class of solutions of interest to us, the plasma density ρ depends only on time. Hence, by virtue of the foregoing assumption, the pressure p depends only on time too. Therefore the pressure gradient ∇p in Equation (16.43) vanishes. So we have equations:

$$\rho \, \frac{d\mathbf{v}}{dt} = - \frac{1}{4\pi} \, \Delta A \, \nabla A \,, \tag{16.46}$$

$$\operatorname{curl} \frac{dA}{dt} = 0 \,, \tag{16.47}$$

$$\frac{d\rho}{dt} + \rho \operatorname{div} \mathbf{v} = 0 \,. \tag{16.48}$$

Let us seek a solution of the set of Equations (16.46)–(16.48) under the following initial conditions.

(a) The plasma density is constant:

$$\rho \, (x, y, 0) = \rho_0 \,, \tag{16.49}$$

(b) The magnetic field is a hyperbolic one (cf. formula (16.2) where put $h_0/2 = a_0$):

$$A \, (x, y, 0) = a_0 \left(x^2 - y^2 \right) , \tag{16.50}$$

(c) The initial velocity depends linearly on the coordinates, so that there is no flow of plasma across the coordinate axes:

$$v_x \, (x, y, 0) = U x \,, \qquad v_y \, (x, y, 0) = V y \,. \tag{16.51}$$

Thus the initial conditions are defined by the four independent quantities ρ_0, a_0, U, and V. We can construct from them three independent combinations with the dimension of time:

$$t_x = \frac{1}{U} \,, \qquad t_y = \frac{1}{V} \,, \qquad t_0 = \frac{(\pi \rho_0)^{1/2}}{\mid a_0 \mid} \tag{16.52}$$

and not even one combination with the dimension of length. We introduce new variables with dimensions equal to a certain power of the length:

$$\tau = \frac{t}{t_0} \,, \quad u_x = t_0 v_x \,, \quad u_y = t_0 v_y \,, \quad a = \frac{A}{a_0} \,, \quad g = \frac{\rho}{\rho_0} \,. \tag{16.53}$$

In terms of these variables, Equations (16.46)–(16.48) take the form

$$\frac{\partial}{\partial x}\frac{da}{d\tau} = 0\,, \qquad \frac{\partial}{\partial y}\frac{da}{d\tau} = 0\,, \tag{16.54}$$

$$g\,\frac{du_x}{d\tau} = -\frac{1}{4}\frac{\partial a}{\partial x}\,\Delta a\,, \qquad g\,\frac{du_y}{d\tau} = -\frac{1}{4}\frac{\partial a}{\partial y}\,\Delta a\,, \tag{16.55}$$

$$\frac{dg}{d\tau} + \left(\frac{\partial u_x}{\partial x} + \frac{\partial u_y}{\partial y}\right) g = 0\,. \tag{16.56}$$

The initial conditions (16.49)–(16.51) then become

$$g\,(x,y,0) = 1\,, \qquad a\,(x,y,0) = x^2 - y^2\,,$$

$$u_x(x,y,0) = \varepsilon_x x\,, \qquad u_y(x,y,0) = \varepsilon_y y\,, \tag{16.57}$$

where

$$\varepsilon_x = U\,\frac{(\pi\rho_0)^{1/2}}{|\,a_0\,|}\,, \qquad \varepsilon_y = V\,\frac{(\pi\rho_0)^{1/2}}{|\,a_0\,|}\,. \tag{16.58}$$

Thus the problem is completely determined by the two dimensionless parameters (16.58) which are similar to the parameter ε in (8.62). As to the choice of the unit of length, Equations (16.54)–(16.56) impose no limitations whatever. So the length unit can be chosen arbitrarily; and both the coordinates x and y, together with all the variables in definition (16.53), can be chosen dimensionless.

Therefore we consider the problem as a self-similar one, more exactly, as the self-similar problem of the *first type* (Zel'dovich and Raizer, 1966, Ch. 12). It means that the set of equations in partial derivatives, (16.54)–(16.56), can be reduced to the set of ordinary differential equations. Let us do it. Substitute in Equations (16.54)–(16.56) the following solution:

$$a\,(x,y,\tau) = a_x(\tau)\,x^2 - a_y(\tau)\,y^2\,, \tag{16.59}$$

$$g\,(x,y,\tau) = g\,(\tau)\,, \tag{16.60}$$

$$u_x(x,y,\tau) = f_x(\tau)\,x\,, \qquad u_y(x,y,\tau) = f_y(\tau)\,y\,. \tag{16.61}$$

We obtain the following set of five ordinary differential equations for the five unknown functions $a_x(\tau)$, $a_y(\tau)$, $g\,(\tau)$, $f_x(\tau)$ and $f_y(\tau)$:

$$\dot{a}_x + 2a_x f_x = 0\,, \qquad \dot{a}_y + 2a_y f_y = 0\,,$$

$$\dot{g} + (f_x + f_y)\,g = 0\,, \tag{16.62}$$

$$\left(\dot{f}_x + f_x^2\right) g = a_x \left(a_y - a_x\right), \qquad \left(\dot{f}_y + f_y^2\right) g = a_y \left(a_x - a_y\right).$$

The dot denotes differentiation with respect to the dimensionless time τ. The initial conditions (16.57) give us the following initial conditions:

$$a_x(0) = 1, \quad a_y(0) = 1, \quad g(0) = 1,$$

$$f_x(0) = \varepsilon_x, \quad f_y(0) = \varepsilon_y. \tag{16.63}$$

Let us eliminate the functions f_x and f_y from the first two and last equations of the set (16.62). As a result we get the equation

$$\frac{\dot{a}_x}{a_x} + \frac{\dot{a}_y}{a_y} - 2\,\frac{\dot{g}}{g} = 0. \tag{16.64}$$

From this, assuming that the functions a_x, a_y and g are not equal to zero and using the initial conditions (16.63), we obtain an integral of the set of ordinary Equations (16.62):

$$g = \left(a_x a_y\right)^{1/2}. \tag{16.65}$$

Since the initial values of these three functions are positive, the subsequent results will pertain to a time interval τ for which these quantities remain positive.

16.4.2 Magnetic collapse at the zeroth point

To illustrate the behaviour of the solutions (16.59)–(16.61), it is convenient to introduce two functions $\zeta_x(\tau)$ and $\zeta_y(\tau)$ such that

$$a_x = \frac{1}{\zeta_x^2}, \qquad a_y = \frac{1}{\zeta_y^2}. \tag{16.66}$$

Without loss of generality, we assume that these new functions are positive.

From the first two equations of the set (16.62) and from the integral (16.65) we obtain formulae for the other three unknown functions:

$$f_x = \frac{\dot{\zeta}_x}{\zeta_x}, \qquad f_y = \frac{\dot{\zeta}_y}{\zeta_y}, \qquad g = \frac{1}{\zeta_x \zeta_y}. \tag{16.67}$$

The set of five equations (16.62) then reduces to two second-order differential equations for $\zeta_x(\tau)$ and $\zeta_y(\tau)$:

$$\ddot{\zeta}_x = -\zeta_y \left(\frac{1}{\zeta_x^2} - \frac{1}{\zeta_y^2}\right), \qquad \ddot{\zeta}_y = \zeta_x \left(\frac{1}{\zeta_x^2} - \frac{1}{\zeta_y^2}\right), \tag{16.68}$$

with the initial conditions

$$\zeta_x(0) = 1 \,, \quad \zeta_y(0) = 1 \,,$$
$$\dot{\zeta}_x(0) = \varepsilon_x \,, \quad \dot{\zeta}_y(0) = \varepsilon_y \,. \tag{16.69}$$

For definiteness, let $\varepsilon_x > \varepsilon_y$. Then a solution of the problem has a singular point which is reached after a finite time τ_0. When $\tau \to \tau_0$ the quantity ζ_x tends to a finite value $\zeta_x(\tau_0)$, and $\zeta_y(\tau) \to 0$. So we retain in Equations (16.68) only the principal terms:

$$\ddot{\zeta}_x = \frac{1}{\zeta_y} \,, \quad \ddot{\zeta}_y = -\frac{\zeta_x}{\zeta_y^2} \,. \tag{16.70}$$

In the region $\tau < \tau_0$ of interest to us, the solution of these equation is

$$\zeta_x(\tau) = \zeta_x(\tau_0) + \ldots \,,$$
$$\zeta_y(\tau) = \left(\frac{9}{2} \, \zeta_x(\tau_0) \right)^{1/3} (\tau_0 - \tau)^{2/3} + \ldots \,. \tag{16.71}$$

Here the terms of higher order of smallness in $(\tau_0 - \tau)$ have been omitted.

Returning to the variables (16.66) and (16.67), we obtain the asymptopic behaviour of the unknown functions near the singularity as $\tau \to \tau_0$:

$$a_x \to a_x(\tau_0) \,, \quad a_y \to \left(\frac{2}{9} \right)^{2/3} (a_x(\tau_0))^{1/3} \, \frac{1}{(\tau_0 - \tau)^{4/3}} \,,$$
$$f_x \to \varepsilon_x(\tau_0) \,, \quad f_y \to -\frac{2}{3(\tau_0 - \tau)} \,, \tag{16.72}$$
$$g \to \left(\frac{2}{9} \right)^{1/3} (a_x(\tau_0))^{2/3} \, \frac{1}{(\tau_0 - \tau)^{2/3}} \,.$$

Here the quantities τ_0, $a_x(\tau_0)$, and $\varepsilon_x(\tau_0)$ depend on the initial conditions (16.57) and can be determined by numerical integrating (Imshennik and Syrovatskii, 1967) the complete set of Equations (16.54)–(16.56).

Let us consider the fraction of the plasma that is located within a circle of radius equal to unity (Figure 16.4) at the initial instant $\tau = 0$. The corresponding Lagrange line is the circle

$$a_x(0) \, x^2 + a_y(0) \, y^2 = 1 \,.$$

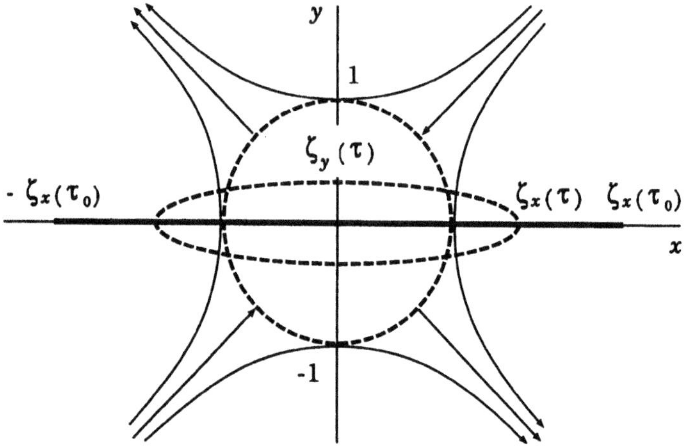

Figure 16.4: Magnetic collapse in the vicinity of a hyperbolic zeroth point.

Therefore at any subsequent instant of time this plasma will be located inside the ellipse

$$a_x(\tau)\, x^2 + a_y(\tau)\, y^2 = \frac{x^2}{\zeta_x^2(\tau)} + \frac{y^2}{\zeta_y^2(\tau)} = 1\,, \tag{16.73}$$

where $\zeta_x(\tau)$ and $\zeta_y(\tau)$ introduced above have the simple meaning of semi-axes of this deforming ellipse.

As follows from the obtained solution, the semi-axis whose direction corresponds to a smaller initial velocity vanishes at the instant τ_0. At the same time, the second semi-axis remains different from zero and bounded. Thus any initial circle is transformed at the instant τ_0 into a segment of the x axis with the ends $x = \pm\, \zeta_x(\tau_0)$ as shown in Figure 16.4.

Let us consider the behaviour of the magnetic field (see definitions (16.50) and (16.59)):

$$\mathbf{B} = h_0 \left\{ -a_y(\tau)\, y,\ -a_x(\tau)\, x,\ 0 \right\}, \tag{16.74}$$

where $h_0 = 2a_0$ is the gradient of the initial field near the zeroth point. In the limit as $\tau \to \tau_0$ the field is equal to

$$\mathbf{B} = h_0 \left\{ \mp \frac{1}{\zeta_y(\tau)}\,,\ -\frac{x}{\zeta_x(\tau)}\,,\ 0 \right\}, \tag{16.75}$$

where the minus and plus signs correspond to the regions $y > 0$ and $y < 0$ respectively. Therefore, when $\tau \to \tau_0$, the magnetic field is always tangent to the x axis segment into which the ellipse (16.73) degenerates, increases in magnitude without limit, and experiences a discontinuity on the x axis:

$$B_x(y=+0) - B_x(y=-0) = -\,\frac{2h_0}{\zeta_y(\tau)} \to \infty\,. \tag{16.76}$$

The appearance of the discontinuity in the magnetic field corresponds to an unbounded increase in the density of the electric current:

$$j_z = \frac{c}{4\pi}\,(\,\mathrm{curl}\,\mathbf{B}\,)_z = -\frac{c}{4\pi}\,\Delta A\,. \tag{16.77}$$

Substituting (16.59) and (16.66) into (16.77), we calculate the current density

$$j_z(\tau) = \frac{ch_0}{4\pi}\left(\frac{1}{\zeta_y^2(\tau)} - \frac{1}{\zeta_x^2(\tau)}\right). \tag{16.78}$$

From this and from the solution (16.71) it follows that when $\tau \to \tau_0$ the current density increases like

$$j_z(\tau) \sim \frac{1}{(\tau_0 - \tau)^{4/3}}\,. \tag{16.79}$$

So, when $\tau \to \tau_0$ a kind of *magnetic collapse* occurs. The x component of the magnetic field and the current density become infinite. The magnetic field is tangential to the x axis everywhere and changes its sign when passing the plane $y = 0$. Therefore

> the magnetic collapse results in the generation of a *neutral* current sheet after a finite amount of time.

As we mentioned above, a similar solution for imcompressible plasma was obtained by Chapman and Kendall (1963). In that solution the quantities ζ_x and ζ_y depend exponentially on time τ. Thus the magnetic collapse in an *incompressible* fluid requires an *infinite* amount of time.

In general, it is difficult to determine the exact conditions under which the derived plasma motion can occur. The most difficult question is that of the realization of the assumed initial linear distribution of velocity (16.51). In practice, such a distribution could be realized as a small perturbation of an stationary initial state. One might therefore assume, as was done by Chapman and Kendall (1963), that the entire process has the same character as an ordinary instability. However, Imshennik and Syrovatskii showed that

> the plasma flow under consideration – magnetic collapse – is caused by external forces and has a cumulative nature

(as we saw in Section 16.1.4). Syrovatskii (1968) showed that the self-similar solutions obtained in both Chapman and Kendall (1963) and Imshennik and Syrovatskii (1967) can be set in correspondence with exact boundary conditions that have a physical meaning. These conditions are a particular case of the conditions considered in Sections 16.1 and 16.2. They correspond to a change of the potential of the external currents producing the hyperbolic field in accordance with a fully defined law (see Syrovatskii 1968).

16.4.3 From collisional to collisionless reconnection

An essential circumstance in magnetic collapse is that the current density (16.79) increases more rapidly than the plasma density and accordingly the particle density

$$n(\tau) \sim g(\tau) \sim \frac{1}{(\tau_0 - \tau)^{2/3}} . \tag{16.80}$$

The specific (per one particle) current density is

$$\frac{j_z}{n} = \frac{ch_0}{4\pi\, n_0} \left(\frac{\zeta_x}{\zeta_y} - \frac{\zeta_y}{\zeta_x} \right) , \tag{16.81}$$

where n_0 is the initial plasma density. In the limit as $\tau \rightarrow \tau_0$

$$\frac{j_z}{n} = \frac{ch_0}{4\pi\, n_0} \left(\frac{2}{9\, a_x(\tau_0)} \right)^{1/3} \left(\frac{1}{\tau_0 - \tau} \right)^{2/3} . \tag{16.82}$$

So the ratio j_z/n tends to infinity when $\tau \rightarrow \tau_0$ within the frame of the solution desribed above. Of course, the solution has no physical meaning near the singularity where a number of quantities increase infinitely.

> When a sufficiently high current density is attained, new effects arise, not accounted from by MHD.

Here they are. First, when the current density

$$j_z \geq \sigma E_{\rm Dr} , \tag{16.83}$$

where $E_{\rm Dr}$ is the Dreicer field (6.65), an intense electric runaway of electrons begins and causes current instabilities inside the reconnecting current sheet. This process leads to a decrease in an effective conductivity of the plasma inside the current sheet (Section 17.3), but still does not impose essential limitations on the applicability of MHD to the description of the macroscopic plasma flows.

If, however,

$$j_z \gg \sigma E_{\rm Dr} \,, \tag{16.84}$$

direct acceleration of the particles by the strong electric field can set in. This is the case of **dynamic dissipation** of the magnetic field, for example, in solar flares (see the estimations in Section 17.1.1). The particle inertia (usually combined with anomalous resistivity due to wave-particle interactions) replaces the classical resistivity in allowing the magnetic reconnection to occur very quickly and practically without any Coulomb collisions.

Fast collisionless reconnection seems to be often observed in a high-temperature, rarefied cosmic plasma in the presence of a strong magnetic field, for example, in the solar corona. At a first sight, to describe the collisionless reconnection process, one may try to use an ordinary resistive MHD with a generalized Ohm's law (Section 7.3) by simply including the electron inertia:

$$E_z = \sigma_{\rm ef}^{-1} \, j_z + \frac{4\pi}{\left(\omega_{pl}^{(\rm e)}\right)^2} \frac{d}{dt} \, j_z \,. \tag{16.85}$$

Here $\sigma_{\rm ef}$ is an anomalous conductivity originated from the wave-particle interaction or the stochasticity of the particle orbits.

The problem will appear soon, however, in such an over-simplified approach because inside actual reconnecting current sheets the magnetic field is not equal to zero. This internal (transversal and longitudinal) magnetic field has a strong influence on the particle acceleration by the strong electric field E_z related to the fast collisionless reconnection. This problem will be discussed in Chapter 18.

16.5 Reconnection in solar flares

16.5.1 The role of magnetic fields

There are three main objections to the hypothesis that the energy of a solar flare can be stored in the form of a magnetic field of one or several reconnecting current sheets.

(a) First, it is claimed that measurements of photospheric magnetic fields do not demonstrate an unambiguous relation between flares and the changes of the magnetic fields. More exactly, the changes in question are those that occur immediately before the flares. These changes were supposed to be the cause but not the consequence of the flare.

(b) The second objection is related to the time of dissipation of the magnetic field in a volume that would contain the energy necessary for the flare. If this time is estimated in a usual way as the diffusion time in a solar plasma of a finite conductivity, then it is too long compared with the observed duration of the flare.

(c) The third objection is the most crucial one: the observers have never seen real reconnecting current sheets in solar flares. What are the answers on the reconnection theory to these objections?

According to contemporary views the principal flare process is contingent on the accumulation of the *free magnetic energy* in the corona and chromosphere. At least, this is one of basic concepts (see also Chapter 22). By 'free' we mean the **surplus energy above that of a potential magnetic field** having the same sources (sunspots, background fields) in the photosphere. In other words, the free magnetic energy is related to the electric currents in the solar atmosphere above the photosphere. The flare correspondes to rapid changes of these currents. So, we distinguish between two processes: the slow accumulation of flare energy and its fast release – a flare.

Let us see these distinctions in the following classical example – the evolution of the quadrupole configuration of sunspots shown in the *two-dimensional* Figure 16.5. Four sunspots of pairwise opposite polarity are shown: N and S represent a bipolar group of sunspots in an active region, n and s model a new emerging flux. All four sunspots are placed along the axis x placed in the photospheric plane Ph at the bottom of the chromosphere Ch.

As in Figure 4.20, **three consequent states of the potential field** are shown. In Figure 16.5a the field line A_1 is the separatrix line of the initial state (a), this field line will reconnect first; X is the neutral point (line along the z axis) of the potential field at the initial state, here the current sheet CS is created at the state (b). The field line A_2 is the separatrix of the final state (c) or the last reconnected field line. In Figure 16.5b three solid arrows show an emergence of the new magnetic flux (the sunspots n and s); the sunspots have been emerged, but the field lines do not start to reconnect.

In general, the redistribution of fluxes appears as a result of the **motions and changes of magnetic field sources** in the photosphere. These changes can be either the emergence of a new flux tube from below the photosphere (Figure 16.5) or many other **flows of photospheric plasma**, in particular the *shear motions* – inhomogeneous horizontal flows along the neutral line of the photospheric magnetic field. For this reason, an actual reconnection in the solar atmosphere is always a three-dimensional process (see next Section). Sometimes the two-dimensional MHD problems still give a simple but

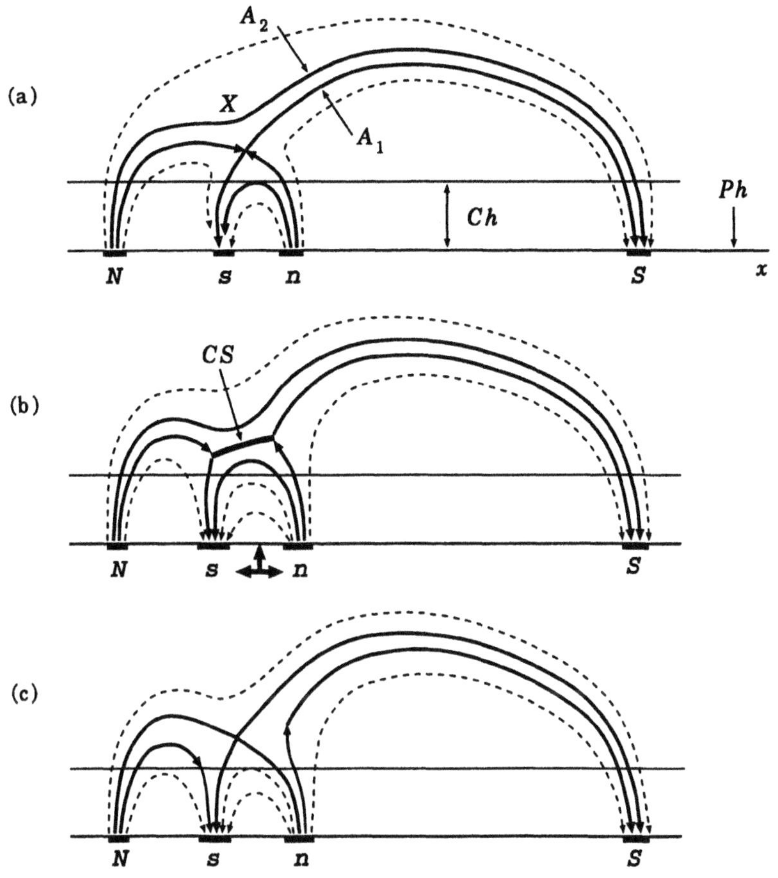

Figure 16.5: Three states of the potential field: (a) the initial state, (b) the *pre-reconnection* state, (c) the final state after reconnection.

realistic illustration of an effect, for example, the formation and dissipation of a current sheet at the X point under action of the photospheric shear (e.g., Kusano and Nishikawa, 1996; Karpen *et al.*, 1998), see also Sections 22.3 and 22.4. The term '2.5-dimensional' frequently refers to such two-dimensional MHD problems (in two spatial variables x and y) to point out the presence of the longitudinal field B_z related to the shear motion.

* * *

Let us come back to the first objection **(a)** to the reconnection theory of solar flares. According to the theory, the free magnetic energy is related to the electric current J inside the current sheet CS. The flare corresponds to rapid changes of this current. It is clear, however, that the magnetic flux through the photospheric plane *Ph* (Figure 16.5) can change only little during this process.

It means that sunspots and other magnetic features in the photosphere are unaffected by the occurrence of a flare because the plasma in the photosphere is almost 10^9 times denser than the plasma in the corona where the flare originates. Therefore it is difficult (but still possible) for disturbances in the tenuous corona and upper chromosphere to affect the extremely massive plasma in the photosphere. Only small MHD perturbations can penetrate into the upper photosphere.

The same is true in particular for the vertical component of the magnetic field, which is usually measured. Therefore

> in the first approximation, the photospheric magnetic field does not change during the solar flare.

As a consequence, it is not surprising that during the flare the large-scale structure in the corona can remain free of noticeable changes, because it is determined essentially by the potential part of the magnetic field above the photospheric sources. More exactly, even being disrupted, the large-scale structure will come to the potential field configuration corresponding to the post-flare position of the photospheric sources.

As for the second objection **(b)** to the hypothesis of accumulation of energy in the form of magnetic field of slowly-reconnecting current sheets in the solar atmosphere, the rapid dissipation of the field necessary for the flare is naturally explained by the theory of current sheets presented in what follows, first of all, in Sections 16.5.4 and 16.5.5.

16.5.2 Three-dimensional reconnection in flares

Gorbachev and Somov (1989, 1990) have developed a three-dimensional model for a potential field in the active region AR 2776 with an extended flare of 1980 November 5. Before discussing the flare, consider, at first, the basic general properties of this model called topological. Four magnetic sources – the 'magnetic charges' e_N and e_S, e_n and e_s, located in the plane Q under the photosphere *Ph* (Figure 16.6) – are used to reproduce the main features of the observed field in the photosphere related to the four most important sunspots: N, S, n and s. As a consequence, the model reproduces only the large-scale features of the actual field in the corona related to these sunspots.

The features are two magnetic surfaces, the boundary surfaces called the *separatrices* S_1 and S_2 (Figure 16.6), that divide the whole space above the

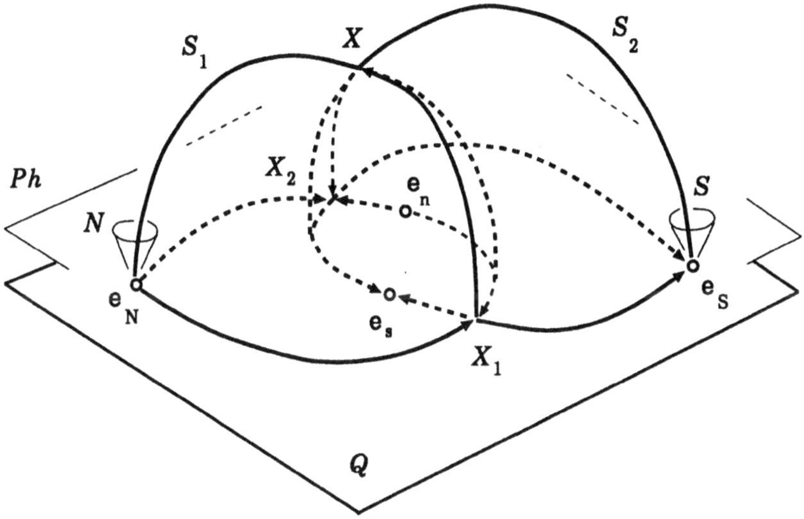

Figure 16.6: The model for the magnetic field of four sunspots of pairwise opposite polarity. The sunspots N and S in the photospheric plane Ph. The separatrixes S_1 and S_2 cross at the separator X_1XX_2 above the plane Q of the effective magnetic 'charges' e_N, e_S, etc.

under-photospheric plane Q into four regions and, correspondingly, the whole field into four magnetic fluxes having different linkages. There is a topologically singular field line X_1XX_2 lying at the intersection of the two separatrices, it belongs to all four fluxes that interact at this line – the 3D magnetic *separator*. So **the separator separates the interacting magnetic fluxes by the separatrices** (see also Lau, 1993).

The linkage of field lines connected to the separator is shown in Figure 16.7. This Figure does not mean, of course, that we assume the existance of real magnetic charges under the photosphere as well as the real X-type zeroth points X_1 and X_2 in the plane Q which does not exist either. We only assume that above the photospheric plane the large-scale magnetic field can be described in terms of such a model. We also assume that the actual conditions for reconnection are better at some point X of the separator rather than at its other points. If the magnetic sources move or/and change, the field also changes.

It is across the separator that the magnetic fluxes are redistributed and reconnected (Figure 16.7) so that the magnetic field could remain potential, if there were no plasma.

In the presence of the solar plasma of high conductivity, the separator

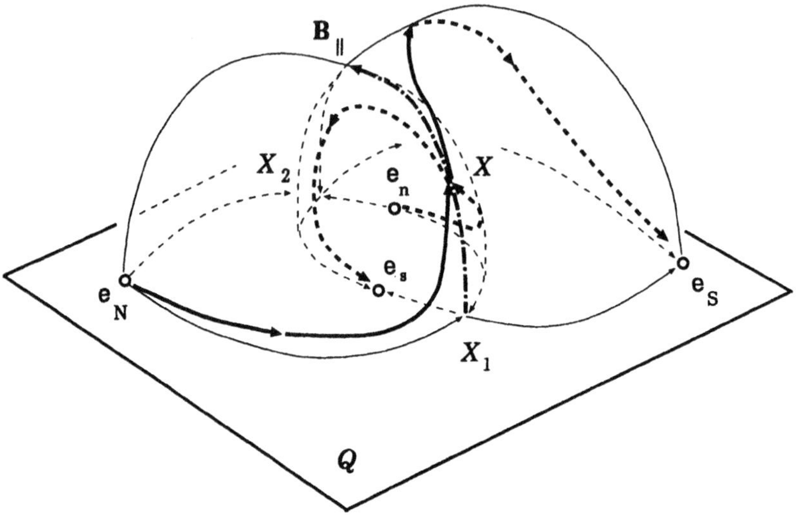

Figure 16.7: The same model for the magnetic field. The field lines located at the separatrices and connected to the separator due to the reconnection process at the point X, the vector $\mathbf{B}_{\|}$ is the longitudinal component of a magnetic field.

plays the same role as the hyperbolic neutral line of magnetic field, familiar from 2D MHD problems (e.g., Syrovatskii, 1966; Sweet, 1969; Brushlinskii et al., 1980; Biskamp, 1986 and 1997). In particular, as soon as the separator appears, the electric field $\mathbf{E}_0$ induced by the varying magnetic field produces an electric current $\mathbf{J}$ along the separator. The current interacts with the potential magnetic field in such a way (Section 4.4.3) that the current assumes the shape of a **thin wide current sheet** (see CS in Figure 16.8).

> In the high-conductivity plasma the current sheet hinders the redistribution of the magnetic fluxes.

This results in an energy being stored in the form of magnetic energy of a current sheet – the *free* magnetic energy.

Therefore the model assumes that the slowly-reconnecting current sheet appears at the separator (Syrovatskii, 1981; Gorbachev and Somov, 1989; Longcope and Cowley, 1996) in a pre-flare stage. If for some reason (see Somov, 1992) the reconnection process becomes fast, then the *free* magnetic energy is rapidly converted into kinetic energy of particles. This is a *flare*. The rapidly-reconnecting current sheet, being in a high-temperature turbulent-current state (see Somov, 1992), provides the flare energy fluxes along the

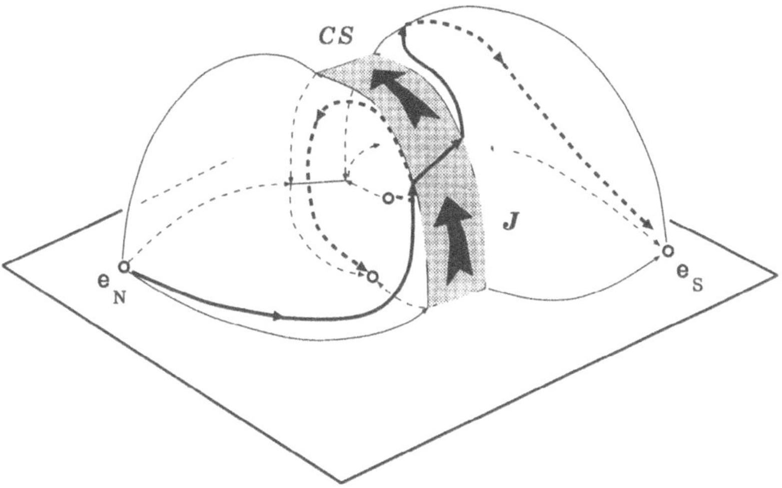

Figure 16.8: The current sheet CS with a total current J at the separator.

reconnected field lines.

* * *

It is important for what follows in Chapters 18, 20, and 22 that

> actual 3D reconnection at the separator proceeds in the presence of an increasing (or decreasing) longitudinal magnetic field $\mathbf{B}_{\parallel}$

(see Figure 16.7), which is parallel to the electric current $\mathbf{J}$ inside the reconnecting current sheet (see Figure 16.8). What factors do determine the increase (or decrease) of the longitudinal field? – The first of them is the global field configuration, i.e. the relative position of the magnetic field sources in an active region. It determines the position of the separator and the value of the longitudinal field at the separator and in its vicinity. This field is not uniform, of course, near the separator.

The second factor is the evolution of the global magnetic configuration, more exactly, the electric field $\mathbf{E}_0$ related to the evolution and responsible for driven reconnection at the separator. The direction of reconnection – with an increase (or decrease) of the longitudinal magnetic field – depends on the sign of the electric field projection on the separator, i.e. on the sign of the scalar product $(\mathbf{E}_0 \cdot \mathbf{B}_{\parallel})$. In general, this sign can be plus or minus with equal probabilities, if there are no preferential configurations of the global field or no preferential directions of the active region evolution. This statement as well as the whole model must be examined by future observations and their analysis.

16.5.3 The solar flare of 1980 November 5

Because the topological model by Gorbachev and Somov uses a minimal number of magnetic sources – four, which is necessary to describe the minimal number of interacting magnetic fluxes – two, we call it the *quadrupole-type* model. This label is not an exact definition (because in general $e_{\rm N} \neq -e_{\rm S}$ and $e_{\rm n} \neq -e_{\rm s}$) but it is convenient for people who know well the *exact-quadrupole* model by Sweet (1969). In fact, the difference – the presence of another separator in the model by Gorbachev and Somov – is not small and can be significant for actual active regions on the Sun. The second separator may be important to give accelerated particles a way to escape out of an active region in interplanetary space.

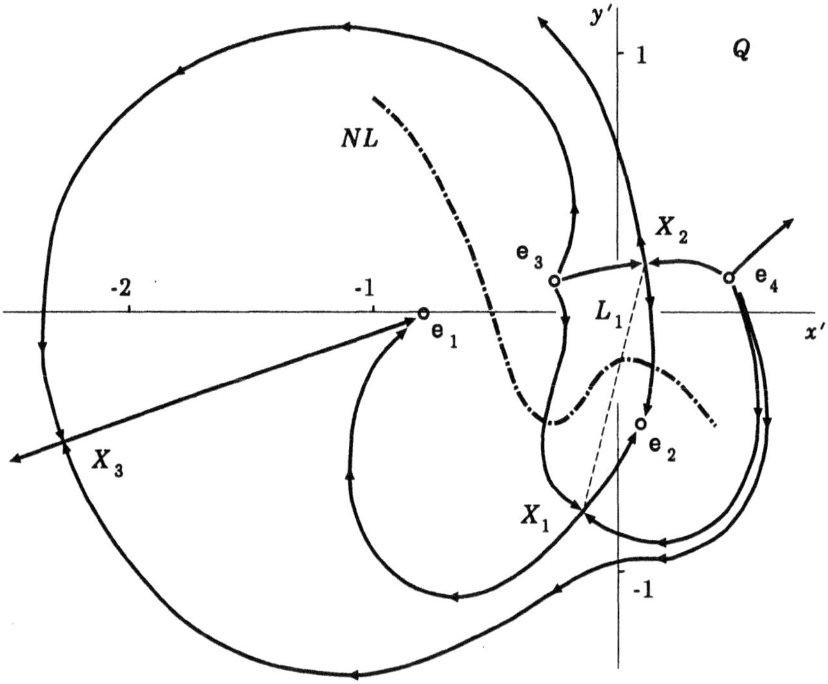

Figure 16.9: The topological portrait of the active region AR 2776 where the solar flare of 1980 November 5 occured.

Figure 16.9 shows the most important field lines in the plane (x', y') which is the plane Q of the effective sources e_1, e_2, e_3, and e_4. They reproduce the main features of the observed magnetic field in the photosphere related to the four largest sunspots in the active region AR 2776 where the extended M4/1B flare of 1980 November 5 was observed by the SMM satellite. Positions and magnitudes of the sources are adjusted to fit the main topological features of the magnetogram (see Figures 1 and 3 in Gorbachev and Somov, 1989).

The field lines shown in Figure 16.9 play the role of separatrices (cf. Figure 16.6) and show the **presence of two separators** in the active region. Two zeroth points X_1 and X_2 are located in the vicinity of the magnetic sources and are connected by the first separator shown by its projection, the thin dasched line L_1. Near this separator, the field and its gradient are strong and determine the flare activity of the region. Another separator starts from the zeroth point X_3 far away from the magnetic sources and goes much higher above the active region. The second separator can be responsible for flares in weaker magnetic fields and smaller gradients high in the corona.

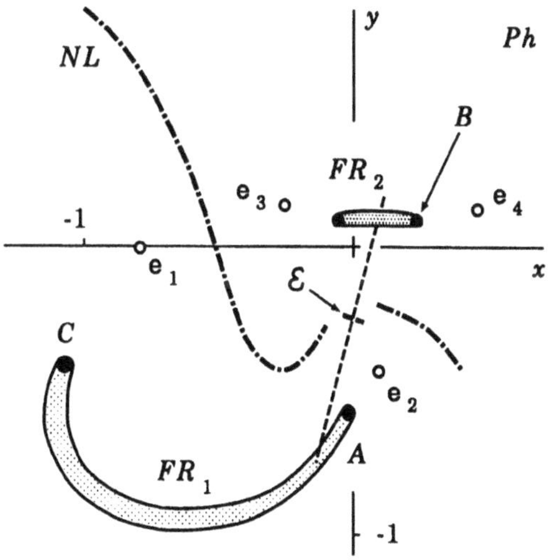

Figure 16.10: The flare ribbons at both sides of the photospheric neutral line NL in the flare of 1980 November 5.

Let us suppose that a part of the flare energy is initially released in some compact region $\mathcal{E}$ near the apex of the main separator X_1X_2. Then energy fluxes $F_{\mathcal{E}}$ will propagate *along* the field lines connecting the energy source with the photosphere. Projections of the energy source $\mathcal{E}$ on the photospheric plane Ph along the field lines are shown as two 'flare ribbons' FR_1 and FR_2 in Figure 16.10. Therefore we identify flare brightenings, mostly in the hydrogen $H\alpha$ line, with the ribbons located at the intersection of the separatrices with the chromosphere which is placed slightly above the photospheric plane (x, y).

The characteristic *saddle* structure of the field in the vicinity of the reconnecting point X at the separator (cf. Figure 9.1) leads to a spatial redistribution of the energy flux $F_{\mathcal{E}}$ of heat and accelerated particles. This flux is efficiently split apart in such a way that it creates the observed long-narrow

Hα ribbons in the chromosphere (see FR_1 and FR_2 in Figure 16.11).

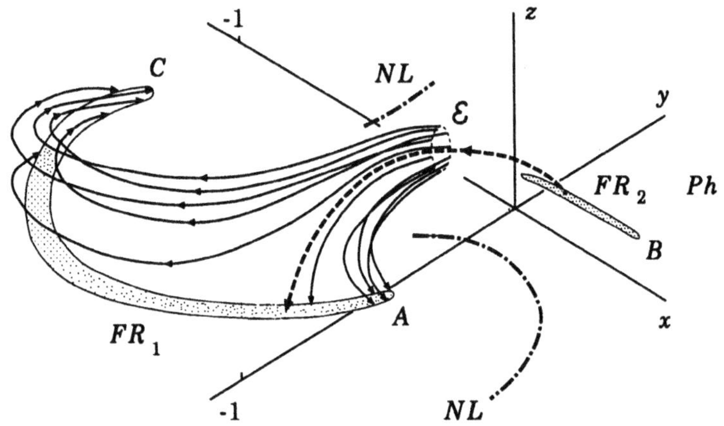

Figure 16.11: A picture of potential field lines crossing the region of primary energy release $\mathcal{E}$, which is situated at the apex of the main separator (boldface dashed curve). The flare ribbons are formed where these field lines cross the photosphere (plane $z = 0$).

For the first time, the model by Gorbachev and Somov (1989, 1990) had reproduced the observed features of the M4/1B flare of 1980 November 5. In particular, the model predicts the simultaneous flaring of the two chromospheric ribbons. Moreover it predicts that a concentration of the field lines that bring energy into the ribbons in the chromosphere is higher at the edges of the ribbons, i.e. at relatively compact regions indicated as A, B, and C. Here the Hα brightenings must be especially bright. This prediction of the model is consistent with observations of Hα 'kernels' in this flare.

The model also predicts another signature of flares. Since in the Hα kernels the flare energy fluxes are more concentrated, the impulsive heating of the chromosphere must create a fast expansion of high-temperature plasma upwards into the corona (e.g., Somov, 1992). This effect is known as the chromospheric 'evaporation' observed in the soft X-ray emission of solar flares. Evaporation lights up the soft X-ray coronal loops in flares.

Moreover the topological model shows that the two flare ribbons as well as the four of their edges with Hα kernels are magnetically connected to the common region of energy release at the separator (see $\mathcal{E}$ in Figure 16.11). Note that Figure 16.11 demonstrates only the field lines connected to one of the ribbons. Through the same region all four Hα kernels are magnetically connected to one another. Therefore the soft X-ray loops look like they are crossing or touching each other somewhere in the region of energy release as shown in Figure 16.12.

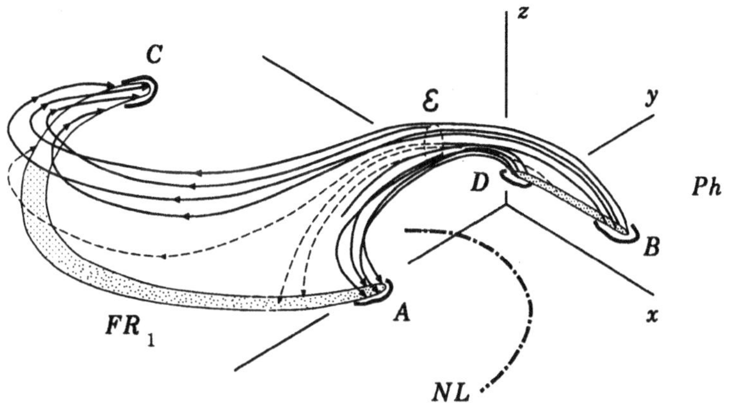

Figure 16.12: Field lines that connect the Hα kernels A, B, C, and D. Chromospheric evaporation creates a picture of the crossing soft X-ray loops predicted by the topological model for a flare in an active region with the quadrupole-type configuration of magnetic sources in the photosphere.

So the quadrupole-type model predicts that the reconnecting magnetic fluxes are distributed in the corona in such a way that the **two soft X-ray loops may look like that they interact with each other**. That is why the soft X-ray observations demonstrating such structures are usually considered as direct evidence in favour of the model of two interacting loops (see review in Sakai and de Jager, 1996). The difference, however, exists in the primary source of energy. High concentrations of electric currents and twisted magnetic fields are created inside the interacting loops by some under-photospheric mechanism. If these currents are mostly parallel they attract each other giving an energy to a flare (Gold and Hoyle, 1960). On the contrary, according to the topological model, the flare energy comes from an interaction of magnetic fluxes that can be mostly potential.

It is interesting to note that the **S-shaped structures**, when they are observed in soft X-rays (e.g., Figure 2 in Pevtsov *et al.*, 1996) or in hydrogen Hα-line, are usually interpreted in favour of non-potential fields. In general, the shapes of coronal loops are signature of the helicity (Section 21.1) of their magnetic fields – i.e., S-shaped loops match flux tubes of positive current helicity, and inverse S-shaped loops match flux tubes of negative current helicity (Pevtsov *et al.*, 1996). As we see in Figure 16.12, the S-shaped structure

$C\mathcal{E}B$ connecting the bright points C and B results from the computations of the potential field in the frame of the topological model.

Perhaps not surprisingly, the potential field produced by four magnetic sources may be even more complicated and may look as a strongly non-potential field. Severely **kinked Ω-type loops**, sometimes connecting two active regions, might be understood in terms of a simple topological model, see Figure 8 in Pevtsov and Longcope (1998).

In the active region AR 2776 where the flare of 1980 November 5 was observed, Den and Somov (1989) had found a considerable **shear of a potential field** above the photospheric neutral line near the region of the brightest flare loop AB. Many authors concluded that an initial energy of flares is stored in magnetic fields with large shear. However, such flares presumably were not the case of potential field having a minimum energy. This means that the presence of magnetic shear is not a sufficient condition for generation of a large flare in an active region.

The topological model by Gorbachev and Somov (1989, 1990) postulated a global magnetic field topology for an active region consisting of a four-flux system in which reconnection between, for example, the upper and lower fluxes transfers a part of the magnetic flux to the two side systems. Antiochos (1998) addresses the following question: 'What is the minimum complexity needed in the magnetic field of an active region so that a similar process can occur in a fully three-dimensional geometry?' He starts with the picture that a highly sheared, newly emerged field near the photospheric neutral line held down by an overlying unsheared field.

Antiochos (1998) concludes that a real active region on the Sun can have much more complexity than very simple configurations. Even so, we expect that the topology of four flux systems meeting along a coronal separator line is the basic topology underlying eruptive activity. It is unlikely that more than four fluxes would share a common boundary. This four-flux topology is precisely what is needed for eruption to occur.

16.5.4 A current sheet as the source of energy

(a) Pre-flare accumulation of energy

Potential field has no free energy. Given common and obvious assumptions, the free energy in the quadrupole-type model described above is simply the magnetic energy of the total electric current J in the current sheet CS in the solar atmosphere (Figure 16.8):

$$\mathcal{E}_f = \frac{1}{2c^2}\, LJ^2. \tag{16.86}$$

Here, according to Syrovatskii (1981),

$$L \approx 2l \ln \frac{2l}{b} \tag{16.87}$$

is the *self-inductance* of the current sheet, l is the distance taken along the separator from the zeroth point X_1 to the point X_2 in Figure 16.7, and b is the half-width of the current sheet.

Since we know the physical properties of a pre-flare current sheet (see Section 17.1.2), we can estimate the total current inside the sheet as well as its free magnetic energy (Syrovatskii, 1976b), the energy of a flare.

If we did not know the properties of the pre-flare reconnection process, we should have considered as an open question the following one: Why can the considerable excess energy, which will be released in a flare, be accumulated in the coronal magnetic field during the pre-flare stage without contradicting the natural tendency that lower energy states are more favourable? We should look for an answer to this question, for example, in a bifurcation structure of force-free fields in the corona (e.g., Kusano and Nishikawa, 1996). However, we may continue our consideration of the pre-flare stage as the creation and existence of the slowly-reconnecting current sheet. In this way, we see that

> slowly-reconnecting current sheets in the solar atmosphere can store the magnetic energy $\mathcal{E}_f$ necessary for flares.

Moreover in a quasi-stationary case (e.g., in the pre-flare state) their output can account for the energetics of the whole active region (Somov and Syrovatskii, 1977; Den and Somov, 1989). We may call such a state the **minimum current corona**.

Note that from (16.86) a simple formula follows for the total current J necessary for a solar flare to release the energy $\mathcal{E}_f$:

$$J = c \left(\frac{2\mathcal{E}_f}{L} \right)^{1/2} \approx (1-6) \times 10^{11}\ \mathrm{A}\,. \tag{16.88}$$

In this estimate the length l is set equal to the characteristic size of a large active region, $l \approx 10^{10}$ cm, and the flare energy to $\mathcal{E}_f \approx (1-3) \times 10^{32}$ erg. The result agrees with the estimates of the total electric current based on measurements of the magnetic field components in the photospheric plane (Moreton and Severny, 1968).

Vector magnetographs determine the transversal field at lower atmospheric levels; the curl of this field yields the vertical current density (e.g., Gopasyuk, 1990; Zhang, 1995; Wang *et al.*, 1996). Distributions of the intensity of the vertical current inferred from the horizontal magnetic field evolve only gradually and demonstrate two possibilities. One is the emergence of a new electric current from the sub-photosphere. The other is the rearrangement of the electric current systems in the solar atmosphere.

(b) Flare energy release

It is known that current sheets in the pre-flare state can suffer many instabilities: thermal instability caused by radiative energy losses (Field, 1965), resistive instability caused by temperature dependence of plasma conductivity, two-steam instabilities of various types, *structural instability* (Chapter 19), tearing instability (Chapter 20) etc. It is assumed that as a result of one of these instabilities the magnetic energy related to the current sheet is rapidly released and a solar flare starts. For example, a flare occurs when the current carried on a separator exceeds some threshold.

At present there are several open questions related to these instabilities: what is the relative importance of each of them, which of them can develop first, and whether an external action upon the current sheet is necessary or whether the sheet gradually evolves towards an unstable equilibrium or a non-equilibrium state by itself. Some attempts to answer these questions using relatively simple models will be demonstrated in what follows. In general, however, answers to these questions depend on the internal structure of the sheet. In its turn this structure depends on the initial and boundary conditions, and on the current sheet evolution during previous stages.

Therefore the investigation of current sheet dynamics is important for cosmic plasma physics. This investigation must include the formation stage, the pre-flare evolution, and the rapid realignment (rupture of the current sheet) with transition to a new state characterized by high temperatures and low conductivity (Chapter 17).

In the process of solving this problem many numerical (e.g., Brushlinskii *et al.*, 1980; Antiochos *et al.*, 1996) and laboratory (e.g., Altyntsev *et al.*, 1977; Stenzel and Gekelman, 1984; Bogdanov *et al.*, 1986) experiments have been performed. The hydrodynamic stage of the rise and evolution of pre-flare current sheets has been studied in detail. Experiments have shown that a thin, extended current sheet can be formed, even in laboratory conditions. To some approximation it has been possible to study the structures of the magnetic field inside the sheet and in the ambient plasma, to find the current distribution, the electron density and other plasma parameters.

Experiments have demonstrated the possibility of a substantial accumulation of magnetic energy and the explosive disruption of the current sheet. The cause of such disruption, which is accompanied by fast reconnection, may be a local resistivity increase related to the development of plasma turbulence.

Future experiments will probably, more than hitherto, concentrate on the study of the conditions for current sheet disruption, of nonlinear interactions in the fast reconnection region, and of particle acceleration (Chapter 18). This would help us to solve the most difficult problem in the reconnection theory and, in particular, give us information necessary to investigate experimentally the characteristics of current sheets as the source of flare energy during the impulsive phase.

The disruptive stage of the evolution cannot be described in hydrodynamic terms only: it requires a kinetic description in the disruption region. The impulsive electric field induced there can efficiently accelerate charged particles (Somov and Syrovatskii, 1975). During this process, plasma turbulence is generated. Its intensity depends on the one hand on the charged particle flux and on the other hand governs plasma conductivity, reconnection rate, and, as a consequence, the electric field intensity. There is thus a nonlinear feedback. Of course, to solve such a self-consistent problem is not easy. We shall, however, bear two limiting cases in mind.

First, low-energy particles interact effectively with the plasma, and most of their energy is rapidly lost by heating the plasma to very high temperatures, the so-called super-hot plasma. Second, in the high-energy region, a part of the accelerated particles enters into the *electric runaway* regime (Section 6.4.2), i.e. it virtually ceases to interact with the plasma (Chapter 18).

16.5.5 A current sheet as a part of an electric circuit

We have not discussed yet another problem of the theory of current sheets as a source of energy for solar flares. This problem has been nicely called *global electrodynamic coupling* (Spicer, 1982; Kan *et al.*, 1983) and it essentially consists in the question about the role of inductance and resistance in an equivalent electric circuit one of whose components is a current sheet in the solar atmosphere. In its simplest form (Baum *et al.*, 1978), the corresponding task can be illustrated by the elementary equation

$$L \frac{d}{dt} J(t) + J(t) R_0 = V(t) . \qquad (16.89)$$

Here $V = V(t)$ is the external *electromotive force* (emf) due to variations of photospheric magnetic fields, or simply the potential difference between the points X_1 and X_2 at the ends of the separator in Figure 16.6. The unknown quantity V depends on the strength of the photospheric sources and in the simplest approach it is treated as a given function of time.

Let us assume that at the initial moment $t = 0$, the current $J(0)$ along the separator was zero. At this point the external emf $V(0)$ was completely used up by acting against the self-induction emf:

$$L \frac{dJ}{dt} + 0 = V(0) . \qquad (16.90)$$

So the current $J(t)$ will appear.

As soon as a nonzero current $J(t)$ appears, the voltage drop on the total separator resistance R_0, according to Equation (16.89), makes the rate of current increase dJ/dt in the circuit smaller, which amounts to decreasing the rate of magnetic energy accumulation prior to a flare. The final steady current J_s depends on the resistance R_0 and the external emf V:

$$J_s = \frac{V}{R_0} . \qquad (16.91)$$

The characteristic time of the process is proportinal to the self-inductance L:

$$\tau_a = \frac{L}{R_0} . \qquad (16.92)$$

Note that $L \sim l$ and $R_0 \sim \sigma^{-1} l$. Therefore $\tau_a \sim \sigma$ does not depend of the length scale l.

The maximum accumulated energy (16.86) is also proportional to the inductance L of the equivalent circuit comprising the separator current sheet:

$$\mathcal{E}_f = \frac{1}{2c^2} \frac{LV^2}{R_0^2} . \qquad (16.93)$$

It is important that the free magnetic energy $\mathcal{E}_f$ and the energy accumulation time τ_a depend also on the total resistance R_0. In the pre-flare state, the current sheet with high Coulomb conductivity has low resistance. For this reason, the accumulated energy can be sufficiently large for a pre-flare active region. The accumulation time is long enough: $\tau_a \approx 3 \times 10^4$ s (Syrovatskii, 1976b). From what we have seen it is evident that

> to release the accumulated energy in a time $\tau_f \approx 10^2 - 10^3$ s corresponding to the solar flare duration, the total current sheet resistance must be increased by 2 to 3 orders of magnitude.

Such an effect can be well the result of the appearance of plasma turbulence (Section 17.3). An alternative possibility (see Chapter 22) is an appearance of one or many local current disruptions which have large enough resistance, *electric double layers*.

Earlier the possibility of formation of the double layers was, for some reason, treated as being alternative or even more in conflict with the concept of reconnection. However, after the laboratory experiment by Stenzel and Gekelman (1984) it became clear that double layers may form inside reconnecting current sheets. The hypothesis of the formation of electric double layers inside the separator-related current sheet can prove useful for the explanation of the extremely rapid energy release observed sometimes during solar flares. However, the concept of collisionless reconnection seems to be a more natural and more realistic alternative.

16.5.6 New topological models

After Gorbachev and Somov (1989, 1990) a series of investigations have sought observational evidence for reconnection in flares (Mandrini *et al.*, 1991, 1993; Mandrini and Machado, 1993; Demoulin *et al.*, 1993; Bagalá *et al.*, 1995; Longcope, 1996; Antiochos, 1998; Longcope and Silva, 1998). The results of these investigations can be summarized as follows: flare brightenings are located at the intersection of the separatrices with the chromosphere and are magnetically connected to one another as well as to a common region close to the separator (cf. Fig. 16.12). In particular, Longcope (1996), Longcope and Silva (1998) demonstrated clearly how

> motions of the photospheric sources (magnetic charges) lead to the build-up of 'ribbon-like' current sheets parallel to the separator

or two separators, as it is in the case of the solar flare on 7 January 1992 (see Section 16.5.3).

The magnitude of the current J at the separator is related through the self-inductance L (see formula (16.87)) to the magnetic flux change which would have occured in a potential field in the corona (Syrovatskii, 1966, 1981). By calculating approximate self-inductances of the separator, the topological model, called now the *minimum current corona*, provides an estimate of the current and the associated free energy from a given displacement of the magnetic sources.

The model developed by Longcope and Silva (1998) applies a topological approach to the magnetic field configuration for 7 January 1992. A new bipole ($\sim 10^{21}$ Mx) emerges amidst a pre-existing active region flux. This emergence gives rise to two current sheets along the separators separating the distinct, new and old, magnetic flux systems. Sudden reconnection across the separators transfers $\sim 10^{20}$ Mx of flux from the bipole into the surrounding flux. The locations of current sheets in the model correspond with observed soft X-ray loops. In addition the footpoints and apexes of the current sheets correspond with observed sources of microwave and hard X-ray emission. The magnitude of the magnetic energy stored by the current sheets compares favourably to the inferred energy content of accelerated electrons.

Sakao *et al.* (1998) have studied the spatial evolution of impulsive flares that clearly show the typical double-source structure at the peak of the M2 band (33-53 keV) emission in the hard X-ray (HXR) images obtained by the Hard X-ray Telescope (HXT) onboard *Yohkoh*. The distance between the sources has been analyzed as a function of time. As a result, two subclasses of flares – *more impulsive* (MI) and *less impulsive* (LI) – have been discovered. We assume that in both subclasses, the three-dimensional reconnection process occurs in the corona at the separator with a longitudinal field.

The difference between the LI and MI flares presumably appears because in the LI flares the reconnection process accompanies an increase of the longitudinal field at the separator (Somov *et al.*, 1998). In contrast, in the MI flares the reconnection proceeds with a decrease of the longitudinal field; hence, the reconnection rate is higher in the MI flares.

To illustrate that the observed variations of the footpoint separation depend on the longitudinal field $\mathbf{B}_{\parallel}$, this field is shown near the separator X in Figure 16.13. The arrows $\mathbf{v}_0$ and $\mathbf{v}_1$ indicate the reconnecting velocity pattern during the impulsive phase of a flare.

Two reconnecting field lines f_1 and f_2 arrive at the separator X and pass through it, the second one after the first. They bring different values of the longitudinal field $\mathbf{B}_{\parallel}$. If the second field line f_2 arrives with a stronger longitudinal field than the first one, i.e. $B_{\parallel 2} > B_{\parallel 1}$, then the length of the line f_2 after reconnection is obviously larger than the length of the line f_1 as shown in Figure 16.13a.

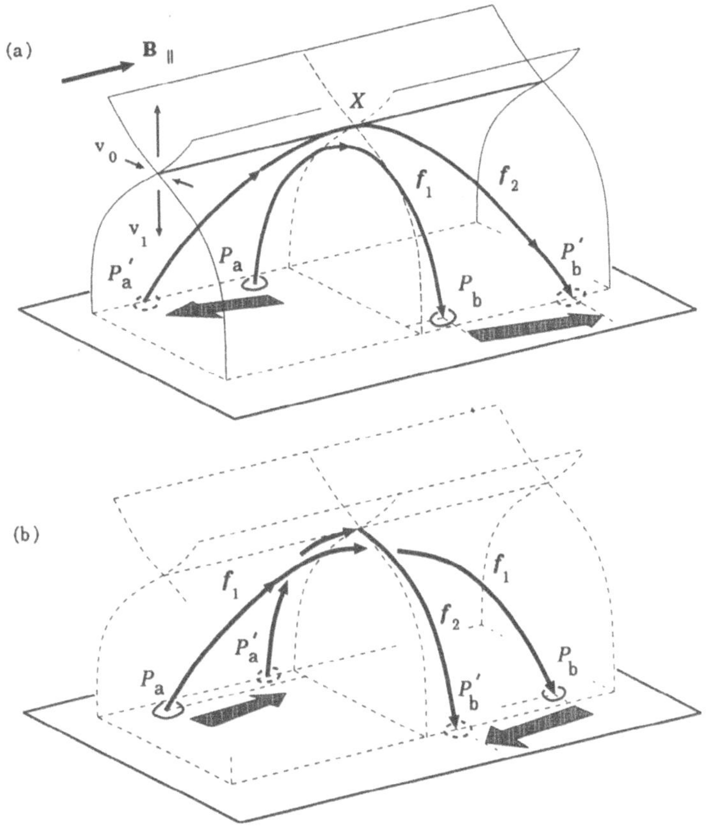

Figure 16.13: An apparent motion of the hard X-ray footpoints during the fast reconnection: (a) the footpoint separation rapidly increases, (b) a decreasing footpoint separation.

Figure 16.13a also shows positions of the footpoints in the chromospheric plane for the same field lines. The footpoints P_{a} and P_{b}, being impulsively heated by accelerated particles, became bright in HXR earlier than the footpoints P_{a}' and P_{b}'. Figure 16.13a demonstrates that, if the longitudinal field becomes stronger at the separator, then the footpoint separation will increase during the fast reconnection. If, on the contrary, the line f_2 brings a weaker longitudinal field, i.e. $B_{\parallel 2} < B_{\parallel 1}$, then the distance between footpoints rapidly becomes shorter as shown in Figure 16.13b.

The topological model makes intelligible the observed decrease (increase) of the separation between the double-footpoint HXR sources in the more impulsive (less impulsive) flares (Somov and Merenkova, 1999). Consider two configurations (a) and (b) in Figure 16.14 for the four magnetic sources in the source plane Q. To a different extent they differ from the ideal configuration when all the four sources are placed along the symmetry axis x. The longitudinal magnetic field at the separator is equal to zero in the ideal symmetrical case.

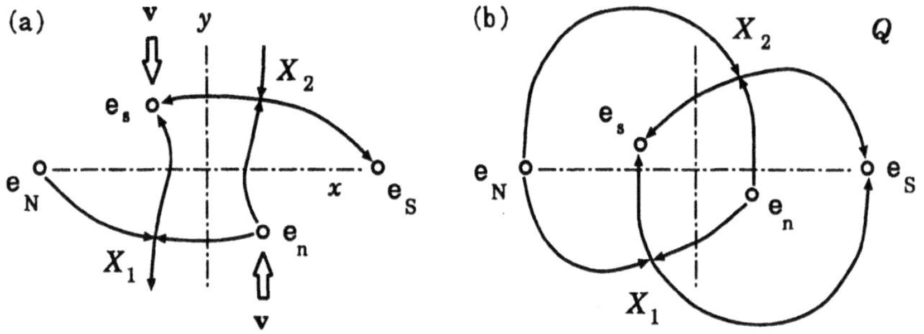

Figure 16.14: Two configurations of magnetic sources in the plane Q.

In general, the initial state (more exactly, the temporary pre-reconnection state which can be considered as initial one) differs from the ideal state, of course. So, the longitudinal field already exists at the separator. This field always presents under condition of actual 3D reconnection in the solar atmosphere, and it will increase (or decrease) depending on the direction of evolution of the magnetic field in an active region. For example, the configuration evolves from the less-ideal initial state (a) to a more-ideal one (b) as shown in Figure 16.14. Under this direction of evolution, indicated by vector $\mathbf{v}$ in Figure 16.14, the reconnection process decreases the longitudinal field at the separator.

Following Gorbachev and Somov (1989, 1990), let us suppose that a part of the flare energy is initially released in some compact region $\mathcal{E}$ near the apex of the separator. Then the energy fluxes will propagate *along* the field lines connecting the energy source with the photosphere. Projections of the energy source $\mathcal{E}$ on the photospheric plane Ph along the field lines are shown as the two 'flare ribbons' FR in Figure 16.15. Therefore we identify flare brightenings, mostly in the hydrogen $H\alpha$ line, with the ribbons located at the intersection of the separatrices with the chromosphere which is placed slightly above the photospheric plane.

The *saddle* structure of the field near the separator leads to a spatial redistribution of the flux of heat and accelerated particles. This flux is split apart in such a way that it creates the long-narrow Hα ribbons in the chro-

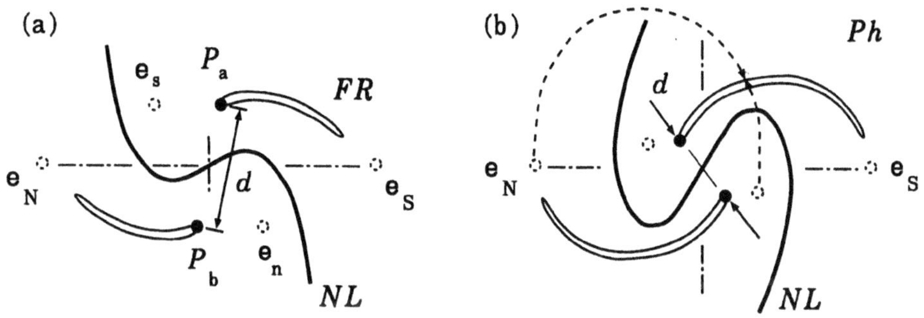

Figure 16.15: The long-narrow Hα ribbons FR and Hα kernels P_a and P_b projected in the photospheric plane Q both sides of the photospheric neutral line NL.

mosphere (see FR in Figure 16.15). Moreover the model predicts that a concentration of the field lines that bring energy into the flare ribbons in the chromosphere is higher at the edges of the ribbons, i.e. at relatively compact regions shown by dark points P_a and P_b in Figure 16.15. Here the Hα brightenings must be especially bright. This prediction of the model is well consistent with observations of Hα kernels in a flare.

Figure 16.15 shows that the foot-point separation, which is the distance d between the points P_a and P_b, decreases if the magnetic configuration evolves from the state (a) to state (b), id. when the longitudinal magnetic field decreases during the reconnection process at the separator. So the reconnection rate is higher in the MI flares of the Sakao type. In contrast, in the LI flares the magnetic configuration evolves from (b) to (a). This means that the reconnection proceeds with an increase of the longitudinal field, more slowly, and with an increase of the foot-point separation.

Therefore we may conclude that, if the evolution of the sunspot configuration goes to a more ideal state with a smaller displacement from the symmetry axies, then the MI flares should occur. This statement as well as the whole model must be examined by future observations and their analysis.

Chapter 17

Stationary Models of Reconnecting Current Sheets

When two oppositely directed magnetic fields are pressed together, the conductive plasma is squeezed out from between them, causing the field gradient to steepen until the resistive dissipation may create a steady state – the stationary reconnection.

17.1 Magnetically neutral current sheets

17.1.1 The simplest MHD model

Let us consider two consequent approximations used to study the reconnection process in current sheets. The first of them was the *neutral* current sheet model (e.g., Sweet, 1969; Parker, 1979; Syrovatskii, 1981). This was initially the simplest 2D configuration of steady reconnection. Two oppositely directed magnetic fields are pushed together into the neutral sheet (see Figure 17.1). The uniform field $\mathbf{B}_0$ immediately outside the sheet is frozen into the uniform plasma inflow with a velocity $\mathbf{v}_0$ perpendicular to the field. The plasma flows out of the neutral sheet through its edges with a large velocity $\mathbf{v}_1$ perpendicular to the velocity $\mathbf{v}_0$.

The strength of the magnetic field, B_0, on the inflow sides of the neutral sheet can be found out, for example, from the analytical solution of the problem for the vertical current sheet in the solar corona above a dipole source of the field in the photosphere (Somov and Syrovatskii, 1972b). This would be just the case of the so-called 'standard model' for a two-ribbon flare

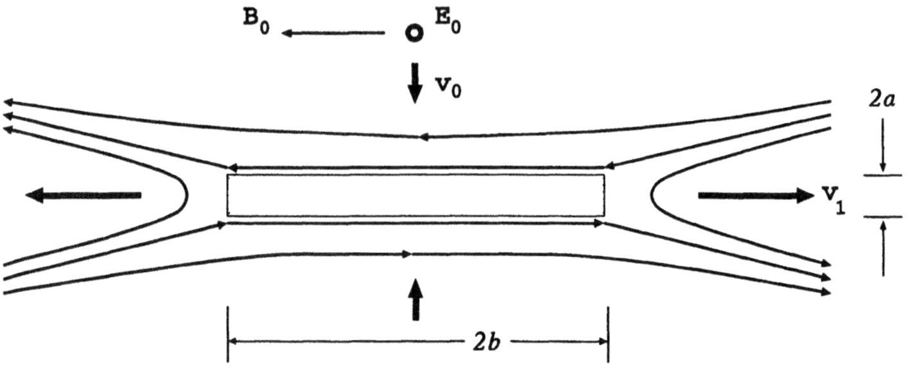

Figure 17.1: A schematic drawing of the field lines undergoing reconnection across the neutral current sheet according to Sweet-Parker model.

(see Tsuneta, 1996, and references there). The strength of the electric field, E_0, near the current sheet can be estimated for a given value of the velocity v_0 for the coronal plasma inflow into the reconnecting current sheet (RCS) and for a given value of the magnetic field B_0.

By definition, there is no magnetic field inside the neutral sheet; that is why it is called a *neutral* or, more exactly, a *magnetically neutral* RCS. This oversimpified approximation seems to be good, however, only for a low-temperature RCS, for example, for cold dense pre-flare current sheets because heat conduction does not play any role in the energy balance for such RCS (Section 17.1.2). Although it is a strong idealization, the approximation of a neutral sheet is still useful for several reasons.

First, the neutral sheet approximation demonstrates the existence of two linear scales corresponding to two different physical processes. (a) The sheet half-thickness

$$a \approx \frac{\nu_m}{v_0} \tag{17.1}$$

is the dissipative scale responsible for the rate of reconnection; here $\nu_m = c^2 \, (4\pi\sigma)^{-1}$ is the magnetic diffusivity. (b) The sheet width $2b$ is responsible for the accumulation of magnetic energy (Syrovatskii, 1976a). The wider the reconnecting sheet, the larger is the energy accumulated in the region of the reconnecting magnetic fluxes interaction.

Second, the neutral sheet approximation indicates that very efficient acceleration of particles can work in the RCS (see Section 4.5). Let us take as the low limits for the magnetic field $B_0 \approx 50$ G and for the inflow velocity $v_0 \approx 20$ km s^{-1}. These values are smaller than those estimated from the *Yokhoh* SXT and HXT observations of the well studied impulsive flare on

1992 January 13 – the magnetic field strengh in the supposed Petschek-type (Problem 11.8) upstream plasma 52 G and the inflow speed range 40-140 km s^{-1}, respectively (Tsuneta et al., 1997). So, the lower limit for the electric field can be estimated as

$$E_0 = \frac{1}{c} \, v_0 B_0 \approx 1 \ \mathrm{V\,cm}^{-1} . \tag{17.2}$$

This field is much stronger than the Dreicer's field – the electric field strength for which the critical runaway speed is equal to the electron thermal velocity (Section 6.4.2):

$$E_{\mathrm{Dr}} = \frac{4\pi e^3}{k_{\mathrm{B}}} \, (\ln \Lambda) \, \frac{n}{T} \approx 10^{-4} \ \mathrm{V\,cm}^{-1} . \tag{17.3}$$

Here we have assumed that the density and temperature of the plasma near the RCS $n_0 \approx 4 \times 10^8 \ \mathrm{cm}^{-3}$ and $T_0 \approx 3 \times 10^6 \ \mathrm{K}$. In fact, near the RCS in solar flares, the magnetic field B_0 can be as high as 100–300 G. So the electric field E_0 can be even stronger by one order of magnitude (Somov, 1981).

Since $E_0 \gg E_{\mathrm{Dr}}$, we neglect collisional energy losses (e.g., Gurevich, 1961) as well as wave-particle interaction of fast particles (Gurevich and Zhivlyuk, 1966). So

> the neutral sheet model predicts very impulsive acceleration of charged particles by the direct strong electric field $\mathbf{E}_0$.

This advantage of the RCS will be discussed in Chapter 18 with account of the fact that real reconnecting sheets are always magnetically non-neutral – they always have an internal magnetic field. The influence of this three-component field inside the RCS on the particle acceleration is considered in Chapter 18.

The main disadvantage of the neutral sheet model is that it does not explain the high power of the energy release in solar flares. The reason will be explained in Section 17.2 by using a less idealized model of the RCS.

17.1.2 The current sheet by Syrovatskii

To establish relations between the parameters of a neutral sheet in *compressible* plasma let us use the equations of continuity and momentum. Under conditions of the strong magnetic field (Section 8.3.3) these equations can be rewritten as the following *order-of-magnitude* relations:

$$n_0 v_0 \, b = n_s v_1 \, a \, , \tag{17.4}$$

$$\frac{B_0^2}{8\pi} = 2n_s k_B T, \tag{17.5}$$

$$2n_s k_B T = \frac{1}{2} M n_s v_1^2 . \tag{17.6}$$

Here n_0 and n_s is plasma density outside and inside the sheet, respectively. T is temperature of the plasma inside the sheet.

It follows from Equations (17.5) and (17.6) that the velocity of outflow from the current sheet

$$\boxed{v_1 = V_{A,s} = \frac{B_0}{\sqrt{4\pi M n_s}} .} \tag{17.7}$$

Note that the value of the magnetic field is taken outside the sheet, for plasma density it is taken *inside* the neutral sheet. So the outflow velocity (17.7) *differs* from the Alfvén speed outside the sheet

$$V_{A,0} = \frac{B_0}{\sqrt{4\pi M n_0}} . \tag{17.8}$$

The downstream flow velocity v_1 of a compressed plasma is *not* equal to the upstream Alfvén speed outside the sheet $V_{A,0}$.

The inflow velocity equals the velocity of the plasma drift to the neutral sheet

$$v_0 = V_d = c\,\frac{E_0}{B_0} . \tag{17.9}$$

Hence we have to add an equation which relates the electric field E_0 with the current sheet parameters. From the Maxwell equation (8.9) and Ohm's law, we find

$$\frac{cB_0}{4\pi a} = \sigma E_0 . \tag{17.10}$$

Here $\sigma = \sigma_0 T^{3/2}$ is the Coulomb conductivity.

Following Syrovatskii (1976b), from Equations (17.4)–(17.6) and (17.10) the neutral sheet half-thickness a, its half-width b, and the plasma density inside the sheet n_s can be expressed in terms of three 'external' (assumed known) parameters n_0, $h_0 = B_0/b$, E_0 and the unknown equilibrium temperature T of the plasma inside the current sheet:

$$a = b\,\frac{c}{4\pi\sigma_0}\left(\frac{h_0}{E_0}\right)\frac{1}{T^{3/2}} , \tag{17.11}$$

$$b = 4\pi \left(\frac{k_{\rm B} \sigma_0^2 M}{4\pi^2} \right)^{1/6} \left(\frac{n_0 E_0^2}{h_0^4} \right)^{1/3} T^{2/3} , \tag{17.12}$$

$$n_s = \left(\frac{\pi \sigma_0^2 M}{4 k_{\rm B}^2} \right)^{1/3} \left(\frac{n_0 E_0^2}{h_0} \right)^{2/3} T^{1/3} . \tag{17.13}$$

To determine the temperature T let us add the energy equation in the following form:

$$\frac{B_0^2}{4\pi} \, V_d \, b = L\,(T)\, n_s^2 \, ab \, . \tag{17.14}$$

It is assumed here that the temperature of the neutral sheet is not high; so the energy transfer from the sheet by plasma outflow and by heat conduction play a secondary role. The principal factors are the influx of magnetic energy into the current sheet (formula (8.90)) and radiative cooling. The radiative loss function $L\,(T)$ can be taken, for example, from Cox and Tucker (1969). More justifications for simple Equation (17.14) follow from the more detailed numerical model by Oreshina and Somov (1999a); see also a comparison between different models in Somov and Oreshina (2000).

Substituting the solution (17.11)–(17.13) in Equation (17.14) we obtain the following equation for the temperature of the plasma inside the current sheet:

$$T = \sigma_0^{2/5} \left(\frac{\pi M}{4 k_{\rm B}^2} \right)^{4/5} \Gamma_{\rm S}^{4/5} L^{6/5}(T) \, . \tag{17.15}$$

Here

$$\Gamma_{\rm S} = \frac{n_0^2 E_0}{h_0^2} \tag{17.16}$$

is the dimensional parameter which characterizes the reconnection conditions. Therefore the values n_0, h_0, and E_0 must be specified in advance. The other quantities can be determined from the solution (Problem 17.1).

Figure 17.2 shows a solution of Equation (17.15) with two unstable branches indicated by dashed curves. On these branches a small deviation of the temperature from equilibrium will cause the deviation to increase with time. It means that the *thermal instability* of the current sheet occurs.

The first appearance of the thermal instability, at $T \approx 2 \times 10^4$ K, is caused by emission in the Lα line of hydrogen. It can hardly be considered significant since the function $L(T)$ was taken from Cox and Tucker (1969) without allowance for the absorption of radiation, which may be important

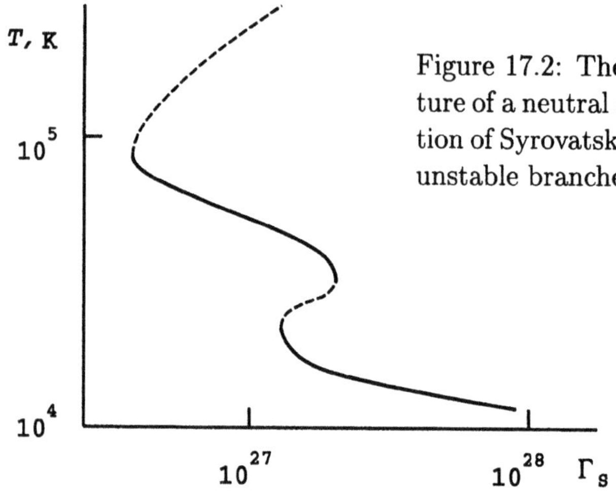

Figure 17.2: The equilibrium temperature of a neutral current sheet as a function of Syrovatskii's parameter $\Gamma_{\rm S}$. Two unstable branches are dashed.

for the hydrogen lines in the solar atmosphere. On the contrary, the second break, at

$$T \approx 8 \times 10^4 \text{ K}, \quad \Gamma_{\rm S} \approx 3.8 \times 10^{26}, \tag{17.17}$$

will necessarily occur because of the maximum in the radiative cooling function $L(T)$. Near this maximum, in the region where $L(T) \propto T^{\alpha}$ with $\alpha < 1$, the condensation mode of the thermal instability (Field, 1965) occurs (see also Somov and Syrovatskii, 1976b and 1982).

Syrovatskii (1976b) assumed that the temperature T of a *cold dense* current sheet in the solar atmosphere gradually increases in the pre-flare stage until the critical values (17.17) are reached. Then the current sheet can no longer stay in equilibrium; the radiative losses cannot balance the Joule heating, and the temperature of the sheet rapidly rises. This leads to a flare. In this way, Syrovatskii suggested to identify the thermal instability of a cold dense current sheet with the onset of the eruptive phase of a solar flare.

Whether such a *thermal trigger* for solar flares occurs or not is unclear yet (Somov and Syrovatskii, 1982). It is clear only that heating of the current sheet leads to the powerful heat-conductive cooling of the plasma electron component. This effect is important for energy balance of high-temperature turbulent-current sheets (HTTCS) discussed in Section 17.3.

17.1.3 Simple scaling laws

In order to determine the parameters of a stationary driven reconnection configuration, the stationary resistive MHD equations must be solved for given boundary conditions. Unfortunately it appears that the problem is too complicated to permit analytical solutions without severe approximations. The severest of them are called the scaling 'laws'.

Let us come back to the Sweet-Parker model of reconnection in *incompressible* plasma. The *order-of-magnitude* relations introduced above become simpler:

$$v_0 \, b = v_1 \, a \, , \tag{17.18}$$

$$v_0 = \frac{\nu_m}{a} \, , \tag{17.19}$$

$$v_1 = V_{A,0} \, . \tag{17.20}$$

These equations follow from (17.4)–(17.13) and give us the ratio of the inflow (upstream) velocity of the incompressible plasma to the upstream Alfvén speed:

$$\frac{v_0}{V_{A,0}} = \left(\frac{\nu_m}{V_{A,0} b} \right)^{1/2} . \tag{17.21}$$

The left-hand side of the relation (17.21) is called the Alfvén-Mach number M_A and is conventionally used as a dimensionless measure of the reconnection rate. The right-hand side is simply related to the magnetic Reynolds number (8.36), more exactly

$$\mathrm{Re}_m(V_{A,0}, b) = \frac{V_{A,0} b}{\nu_m} \equiv N_L \, . \tag{17.22}$$

Here N_L is called the Lundquist number. Therefore the Sweet-Parker reconnection rate

$$\boxed{M_A = N_L^{-1/2} \, .} \tag{17.23}$$

Order-of-magnitude relations similar to (17.23) are often called scaling 'laws'. They certainly do not have a status of any law but are useful since they simply characterize the *scaling properties* of stationary reconnecting configurations as a proper dimensionless parameter.

Since in formula (17.22) the linear scale L is taken to be equal to the large half-width b of the Sweet-Parker neutral sheet, the Lundquist number (17.23) is rather a *global* parameter of the reconnection problem. In the most cases of practical interest the Lundquist number is too large, typically $10^{14} - 10^{15}$ in the solar corona (Problem 17.1), such that the Sweet-Parker rate would lead to reconnection times many orders of magnitude longer than observed in flares. This means that slowly-reconnecting current sheets can exist in the solar corona for a long time.

In general, scaling relations are useful to summarize and classify different regimes and configurations of reconnection as they are observed, for example, in numerical simulations (see Ch. 6 in Biskamp, 1993; Horiuchi and Sato, 1994).

17.2 Magnetically non-neutral RCS's

Magnetic neutrality of the RCS, as assumed in the previous Section, means that there is no penetration of magnetic field lines through the sheet (the transversal field $\mathbf{B}_\perp = 0$) as well as no longitudinal magnetic field parallel to the electric current inside the RCS (the longitudinal field $\mathbf{B}_\parallel = 0$). In general, both assumptions are incorrect (for a review see Somov, 1992). The first of them is the most important for what follows in this Chapter.

17.2.1 Transversal magnetic fields

As it reconnects, every field line penetrates through the current sheet as shown in Figure 17.3. So, the reconnecting sheet is magnetically non-neutral by definition because of physical meaning of the reconnection process. In many real cases (for example, the magnetospheric tail or interplanetary sectorial current sheets) a small transversal component of the magnetic field is well observed. This is also the case of laboratory and numerical experiments (e.g., Hesse et al., 1996; Ono et al., 1996; Horiuchi and Sato, 1997).

We characterize the penetration of the magnetic field into the current sheet by the parameter $\xi_\perp = B_\perp / B_0$ which is the relative value of the transversal component $\mathbf{B}_\perp$. As distinguished from the neutral-sheet approximation, we assume that $\xi_\perp \neq 0$ and satisfies the inequality

$$a/b \ll \xi_\perp \ll 1 \,. \tag{17.24}$$

What are the consequences of such a penetration?

The penetration of even a very small transversal field into the high-temperature sheet essentially increases the outflows of energy and mass from the sheet along the field lines. The effective cross-section for the outflows of energy and mass is proportional to the outflow scale

$$a^{out} \approx \xi_\perp b \gg a \,. \tag{17.25}$$

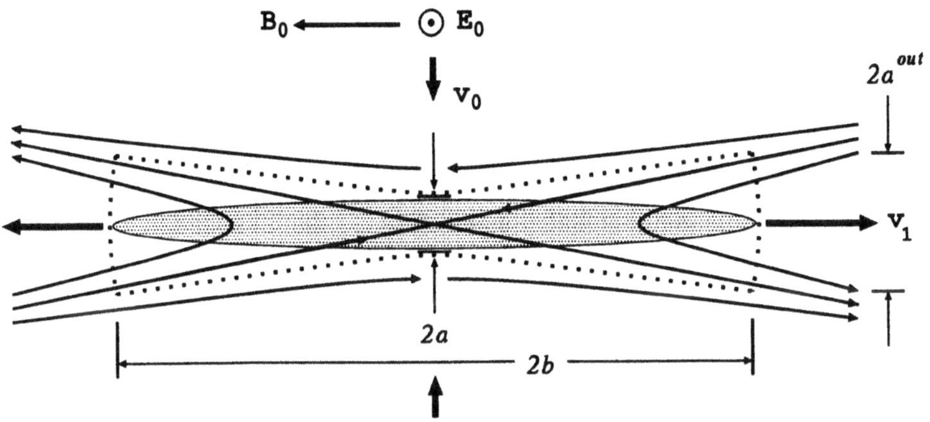

Figure 17.3: A magnetically non-neutral reconnecting sheet: the electric current distribution is schematically shown by the shadow, the dotted boundary indicates the field lines going through the current sheet.

Hence, corresponding to three different physical processes, the magnetically **non-neutral current sheet is characterized by three different linear scales:** $2a$ is a small dissipative thickness of the sheet, $2b$ is the scale responsible for the energy accumulation process, and $2a^{out}$ is the linear scale which determines the outflow of energy and mass along the field lines into the surrounding plasma.

As we see in Section 17.3, even a very small (like $\xi_{\perp} \approx 10^{-3}$) transversal field $\mathbf{B}_{\perp}$ **significantly increases the plasma outflows** as well as the heat-conductive cooling of the non-neutral high-temperature turbulent-current sheet (HTTCS). As a result, its energy output is much larger than that of the neutral HTTCS. (In the neutral-sheet approximation $a^{out} = a$.) The last reason will enable us to consider the HTTCS with a small transversal component of the magnetic field as the source of energy in flares.

17.2.2 Longitudinal magnetic fields

As we saw in Section 16.5, the reconnection process under the actual conditions in the solar atmosphere is released at the separator, which differs from the X-type neutral line in that the separator has a longitudinal field $\mathbf{B}_{\parallel}$. In this context, it is necessary to understand the physical effects that are created by the longitudinal field inside the RCS and its vicinity.

It is intuitively clear that the longitudinal field at the separator decreases the reconnection rate

$$\mathbf{v}_0 = c\,\frac{\mathbf{E}_0 \times (\mathbf{B}_0 + \mathbf{B}_{\parallel 0})}{B_0^2 + B_{\parallel 0}^2} = c\,\frac{\mathbf{E}_0 \times \mathbf{B}_0}{B_0^2\left[1 + (B_{\parallel 0}/B_0)^2\right]} \,. \tag{17.26}$$

Here $\mathbf{B}_0$ and $\mathbf{B}_{\parallel 0}$ are the strengths of the reconnecting component and of the longitudinal component of the magnetic field on the inflow side of the RCS, respectively; they are not *free* parameters, they have to be determined from a self-consistent solution of the problem on the RCS properties.

The appearance of the longitudinal field changes, first of all, the balance of forces across the sheet. The pressures of the plasma and the magnetic field outside the RCS should balance not only the plasma pressure but also the magnetic pressure of the longitudinal field inside it:

$$2n_0 k_{\mathrm{B}} T_0 + \frac{B_0^2}{8\pi} + \frac{B_{\parallel 0}^2}{8\pi} = 2n_s k_{\mathrm{B}} T + \frac{B_{\parallel s}^2}{8\pi} \,. \tag{17.27}$$

Here n_0 and n_s are the plasma densities outside and inside the RCS. T_0 is the temperature of inflowing plasma outside the sheet, T is the temperature of plasma inside the RCS. In the right-hand side of Equation (17.30) $B_{\parallel s}$ is the strength of the longitudinal field inside the RCS.

If the longitudinal field could be effectively accumulated inside the current sheet, its pressure would impose strong limitations on the sheet compression and, hence, on the rate of reconnection. In terms of the ideal MHD approximation, the longitudinal field must increase proportionally to the plasma density n_s inside the sheet because the field is frozen in the plasma:

$$B_{\parallel s} = B_{\parallel 0}\,\frac{n_s}{n_0} \,. \tag{17.28}$$

On the contrary, in a real finite-conductivity plasma, the increase of the longitudinal field is accompanied by dissipative effects. As soon as the longitudinal field inside the sheet becomes stronger than outside the sheet, a gradient of the longitudinal field $\mathbf{B}_{\parallel}$ will appear and give rise to an electric current. In turn, the dissipation of this current produced by the field compression affects the $\mathbf{B}_{\parallel}$ field value. Thus the compression of the longitudinal field seems to facilitate its dissipation. In reality, however, this problem proves to be more delicate; see Somov and Titov (1985), Somov (1992).

The essence of the effect is that any compression of the longitudinal field $\mathbf{B}_{\parallel}$ within a current sheet does create a gradient of the longitudinal field, $\nabla \mathbf{B}_{\parallel}$. By so doing, compression generates an associated electric current $\mathbf{J}_{\perp}$ which circulates in the transversal (relative to the main current $\mathbf{J}$ in the sheet) plane. The ohmic dissipation of the current $\mathbf{J}_{\perp}$, circulating around the sheet, gives rise to an outward diffusion of the longitudinal field from the current sheet and to the Joule heating of the plasma. It is of importance that **the total flux of the longitudinal field is conserved**, while

> the Joule heating due to the $\mathbf{B}_{\parallel}$ field compression is produced by the dissipation of the reconnecting magnetic field $\mathbf{B}_0$.

This effect is certainly valid for collisionless reconnection in the RCS.

So, on the one hand, the magnetic field compression decreases the velocity v_0 of plasma inflows. On the other hand, due to the large magnetic diffusion in the small scale of the current sheet thickness $2b$, the longitudinal field $\mathbf{B}_{\parallel}$ does not have an overwhelming effect on the parameters of the current sheet and the reconnection rate. For this reason, we regard as likely that

> the longitudinal field $\mathbf{B}_{\parallel}$ at the separator changes the reconnection rate in the RCS not too strongly.

This can be especially true if the compression of the plasma inside the RCS, n_s/n_0, is not high, for example, in high-temperature turbulent-current sheets (HTTCS) of solar flares. Therefore, in the first approximation, we hope that we can neglect the longitudinal magnetic field in the next Section.

17.3 Basic physics of the HTTCS

17.3.1 A general formulation of the problem

Coulomb collisions do not play any role in the HTTCS. So, the plasma inside the HTTCS has to be considered as essentially collisionless (see Chapter 3 in Somov, 1992). The concept of an anomalous resistivity, which originates from wave-particle interactions, is then useful to describe the fast conversion from field energy to particle energy. Some of the general properties of such a **collisionless reconnection** can be examined in a frame of a self-consistent model which makes it possible to estimate the main parameters of the HTTCS. Basing on the mass, momentum and energy conservation laws, we write the following relations (valid for a quarter of the current sheet and a unit length along the electric current):

$$n_0 v_0 \, b = n_s v_1 \, a^{out}, \tag{17.29}$$

$$2 n_0 k_{\rm B} T_0 + \frac{B_0^2}{8\pi} = n_s k_{\rm B} T \left(1 + \frac{1}{\theta} \right), \tag{17.30}$$

$$n_s k_B T \left(1 + \frac{1}{\theta}\right) = \frac{1}{2} M n_s v_1^2 + 2 n_0 k_B T_0 \,, \tag{17.31}$$

$$\chi_{\rm ef}\, \mathcal{E}_{mag}^{\,in} + \mathcal{E}_{th,e}^{\,in} = \mathcal{E}_{th,e}^{\,out} + \mathcal{C}_{\parallel}^{\,an} \,, \tag{17.32}$$

$$(1 - \chi_{\rm ef})\, \mathcal{E}_{mag}^{\,in} + \mathcal{E}_{th,i}^{\,in} = \mathcal{E}_{th,i}^{\,out} + \mathcal{K}_i^{\,out} \,. \tag{17.33}$$

Here n_0 and n_s are the plasma densities outside and inside the HTTCS. T_0 is the temperature of inflowing plasma outside the sheet, $T = T_e$ is an effective electron temperature (the mean kinetic energy of chaotic motion per single electron) inside the HTTCS, the ratio $\theta = T_e/T_i$, T_i is an effective temperature of ions inside the HTTCS.

$$v_0 = V_d = c\,\frac{E_0}{B_0} \tag{17.34}$$

is the velocity of the plasma drift to the current sheet, and

$$v_1 = V_{\rm A,S} = \frac{B_0}{\sqrt{4\pi M n_s}} \tag{17.35}$$

is the velocity of the plasma outflow from the sheet. Compare this approximate formula with (17.7).

The continuity Equation (17.29) as well as the energy Equations (17.32) and (17.33) are of integral form for a quarter of the current sheet assumed to be symmetrical and for a unit length along the electric current.

The left-hand sides of the energy equations for electrons (17.32) and ions (17.33) contain the magnetic energy flux (see formula (8.90))

$$\mathcal{E}_{mag}^{\,in} = \frac{B_0^2}{4\pi}\, v_0\, b \,, \tag{17.36}$$

which coincides with the direct heating of the ions and electrons due to their interactions with waves inside the HTTCS. A relative fraction $\chi_{\rm ef}$ of the heating is consumed by electrons, while the remaining fraction $(1 - \chi_{\rm ef})$ goes to the ions.

The electron and ion temperatures of the plasma inflowing to the sheet are the same. Hence, the fluxes of the electron and ion thermal energies are also the same:

$$\mathcal{E}_{th,e}^{\,in} = \mathcal{E}_{th,i}^{\,in} = \frac{5}{2}\, n_0 k_B T_0 \cdot v_0 b \,. \tag{17.37}$$

Because of the difference between the effective temperatures of electrons and ions in the outflowing plasma, the electron and ion thermal energy outflows also differ:

$$\mathcal{E}_{th,e}^{out} = \frac{5}{2} n_s k_{\rm B} T \cdot v_1 a^{out} , \qquad \mathcal{E}_{th,i}^{out} = \frac{5}{2} n_s k_{\rm B} \frac{T}{\theta} \cdot v_1 a^{out} . \tag{17.38}$$

The ion kinetic energy flux from the HTTCS

$$\mathcal{K}_i^{out} = \frac{1}{2} M n_s v_1^2 \cdot v_1 a^{out} \tag{17.39}$$

is important in the energy balance (17.33). As to the electron kinetic energy, it is negligible and disregarded in (17.32). However, electrons play the dominant role in the heat conductive cooling of the HTTCS:

$$\mathcal{C}_{\parallel}^{an} = f_{\rm M}(\theta) \, \frac{n_s (k_{\rm B} T)^{3/2}}{M^{1/2}} \, a^{out} . \tag{17.40}$$

Here

$$f_{\rm M}(\theta) = \begin{cases} \frac{1}{4} \left(\frac{M}{m} \right)^{1/2} & \text{at } 1 \le \theta \le 8.1 , \\ \left(\frac{M}{m} \right)^{1/2} \theta^{3/2} \left[\left(1 + \frac{3}{\theta} \right)^{1/2} - \frac{1}{\theta^{1/2}} \right] \times & \\ \times \exp \left[- \frac{2(\theta + 3)}{5} \right] + \left(1 + \frac{3}{\theta} \right)^{1/2} & \text{for } \theta > 8.1 \\ & \text{or } \theta < 1 . \end{cases} \tag{17.41}$$

is the Manheimer function which allows us to consider the anomalous magnetic-field-aligned thermal flux depending on the the effective temperature ratio θ.

Under the coronal conditions derived from the *Yohkoh* data, especially in flares, contributions to the energy balance are not made either by the energy exchange between the electrons and the ions due to collisions, the thermal flux across the magnetic field, and the energy losses for radiation. The magnetic-field-aligned thermal flux becomes anomalous and plays the dominant role in the cooling of electron component inside the HTTCS. All these properties are typical for collisionless 'superhot' plasma.

Under the same conditions, the effective anomalous conductivity $\sigma_{\rm ef}$ in the Ohm's law

$$\frac{c B_0}{4 \pi a} = \sigma_{\rm ef} E_0 , \tag{17.42}$$

as well as the relative fraction $\chi_{\rm ef}$ of the direct heating consumed by electrons, are determined by the wave-particle interaction inside the HTTCS and depend on the type of plasma turbulence and its regime (see Chapter 3 in Somov, 1992). For example, if the resistivity was caused by Coulomb collisions, it would depend on the electron temperature only. However, when the plasma is in a **collisionless turbulent state**, the electrons carring the current and the ions interact with the field fluctuations in the waves, which changes the resistivity and other transport coefficients of the plasma in a way that depends on the type of waves that grow.

17.3.2 Problem in the strong field approximation

Let the conditions of a strong magnetic field (Section 8.3.3) be satisfied. Then, the set of Equations (17.29)–(17.33) takes the following form:

$$n_0 V_d = n_s V_{\rm A,S}\, \xi_{\perp} , \tag{17.43}$$

$$\frac{B_0^2}{8\pi} = n_s k_{\rm B} T \left(1 + \frac{1}{\theta} \right) , \tag{17.44}$$

$$n_s k_{\rm B} T \left(1 + \frac{1}{\theta} \right) = \frac{1}{2} M n_s V_{\rm A,S}^2 , \tag{17.45}$$

$$\chi_{\rm ef} \, \frac{B_0^2}{4\pi} \, V_d = \frac{5}{2} n_s k_{\rm B} T \cdot V_{\rm A,S}\, \xi_{\perp} + f_{\rm M}(\theta) \, \frac{n_s \, (k_{\rm B} T)^{3/2}}{M^{1/2}} \, \xi_{\perp} , \tag{17.46}$$

$$(1 - \chi_{\rm ef}) \, \frac{B_0^2}{4\pi} \, V_d = \left(\frac{5}{2} n_s k_{\rm B} \frac{T}{\theta} + \frac{1}{2} M n_s V_{\rm A,S}^2 \right) V_{\rm A,S}\, \xi_{\perp} . \tag{17.47}$$

In Ohm's law (17.42) it is convenient to replace the effective conductivity $\sigma_{\rm ef}$ by effective resistivity $\eta_{\rm ef}$; so,

$$\frac{cB_0}{4\pi a} = \frac{E_0}{\eta_{\rm ef}} . \tag{17.48}$$

In general, the partial contributions to the effective resistivity may be made simultaneously by several processes of electron scattering by different sorts of waves, so that the resistivity proves to be merely a sum of the contributions:

$$\eta_{\rm ef} = \sum_k \eta_k . \tag{17.49}$$

The relative share of the electron heating $\chi_{\rm ef}$ is also presented as a sum of the respective shares χ_k of the feasible processes taken, of course, with the weight factors $\eta_k/\eta_{\rm ef}$ which defines the relative contribution from one or another process to the total heating of electrons inside the HTTCS:

$$\chi_{\rm ef} = \sum_k \frac{\eta_k}{\eta_{\rm ef}} \chi_k \, . \tag{17.50}$$

In usual practice (e.g., Chapter 3 in Somov, 1992), the sums (17.49) and (17.50) consist of no more than two terms, either of which corresponds to one of the turbulent types or states. Note also that more detailed numerical results (e.g., Somov and Oreshina, 2000) confirm validity of the assumptions made above.

17.3.3 Basic local parameters of the HTTCS

We shall assume that the magnetic field gradient h_0 locally characterizes the potential field in the vicinity of the separator or X-type neutral line. It means that we consider a less specific configuration of reconnecting magnetic fluxes in comparison with the 2D MHD 'standard model' mentioned in Section 17.1.1. We shall also assume that, at distances larger than the current sheet width $2b$, the magnetic field structure becomes, as it should be, the same as the structure of the potential field of 'external sources', for example, of sunspots in the solar photosphere. So the gradient h_0 is the local parameter which 'remembers' the global structure of the potential field.

Under the assumptions made, the field B_0 on the inflow sides of the current sheet may be estimated as

$$B_0 = h_0 b \, . \tag{17.51}$$

The second local parameter of the reconnection region is the inflow velocity v_0 or, alternatively, the electric field E_0 determined by formula (17.2). We shall use E_0 in what follows.

In the approximation of a strong magnetic field, the pressure (or temperature T_0) of inflowing plasma is negligible, but its density n_0 certainly has to be prescribed as a local parameter of the reconnection region. In fact, as we shall see below, all characteristics of the HTTCS depend on n_0.

The dimensionless parameter $\xi_\perp$ could be, in principle, obtained as a result of the solution of the more self-consistent problem on the current sheet structure (Section 3.4 in Somov, 1992). However, to keep the problem under consideration as simple as possible, here we shall consider the small (see Inequalities (17.24)) parameter $\xi_\perp$ as the specified one.

Summarizing the formulation of the problem, we see that the set of Equations (17.43)–(17.48) becomes closed if the particular expressions (17.49) and (17.50) are added to this set. This allows us to find the following parameters of the HTTCS: a, b, n_s, T, and θ.

17.3.4 The general solution of the problem

The input set of Equations (17.43)–(17.47) exibits a remarkable property which facilitates the solution of the problem as a whole. The property consists of the fact that the first three Equations (17.43)–(17.45) allow us to transform the last two Equations (17.46) and (17.47) into a simpler form:

$$2\,\chi_{\rm ef}\,\frac{n_s}{n_0} = \frac{2.5}{1+\theta^{-1}} + \frac{f_{\rm M}(\theta)}{\sqrt{2}\,(1+\theta^{-1})^{3/2}}\,, \tag{17.52}$$

$$2\,(1-\chi_{\rm ef})\,\frac{n_s}{n_0} = 1 + \frac{2.5}{1+\theta}\,. \tag{17.53}$$

From these two Equations we find the plasma compression and the relative share of the total heating of the electrons in the current sheet:

$$\frac{n_s}{n_0} = N(\theta) = 1.75 + \frac{f_{\rm M}(\theta)}{\sqrt{8}\,(1+\theta^{-1})^{3/2}}\,, \tag{17.54}$$

$$\chi_{\rm ef} = f_\chi(\theta) = 1 - \frac{3.5+\theta}{2N(\theta)\,(1+\theta)}\,. \tag{17.55}$$

Now we use Equations (17.43)–(17.45) together with (17.48) to find the general solution of the problem, which determines the following parameters of the HTTCS: the current-sheet half-thickness

$$a = \frac{c\,m^{1/2}}{e\,(2\pi)^{1/2}} \left[\left(\frac{1+\theta^{-1}}{N(\theta)} \right)^{1/2} \frac{1}{U_k\,(\theta)} \right] \times \frac{1}{n_0^{1/2}}\,, \tag{17.56}$$

its half-width

$$b = (2c)^{1/2}\,(\pi M)^{1/4} \left[\frac{1}{N(\theta)} \right]^{1/4} \times n_0^{1/4}\,\frac{1}{h_0} \left(\frac{E_0}{\xi_\perp} \right)^{1/2}, \tag{17.57}$$

the effective temperature of electrons inside the HTTCS

$$T = \frac{c M^{1/2}}{4k_{\rm B}\,\pi^{1/2}} \left[\frac{1}{(1+\theta^{-1})\,N^{3/2}(\theta)} \right] \times \frac{1}{n_0^{1/2}} \left(\frac{E_0}{\xi_\perp} \right), \tag{17.58}$$

the effective anomalous resistivity

$$\eta_{\rm ef} = \frac{2\,m^{1/2}\,\pi^{1/4}}{e\,c^{1/2}M^{1/4}} \left[\frac{(1+\theta^{-1})^{1/2}}{N^{1/4}(\theta)\,U_k\,(\theta)} \right] \times \frac{1}{n_0^{3/4}}\,(\xi_\perp E_0)^{1/2}\,. \tag{17.59}$$

Thus to complete the solving this problem, we have to find a form of the function $U_k\,(\theta)$ which depends on the regime of the plasma turbulence. This will be done in Section 17.3.5.

In addition, from definitions (17.51), (17.34), (17.35), and (17.36), by using the obtained solutions (17.56)–(17.59), we have the following formulae: the magnetic field near the HTTCS

$$B_0 = (2c)^{1/2}\,(\pi M)^{1/4} \left[\frac{1}{N(\theta)} \right]^{1/4} \times n_0^{1/4} \left(\frac{E_0}{\xi_\perp} \right)^{1/2}, \tag{17.60}$$

the reconnection velocity

$$v_0 = \frac{c^{1/2}}{2^{1/2}\,\pi^{1/4} M^{1/4}}\,[N(\theta)]^{1/4} \times \frac{1}{n_0^{1/4}}\,(\xi_\perp E_0)^{1/2}\,, \tag{17.61}$$

the outflow velocity

$$v_1 = \frac{c^{1/2}}{2^{1/2}\,\pi^{1/4} M^{1/4}} \left[\frac{1}{N(\theta)} \right]^{3/4} \times \frac{1}{n_0^{1/4}} \left(\frac{E_0}{\xi_\perp} \right)^{1/2}, \tag{17.62}$$

the power of energy release per unit length along the current sheet l_j

$$\frac{P_s}{l_j} = \frac{B_0^2}{4\pi}\,v_0\,4b = \frac{2c^2 M^{1/2}}{\pi^{1/2}} \left[\frac{1}{N(\theta)} \right]^{1/2} \times n_0^{1/2}\,\frac{1}{h_0} \left(\frac{E_0^2}{\xi_\perp} \right), \tag{17.63}$$

the rate of high-temperature plasma production by the HTTCS per unit length along the current sheet l_j

$$\frac{\dot N}{l_j} = n_s v_1\,4a^{out} = n_0 v_0\,4b = 4c \times n_0\,\frac{1}{h_0}\,E_0\,. \tag{17.64}$$

Formula (17.64) demonstrates a high level of self-consistency for the HTTCS model under consideration. It shows that the total flux of plasma through the reconnecting current sheet depends only on the plasma density n_0 on the inflow sides of the HTTCS, the driving electric field E_0, and the gradient h_0 of potential magnetic field in the vicinity of the X-type neutral point. It is remarkable that other parameters, like the dimensionless parameter $\xi_\perp$, as well as the assumptions on the plasma turbulence inside the HTTCS, discussed in the next Section, do not influence the total flux of plasma passing through the current sheet.

17.3.5 Plasma turbulence inside the HTTCS

In the case of the marginal regime (e.g. Duijveman et al., 1981), the electron current velocity

$$u = \frac{E_0}{e n_s \eta_{\rm ef}} \tag{17.65}$$

coincides with the critical velocity u_k of the k-type wave excitation. Hence, in formulae (17.56) and (17.59), the unknown function

$$U_k(\theta) = U_k^{mar}(\theta) = \frac{u_k}{V_{\rm Te}}\,. \tag{17.66}$$

For example, the ion-cyclotron instability becomes enhanced when the electron current velocity u is not lower than the critical value u_{ic} of the ion-cyclotron (ic) waves. In the marginal regime of the ion-cyclotron instability

$$U_{ic}^{mar}(\theta) = \frac{u_{ic}}{V_{\rm Te}}\,. \tag{17.67}$$

As long as the ion-cyclotron waves are not saturated, the electron current velocity u remains approximately equal to u_{ic} and thus it is possible to calculate the effective resistivity $\eta_{\rm ef}$ from Equation (17.65).

In the saturated turbulence regime, $U_k(\theta)$ must be replaced by certain functions $U_{ic}^{sat}(\theta)$ and $U_{ia}^{sat}(\theta)$ for the ion-cyclotron and ion-acoustic turbulence, respectively (see, for example, Section 3.3.2 in Somov, 1992).

17.3.6 Formulae for the basic parameters of the HTTCS

So we rewrite the general solution (17.56)–(17.59) as follows: the current-sheet half-thickness

$$a = 7.5 \times 10^5 \, f_a(\theta) \times \frac{1}{n_0^{1/2}}\,, \ \text{cm}\,; \tag{17.68}$$

the half-width of the HTTCS

$$b = 3.7 \times 10^{-1} \, f_b(\theta) \times n_0^{1/4} \, \frac{1}{h_0} \left(\frac{E_0}{\xi_\perp} \right)^{1/2}, \ \text{cm}\,; \tag{17.69}$$

the effective temperature of electrons inside the HTTCS

$$T = 4.0 \times 10^{13} \, f_T(\theta) \times \frac{1}{n_0^{1/2}} \left(\frac{E_0}{\xi_\perp} \right), \ \text{K}\,; \tag{17.70}$$

the effective anomalous resistivity

$$\eta_{\text{ef}} = 8.5 \times 10^{-4} \, f_{\eta}(\theta) \times \frac{1}{n_0^{3/4}} \, (\xi_{\perp} E_0)^{1/2} \ , \ \text{s} . \tag{17.71}$$

Here we write separatelly the functions which are determined by the plasma turbulence inside the HTTCS:

$$f_a(\theta) = \left(\frac{1+\theta^{-1}}{N(\theta)} \right)^{1/2} \frac{1}{U_k(\theta)} \approx 2.9 \, , \tag{17.72}$$

$$f_b(\theta) = \frac{1}{N^{1/4}(\theta)} \approx 6.8 \times 10^{-1} \, , \tag{17.73}$$

$$f_{T}(\theta) = \frac{1}{(1+\theta^{-1}) \, N^{3/2}(\theta)} \approx 8.2 \times 10^{-2} \, , \tag{17.74}$$

$$f_{\eta}(\theta) = \frac{(1+\theta^{-1})^{1/2}}{N^{1/4}(\theta) \, U_k(\theta)} \approx 4.3 \, . \tag{17.75}$$

Bearing in mind the discussion of solar flares in Section 17.4, we have calculated the right-hand sides of functions (17.72)–(17.75) in the marginal regime of the ion-acoustic turbulence: $\theta \approx 6.5$, $N \approx 4.8$, $U_k = U_{ia}^{mar} \approx 0.17$ (see Section 3.3 in Somov, 1992).

The magnetic field on the inflow sides of the current sheet can be found from formula (17.60):

$$B_0 = 3.7 \times 10^{-1} \, f_b(\theta) \times n_0^{1/4} \left(\frac{E_0}{\xi_{\perp}} \right)^{1/2} \ , \ \text{G} . \tag{17.76}$$

From (17.61) it follows that the reconnection velocity

$$v_0 = 8.1 \times 10^5 \, N^{1/4}(\theta) \times \frac{1}{n_0^{1/4}} \, (\xi_{\perp} E_0)^{1/2} \ , \ \text{km s}^{-1} . \tag{17.77}$$

From (17.62) and (17.63) we obtain the outflow velocity

$$v_1 = 8.1 \times 10^5 \, N^{-3/4}(\theta) \times \frac{1}{n_0^{1/4}} \left(\frac{E_0}{\xi_{\perp}} \right)^{1/2} \ , \ \text{km s}^{-1} , \tag{17.78}$$

and the power of energy release per unit length along the current sheet l_j

$$\frac{P_s}{l_j} = 6.0 \times 10^8 \, N^{-1/2}(\theta) \times n_0^{1/2} \, \frac{1}{h_0} \left(\frac{E_0^2}{\xi_{\perp}} \right) \ , \ \text{erg s}^{-1} \, \text{cm}^{-1} . \tag{17.79}$$

The rate of high-temperature plasma production by the HTTCS is found from (17.64):

$$\frac{\dot{N}}{l_j} = 1.2 \times 10^{11} \times n_0 \frac{1}{h_0} E_0 \,, \ \mathrm{s^{-1}\,cm^{-1}} \,. \tag{17.80}$$

The applicability scope of the HTTCS model has been considered in Somov (1992) with account of the ion-acoustic and ion-cyclotron instabilities in marginal and saturated regimes. It follows from this consideration that the best agreement between the average quantities predicted by the model and those observed in solar flares can be achieved in the marginal regime of ion-acoustic turbulence. A small parameter of the model, $\xi_\perp$, is really small; on average $\xi_\perp \leq 3 \times 10^{-3}$. With this value taken into account, we finally have the following approximate formulae: the current-sheet half-thickness

$$a = 2.2 \times 10^6 \times \frac{1}{n_0^{1/2}} \,, \ \mathrm{cm} \,; \tag{17.81}$$

the half-width of the HTTCS

$$b = 4.6 \times n_0^{1/4} \frac{1}{h_0} E_0^{1/2} \,, \ \mathrm{cm} \,; \tag{17.82}$$

the effective temperature of electrons inside the HTTCS

$$T = 1.1 \times 10^{15} \times \frac{1}{n_0^{1/2}} E_0 \,, \ \mathrm{K} \,; \tag{17.83}$$

the effective anomalous resistivity

$$\eta_{\mathrm{ef}} = 2.0 \times 10^{-4} \times \frac{1}{n_0^{3/4}} E_0^{1/2} \,, \ \mathrm{s} \,; \tag{17.84}$$

the magnetic field on the inflow sides of the current sheet

$$B_0 = 4.6 \times n_0^{1/4} E_0^{1/2} \,, \ \mathrm{G} \,; \tag{17.85}$$

the reconnection velocity

$$v_0 = 6.6 \times 10^4 \times \frac{1}{n_0^{1/4}} E_0^{1/2} \,, \ \mathrm{km\ s^{-1}} \,; \tag{17.86}$$

the outflow velocity of high-temperature plasma

$$v_1 = 4.6 \times 10^6 \times \frac{1}{n_0^{1/4}} E_0^{1/2} \,, \ \mathrm{km\ s^{-1}} \,; \tag{17.87}$$

the power of energy release per unit length along the current sheet l_j

$$\frac{P_s}{l_j} = 2.0 \times 10^{11} \times n_0^{1/2} \frac{1}{h_0} E_0^2 \,, \text{ erg s}^{-1} \text{cm}^{-1} \,; \tag{17.88}$$

and the rate of high-temperature plasma production by the HTTCS

$$\frac{\dot{N}}{l_j} = 1.2 \times 10^{11} \times n_0 \frac{1}{h_0} E_0 \,, \text{ s}^{-1} \text{cm}^{-1} \,. \tag{17.89}$$

Formulae (17.81)–(17.89) depend on **three principal parameters of the reconnection region**: the gradient of the magnetic field h_0 in the vicinity of separator, the value of the inductive electric field E_0 and the plasma density n_0.

For applications to the solar flares in the next Section, we also introduce the *heating time* t_h which is the time for a given field line to be connected to the HTTCS. In other words, during the time t_h, the thermal flux from the HTTCS along the field line heats the high-temperature plasma flowing out of the current sheet along this field line. Let us take by definition

$$t_h = \frac{2b}{v_1} = 4(\pi M)^{1/2} \, [\, N(\theta) \,]^{1/2} \times n_0^{1/2} \frac{1}{h_0} = 2.0 \times 10^{-11} \times n_0^{1/2} \frac{1}{h_0} \,, \text{ s}. \tag{17.90}$$

In all these formulae all the quantities, except the temperature, are measured in CGS units; the temperature is given in degrees Kelvin.

17.4 HTTCS in solar flares

17.4.1 Why are flares so different?

(a) Magnetic-field gradient effects

Even if one considers the flares driven by reconnection in the HTTCS with the same kind of plasma turbulence, one can see from the solution described above that very different physical processes will dominate in a flare depending on physical conditions. The advantage which this solution gives us is that we can estimate the most important parameters which determine the physical difference in solar flares.

Let us distinguish *impulsive* and *gradual* flares in the following way. If the difference in the time scale of a flare t_f would be mainly determined by the difference in its linear size l_f, then the impulsive flares should have the stronger gradient h_0 near the separator of the potential field in an active region (see Section 16.5.2). By thinking so, we would believe that the impulsive flares are the compact flares in strong magnetic fields, for example, flares in the low corona not far from sunspots. On the contrary, the gradual or long-duration flares may occur in a large scale region placed high in the corona at a significant distance above the strong sunspots.

For definiteness, let us put $l_f \approx 3 \times 10^9$ cm as a typical value at an imaginary boundary between compact (impulsive) and large-scale (long-duration or gradual) flares. In that case, the typical value of the field gradient $h_f = B_f/l_f$, where B_f is a typical value of the external (with respect to the reconnecting current sheet) magnetic field in the photosphere. Since in sunspots $B_f \approx 10^3$ G, we take

$$h_f = \frac{B_f}{l_f} \approx 3 \times 10^{-7}\,\mathrm{G\ cm^{-1}} \tag{17.91}$$

as a boundary value of the field gradient. Therefore, by our conventional definition, which is not always true, in impulsive flares $h_0 > h_f$ but in gradual flares $h_0 < h_f$.

Note that the half-thickness a of the current sheet, its temperature T and effective anomalous resistivity η_{ef}, the magnetic field B_0 on the inflow sides of the current sheet, the inflow and outflow velocities v_0 and v_1 do *not* depend on the gradient h_0. This remarkable feature follows from formulae (17.81), (17.83)–(17.87), respectively. Perhaps, that is why there still exists some similarity between solar flares, in spite of the great difference in their scales and shapes.

On the contrary, the current-sheet half-width b and, as a consequence, the power of energy release per unit length along the current P_s/l_j and the rate of high-temperature plasma production by the HTTCS $\dot{N}/l_j$ are inverse proportional to the field gradient h_0, see formulae (17.82), (17.88) and (17.89). The plasma production rate is proportional to the electric field E_0, which is typical for driven reconnection.

(b) The role of the plasma density

Also conventionally, we shall distinguish *thermal* and *non-thermal* flares. Plasma heating is an unavoidable phenomenon in all flares. The relative role of the thermal part of a flare certainly depends on collisional relaxation processes mainly in the secondary (Chapter 2 in Somov, 1992) transformations of the flare energy. It is natural to assume that the plasma density n_0 determines the importance of collisions in flares: the higher the density, the faster is the thermalization.

Thermal flares have to produce very efficient heating but inefficient acceleration. The opposite seems to be true for the non-thermal flares.

The solutions (17.56)–(17.63) show that all parameters of the HTTCS depend on the density n_0. Generally, this dependence is not strong ($n_0^{1/2}$, $n_o^{1/4}$ etc.), but the difference in density can be large. This is important for what follows. For example, Figure 17.4 shows the effective temperature of electrons (17.83) as a function of the plasma density n_0 and electric field E_0. As we

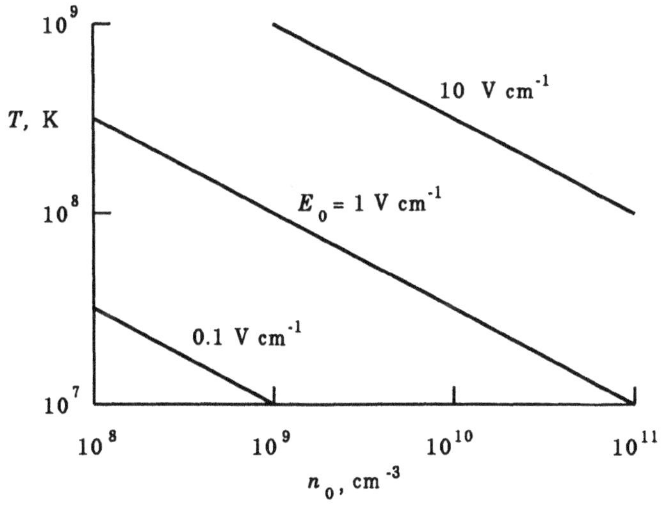

Figure 17.4: The effective temperature of electrons inside the HTTCS as a function of the plasma density n_0 and the driving electric field E_0.

see, temperatures greater than 10^8 K can be easily reached in the HTTCS of flares. Moreover, the effective temperature of electrons does not depend on the field gradient h_0. So the HTTCS may well exist in both impulsive and gradual flares.

In the conditions of the 'main' or 'hot' phase of solar flares the characteristic parameters of such collisionless current sheets, computed in the frame of the model described above (see also Table 3.3.3 in Somov, 1992), are the followings. (a) The effective electron temperature inside the sheet $T_e \approx 100 - 200$ MK, the temperature ratio $\theta = T_e/T_p \approx 6.5$, the plasma compression $n_s/n_0 \approx 4.8$. (b) The dissipative thickness of the current sheet $2a \approx 20$ cm is very small but its width $2b \approx (1-2) \times 10^9$ cm is large, for this reason the linear scale (17.25) for the outflows of energy and mass $2a^{out} \approx (3-6) \times 10^6$ cm is not small.

A high anomalous resistivity $\eta \approx (3-10) \times 10^{13}\ \mathrm{s}^{-1}$ is induced by the ion-acoustic turbulence in a marginal regime inside the HTTCS. Under this condition, the energy release power per unit sheet length l_j (along the third dimension – the direction of electric current inside the sheet) is $P_s/l_j \approx (1-7) \times 10^{19}\ \mathrm{erg\ (s\ cm)}^{-1}$, if the plasma inflow velocity $v_0 \approx 10-30\ \mathrm{km\ s}^{-1}$. Hence, if the current sheet length $l_j \approx 3 \times 10^9$ cm, then the power of energy release $P_s \approx 3 \times 10^{28} - 2 \times 10^{29}\ \mathrm{erg\,s}^{-1}$. The outflow velocity equals $v_1 \approx 1400-1800\ \mathrm{km\ s}^{-1}$.

17.4.2 Superhot plasma production

How much high-temperature plasma can be generated by the HTTCS? – According to formula (17.89), for the impulsive flares with the field gradient $h_0 \approx 5 \times 10^{-7}\ \mathrm{G\ cm}^{-1}$, the rate of high-temperature plasma production by the HTTCS (per unit length along the current sheet l_j) is $\dot{N}/l_j \approx 2 \times 10^{17}\, n_0\, E_0\,,\ \mathrm{s}^{-1}\,\mathrm{cm}^{-1}$. If we take the maximum value of the electric field $E_0 \approx 10\ \mathrm{V\,cm}^{-1}$ and plasma density $n_0 \approx 10^9 - 10^{10}\ \mathrm{cm}^{-3}$, then we estimate the rate of plasma production as $\dot{N}/l_j \approx 10^{25} - 10^{26}\ \mathrm{s}^{-1}\,\mathrm{cm}^{-1}$.

Let us take the characteristic value of the current sheet length $l_j \sim l_f \approx 3 \times 10^9$ cm and the characteristic value of the impulsive phase duration $\tau \approx 30$ s. Then the amount of high-temperature plasma produced by the HTTCS can be estimated as

$$N = \frac{\dot{N}}{l_j} \times l_j\, \tau \approx \left(10^{36} - 10^{37}\right) \text{particles}\,. \tag{17.92}$$

This amount of high-temperature particles seems to be well comparable with the total number of accelerated electrons having energies larger than ≈ 10 keV during the impulsive phase of a typical flare. So, in principle, the HTTCS can produce an observable amount of the superhot plasma (Section 13.4).

Let us estimate the emission measure of the superhot plasma. 2D distributions of temperature and pressure, that follow from the *Yohkoh* SXT and HXT observations (Tsuneta *et al.*, 1997), do not allow us to estimate the volume V_{sh} occupied by superhot plasma. So we have to start from a rather arbitrary assumption frequently used in this situation as a first approximation. If this plasma would be distributed uniformly over the large volume of a flare $V_f = l_f^{\,3}$, then the emission measure should be

$$EM_{\min} = \frac{N^2}{l_f^{\,3}} \approx 3 \times \left(10^{43} - 10^{45}\right) \text{ cm}^{-3} . \tag{17.93}$$

This is not the case. The emission measure can be much higher because the superhot plasma is concentrated in a much smaller volume, for example, in a compact source above the soft X-ray (SXR) loops (see Figures 13.8 and 13.9). So the value (17.93) is only a *lower limit* to the emission measure of the superhot plasma in real flares. A reasonable value of the volume filling factor V_{sh}/V_f, which we may assume, is of about $3 \times 10^{-4} - 10^{-3}$. That is why the superhot plasma can be observed in solar flares by the HXT on board *Yohkoh*.

* * *

Before *Yohkoh*, a little indirect evidence of the superhot plasma was known. First, the high-resolution ($\approx$ 1 keV FWHM) spectral measurements (Lin *et al.* 1981) from 13 to 300 keV of a flare on June 27, 1980 have shown, at energies below $\approx$ 35 keV, an extremely steep spectrum which fits closely to that from the Maxwellian distribution with an electron temperature $T_e \approx$ 34 MK and an emission measure $EM \approx 3 \times 10^{48}\,\text{cm}^{-3}$. Second, statistical properties of a large number of solar flares detected with the Hard X-Ray Burst Spectrometer (HXRBS) on the *SMM* satellite allowed to confirm the existence of superhot thermal flares (Type A) with temperatures 30-40 MK (Dennis, 1985, 1988).

Third, the 2D distributions of electron temperature and emission measure of the 'hot' (say $10 \le T_e \le 30$ MK) and superhot plasma (Den and Somov, 1989) were calculated for the 1B/M4 flare on November 5, 1980 on the basis of data obtained with the Hard X-ray Imaging Spectrometer (HXIS) on board *SMM*. It was shown that the large and small SXR 'interacting loops' do not coincide with the location of super-hot ($T \geq 30$ MK) plasma in a long structure ($\approx$ 1 arc min) during the long after-impulsive phase. The total emission measure of the superhot plasma was of about 10^{47} cm^{-3}. In two maxima, the electron temperature reaches enormous values, $\approx$ 50-60 MK, determined with accuracy better than 20 %.

Hard X-ray imaging telescopes on *Hinotori* observed a superhot plasma of 30-35 MK with an emission measure of the order of 10^{49}cm^{-3} (Tsuneta *et al.*, 1984, Tanaka 1987). The same superhot plasma was detected by the Bragg-type spectrometer (Tanaka, 1987).

Fast flows of the hot plasma can produce a symmetrical broadening of the optically thin SXR lines observed during solar flares. This broadening is larger than the thermal one. A comparison of the observed profiles of the Fe XXV emission lines with the predictions of the HTTCS model suggests that the presence in the flare region of several small-scale or one (or a few) large-scale curved HTTCS (Antonucci *et al.*, 1996).

* * *

The *Yohkoh* data obtained simultaneously with the HXT, SXT, and BCS offer a wonderful opportunity for a detailed analysis which is necessary to distinguish the superhot plasma components of different origins in different classes of flares as well as at different phases of the flare development.

Fast outflows of superhot plasma create complicated dynamics of plasma in an external (relative to the RCS) region (see Section 13.4.1). If the distance between the HTTCS and the magnetic obstacle is not large, then the outflow becomes wider but does not relax in the coronal plasma before reaching the obstacle. Moreover, if the plasma velocity still exceeds the local fast-magnetoacoustic-wave velocity, a fast MHD shock wave appears ahead the obstacle (see Figure 13.6).

If, on the contrary, the distance is large, the outflow of superhot plasma relaxes gradually with (or even without) a collisinal shock depending on the height and the conditions in an active region where a flare occurs (e.g., Tsuneta, 1996). For example, collisional relaxations can be fast just near the HTTCS if the plasma density is relatively high but its temperature inside the current sheet is relatively low.

We do not discuss in this Chapter an existance of slow or fast MHD shocks (or other MHD discontinuities) which may be attached to external edges of the collisionless HTTCS. This will be reasonable to discuss as a part of the current sheet evolutionarity problem in Chapter 19, see also Problem 11.8.

17.4.3 Concluding comments

The collisionless transformation of the magnetic energy into kinetic energy of particles inside the non-steady 2D reconnecting current sheet was introduced by Syrovatskii (1966a) as a *dynamic dissipation*. An essential peculiarity of the dynamic dissipation is that

> the inductive electric field $\mathbf{E}_0$ is directed along the current in the reconnecting current sheet; this field does positive work on charged particles, thus increasing their energy.

Naturally, some instabilities are excited in the plasma-beam system in the RCS. Wave-particle interactions transform a part of this work into direct heating of ions and electrons.

Three-component collisionless reconnection (e.g., Ono et al., 1996; Horiuchi and Sato, 1997) includes several natural complications. For example, large ion viscosity possibly contributes to the thermalization process of the ion kinetic energy. However, the general inference as to the possibility of particle acceleration and heating inside the collisionless RCS (i.e. dynamic dissipation of the magnetic field) remains valid and is used in the HTTCS model. This allows us to consider the HTTCS as the primary source of flare energy and, at least, the first-step acceleration mechanism.

17.5 Practice: Problems and Answers

Problem 17.1. Evaluate the characteristic value of the global Lundquist number (17.22) for a current sheet with the classical Coulomb conductivity in the solar corona before an impulsive flare. Compare a predicted reconnection rate with the real one.

Answer. First, let us formally apply the Sweet-Parker scaling property (17.23) to the Syrovatskii current sheet (see Section 17.1.2). Consider the main parameters of the neutral sheet at the limit of thermal stability (17.17). The values $n_0 \approx 5 \times 10^8$ cm^{-3}, $h_0 \approx 5 \times 10^{-7}$ Gauss cm^{-1}, and $E_0 \approx 1.2 \times 10^{-1}$ V cm^{-1} have been specified in advance. The other quantities have been determined from the Syrovatskii solution. For example, the half-width of the current sheet $b \approx 7 \times 10^8$ cm, the magnetic field near the sheet $B_0 = h_0 b \approx 340$ Gauss, the plasma density inside the neutral sheet $n_s \approx 2 \times 10^{14}$ cm^{-3}.

The upstream Alfvén speed (17.8), see also the non-relativistic formula (8.84),

$$V_{\mathrm{A,0}} = 2.18 \times 10^{11} \, \frac{B_0}{\sqrt{n_0}} \approx 3 \times 10^9 \text{ cm s}^{-1} \approx 0.1 \, c \, . \tag{17.94}$$

Here c is the light speed.

The global Lundquist number (17.22), with account of (8.82),

$$N_{\mathrm{L}} = \frac{V_{\mathrm{A,0}} b}{\nu_{\mathrm{m}}} \approx 2.3 \times (10^{14} - 10^{15}) \, .$$

Therefore the Sweet-Parker reconnection rate (17.23) predicted for the Syrovatskii neutral sheet is extremely low:

$$M_{\rm A} = N_{\rm L}^{-1/2} \approx (2.1 - 6.7) \times 10^{-8} .$$

Let us compare this rate with the one which directly corresponds to the Syrovatskii model. According to formula (17.9) the inflow velocity

$$v_0 = c\,\frac{E_0}{B_0} \approx 3.5 \times 10^4 \text{ cm s}^{-1} = 0.35 \text{ km s}^{-1} .$$

Hence an actual reconnection rate in the Syrovatskii neutral sheet

$$M_{\rm A,S} = \frac{v_0}{V_{\rm A,0}} \approx 1.1 \times 10^{-5} \gg M_{\rm A} .$$

Obviously a difference in the reconnection rate is related to the compressibility of the plasma in the Syrovatskii model. With account the plasma compressibility inside the reconnecting current sheet, the actual reconnection rate

$$\boxed{M_{\rm A,S} = \frac{v_0}{V_{\rm A,0}} = \left(\frac{n_s}{n_0}\right)^{1/2} N_{\rm L}^{-1/2} .} \qquad (17.95)$$

In the frame of Syrovatskii's model for the neutral sheet

$$\left(\frac{n_s}{n_0}\right)^{1/2} > 10^2 .$$

So, the plasma compressibility is really very important factor in the magnetic reconnection theory.

Problem 17.2. Evaluate the characteristic value of the parameters ε^2 and γ^2 for a plasma flow into the HTTCS during the 'main' phase of an impulsive solar flare.

Answer. Let us consider, first, the reconnection velocity v_0 of plasma in the vicinity of the HTTCS. According to formula (17.86), v_0 does not depend on the magnetic-field gradient h_0. For given values of n_0 and E_0, the reconnection velocity is shown in Figure 17.5. On average, the characteristic value of the reconnection velocity is $v_0 \sim 10$ km s^{-1}.

So, the reconnection velocity during the 'main' phase of solar flares is much higher than in the pre-flare state (cf. Problem 17.1).

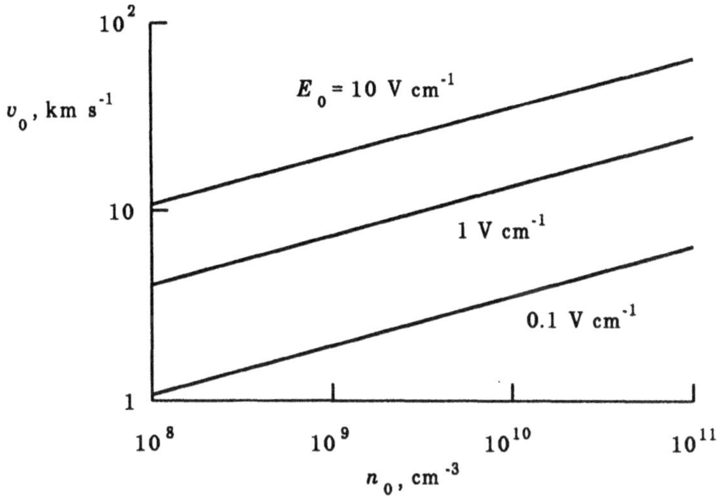

Figure 17.5: The reconnection velocity v_0 in the vicinity of the HTTCS as a function of the plasma density n_0 and the electric field E_0.

Second, if the characteristic value of the upstream Alfvén speed in the undisturbed solar corona $V_{A,0} \approx 3 \times 10^4$ km s^{-1} (see (17.94)), then the parameter $\varepsilon \approx 3 \times 10^{-4}$. Hence the parameter $\varepsilon^2 \approx 10^{-7}$ is really very small. Therefore the approximation of a strong magnetic field (see Section 8.3.3) is well applicable to the HTTCS in solar flares. Except, the parameter γ^2 is small but not so small as ε^2:

$$\gamma^2 \approx \frac{V_s^2}{V_{A,0}^2} \sim 10^{-4} \gg \varepsilon^2$$

(see Problem 10.1). So, the condition (8.68) would be well satisfied in the undisturbed corona near the HTTCS.

This means that, in a first approximation, the parameter γ^2 is more important than the ε^2 (see Equation (8.70)). Hence, we cannot neglect the gas-pressure-gradient effects in the vicinity of the HTTCS. We have to take into account a compression of the plasma by a magnetic field near the HTTCS. That is why we use in the HTTCS model the plasma density $n_0 \sim 10^9 - 10^{11}$ cm^{-3} which is different from the plasma density in the undisturbed corona.

Chapter 18

Particle Acceleration in Current Sheets

The inductive electric field is directed along the current inside the collisionless reconnecting current sheet; this strong field does positive work on charged particles, thus increasing their energy impulsively.

18.1 Magnetically non-neutral RCS's

18.1.1 An introduction in the problem

Reconnection determines many phenomena in cosmic plasma (for a review of pioneering works see Sweet, 1969; Syrovatskii, 1981, 1982). The theory of reconnection in high-temperature turbulent-current sheets (Section 17.3) explains the total amount of energy accumulated in solar flares, the power of energy released and some other parameters of flares (Section 17.4). In particular, it has been shown (Litvinenko and Somov, 1991b) that acceleration by the electric field and scattering of particles by ion-acoustic turbulence in an HTTCS lead to the appearance of about $10^{35} - 10^{36}$ electrons with a power-law spectrum and with energies of the order of tens of keV. Future development of the theory should result in models for the total number of accelerated particles, their maximum energy and the rate of particle acceleration (see Bai and Sturrock, 1989; Somov, 1992; Hudson and Ryan, 1995).

In this section we return to the question of the maximum energy of particles accelerated in a current sheet, which has been formulated in Section 4.5. Three points are important here.

(a) The problem of particle motion in a magnetic field which changes the sign of its direction and in the electric field related to reconnection has been considered several times. Speiser (1965) found particle trajectories near the neutral plane where the magnetic field is zero (Section 4.4). The physical meaning of the Speiser solution is in the following. Formally speaking,

> a charged particle can spend an *infinite* time near such a neutral plane and can take an infinite energy from the electric field.

However, under real conditions, the probability of such a situation is small; usually the magnetic field in the 'reconnecting plane', i.e. the current sheet, has non-zero transversal and longitudinal components. Therefore actual current sheets are *magnetically* non-neutral RCSs. This is of importance for their energetics (Chapter 17), stability (Chapter 20), and for the mechanism of acceleration that will be considered in the present Chapter.

(b) Speiser (1965) showed also that

> even a small transversal field changes the particle motion in such a way that the particle leaves the current sheet after a *finite* time,

the particle energy being finite. In what follows we show that this time is small and the energy is not sufficient in the context of solar flares.

(c) Can we increase the time spent by the particle inside the current sheet? – In the following it will be shown that (Somov and Litvinenko, 1993)

> the longitudinal field increases the acceleration time and, in this way, strongly increases the efficiency of particle acceleration

thus allowing us to explain the first step of acceleration in flares. An iterative method will be presented which gives an approximate general solution of the problem.

18.1.2 Dimensionless parameters and equations

Let us consider a current sheet placed in the (x, z) plane in Figure 18.1. The electric field $\mathbf{E}$ and current density $\mathbf{j}$ are parallel to the z axis; so the associated magnetic field components are parallel to the x axis and change their sign in the plane $y = 0$. Therefore we prescribe the electric and magnetic fields inside the current sheet as follows:

$$\mathbf{E} = \{\, 0,\, 0,\, E_0 \,\}\,, \quad \mathbf{B} = \left\{ -y/a\,,\, \xi_{\perp},\, \xi_{\parallel} \right\} B_0\,. \tag{18.1}$$

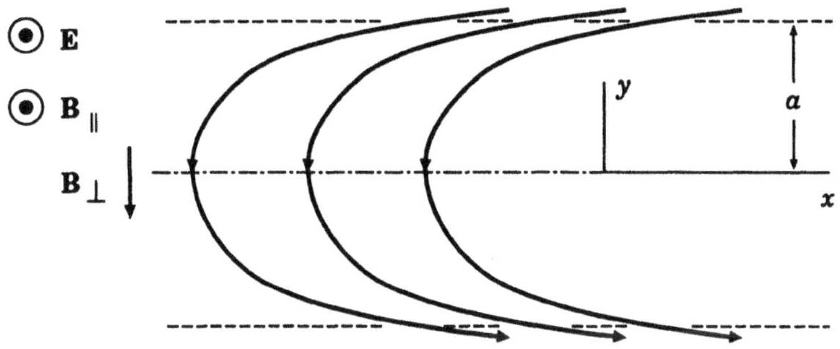

Figure 18.1: The projection of field lines inside the current sheet to the plane (x, y); $\mathbf{B}_{\|}$ is the longitudinal magnetic field. $\mathbf{E}$ is the electric field related to reconnection.

The non-relativistic equation of motion for a particle with mass m and charge $q = Ze$ is

$$m \, \frac{\partial \mathbf{v}}{\partial t} = q \left(\mathbf{E} + \frac{1}{c} \, \mathbf{v} \times \mathbf{B} \right) . \tag{18.2}$$

Let us take the half-thickness a of the sheet as a unit of length and the inverse gyro-frequency $\omega_{_B}^{-1} = mc/qB_0$ as a unit of time. Then Equation (18.2) can be rewritten in the dimensionless form:

$$\frac{\partial^2 x}{\partial t^2} = \xi_{\|} \, \frac{\partial y}{\partial t} - \xi_{\perp} \, \frac{\partial z}{\partial t} \, , \tag{18.3}$$

$$\frac{\partial^2 y}{\partial t^2} = - \xi_{\|} \, \frac{\partial x}{\partial t} - y \, \frac{\partial z}{\partial t} \, , \tag{18.4}$$

$$\frac{\partial^2 z}{\partial t^2} = \varepsilon + \xi_{\perp} \, \frac{\partial x}{\partial t} + y \, \frac{\partial y}{\partial t} \, . \tag{18.5}$$

Here the dimensionless electric field

$$\varepsilon = \frac{mc^2 E_0}{aqB_0^2} \, . \tag{18.6}$$

The influence of plasma turbulence on particle motions is ignored in (18.2). This is justified provided the time spent by a particle inside the current sheet is less than the inverse frequency of the wave-particle interactions $\nu\,(v)$. For the typical case, like the ion-acoustic turbulence,

$$\nu\,(v) = \nu_{\text{eff}} \left(\frac{\sqrt{k_{_B} T / m}}{v} \right)^3 , \tag{18.7}$$

T being the temperature in the sheet. For typical parameters of HTTCS (Chapter 17), the effective collision frequency can be estimated as

$$\nu_{\text{eff}} \approx \xi_{\perp} \omega_{\text{B}} \approx 10^6 \ \text{s}^{-1} .$$

Hence the turbulence can be ignored for suprathermal particles, once the time spent by a particle inside the HTTCS does not exceed

$$\tau_{\text{eff}} = (\xi_{\perp} \omega_{\text{B}})^{-1} \approx 10^{-6} \ \text{s} .$$

On integrating Equations (18.3) and (18.5) and substituting in (18.4), the set of Equations (18.3)–(18.5) becomes

$$\frac{\partial x}{\partial t} = \xi_{\parallel} y - \xi_{\perp} z + c_1 , \tag{18.8}$$

$$\frac{\partial^2 y}{\partial t^2} + \xi_{\parallel}^2 y = - \left(\varepsilon t + \xi_{\perp} x + \frac{1}{2} y^2 + c_2 \right) y + + \xi_{\parallel} \left(\xi_{\perp} z - c_1 \right) , \tag{18.9}$$

$$\frac{\partial z}{\partial t} = \varepsilon t + \xi_{\perp} x + \frac{1}{2} y^2 + c_2 . \tag{18.10}$$

Let x_0, y_0, and z_0 be the initial coordinates of the particle. Its initial velocity is assumed to be negligible. In this case the constants of integration are as follows:

$$c_1 = -\xi_{\parallel} y_0 + \xi_{\perp} z_0 , \quad c_2 = -\xi_{\perp} x_0 - \frac{1}{2} y_0^2 . \tag{18.11}$$

So, in principle, the problem can be solved.

18.1.3 An iterative solution of the problem

The simple-looking set of ordinary differential Equations (18.3)–(18.5) for the single particle motion inside the current sheet is still complex, because the equations are not linear in the variables. As surprising as it may seem, we cannot solve these equations exactly, except for very special cases or with some simplifications.

Until the particle leaves the current sheet, the value of $y(t)$ is small, since the sheet is supposed to be thin. The behaviour of the functions $x(t)$ and $z(t)$ does not depend strongly on the exact form of the solution $y(t)$. For this reason the Equations (18.8) and (18.10) can be solved by the following iterative procedure. First, let us prescribe some function $y(t) = y^{(0)}(t)$.

Second, using this function, calculate $x^{(0)}(t)$ and $z^{(0)}(t)$ from Equations (18.8) and (18.10). Third, let us use these functions to find a small correction $y^{(1)}(t)$ from Equation (18.9).

In zeroth approximation Equation (18.9) takes the simplest form

$$\frac{\partial^2 y^{(0)}}{\partial t^2} + \xi_{\parallel}^2 \left(y^{(0)} - y_0 \right) = 0 \,, \tag{18.12}$$

whence $y^{(0)} = y_0 = \text{const}$. Now, from Equations (18.8) and (18.10) we find the zeroth order functions:

$$x^{(0)}(t) = x_0 + (\sin \xi_{\perp} t - \xi_{\perp} t)\, \varepsilon / \xi_{\perp}^2 \,,$$

$$z^{(0)}(t) = z_0 + (1 - \cos \xi_{\perp} t)\, \varepsilon / \xi_{\perp}^2 \,. \tag{18.13}$$

In this approximation the projection of the particle's trajectory on the plane (x, z) is a cycloid curve whose shape does not depend on the longitudinal field $B_z = \xi_{\parallel} B_0$. Physically, formulae (18.13) describe the particle drift in the perpendicular fields $B_y = \xi_{\perp} B_0$ and $E_z = E_0$ (Section 4.1), the influence of the B_z component being neglected.

Now let us write an equation which will allow us to find a correction to $y^{(0)}(t)$. Making use of (18.9) and (18.13), we obtain

$$\frac{\partial^2 y}{\partial t^2} + \left(\xi_{\parallel}^2 + \varepsilon \, \frac{\sin \xi_{\perp} t}{\xi_{\perp}} \right) y = \xi_{\parallel}^2 \, y^{(0)} + (1 - \cos \xi_{\perp} t)\, \varepsilon \, \frac{\xi_{\parallel}}{\xi_{\perp}} \,. \tag{18.14}$$

So the character of the particle motion is determined by two dimensionless parameters: $\xi_{\parallel}$ and $\xi_{\perp}$. Depending on them, two cases can be considered.

(a) No longitudinal field

The case $\xi_{\parallel} = 0$ means that there is no longitudinal magnetic field inside a current sheet. Equation (18.14) becomes

$$\frac{\partial^2 y}{\partial t^2} + \left(\varepsilon \, \frac{\sin \xi_{\perp} t}{\xi_{\perp}} \right) y = 0 \,. \tag{18.15}$$

This is the equation of a one-dimensional oscillator with a time-dependent frequency. From (18.15), together with (18.13), Speiser's results follow. In particular, a particle can remain inside the current sheet only for the time

$$\tau = \frac{\pi}{\xi_{\perp}} \,. \tag{18.16}$$

When $t > \tau$, the particle quickly moves out of the current sheet, since the frequency formally becomes an imaginary value. At this instant,

$$\frac{\partial x(\tau)}{\partial t} = -\frac{2\varepsilon}{\xi_\perp}, \qquad \frac{\partial z(\tau)}{\partial t} = 0. \tag{18.17}$$

Note that in the case of a neutral sheet $\xi_\perp = 0$ and the particle acceleration along the z axis is not restricted. According to (18.16), $\tau \to \infty$; the non-relativistic kinetic energy increases as $\mathcal{K} \sim z \sim \tau^2$, while the oscillation amplitude decreases as $A_y \sim \tau^{-1/4}$ (formula (4.81)).

If $\xi_\perp \neq 0$ and the electric field is small enough,

$$\varepsilon < \frac{1}{2}\,\xi_\perp^3, \tag{18.18}$$

then small oscillations near the plane $y = 0$ are stable, and particles are not pushed out of the current sheet. However, in the HTTCS model pertaining to solar flare conditions (Chapter 17), $\xi_\perp \sim 10^{-3}$ and $\varepsilon \sim 10^{-5}$. Therefore the inequality (18.18) cannot be satisfied and particles go out of the current sheet without being accelerated.

(b) Stabilization by the longitudinal field

The case $\xi_\parallel \neq 0$, the current sheet with a longitudinal magnetic field. Equation (18.14) describes an oscillator the frequency of which changes with time and which is also subject to the action of an external periodic force. Hence the oscillating system represented by Equation (18.14) is not closed and may have resonance increases of $y = y(t)$. This corresponds to the particle going out of the sheet.

It is important, however, that the particle's motion becomes *stable* provided $\xi_\parallel$ is large enough. In this case the particle remains in the vicinity of the current sheet plane, $y = 0$. For the resonance effects to be absent, the oscillation frequency must always be real:

$$\xi_\parallel^2 > \frac{\varepsilon}{\xi_\perp}. \tag{18.19}$$

Once the inequality (18.19) is valid, **particles do not leave the current sheet** due to unstable trajectories. Were it not for the turbulence, the particles would **simply drift along the current sheet, gaining energy**. The ion-acoustic turbulence in HTTCS (cf. formula (18.7)) makes the particle motion more complex.

18.1.4 The maximum energy of an accelerated particle

An issue of great concern is what is the maximum energy to which a particle can be accelerated by the reconnecting current sheet?

For the case of a strong longitudinal magnetic field, the maximum velocity can be evaluated as

$$v_{\max} \approx \xi_{\parallel} . \tag{18.20}$$

Here a unit of velocity (Section 18.1.2) is

$$V_1 = a\,\omega_{\rm L} = \frac{aqB_0}{mc} . \tag{18.21}$$

Therefore the longitudinal field qualitatively changes the character of particle motion inside the current sheet. As an example, let us consider electron acceleration in HTTCS during solar flares.

The HTTCS model allows us to express the characteristics of a current sheet through the external parameters of a reconnection region: the concentration of plasma n_0 outside the sheet, the electric field E_0, the magnetic field gradient h_0 and the relative value $\xi_{\perp}$ of a transversal magnetic field (Chapter 17). In the case $\xi_{\parallel} = 0$ (no longitudinal field), i.e. (18.17), the maximum electron energy is given by

$$\mathcal{E}_{\max} = 2\,mc^2 \left(\frac{E_0}{\xi_{\perp} B_0} \right)^2 \tag{18.22}$$

or, using the HTTCS model,

$$\mathcal{E}_{\max}\,(\text{keV}) \approx 5 \times 10^{-9}\, T\,(\text{K}) . \tag{18.23}$$

Formula (18.23) shows that acceleration in the current sheet without a longitudinal field is not efficient: for the temperature inside the current sheet $T \approx 10^8$ K, the maximum energy of accelerated electrons is only 0.5 keV.

Let us consider now the case of a non-zero longitudinal field. The stabilization condition (18.19) can be rewritten in dimensional units as follows:

$$\left(\frac{B_{\parallel}}{B_0} \right)^2 > \frac{mc^2 E_0}{aq\, B_{\perp} B_0} . \tag{18.24}$$

In the frame of the HTTCS model the last inequality becomes especially simple:

$$B_{\parallel} > 0.1\, B_0 . \tag{18.25}$$

Thus the longitudinal component can be one order of magnitude smaller than the reconnecting components related to the electric current in the current sheet.

The maximum energy (written in dimensional units) of accelerated electrons in the current sheet is

$$\mathcal{E}_{\max} = \frac{1}{2m} \left(\frac{qa\, B_{\parallel}}{c} \right)^2 \tag{18.26}$$

or, in the HTTCS model,

$$\mathcal{E}_{\max}\,(\text{keV}) \approx 10^{-5}\, \xi_{\parallel}^{\,2}\, T\,(\text{K})\,. \tag{18.27}$$

If the current-sheet temperature $T \approx 10^8$ K and $\xi_{\parallel}^{2} \approx 0.1$, formula (18.27) gives $\mathcal{E}_{\max} \approx 100$ keV. Therefore

> the longitudinal magnetic field increases the acceleration efficiency to such a degree that it becomes possible to interpret the *first stage* or the *first step* of electron acceleration in solar flares

as the particle energization process in a non-neutral HTTCS.

The results obtained are very clear. On the one hand, the transversal magnetic field turns a particle trajectory in the current sheet plane (the plane (x, z) in Figure 18.1). At some point, where the projection of particle velocity v_z on the electric field direction changes its sign, the Lorentz force component associated with the field component $B_x = (-y/a)\, B_0$ pushes the particle out of the current sheet. This process is described by Equation (18.4) with $\xi_{\parallel} = 0$, or by Equation (18.15). On the other hand, a non-zero longitudinal magnetic field tries to turn the particle back to the current sheet. This effect is related to the first term on the right-hand side of Equation (18.4). That is why the maximum velocity of a particle is proportional to the gyro-frequency in the longitudinal field.

18.1.5 The non-adiabatic thickness of current sheet

The condition (18.24) can also be simply understood from the physical point of view. In the absence of a longitudinal magnetic field, there exists a region near the neutral plane (x, z), where the adiabatic approximation is not valid (see Section 4.5.2). So we had to solve Equation (18.2) to determine the character of the particle motion. The thickness of this region which is called the *non-adiabatic thickness* of a current sheet equals

$$d = (r_{\rm L} a)^{1/2} = \left(\frac{mc\,va}{qB_0} \right)^{1/2} . \tag{18.28}$$

Here the maximum velocity $v \approx cE_0/\xi_\perp B_0$ is substituted in the formula (4.14) of the Larmor radius $r_{\rm L}$. The longitudinal field tends to keep particles 'frozen' and to confine them inside the current sheet. Obviously such a confinement becomes efficient, once

$$r_{\rm L}\left(B_{\|}\right) < d\,, \tag{18.29}$$

where

$$r_{\rm L}\left(B_{\|}\right) = \frac{mcv}{qB_{\|}} = \frac{r_{\rm L}}{\xi_{\|}}\,. \tag{18.30}$$

This last expression coincides with condition (18.24).

* * *

To conclude, let us remind that, in the solar atmosphere, reconnection usually takes place at the separators with the non-zero transversal and longitudinal components of the magnetic field (Section 16.5). This effect was already considered in the MHD approximation from the viewpoint of current sheet energetics (Chapter 17). The longitudinal and transversal components of the magnetic field are also important for the current sheet stability (Chapter 20). As was shown in this section, the longitudinal field has strong influence on the kinetics of suprathermal particles: the magnetically non-neutral HTTCS does efficient work as an electron accelerator and, at the same time, as a trap for fast electrons in solar flares.

18.2 Regular versus chaotic acceleration

Considerable attention is focused on the phenomenon of *dynamic chaos*. The stochastic behaviour of a dynamic system is due to its intrinsic non-linear properties rather than some external noise (see Lichtenberg and Lieberman, 1983). A particular example of such a system is a particle moving in a reconnecting current sheet.

So far both numerical (Chen and Palmadesso, 1986) and analytic (e.g., Büchner and Zelenyi, 1989) treatments of the particle's motion have concentrated on a current sheet with a small magnetic field component perpendicular to the sheet. This small transversal component has been shown to give rise to chaotic particle behaviour. However, current sheets in the solar atmosphere usually have not only transversal but also longitudinal (parallel to the electric field inside the sheet) magnetic field components. The purpose of this section is to illustrate the influence of the longitudinal field on the character of charged particle motion in non-neutral current sheets.

18.2.1 Reasons for chaos

Consider the current sheet with the electric and magnetic fields (18.1). An approximate solution to Equations (18.3)–(18.5) of particle motion in such current sheet was discussed above. Now we consider some general properties of this set of equations, starting from the fact that it possesses three exact constants of motion – the invariants of particle motion:

$$C_x = \dot{x} - \xi_{\parallel}\, y + \xi_{\perp} z\,, \tag{18.31}$$

$$C_z = \dot{z} - \xi_{\perp} x - \frac{1}{2}\, y^2 - \varepsilon\, t\,, \tag{18.32}$$

$$H = \frac{1}{2}\left(\dot{x}^{\,2} + \dot{y}^{\,2} + \dot{z}^{\,2}\right) - \varepsilon\, z\,. \tag{18.33}$$

Here H is the usual Hamiltonian (e.g., Landau and Lifshitz, *Mechanics*, 1960, Ch. 7, § 40).

Rewrite the set of master Equations (18.3)–(18.5) in the Hamiltonian form. The usual way to do this is to introduce the four generalized coordinates

$$Q = \{\, t\,,\, x\,,\, y\,,\, z\,\} \tag{18.34}$$

and the generalized momenta

$$P = \left\{ -H,\ \dot{x} - \xi_{\parallel}\, y\,,\ \dot{y}\,,\ \dot{z} - \xi_{\perp} x - \frac{1}{2}\, y^2 \right\}. \tag{18.35}$$

Then the equations of motion take the form

$$\dot{Q}_i = \frac{\partial \mathcal{H}}{\partial P_i}\,, \qquad \dot{P}_i = -\,\frac{\partial \mathcal{H}}{\partial Q_i} \qquad (\, i = 0, 1, 2, 3\,)\,, \tag{18.36}$$

where

$$\mathcal{H} = H(P, Q) + P_0\,. \tag{18.37}$$

The *transformed* Hamiltonian $\mathcal{H}$ is formally time-independent since t is treated as another coordinate variable. The constants of motion are now as follows:

$$C_x = P_x + \xi_{\perp} z\,, \tag{18.38}$$

$$C_z = P_z - \varepsilon\, Q_0\,, \tag{18.39}$$

$$\mathcal{H} = \frac{1}{2}\left(P_x + \xi_{\parallel}\, y\right)^2 + \frac{1}{2}\, P_y^2 + \frac{1}{2}\left(P_z + \xi_{\perp}\, x + \frac{1}{2}\, y^2\right)^2 - \varepsilon\, z + P_0\,. \tag{18.40}$$

The Hamiltonian system (18.36) is integrable if the three constants of motion are in *involution*, i.e. their Poisson brackets are zero

(see Landau and Lifshitz, *Mechanics*, 1960, Ch. 7, § 42). Otherwise the system is likely to demonstrate *chaotic* behaviour, i.e. the particle trajectory inside the current sheet is unpredictable.

Straightforward calculation, based on the definition (1.49) for the Poisson brackets, shows that

$$[\mathcal{H}, C_x] = 0 \quad \text{and} \quad [\mathcal{H}, C_z] = 0 .$$

However, for C_x and C_z we find

$$\boxed{[C_x, C_z] = \xi_\perp ,} \tag{18.41}$$

so that the constants C_x and C_z are not in involution.

Chen and Palmadesso (1986) have obtained this result for the case $\xi_\parallel = 0$ and numerically showed the particle trajectory to be chaotic. In what follows our attention will be drawn to the fact that a non-zero longitudinal magnetic field leaves the result (18.41) unchanged. This means that **the chaos is entirely due to the transversal field** which is proportional to $\xi_\perp$ inside the reconnecting current sheet (Litvinenko, 1993).

Moreover, as will be proved below,

the longitudinal magnetic field tends to make the particle trajectory bounded and integrable inside the current sheet.

Therefore an additional constant of motion must be present in the set of equations under consideration for a sufficiently large value of the parameter $\xi_\parallel$ (Litvinenko, 1993). Seemingly, this additional constant of motion cannot be expressed in terms of elementary functions.

18.2.2 The stabilizing effect of the longitudinal field

Because of the presence of three constants of motion, the phase trajectory – the particle trajectory inside a six-dimensional phase space X (see definition in Section 1.2.1 and Figure 1.3) – is restricted to a three-dimensional surface. It follows from Equations (18.31)–(18.33) that the particle coordinate and velocity components are subject to the relation

$$H = \frac{1}{2}\dot{y}^2 + \frac{1}{2}\left(\xi_\parallel y - \xi_\perp z\right)^2 + \frac{1}{2}\left(\varepsilon t + \xi_\perp x + \frac{1}{2}y^2\right)^2 - \varepsilon z = \text{const} , \tag{18.42}$$

where zero initial conditions are assumed for simplicity.

A useful way to study the character of the particle motion is to calculate the curvature of the *energy surface* $H = H(P, Q)$.

> The negative curvature K implies the exponentially fast divergence with time of initially close trajectories.

In its turn, that gives rise to *mixing* of trajectories in phase space and hence to chaos. Analogous inferences can be drawn concerning the particle motion in the usual coordinate space (see Anosov, 1967). Provided the curvature $K \leq 0$, the asymptotic (for large t) behaviour of the trajectory is indistinguishable from that of random motion, which corresponds to stochasticity.

As was shown by Speiser (1965, 1968), particle motions in the current sheet plane and across it occur almost independently. Thus, while studying the instability in the y direction, it is justifiable to consider the two-dimensional energy surface $H = H(y, \dot{y})$, treating x and z as some time-dependent constants. Attention must be centred on the motion along the y axis, which is known to possess the strongest instability (Speiser, 1965). Therefore the quantity to be calculated is

$$K = \frac{H_{\dot{y}\dot{y}} H_{yy} - {H_{\dot{y}y}}^2}{(1 + {H_{\dot{y}}}^2 + {H_y}^2)^2} \, . \tag{18.43}$$

Assuming that $\xi_{\parallel}^2 \ll 1$ and that the particle is near the current sheet plane (i.e., $y \ll 1$), we show that the denominator of formula (18.43) approximately equals unity. Anyway, being positive, it does not influence the sign of K. The curvature of the energy surface is calculated to be

$$K(t) \approx \xi_{\parallel}^2 + \varepsilon t + \xi_{\perp} x + \frac{3}{2} y^2 , \tag{18.44}$$

or on making use of the invariant (18.32),

$$K(t) \approx \xi_{\parallel}^2 + \dot{z}(t) + y^2(t) \, . \tag{18.45}$$

It is known that $\dot{z} \geq -\varepsilon / \xi_{\perp}$ (Speiser, 1965). Thus **strong chaos is expected** in the vicinity of the neutral plane $y = 0$, provided $\xi_{\parallel} = 0$. In this case the model of Büchner and Zelenyi (1989) is applicable. On the other hand, inside the reconnecting current sheet and in its vicinity,

> a sufficiently strong longitudinal magnetic field tends to suppress chaos and make the particle motion regular.

The necessary condition for such a suppression is $K > 0$, that is

$$\xi_{\parallel} > \left(\frac{\varepsilon}{\xi_{\perp}} \right)^{1/2} . \tag{18.46}$$

So, in another way, we arrive at an inequality which coincides with (18.19). The inequality (18.46) gives $\xi_{\parallel} > 0.1$ for typical solar flare conditions if the particles under consideration are electrons (Somov, 1992; Somov et al., 1998; Somov and Merenkova, 1999). Litvinenko and Somov (1993) have been the first to pay attention to this important property of the magnetically non-neutral current sheet.

18.2.3 Characteristic times of processes

It might seem surprising that $\xi_{\parallel}$ in inequality (18.46) should tend to infinity for $\xi_{\perp} \to 0$. However, it is incorrect to consider such a limiting case. The point is that the time needed for the instability to start developing is of the order of ${\xi_{\perp}}^{-1}$ (Speiser, 1965). Hence, while being formally unstable, the particle's motion in the limit of small $\xi_{\perp}$ is regular for all reasonable values of time.

The result (18.46) is easy to understand from the physical viewpoint. A typical time for destabilization of the y-motion, i.e. the time for divergence of initially close trajectories inside the current sheet, is (in dimensional units)

$$t_{\perp} = \left(\frac{am}{F} \right)^{1/2} , \tag{18.47}$$

where the Lorentz force component is evaluated to be

$$F \approx \frac{1}{c}\, q v B_0 = \frac{1}{c}\, q \, \frac{cE}{B_{\perp}}\, B_0 = \frac{qE}{\xi_{\perp}} \tag{18.48}$$

and some typical value of $v = cE/B_{\perp}$ is assumed; $q = Ze$. The instability creating the chaos becomes suppressed once it has no time for developing, i.e.

$$t_{\perp} > t_{\parallel} , \tag{18.49}$$

$t_{\parallel}$ being the time scale introduced by the longitudinal magnetic field:

$$t_{\parallel} = \frac{mc}{qB_{\parallel}} = \frac{mc}{\xi_{\parallel} q B_0} . \tag{18.50}$$

Once (18.49) is valid, the particle becomes magnetized inside the current sheet and its trajectory is no longer chaotic. Clearly the inequality (18.49) is equivalent to condition (18.46).

18.2.4 Dynamics of accelerated electrons in solar flares

A question at this point is: What observational data can be used to verify the above-presented results? To put it another way: What are the observational consequences of chaotic particle dynamics? – Such consequences do exist.

Consider electron acceleration in solar flares. The accelerated electrons spiral in the coronal magnetic field and produce flare radio emission. Using the data on radio pulsations, Kurths and Herzel (1986); Kurths, Benz, and Aschwanden (1991); Isliker (1992) have calculated the dimension of the pseudo-phase space related to the electron source. The technique for reconstructing phase space from a one-dimensional data array is described, e.g., by Schuster (1984), where also the references to original works can be found.

> The dimension of the pseudo-phase space can serve as a measure of chaos: the larger the dimension, the more chaotic is the system.

Using the data on ms-spikes, Isliker (1992) has found that the degree of chaos varied from flare to flare and during the course of a flare. He conjectured that such behaviour was due to some *exterior* (to the electron source) parameter which could change with time. Based on the above discussion, the role of this parameter may be ascribed to the value of the longitudinal magnetic field.

This conclusion is in agreement with previous findings. From the theoretical viewpoint, the longitudinal field is determined by the photospheric sources and does change in time. It is this change that can be responsible for flare onset, i.e., the longitudinal field can be the 'topological trigger' of a solar flare (Section 16.5.2). As far as observations are concerned, the electron acceleration during flares is likely to occur at the separators with a strong longitudinal field, where magnetically non-neutral current sheets are formed (Section 16.5). As indicated above, the relative value of this field, $\xi_{\parallel} = B_{\parallel} / B_0$, determines whether the acceleration occurs in a regular or stochastic manner.

To summarize, the motion of electrons in magnetically non-neutral current sheets of solar flares becomes **regular rather than chaotic**, once the relative value of the longitudinal magnetic field $\xi_{\parallel} > 0.1$. This fact can have important implications for the dynamics of the particle acceleration in solar flares. It would be also of interest to perform calculations analogous to those of Isliker (1992), in the context of the geomagnetic tail.

Recommended Reading: Froyland (1992).

18.2.5 Particle simulations of collisionless reconnection

A particle simulation study (e.g., Horiuchi and Sato, 1997) has investigated collisionless driven reconnection in a sheared magnetic field by modeling the response of a collisionless plasma to an external driving flow. They specifically studied the effects of the transversal and longitudunal magnetic fields on the rate of reconnection and the acceleration of electrons.

Litvinenko (1997) has used our model for electron acceleration in a magnetically non-neutral current sheet to interpret the results of the simulation. He explained the electron energization in both two-dimensional ($\xi_\perp \neq 0$, $\xi_\parallel = 0$) and three-dimensional ($\xi_\perp \neq 0$, $\xi_\parallel \neq 0$) magnetic fields. An agreement was obtained between the analytical predictions and the numerical results for the electron energy gain, the acceleration time, the longitudinal field diving rise to adiabatic particle motion, and the scaling with $B_\parallel$ of the collisionless resistivity due to particle escape from the current sheet.

The particle simulation, therefore, has substantiated our theoretical modeling presented in Section 18.1. This is important both for future more general analytical models of particle acceleration and for the application of the existing models, for example, to the electron acceleration in solar flares (Sections 18.1.4 and 18.2.4).

Although the particle simulation (Horiuchi and Sato, 1997) had not been run for a sufficient time to study the acceleration of protons, it did show that the question of proton acceleration is more complicated. Their motion, as we shall see in the next Section, is influenced by the polarization electric field arising due to charge separation. Because it is much more difficult to magnetize an ion than an electron, the protons tend to escape the current sheet across its border even when the electrons are well magnetized by the longitudinal field $B_\parallel$. This leads to the generation of a transversal electric field $E_\perp$ directed towards the plane of the current sheet. This field may have important consequences for the proton motion as we discuss below.

18.3 Ion acceleration in current sheets

18.3.1 Ions are much heavier than electrons

In Section 18.1 we considered the particle acceleration in a current sheet, taking into account not only the reconnecting field $\mathbf{B}_0$, parallel to the x axis, but also a small transversal field component $B_\perp = \xi_\perp B_0$, parallel to the y axis as shown in Figure 18.1. A typical relative value of the transversal field is $\xi_{\perp} \sim 10^{-3} \div 10^{-2}$ (see Somov, 1992). In what follows we adopt the value of $\xi_{\perp} \approx 3 \times 10^{-3}$ for our estimates. The basic Speiser's (1965) result is that both the energy gain $\delta\mathcal{E}$ and the time that the particles spend in the magnetically non-neutral RCS, $\delta t_{\,\rm in}$, are finite.

> The transversal magnetic field makes the particle turn in the plane of the sheet, and then a component of the Lorentz force expels it from the RCS plane almost along the field lines

(see Figure 3 in Speiser, 1965). The distance that the particle can travel along the sheet equals the Larmor diameter determined by the transversal field and a typical speed of the particle.

Litvinenko and Somov (1993) generalized the results of Speiser (1965) by including into consideration the *longitudinal* (parallel to the main electric field **E** in Figure 18.1) magnetic field $\mathbf{B}_{\parallel}$ in the sheet.

> The longitudinal field efficiently magnetizes fast electrons in the reconnecting current sheet, but it cannot influence the motion of the accelerated protons and heavier ions.

The Larmor radius (formula (4.14)) of ions is much larger than the Larmor radius of electrons having the same velocity because ions are much heavier than electrons.

As a consequence of this fact, the critical longitudinal field, necessary to magnetize a particle and to accelerate it, is proportional to the square root of the particle mass (see (18.24)). Hence we can use, first, the Speiser's non-relativistic formulae, derived for the case when an ion of mass m and charge $q = Ze$ enters the RCS with a negligible velocity:

$$\delta\mathcal{E} = 2\,mc^2 \left(\frac{E_0}{B_{\perp}} \right)^2 , \tag{18.51}$$

$$\delta t_{\,\rm in} = \frac{\pi\, mc}{q B_{\perp}} \; . \tag{18.52}$$

Generalizations of these formulae to particles with nonzero initial velocities are given in the Section 18.3.3.

Thus, on the one hand, electrons can acquire even relativistic energies in current sheets with a nonzero longitudinal field $B_{\parallel}$ (Litvinenko and Somov, 1993). On the other hand, application of formulae (18.51) and (18.52) to the RCS, formed, for example, behind a rising coronal mass ejection – CME (Litvinenko and Somov, 1995), shows that a nonzero field $B_{\perp}$ radically restricts the energy of heavier particles: $\delta\mathcal{E}$ for protons cannot exceed 20 MeV if a typical value of $\xi_\perp = 3 \cdot 10^{-3}$ ($B_\perp = 0.3$ G) is assumed.

Therefore the relativistic energies cannot be reached after a single 'interaction' of a proton with the sheet (cf. Martens, 1988). To overcome this difficulty, Martens conjectured that the relativistic acceleration could take place in RCS regions where $B_\perp \to 0$ (the neutral sheet approximation), and the protons are freely accelerated by the electric field. This conjecture, however, does not seem to be adequate for actual RCSs, where reconnection always occurs in the presence of a transversal magnetic field. Though we expect the latter to vary somewhat along the RCS (Somov, 1992), the region with a vanishing $B_\perp$ is so small that a particle will quickly leave the region (and hence the RCS) before being accelerated. Thus we are led to modify the classic Speiser's model significantly.

Let us propose that a proton (or another ion) interacts with the RCS more than once, each time gaining a finite, relatively small amount of energy. The effect could be the required relativistic acceleration. A similar model was considered in the context of acceleration in the geomagnetic tail (see Section 2.4 in Shabansky, 1971). However, the magnetic structures in the solar atmosphere are quite different from that of the geomagnetic tail; and conditions also differ. Therefore formulae given by Shabansky are inapplicable to the problem at hand. For this reason, we have to consider another model in application to the solar atmosphere.

18.3.2 Electrically non-neutral current sheets

The factor that makes positively charged particles return to the RCS is the *transversal* electric field $\mathbf{E}_\perp$, which is parallel to the y axis in Figure 18.2 and directed toward the current sheet plane from both sides (cf. Figure 18.1). What is the origin of this electric field?

As we have seen in the previous Section, protons and other ions, having much larger masses than the electron mass, have significanly larger Larmor radii. Both electrons and protons try to escape from magnetic confinement inside the current sheet. They are deflected by the magnetic field when they move out of the sheet. However the trajectories of electrons are bent to a much greater degree owing to their smaller mass. As for the much heavier protons and ions, they stream out of the RCS almost freely. Hence the charge separation arises, leading to the electric field $\mathbf{E}_\perp$ at both sides of the RCS. This field detains the protons and ions in the RCS region, more exactly, in the vicinity of the electron current sheet (Harris, 1962; see also Ch. 5 in

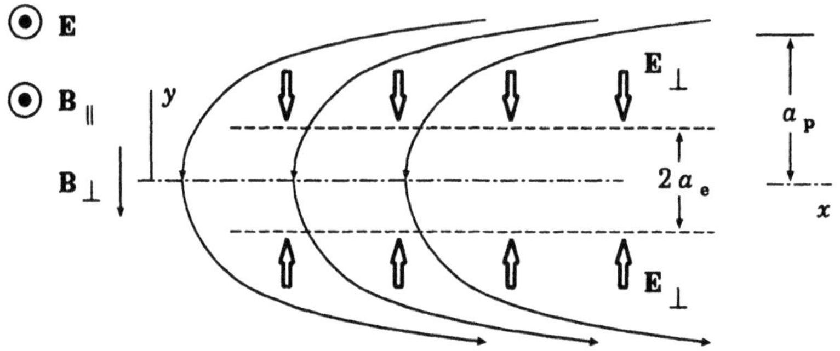

Figure 18.2: An electrically non-neutral current sheet: $\mathbf{E}_\perp$ is the transversal component of the electric field. $\mathbf{E}$ is the electric field related to the reconnection process.

Longmire, 1963; Hoh, 1966; Dobrowolny, 1968).

In an exact self-consistent one-dimensional model of the *electrically* non-neutral current sheet due to Harris (1962), this field equals

$$E_\perp = 2\pi \, \sigma^q . \tag{18.53}$$

Here the magnitude of the electric charge density integrated over the sheet thickness is

$$\sigma^q = \left(\frac{u}{c} \right)^2 nea , \tag{18.54}$$

u is the current velocity of electrons in the RCS.

Let us estimate the velocity u from the Maxwell Equation (1.1) as

$$u = \frac{c}{4\pi} \frac{B_0}{nea} . \tag{18.55}$$

On substituting (18.55) and (18.54) in (18.53), we obtain

$$E_\perp \approx \frac{k_B T}{e a} , \tag{18.56}$$

where the equation $B_0^2 / 8\pi \approx n k_B T$ has been used, T being the plasma temperature in the RCS.

It is not obvious *a priori* that Harris's solution applies to actual reconnecting current sheets with nonzero $\xi_\perp$ and finite conductivity σ. It should be valid, however, for small $\xi_\perp$, at least as a first approximation. In fact all we need for our calculations is the electric potential

$$\phi = \int E_\perp \, \mathrm{d}y , \tag{18.57}$$

which we take to equal $k_B T/e$, the usual value owing to spread of a 'cloud' of charged particles.

The following point is worth emphasizing here. The charge separation that gives rise to the potential ϕ mainly stems from the motion of protons perpendicular to the RCS plane. At the same time, some protons are known to leave the RCS almost along its plane. This property is a characteristic feature of the Speiser's mechanism of acceleration. It seems obvious that even a modest transversal electric field will considerably influence the motion of these particles because they always move almost perpendicular to this field. Having made this qualitative remark, we now proceed to calculating the energy gain rate and maximum energy for the protons being accelerated in the RCS, taking into account both the main components of electromagnetic field ($\mathbf{B}_0$ and $\mathbf{E}_0$) and the transversal ones ($\mathbf{B}_\perp$ and $\mathbf{E}_\perp$).

18.3.3 Maximum particle energy and acceleration rates

According to the model delineated above, a positively charged particle ejected from the RCS may be quickly reflected and moves back to the RCS. The reason for this is the electric field $\mathbf{E}_\perp$, directed perpendicular to the current sheet, which always exists outside the RCS (Harris, 1962). It is of importance for what follows that the accelerated protons and other ions are ejected from the RCS almost *along* the field lines (Speiser, 1965). The transversal electric field efficiently locks the particles in the RCS because they always move almost in the plane of the sheet. On getting into the sheet again, the particles are further accelerated and the cycle repeats itself.

In order to find the properties of the acceleration mechanism, we need to dwell at some length on the particle motion outside the RCS. Consider a proton leaving the RCS plane with energy $\mathcal{E}$ and momentum $\mathbf{p}$. According to Speiser (1965), the component of the momentum perpendicular to the sheet is

$$p_\perp \approx \xi_\perp p \ll p \tag{18.58}$$

for such a proton. The perpendicular component of the equation of motion for the particle outside the RCS is

$$\frac{\mathrm{d}}{\mathrm{d}t}\, p_\perp(t) = -\, qE_\perp \,. \tag{18.59}$$

Here we neglect the magnetic force, in order not to obscure the essential physical point made in this Section. Equation (18.59) allows us to estimate the time spent by the proton between two successive interactions with the RCS,

$$\delta t_{\text{out}} = \frac{2 p_{\perp}}{q E_{\perp}} \approx \frac{2 \xi_{\perp} p}{q E_{\perp}} \, . \tag{18.60}$$

The largest energy attainable is determined by the condition that the potential (18.57) is just enough to prevent the proton from leaving the RCS. In other words, the field $\mathbf{E}_{\perp}$ must cancel the perpendicular momentum $\mathbf{p}_{\perp}$. The energy conservation gives:

$$\mathcal{E}_{\max} = \left(\mathcal{E}_{\max}^{2} - p_{\perp}^{2} c^{2} \right)^{1/2} + q\phi \, , \tag{18.61}$$

where

$$p_{\perp}^{2} c^{2} = \xi_{\perp}^{2} \left(\mathcal{E}_{\max}^{2} - (mc^{2})^{2} \right) . \tag{18.62}$$

Eliminating the unknown $p_{\perp}$ between (18.61) and (18.62), we get the maximum energy

$$\mathcal{E}_{\max} = q\phi \, \frac{1}{\xi_{\perp}^{2}} \left[1 + \left(1 - \xi_{\perp}^{2} + \frac{\xi_{\perp}^{4} (mc^{2})^{2}}{q^{2} \phi^{2}} \right)^{1/2} \right] . \tag{18.63}$$

According to formulae (18.56) and (18.57), here the electric field potential $\phi \approx k_{\rm B} T / e$. Formula (18.63) shows that

> protons can actually be accelerated to GeV energies in the high-temperature turbulent-current sheets (HTTCS) in solar flares

(Chapter 17): for instance $\mathcal{E}_{\max} \approx 2.4$ GeV provided $T_{\rm e} \approx 10^{8}$ K. Even larger energies can be reached in RCS regions with a smaller transversal magnetic field.

Note in passing that if a particle leaves the sheet with the velocity that is perpendicular to the magnetic field lines outside the RCS, the magnetic reflection is very efficient too. In this case it occurs in a time of order the inverse gyrofrequency in the field $\mathbf{B}_{0}$.

The resulting acceleration rate can be estimated as

$$\frac{\mathrm{d}\mathcal{E}}{\mathrm{d}t} \approx \frac{\langle \delta \mathcal{E} \rangle}{\delta t_{\text{in}} + \delta t_{\text{out}}} \, . \tag{18.64}$$

Here

$$\langle \delta \mathcal{E} \rangle = 2 \mathcal{E} \left(\frac{E_{0}}{B_{\perp}} \right)^{2} \tag{18.65}$$

is the relativistic generalization of the Speiser formula (18.51) for the average energy gain. The averaging needs to be introduced because, in general, a term linear in a component of the particle momentum appears in the expression for $\delta\mathcal{E}$ (cf. Speiser and Lyons, 1984).

In much the same way

$$\delta t_{\text{in}} = \frac{\pi \mathcal{E}}{c q B_{\perp}} \tag{18.66}$$

is the relativistic generalization of the Speiser formula (18.52). The approach using the differential equation (18.64) is quite justified once the inequality $\langle \delta\mathcal{E} \rangle \ll \mathcal{E}_{\max}$ holds.

Equation (18.64), with account taken of the formulae (18.60), (18.65), and (18.66), can be integrated in elementary functions. To simplify the problem further, we note that

$$\frac{\delta t_{\text{in}}}{\delta t_{\text{out}}} = \frac{\pi E_{\perp}}{2 \xi_{\perp} B_{\perp}} \left(\frac{\mathcal{E}}{pc} \right) \approx 10^3 \frac{\mathcal{E}}{pc} \gg 1 . \tag{18.67}$$

Hence it is justifiable to ignore the second term in the denominator of Equation (18.64). The simplified equation is integrated to give the kinetic particle energy

$$\mathcal{E}_{kin}(t) \equiv \mathcal{E} - mc^2 = \frac{2}{\pi} c q E_0 \left(\frac{E_0}{B_{\perp}} \right) t , \tag{18.68}$$

whence the time of the particle acceleration is

$$t_{\text{ac}}(\mathcal{E}_{kin}) \approx 0.03 \left(\frac{\mathcal{E}_{kin}}{1\,\text{GeV}} \right) \text{ s} . \tag{18.69}$$

This result demonstrates the possibility of very efficient acceleration of protons and other ions by the direct electric field in the RCS. At the same time, taking care of the actual magnetic field structure has considerably diminished (by a factor of $E_0 / B_{\perp} = V / (\xi_{\perp} c) \approx 10^{-1}$) the magnitude of the energy gain rate, as compared with the case $B_{\perp} = 0$.

Alternatively, we could rewrite formula (18.68) to obtain the energy $\mathcal{E}$ as a function of the number of particle entries to the RCS, N_{int}:

$$\mathcal{E}(N_{int}) = mc^2 \exp \left[2 \left(\frac{E_0}{B_{\perp}} \right)^2 N_{int} \right] . \tag{18.70}$$

Therefore the particle must interact with the RCS

$$N_{\max} \approx \left(\frac{B_{\perp}}{E_0} \right)^2 \approx 10^2 \tag{18.71}$$

times in order to reach a relativistic energy. As was shown above (see Equation (18.63)), the transversal electric field outside the RCS is actually capable of providing this number of reentries into the current sheet.

In principle, the protons or other ions could leave the RCS along its plane rather than across it. This is not likely, however, because of a very short acceleration time t_{ac} ; the distance a proton can travel along the RCS when being accelerated is less than $c\,t_{ac} \approx 10^9$ cm, that does not exceed a typical RCS width and length $10^9 \div 10^{10}$ cm.

Therefore we have estimated the efficiency of the acceleration process in the frame of the simple RCS model which contains several taciturn assumptions. The most important of them is a modification of the steady two-dimensional model for high-temperature turbulent-current sheet (Chapter 17) with account of the Harris type equilibrium across the sheet. Such a possibility does not seem surprising one *a priory*, but it certainly has to be considered in detail somewhere else.

Another assumpion is that the initially assumed conditions of current sheet equilibrium are not changed due to the acceleration, more exactly, during the characteristic time of the acceleration of a particle. In fact, we consider the number of particles accelerated to high energies as a small one in comparison with the number of current driving thermal electrons inside the RCS. However, generally speaking, it remains to be seen that this assumption can be well justified without careful numerical modelling of the real plasma processes in the region of reconnection and particle acceleration.

18.3.4 Early and late acceleration in solar flares

It was widely believed that the longest-lasting and most-energetic *solar energetic particle* events (SEPs) observed in interplanetary space result from acceleration by the bow shocks of coronal mass ejections (CMEs). However, using gamma-ray, X-ray and radio diagnostics of interacting (with the solar plasmas and magnetic fields) particles and spaceborne and ground-based detection of ≥ 20 MeV protons at 1 AU during two large events (1989 September 29 and October 19), Klein *et al.* (1999) demonstrate that time-extended acceleration processes in the low and middle corona, far behind the CME, leave their imprints in the proton intensity time profiles in interplanetary space for one or several hours after the onset of the solar flare. So, the CME may play the role of a trigger or even contribute to the buildup of magnetic stresses in the corona, but its bow shock is not the main accelerator of the high-energy protons.

Litvinenko and Somov (1995) have suggested that the time-extended (or late, or second) acceleration of protons and perhaps heavier ions to relativistic energies during the late phase of large-scale solar flares (e.g., Akimov *et al.*, 1996) occurs in a 'vertical' RCS. Here the field lines are driven together and forced to reconnect below erupting loop prominences or coronal streamers. The time of RCS formation corresponds to the delay of the **second phase of acceleration** after the first (or early), impulsive phase. The mechanism invoked – the direct electric field acceleration – is, in fact, quite ordinary in studies of the impulsive phase (see review in Syrovatskii, 1981; Chupp, 1996). There are good reasons to believe that the same mechanism also efficiently operates during the second phase of the acceleration in large-scale flares occuring high in the corona.

First, already early radio observations of solar flares (Palmer and Smerd, 1972; Stewart and Labrum, 1972) were indicative of particle acceleration at the cusps of helmet magnetic structures in the corona. These are exactly the structures where RCSs are expected to form according to the *Yohkoh* observations in soft and hard X-rays (see review in Kosugi and Somov, 1998).

Note that the acceleration by Langmuir turbulence inside the RCS in the helmet structure, invoked by Zhang and Chupp (1989) to explain the electron acceleration in the flare of April 27, 1981, is too slow to account for the generation of relativistic protons and requires an unreasonably high turbulence level.

Specific models have been designed to explain the physics of particle acceleration in magnetic cusp geometry, in particular the two-step acceleration model with a reconnecting current sheet and magnetic collapsing trap, described in Section 13.4.

Second, gamma-emission during large flares consists of separate peaks with a characteristic duration of 0.1–0.3 s, down to 0.04 s (Gal'per *et al.*, 1994; Akimov *et al.*, 1996). If this behaviour is interpreted in terms of a succession of separate acts of the acceleration, then the shock mechanism is also too slow since the acceleration time would be

$$t_{ac} = 50 \left(\frac{100\,\mathrm{G}}{B_0} \right) \left(\frac{\mathcal{E}}{1\,\mathrm{GeV}} \right) \mathrm{s} \approx 50\ \mathrm{s} \qquad (18.72)$$

(Colgate, 1988). By contrast, as we saw above,

> the direct electric field inside the RCS provides not only the necessary maximum energy but also the necessary energy gain rate

(see formula (18.69)). High velocities (up to the coronal Alfvén speed) of erupting filaments and other CMEs imply a large direct electric field in the

RCS. This is the reason why the acceleration mechanism considered is so efficient in *fast transient phenomena* in the solar corona (Somov, 1981). Strong variability of gamma-emission may reflect the regime of impulsive, bursty reconnection in the RCS.

An interesting feature of the mechanism considered is that neither the maximum energy nor the acceleration rate depend upon the particle mass. Hence the mechanism may play a role in the preferential acceleration of heavy ions during solar flares.

Recall that Martens (1988) applied the Speiser (1965) model when considering relativistic acceleration of protons during the late phase of flares. However, it turned out necessary to assume an idealized geometry of the magnetic field in the RCS, viz. $\mathbf{B}_{\perp} \to 0$, in order to account for the relativistic acceleration. We have seen that the difficulty can be alleviated by allowing for the transversal electric field $\mathbf{E}_{\perp}$ outside the sheet. This field necessarily arises in the vicinity of the RCS (Harris, 1962).

To conclude, though MHD shocks are usually thought to be responsible for the relativistic generation of protons during the late phase of extended (gradual) gamma-ray/proton flares (Bai and Sturrock, 1989), another mechanism – the direct electric field acceleration in RCS – can explain the proton acceleration to the highest energies observed, at least in flares with strong variability of gamma-emission. Of course, the same sudden mass motions that lead to formation of current sheets also give rise to strong shock waves, so the two mechanisms of acceleration can easily coexist in a single solar flare.

Chapter 19

Structural Instability of Reconnecting Current Sheets

The interrelation between the stability and the structure of current sheets governs their nonlinear evolution and determines a reconnection regime.

19.1 Properties of reconnecting current sheets

19.1.1 Current sheet splitting

The continuous MHD flow of a perfectly conducting medium is impossible in the zeroth point of a magnetic field, in which the electric field differs from zero. In the vicinity of this peculiar point the frozen-in condition breaks down (Section 9.3), and the current sheet (CS in Figure 19.1) – the discontinuity dividing magnetic fields of opposite directions – forms there in compliance with the statement of Syrovatskii (1971). Later on Brushlinskii, Zaborov, and Syrovatskii (1980), Podgornii and Syrovatskii (1981), Biskamp (1986, 1997) observed the splitting of the reconnecting current sheet into other MHD discontinuities in their numerical experiments.

This splitting is usually discussed in relation to the configuration suggested by Petschek (1964), which appears in particular during the reconnection of uniform magnetic fluxes (see Problem 11.8). It consists of a system of MHD discontinuities, crossing in the small central diffusion region D.

As distinct from Petschek's configuration, the thin wide current sheet forms in the vicinity of a hyperbolic zeroth point of a strong magnetic field as shown in Figure 19.2. Just this case (and more complicated ones) has

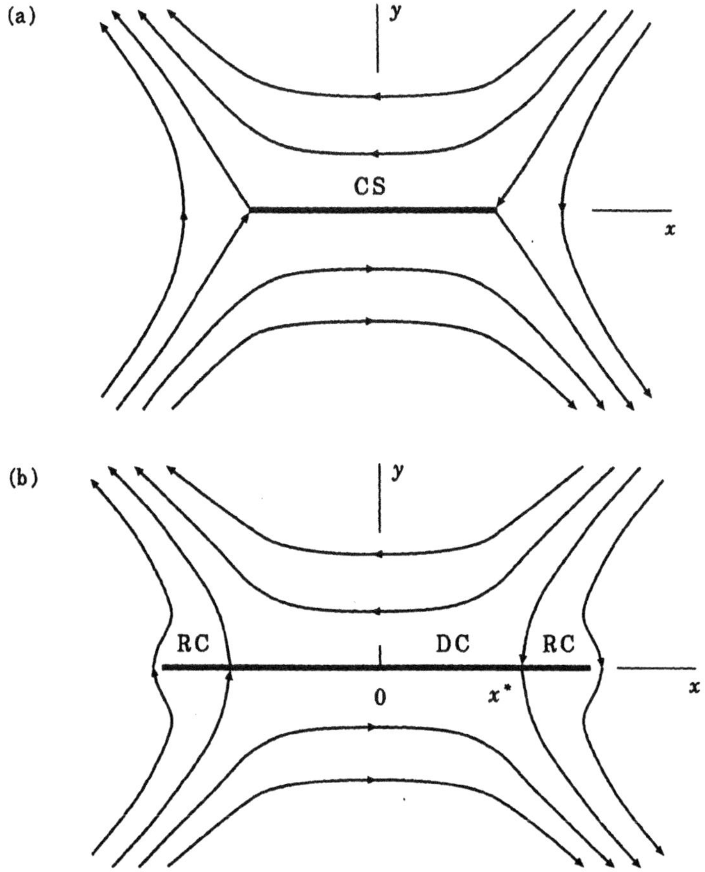

Figure 19.1: Thin current sheets: (a) without reverse electric currents, and (b) with two reverse currents (RC), DC is a region of direct current.

been realized in the numerical MHD experiments carried out by Brushlinskii *et al*. (1980), Podgornii and Syrovatskii (1981), Biskamp (1986), Antiochos *et al*. (1996), Karpen *et al*. (1998) and will be considered below.

The splitting of the current sheet means a change of the regime of magnetic reconnection, since the distribution of electric current becomes two-dimensional. In the present chapter we consider the conditions under which the splitting takes place and point out its possible reason. This reason is the *non-evolutionarity* of the reconnecting current sheet as a discontinuity or its *structural* instability, as people sometimes say.

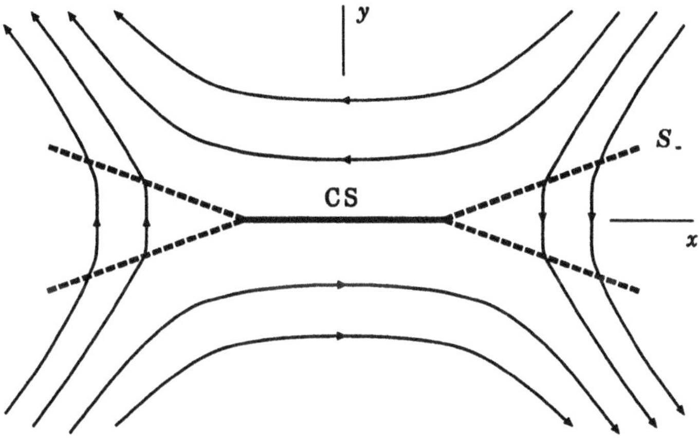

Figure 19.2: A splitted current sheet (CS) with the attached MHD discontinuities – the four slow shock waves (S_-).

19.1.2 Evolutionarity of reconnecting current sheets

One-dimensional equations of ideal MHD have discontinuous solutions: fast and slow shock waves, tangential, contact and Alfvén discontinuities, peculiar shocks (Chapter 11). As was shown, a steady discontinuity may exist in a real plasma only if it is stable with respect to the break up into other discontinuities or the transition to some unsteady flow (Chapter 12).

Let the MHD quantities be subjected to an infinitesimal perturbation at the initial instant of time. Then a linear passage of waves out from the discontinuity occurs. If the amplitudes of these waves and the displacement of the discontinuity are uniquely determined from the linearized boundary conditions, then the problem of the *time evolution* of the initial perturbation has a single solution. If this problem does not have a single solution, then the supposition that the initial perturbation is small is not valid. In this case

> the infinitesimal perturbation results in an instant (in the approximation of an ideal medium) non-linear change of the original flow.

This is a *non-evolutionary* discontinuity. Note that, as distinct from a non-evolutionary discontinuity, the perturbation of an unstable evolutionary discontinuity remains infinitesimal during a small enough period of time.

The criterion of evolutionarity results from the comparison of two numbers. N_w is the number of the independent unknown parameters: the amplitudes of outgoing, i.e. reflected and refracted, waves and the displacement of the discontinuity, describing infinitesimal perturbation. And N_e is the number of independent boundary conditions (equations) which infer the unknown parameters by the amplitudes of the incident waves. If these numbers are equal, then the discontinuity satisfies the requirement of evolutionarity. Otherwise the problem of the time evolution of an initial infinitesimal perturbation does not have a solution, or else it has an infinite amount of solutions. Such a discontinuity cannot exist in a real medium.

As the direction of the propagation of a wave depends on the relationship between its group velocity and the flow velocity,

> the requirement of evolutionarity gives the restriction on the unperturbed MHD quantities on both sides of the discontinuity.

In particular, the shock waves turn out to be evolutionary when either the upflow and the downflow velocities are larger than the Alfvén speed (fast shocks) or smaller than it (slow shocks).

The reconnecting current sheet cannot be reduced to a one-dimensional flow, since the inhomogeneity of velocity in it is two-dimensional, and is characterized by two spatial parameters. The thickness of the sheet, i.e. the distance $2a$ between the reconnecting magnetic fluxes (see Figure 4.19), determines the rate of magnetic field dissipation in it, but the width $2b$ characterizes the storage of magnetic energy in the domain of the flux interaction.

In what follows we obtain the conditions under which, in a plasma of high conductivity, infinitesimal perturbations interact with the current sheet as with a discontinuity, and the problem of its evolutionarity with respect to such perturbations can be solved (Markovskii and Somov, 1996a).

19.1.3 Magnetic field near the current sheet

Consider the thin current sheet, appearing in the vicinity of the zeroth point of a magnetic field $\mathbf{B}_0 = (h_0 y, h_0 x, 0)$, at which the electric field $\mathbf{E} = (0, 0, E)$ differs from zero. The field lines, frozen into the plasma, drift along the y axis into the sheet, where the frozen-in condition breaks down, reconnect in it, and flow out along the x axis. Syrovatskii (1971) represented the coordinate dependence of the field $\mathbf{B}$ outside the sheet in a complex form, supposing that the half-thickness of the current sheet a (size along the y axis) equals zero (see Figure 19.1),

$$B_y + \mathrm{i}B_x = h_0 \left(\zeta^2 - (x^*)^2\right) \left(\zeta^2 - b^2\right)^{-1/2} \qquad (19.1)$$

(see also Ch. 3 in Somov and Syrovatskii, 1976a). Here the complex variable $\zeta = x + \mathrm{i}\,y$, b is the half-width of the sheet (size along the x axis), c is the speed of light, and I is the total current in the sheet. The quantity I is varied through the range $0 \le I \le c h_0 b^2/4$. At the points

$$x^* = \pm \sqrt{\frac{1}{2}\, b^2 + \frac{2I}{c h_0}} \tag{19.2}$$

the magnetic field changes its sign (see formula (19.1) and Figure 19.1b). For $\mid x \mid < \mid x^* \mid$ the direction of the current coincides with the direction of the electric field. This is direct (DC) or forward current in Figure 19.1b. However for $\mid x^* \mid < \mid x \mid < b$ it has the opposite direction (*reverse* currents RC). If $x \sim b$ and $b - \mid x^* \mid \sim b$, then the reverse current is comparable with the forward one. Suppose that precisely this configuration appears. In so doing all MHD quantities outside (but near) the current sheet may be treated as quasi-homogeneous everywhere, except in some neighborhood of the points $x = x^*$ and $x = \pm\, b$, which are excluded from the further consideration.

Given the plasma conductivity σ is infinite the quantity b increases indefinitely with time. If σ is limited, then the finite width $2b$ settles in finite time (Syrovatskii, 1976a) and $a/b \neq 0$, although $a \ll b$. In this case, as distinct from (19.1), $B_y \neq 0$ on the surface of the current sheet. However, when σ is large enough, $B_x \gg B_y$ outside some neighborhood of the points (19.2). Later on B_y is assumed to be zero. More general formulation of the problem is given in Section 3.4.3 in Somov (1992).

19.1.4 Current sheet flows

Let the flow of the plasma satisfy the MHD approximation. Provided $a \ll b$, all quantities except the velocity $\mathbf{v}$ are quasi-homogeneous along the x axis inside the sheet. As for the inhomogeneity of the velocity, it is two-dimensional, since it follows from the mass conservation equation that at the point $x = 0$, $y = 0$

$$\frac{\partial v_x}{\partial x} = - \frac{\partial v_y}{\partial y}$$

because of the flow symmetry. Therefore the reconnecting sheet cannot be reduced to a one-dimensional flow. This is obvious because

> two reconnecting magnetic fluxes move towards each other and the plasma flow inside the current sheet is thus two-dimensional.

If the conductivity is infinite it becomes a tangential discontinuity in the limit $t \to \infty$.

Let us consider a settled current sheet. Then the electric field $\mathbf{E}$ is independent of time. This being so the ratio a/b was estimated by Syrovatskii (1976a) from the steady-state Ohm's law

$$\frac{a}{b} \sim \frac{\nu_m h_0}{cE}, \tag{19.3}$$

where ν_m is the magnetic diffusivity. Besides, in the stationary model, the electric field is independent of the coordinates. Hence

> in the region of forward current the plasma flows into the sheet, but in the regions of reverse currents it flows out along the y axis.

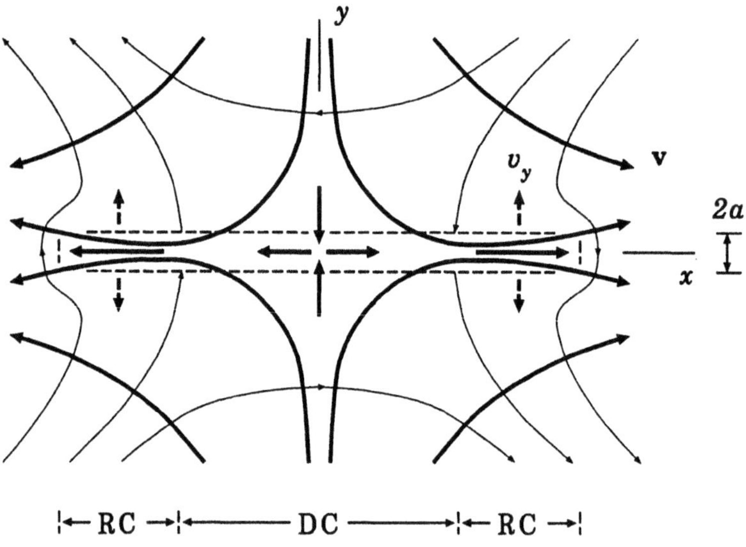

Figure 19.3: Plasma motions inside the reconnecting current sheet and in its vicinity.

Such character of the conductive plasma flows is shown schematically in Figure 19.3. The velocity component v_y changes the sign when the plasma flows from the region DC of direct current into two regions RC of reverse current, which are the same regions as in Figure 19.1b. This is important for counting the number N_w of the outgoing small-amplitude waves.

19.1.5 Additional simplifying assumptions

Suppose that all dissipative factors except the magnetic diffusivity ν_m equal zero, but ν_m is so small that

$$\frac{cE}{h_0 b} \ll \frac{h_0 b}{\sqrt{4\pi\rho}} . \tag{19.4}$$

The left side of this inequality represents the characteristic value of the drift velocity directed to the current sheet v_y, the right side gives the value of the Alfvén speed V_{A}.

Consider also that

$$\rho^{in} \sim \rho^{ex} . \tag{19.5}$$

Here the indexes '*in*' and '*ex*' denote the quantities inside and outside the sheet. Such a distribution was, for example, in the numerical experiment by Brushlinskii, Zaborov, and Syrovatskii (1980).

On the surface of the current sheet the magnetic field increases without bound but the drift velocity tends to zero, if the conductivity is infinite. At the same time the quantity of the pressure p outside the current sheet is close to its value for $\zeta = \infty$ and does not equal zero or infinity for all σ. On this basis it may be thought that, outside the neighborhood of the point (19.2), the sound velocity V_s satisfies the condition

$$\boxed{v_y^{ex} \ll V_s^{ex} \ll V_{\mathrm{A}}^{ex} ,} \tag{19.6}$$

when the conductivity is large enough. Inequalities (19.6) are well consistent with the magnetostatic approximation (8.68).

Taking the characteristic values of these quantities for an active region in the solar corona: $v_y \sim 10$ km/s, $V_s \sim 100$ km/s, $V_{\mathrm{A}} \sim 1000$ km/s, we see that the approximation (19.6) well holds there.

As far as the component of the velocity v_x is concerned, its modulus grows from zero for $x = 0$ to

$$\mid v_x^{in} \mid \sim \frac{h_0 b}{\sqrt{4\pi\rho}} \tag{19.7}$$

for $x = x^*$ (Syrovatskii, 1981) and then reduces to zero for $\mid x \mid = b$. Outside, the component v_x also does not exceed the characteristic Alfvén speed.

Let us now investigate the infinitesimal perturbation of the reconnecting current sheet (RCS) using the outlined properties of the plasma flow.

19.2 Small perturbations outside the RCS

19.2.1 Basic assumptions

Assume that the MHD quantities Q are subjected to an infinitesimal perturbation δQ. Suppose that $\delta v_z \equiv 0$ and $\delta B_z \equiv 0$, and outside the current sheet the perturbation satisfies the WKB approximation. Then, its wave vector $\mathbf{k}$, in the zeroth order in terms of the small parameter $1/kb$, is determined from the dispersion equation

$$\omega_0 \left[\mathrm{i}\, k^2 V_s^2 \, (\mathbf{k}\mathbf{V}_{\mathrm{A}})^2 - V_s^2 \, k^2 \, \omega_0 \left(\mathrm{i}\, \omega_0 - \nu_m k^2 \right) - \right.$$

$$\left. - \mathrm{i}\, k^2 \, V_{\mathrm{A}}^2 \, \omega_0^2 + \omega_0^3 \left(\mathrm{i}\, \omega_0 - \nu_m k^2 \right) \right] = 0 \,, \tag{19.8}$$

where $\omega_0 = \omega - \mathbf{k}\mathbf{v}$.

Let us impose the following restriction on the frequency ω:

$$\boxed{\frac{v_y}{a} \ll \omega_{\|} \ll \frac{V_s}{a} \,,} \tag{19.9}$$

where

$$\omega_{\|} = \omega - k_x \, v_x \,. \tag{19.10}$$

Besides, for the sake of simplicity, we will put

$$v_y \sim \frac{V_s^{\,3}}{V_{\mathrm{A}}^{\,2}} \,. \tag{19.11}$$

We will show in Section 19.5.3 that precisely this velocity appears in the criterion of evolutionarity for the reconnecting current sheet.

19.2.2 The propagation of perturbations normal to the RCS

At first, consider the case of the propagation of the perturbations normal to the current sheet, i.e. the perturbations with $k_x = 0$. In the zeroth order in terms of the small parameters, given by inequality (19.9), the solutions of Equation (19.8) take the form

$$k_y^{\,d} = -\,\mathrm{i} \, \frac{v_y}{\nu_m} \, \frac{V_{\mathrm{A}}^{\,2}}{V_s^{\,2}} \,, \tag{19.12}$$

$$k_y^0 = \frac{\omega}{v_y}, \tag{19.13}$$

$$k_y^- = \frac{\omega}{v_y}, \tag{19.14}$$

$$k_y^+ = \pm \frac{\omega}{V_A}. \tag{19.15}$$

Here the root (19.14) is twofold.

The WKB approximation holds for these perturbations if $1/k_y^+ b \ll 1$, since $| \, k_y^+ \, |$ is the least wave number. This is equivalent to the following condition for the frequency ω:

$$\omega \gg \frac{h_0}{\sqrt{4\pi\rho}}. \tag{19.16}$$

When condition (19.16) is true, the derivatives of the unperturbed quantities over the coordinates in the linear MHD equations are negligible and the dispersion Equation (19.8) is valid.

To obtain the criterion of evolutionarity it is necessary to classify the perturbations according to whether they are incoming to the current sheet or outgoing from it. Generally, such a classification has to be made by the sign of the sum of the projections of the velocity **v** of the medium and the *group* velocity (see formula (10.12)) on the normal to the sheet. However, as it was mentioned by Kontorovich (1959), in the case of normal propagation it is sufficient to determine only the sign of the phase velocity, since in the absence of frequency dispersion the latter coincides with the projection of the group velocity on the direction of the vector **k** in the system of coordinates, where the plasma is at rest.

The perturbation with the wave vector k_y^0 from formula (19.13) corresponds to an entropy wave (Section 10.2), but k_y^- from (19.14) corresponds to the slow magnetoacoustic wave propagating perpendicularly to the magnetic field. In the system of coordinates, where the plasma is at rest, their phase velocities equal zero, but in the laboratory system they coincide with the plasma velocity **v**. This being so,

> both perturbations are incoming to the sheet when the plasma flows into it, and are outgoing ones when the plasma flows out.

Besides, by virtue of the left side of inequality (19.9), we have conditions

$$k_y^0 \gg 1/a \quad \text{and} \quad k_y^- \gg 1/a.$$

Hence the current sheet is not a discontinuity for the perturbations (19.13) and (19.14).

The perturbation with the wave vector k_y^+ from (19.15) represents fast magnetoacoustic waves. Their phase velocity ω / k_y^+ satisfies the condition $V_{ph}^+ \gg v_y$ (see (19.6) and (19.15)) and is aligned with the normal to the current sheet or opposed to it. So, one of them is always incoming to the sheet and the other is outgoing from it, regardless of the sign of v_y. As distinct from k_y^0 and k_y^-, the quantity $k_y^+ \ll 1/a$, and the waves (19.15) interact with the reconnecting current sheet as with a discontinuity.

Finally, the perturbation k_y^d from (19.12) is a dissipative wave and it damps within a distance which is much smaller than the sheet half-thickness a. Consequently, as was pointed out by Roikhvarger and Syrovatskii (1974), its amplitude does not appear in the boundary conditions on the surface of a discontinuity. This being so, the dissipative effects outside the current sheet are negligible.

Thus, in the case of normal propagation,

> there is one outgoing wave on each side of the current sheet when the plasma flows into it (in the region DC of forward current),

and there are four of such waves, when the plasma flows out (in the domains RC of the reverse currents).

19.2.3 The inclined propagation of small perturbations

Let us now turn to the inclined propagation. To solve the problem of the evolutionarity of the current sheet as a discontinuity, it is necessary to obtain the solution of Equation (19.8) with common ω and k_x. Kontorovich (1959) showed that for a given flow the number of waves incoming to the x axis and outgoing from it, with common ω and k_x, is independent of k_x, i.e. of the angle of propagation (see also Ch. 3 in Anderson, 1963). Thus it is sufficient to determine the number of such waves for $k_x = 0$. From the preceding it follows that, when the plasma flows into the current sheet (the region DC of the forward current in Figure 19.1b), there is one outgoing wave on each side of it. But when the plasma flows out there are four of them.

However, for the reconnecting current sheet, under condition (19.9), the number of the perturbations with $k_y \ll 1/a$ – i.e. those for which the amplitudes are discontinuous across it – depends on k_x. If $k_x = 0$, then there are two of such perturbations, determined by the wave vector k_y^+ from (19.15).

As will be shown below, there are three for the inclined propagation. This fact is important in our further considerations.

The wave vector of a slow magnetoacoustic wave is given by the formula

$$| \mathbf{k}^- | = \frac{\omega}{v_y \sin\theta + v_x \cos\theta \pm | V_{ph}^- |}, \tag{19.17}$$

where V_{ph}^- is the phase velocity, and θ is the angle between $\mathbf{k}^-$ and the x axis. Here the scalar product $\mathbf{kv}$ is represented in the form

$$\mathbf{kv} = | \mathbf{k}^- | \times (v_y \sin\theta + v_x \cos\theta) .$$

With $V_s \ll V_{\rm A}$ the following expression for $| V_{ph}^- |$ is valid:

$$| V_{ph}^- | = \frac{V_{\rm A} V_s}{V_\perp} | \cos\theta | \left[1 + \frac{1}{2} \frac{V_{\rm A}^2 V_s^2}{V_\perp^4} \cos^2\theta + o\left(\frac{V_{\rm A}^2 V_s^2}{V_\perp^4} \right) \right], \tag{19.18}$$

where $V_\perp^2 = V_{\rm A}^2 + V_s^2$.

Let us choose the angle θ_0 in such a way that $| V_{ph}^- | \sim V_s$, i.e. $| \cos\theta_0 |$ is not small, and find the solutions of Equation (19.8) for fixed ω and

$$k_x = | \mathbf{k}^- | \cos\theta_0 . \tag{19.19}$$

For this purpose let us separate out the unknown variable k_y

$$\begin{aligned} (\omega_\| - k_y v_y) \Big[\left(\nu_m v_y V_s^2 \right) k_y^5 + \left(\mathrm{i} v_y^2 V_\perp^2 - \nu_m \omega_\| V_s^2 \right) k_y^4 - \\ - \left(2\mathrm{i} \, \omega_\| v_y V_\perp^2 \right) k_y^3 + \mathrm{i} A k_y^2 - \\ - \left(2\mathrm{i} \, \omega_\| v_y \left(V_\perp^2 k_x^2 - 2 \omega_\|^2 \right) \right) k_y + \mathrm{i} k_x^2 A - \mathrm{i} \omega_\|^4 \Big] = 0 . \end{aligned} \tag{19.20}$$

Here the coefficient

$$A = \omega_\|^2 V_\perp^2 - k_x^2 V_{\rm A}^2 V_s^2$$

and condition (19.9) is used. In the zeroth order in terms of the small parameters, given by inequality (19.9), this equation has the following solutions: (19.12) and

$$k_y^0 = \frac{\omega_\|}{v_y}, \tag{19.21}$$

$$k_y^{1-} = \frac{2 \omega_\|}{v_y}, \tag{19.22}$$

$$k_y^{2-} = k_x \tan\theta_0 \,, \tag{19.23}$$

$$k_y^s = \frac{1}{2}\left[\frac{\omega_{\|} V_s^2 \cos^2\theta_0}{2v_y V_{\rm A}^2} \pm \left(-\frac{4\omega_{\|}^2}{V_s^2} + \right.\right.$$

$$\left.\left. + \frac{\omega_{\|}^2 V_s^4 \cos^4\theta_0}{4v_y^2 V_{\rm A}^4} \pm 2\sin\theta_0 \mid \cos\theta_0 \mid \frac{\omega_{\|}^2 V_s}{v_y V_{\rm A}^2}\right)^{1/2}\right]. \tag{19.24}$$

The sign in the round brackets in (19.24) coincides with the sign in front of $\mid V_{ph}^- \mid$ in formula (19.17), but that in front of the round brackets specifies two different solutions of Equation (19.20). From inequality (19.9) it follows that for the perturbations (19.21) and (19.22) $k_y \gg 1/a$, but for (19.23) and (19.24), on the contrary, $k_y \ll 1/a$.

The waves k_y^{1-} and k_y^{2-} are slow magnetoacoustic ones, here with the angle between $\mathbf{k}^{2-}$ and the x axis equals θ_0 for k_x from (19.19). As for the waves k_y^s, they may be either slow magnetoacoustic or the *surface* ones, depending on the ratio $v_y V_{\rm A}^2/V_s^3$. Recall that if the perturbations are characterized by a common θ, but not k_x, as in the present case, then there are always two slow waves, but the rest are fast magnetoacoustic waves.

If the expression in the round brackets in formula (19.24) is negative, then k_y^s has an imaginary part and the corresponding perturbations increase or decrease exponentially with the characteristic length, which is much smaller than a, while propagating away from the surface.

Investigation of the polynomial of the second degree in v_y in the round brackets in formula (19.24) shows that it equals zero at the points

$$v_y = \frac{V_s^3}{4V_{\rm A}^2} \mid \cos\theta_0 \mid \times (\pm \sin\theta_0 \pm 1)\,. \tag{19.25}$$

Here the sign in front of $\sin\theta_0$ is given by the sign in formula (19.17). Two signs in front of 1 determine two ends of the length on the axis of v_y, within which the perturbations (19.24) are slow magnetoacoustic waves. Outside this length they become surface waves. The one of them which increases, while propagating away from the surface, should be rejected as it does not satisfy the boundary condition at infinity. As was stated by Kontorovich (1959), the decreasing perturbation should be classified as outgoing from the discontinuity surface.

Below we will use the fact that for large enough velocities, v_y, the waves (19.24) are surface ones, independent of θ_0. It may be shown that the function

$v_y(\theta_0)$, determined by formula (19.24), is restricted by modulus from above by the quantity

$$v_y^{max} = \frac{3\sqrt{3}}{16} \frac{V_s^3}{V_A^2}, \tag{19.26}$$

here the maximum value (19.26) is reached for $\theta_0 = \pi/6$. If

$$\mid v_y \mid > v_y^{max}, \tag{19.27}$$

the waves (19.24) are surface ones for all θ_0.

The surface perturbation, which decreases with distance from the x axis, does not transfer energy away from the sheet surface, because its amplitude equals zero at $y = \infty$. However, this

> surface wave enters into the total perturbation of the current sheet and its amplitude must be determined from the boundary conditions. In this sense the wave is classified as an outgoing one

(Kontorovich, 1959). As for the increasing perturbation, it is formally an incoming wave, but it must be discarded, since it tends to infinity as $y \to \infty$. Note that for this reason in the domain of the plasma outflow, where only one incoming wave is possible, the incoming waves are absent, for a given θ_0, when $\mid v_y \mid > v_y^{max}$.

Note that v_y^{max} coincides with the maximum value of the projection of the group velocity of a slow magnetoacoustic wave on the y axis, which in the approximation $V_s \ll V_A$ has the form

$$(V_{gr}^-)_y = \frac{V_s^3}{V_A^2} \sin\theta \cos^3\theta \, . \tag{19.28}$$

Moreover, this value is also reached for the angle $\theta = \pi/6$. So inequality (19.27) means that

> all slow waves are either incoming or outgoing, provided the plasma flows into or out of the reconnecting current sheet.

To solve the problem of evolutionarity of the current sheet we now have to derive boundary conditions. They relate the amplitudes of the perturbations with $k_y \ll 1/a$ (that interact with the sheet as with a discontinuity) on two sides of the surface.

However, as distinct from a one-dimensional discontinuity, the waves with $k_y \ll 1/a$ outside the current sheet may lead to the perturbations for which the inverse inequality is valid in the interior. Furthermore, since inside the sheet the dissipative effects are essential, the wave numbers of these perturbations have imaginary parts that tend to infinity in the limit $a/b \to 0$. This means that the magnitude of the perturbation increases without bound, and we cannot neglect the integrals in Equation (12.11), so

> the linearized one-dimensional boundary conditions generally do not hold at the reconnecting current sheet

(Markovskii and Somov, 1996a). This fact can be understood in the next Section from the analysis of the perturbations inside the current sheet.

19.3 Perturbations inside the RCS

19.3.1 Linearized dissipative MHD equations

Let us deduce the equations for the perturbed MHD quantities δQ inside the reconnecting current sheet, i.e. for $Q \sim Q^{in}$. In this case $y \lesssim a$.

Let us linearize the dissipative MHD equations. For $Q_z \equiv 0$ and $\partial\, \delta Q / \partial z \equiv 0$ the equations for δv_z and δB_z, which we put equal to zero, are separated from the equations for the other small quantities. In the latter we may neglect the derivatives $\partial p / \partial x$, $\partial\, \mathbf{B} / \partial x$, and $\partial \rho / \partial x$ in the approximation $a \ll b$. The left side of inequality (19.9) allows us also to neglect the derivative $\partial v_x / \partial x$.

Consider, for example, the linear equation of mass conservation

$$\frac{\partial\, \delta\rho}{\partial t} + \delta\rho \, \frac{\partial v_x}{\partial x} + \rho \, \frac{\partial\, \delta v_x}{\partial x} + \delta v_x \, \frac{\partial\, \rho}{\partial x} + v_x \, \frac{\partial\, \delta\rho}{\partial x} +$$

$$+ v_y \, \frac{\partial\, \delta\rho}{\partial y} + \delta\rho \, \frac{\partial v_y}{\partial y} + \delta v_y \, \frac{\partial \rho}{\partial y} + \rho \, \frac{\partial\, \delta v_y}{\partial y} = 0 \, . \tag{19.29}$$

Since, inside the reconnecting current sheet, the inhomogeneity of the velocity is two-dimensional then, together with the terms proportional to $\partial v_x / \partial x$, we have to neglect the terms with $\partial v_y / \partial y$. We can try to do it.

Let us choose the sign in formula (19.17) coinciding with the sign of v_x. Inside the current sheet $\mid v_x \mid$ is a growing function of $\mid y \mid$, but k_x is constant. So from formulae (19.10) and (19.17) it follows that $\mid \omega_{\parallel} \mid$ increases, while $\mid y \mid$ decreases, and satisfies the condition

$$\mid \omega_{\parallel} \mid > \mid \omega_{\parallel}^{ex} \mid . \tag{19.30}$$

Estimating

$$\frac{\partial\,\delta\rho}{\partial t} + v_x \frac{\partial\,\delta\rho}{\partial x} \sim \omega_{\parallel}\,\delta\rho\,, \qquad \frac{\partial v_y}{\partial y} \sim \frac{v_y^{ex}}{a}\,,$$

we get from (19.30) and the left side of (19.9) that

$$\frac{\partial\,\delta\rho}{\partial t} + v_x \frac{\partial\,\delta\rho}{\partial x} \gg \delta\rho\,\frac{\partial v_y}{\partial y}\,, \qquad \text{q.e.d.}$$

If the other sign in (19.17) is chosen, then a value of y exists for which $\omega_{\parallel} = 0$ and this inequality does not hold.

Similar reasoning is valid for the other equations. Hence $\partial Q/\partial x = 0$ in the zeroth order in terms of the small parameters given by relation (19.9). Besides, we put $\partial Q/\partial t = 0$ in all equations.

Following Syrovatskii (1956), let us substitute $\partial\,\delta Q/\partial t$ by

$$-\,\mathrm{i}\,\omega\left(\delta Q - \xi\,\frac{\partial Q}{\partial y}\right) \equiv -\,\mathrm{i}\,\omega\,\hat{\mathcal{D}}Q\,, \tag{19.31}$$

and $\partial\,\delta Q/\partial x$ by $\mathrm{i}\,k_x\,\hat{\mathcal{D}}Q$, where ξ is the displacement of the current sheet as a unit. Then we obtain the set of *linear ordinary* differential equations with respect to y

$$\mathrm{i}\,\omega_{\parallel}\,\hat{\mathcal{D}}\rho = \mathrm{i}\,k_x\,\rho\,\hat{\mathcal{D}}v_x + (\rho\,\delta v_y)' + v_y\,\delta\rho'\,, \tag{19.32}$$

$$\mathrm{i}\,k_x\,\hat{\mathcal{D}}B_x + \delta B_y' = 0\,, \tag{19.33}$$

$$\mathrm{i}\,\omega_{\parallel}\,\rho\,\hat{\mathcal{D}}v_x = \mathrm{i}\,k_x\,\hat{\mathcal{D}}p + \rho\,v_y\,\delta v_x' - \frac{B_x'\,\delta B_y}{4\pi} + v_x'\,\rho\,\delta v_y\,, \tag{19.34}$$

$$\mathrm{i}\,\omega_{\parallel}\,\rho\,\delta v_y = \delta\left(p + \frac{B_x^2}{8\pi}\right)' + \rho\,v_y\,\delta v_y' - \mathrm{i}\,k_x\,\frac{B_x\,\delta B_y}{4\pi}\,, \tag{19.35}$$

$$\mathrm{i}\,\omega_{\parallel}\,\hat{\mathcal{D}}p = \mathrm{i}\,k_x\,\gamma p\,\hat{\mathcal{D}}v_x + \gamma p\,\delta v_y' +$$

$$+\,\delta\,(p' v_y) - \frac{(\gamma-1)}{2\pi}\,\nu_m\,B_x'\,\delta B_x'\,, \tag{19.36}$$

$$\mathrm{i}\,\omega_{\parallel}\,\hat{\mathcal{D}}B_x = (B_x\,\delta v_y)' + v_y\,\delta B_x' - v_x'\,\delta B_y - \nu_m\,\delta B_x''\,, \tag{19.37}$$

where the prime denotes the differentiation with respect to y. Here we make use of the equality

$$p + \frac{B_x^2}{8\pi} = \text{const}\,, \tag{19.38}$$

which follows from the y component of the unperturbed momentum equation.

19.3.2 Boundary conditions

Under certain restrictions on the unperturbed MHD quantities Q and the frequency ω the *boundary conditions* (conservation laws), which relate the amplitudes of the small perturbations on both sides of the current sheet, may be deduced from the system of linear Equations (19.32)–(19.37).

For a one-dimensional discontinuity these conditions are obtained as a result of integrating the linear equations over the thickness of the domain in which the unperturbed quantities change substantially, and allowing this thickness (the thickness $2a$ of the current sheet in Fig. 19.3) to tend to zero.

Let us integrate, for example, the induction Equation (19.37), substituting $v_x' = -\omega_{\parallel}' / k_x$ (see definition (19.10)) and δB_y from Equation (19.33)

$$\mathrm{i}\,\omega_{\parallel}^{ex} \int_{-a}^{+a} \delta B_x \, dy = \left\{ B_x \left(\delta v_y + \mathrm{i}\,\omega_{\parallel} \xi \right) \right\} +$$

$$+ \int_{-a}^{+a} v_y \, \delta B_x' \, dy - \nu_m \left\{ \delta B_x' \right\}. \tag{19.39}$$

Here and below, as in Sections 11.1 and 11.2, the braces denote the jump of a quantity over a discontinuity. Supposing that δQ varies only slightly inside the discontinuity, if $k_y^{ex} a \ll 1$ outside it, we can, in particular, estimate the integral proportional to $\omega_{\parallel}^{ex}$:

$$\omega_{\parallel}^{ex} \int_{-a}^{+a} \delta B_x \, dy \sim \omega_{\parallel}^{ex} \, \delta B_x^{ex} \, a \, .$$

Let us compare this expression with the jump

$$\left\{ B_x \, \delta v_y \right\} \sim B_x^{ex} \, \delta v_y^{ex} .$$

In the case under study the requirement $k_y^{ex} a \ll 1$ is satisfied for the waves (19.23) and (19.24). The relationship between the perturbations δQ in such waves, in approximation (19.6) and (19.9), is given by the formulae:

$$\delta p \sim V_s^2 \, \delta \rho \, , \qquad \delta v_x \sim V_s \, \frac{\delta \rho}{\rho} \, , \qquad \delta B_x \sim B_x \left(\frac{V_s}{V_{\mathrm{A}}} \right)^2 \frac{\delta \rho}{\rho} \, ,$$

$$\delta v_y \sim V_s \left(\frac{V_s}{V_{\mathrm{A}}} \right)^2 \frac{\delta \rho}{\rho} \, , \quad \text{and} \quad \delta B_y \sim B_x \left(\frac{V_s}{V_{\mathrm{A}}} \right)^2 \frac{\delta \rho}{\rho} \, . \tag{19.40}$$

Taking (19.40) into account, we find that the condition

$$\omega_{\parallel}^{ex} \int_{-a}^{+a} \delta B_x \, dy \ll \{ B_x \, \delta v_y \}$$

coincides with the inequality $k_y^{ex} a \ll 1$, i.e. with the right side of (19.9). Similar reasoning for the other terms in Equation (19.37) leads to the following boundary condition

$$\left\{ B_x \left(\delta v_y + \mathrm{i} \omega_{\parallel} \xi \right) \right\} = 0 . \tag{19.41}$$

The application of this approach to Equation (19.33) gives

$$\{ \delta B_y - \mathrm{i} \, k_x \, B_x \, \xi \} = 0 . \tag{19.42}$$

As in the magnetoacoustic waves, in approximation (19.9)

$$\delta v_y = - \frac{\omega_{\parallel} \, \delta B_y}{k_x \, B_x} , \tag{19.43}$$

Equations (19.41) and (19.42) are satisfied if

$$\delta B_y = \mathrm{i} \, k_x \, \xi \, B_x^{ex} , \tag{19.44}$$

and, consequently,

$$\delta v_y = - \mathrm{i} \, \omega_{\parallel}^{ex} \, \xi . \tag{19.45}$$

As distinct from a one-dimensional discontinuity, δQ changes substantially inside the reconnecting current sheet. It will be shown below that the perturbation with $k_y^{ex} \ll 1/a$ outside the current sheet may lead to perturbations inside it, for which $k_y^{in} \gg 1/a$ and k_y^{in} has an imaginary part. These perturbations increase or decrease exponentially on the characteristic length, which is much smaller than a. So the above estimations of the terms in Equation (19.37) are generally not valid.

19.3.3 Dimensionless equations and small parameters

To deduce the boundary conditions on the current sheet as on the surface of a discontinuity, let us obtain the solutions of the set (19.32)–(19.37) inside the sheet for given ω and k_x. Assume that outside the sheet only the amplitudes of the waves with $k_y^{ex} \ll 1/a$ differ from zero. Let us bring Equations (19.32)–(19.37) to a dimensionless form by the following substitution of variable and unknown functions:

$$y = a \, \tilde{y} , \quad Q = Q^{ex} \, \tilde{Q} , \quad \delta Q = \delta Q^{ex} \, \delta \tilde{Q} , \tag{19.46}$$

$$\xi = \frac{\delta v_y^{ex}}{\omega_{\parallel}^{ex}}\,\tilde{\xi}\,, \qquad k_x = \frac{\omega_{\parallel}^{ex}}{V_s^{ex}}\,\tilde{k}_x\,, \tag{19.47}$$

$$\delta v_y = -\mathrm{i}\,\xi\,\omega_{\parallel} + \frac{a\,\omega_{\parallel}^{ex}}{V_s^{ex}}\;\delta v_y^{ex}\,\tilde{\omega}_{\parallel}\,\delta\tilde{v}_y\,, \tag{19.48}$$

$$\delta B_y = \mathrm{i}\,k_x\,\xi B_x + \frac{a\,\omega_{\parallel}^{ex}}{V_s^{ex}}\;\delta B_y^{ex}\,\delta\tilde{B}_y\,. \tag{19.49}$$

Here the quantities δQ^{ex} are related by formula (19.40), the tilde denotes the dimensionless functions and the expressions for δv_y and δB_y contain the boundary values (19.44) and (19.45) in an explicit form.

Let us insert expressions (19.46)–(19.49) into Equations (19.32)–(19.37) and introduce the following four small parameters in accordance with the basic assumptions (19.9) and (19.11):

$$\varepsilon_0 = \frac{v_y^{ex}}{a\,\omega_{\parallel}^{ex}}\,, \quad \varepsilon_1 = \frac{a\,\omega_{\parallel}^{ex}}{V_s^{ex}}\,, \quad \varepsilon_2 = \frac{v_y^{ex}}{V_s^{ex}}\,, \quad \varepsilon_3 = \left(\frac{V_s^{ex}}{V_A^{ex}}\right)^2. \tag{19.50}$$

As a result, we obtain equations describing the dimensionless functions,

$$\mathrm{i}\,\tilde{\omega}_{\parallel}\,\delta\tilde{\rho} = \mathrm{i}\,\tilde{k}_x\,\tilde{\rho}\,\delta\tilde{v}_x + \varepsilon_3\,(\tilde{\rho}\,\tilde{\omega}_{\parallel}\,\delta\tilde{v}_y\,)' + \varepsilon_0\,\tilde{v}_y\,\delta\tilde{\rho}'\,, \tag{19.51}$$

$$\mathrm{i}\,\tilde{k}_x\,\delta\tilde{B}_x + \delta\tilde{B}_y' = 0\,, \tag{19.52}$$

$$\begin{aligned} \mathrm{i}\,\tilde{\omega}_{\parallel}\,\tilde{\rho}\,\delta\tilde{v}_x &= \mathrm{i}\,\tilde{k}_x\,\delta\tilde{p} - \frac{1}{\tilde{k}_x}\,\varepsilon_3\,\tilde{\omega}_{\parallel}\,\tilde{\omega}_{\parallel}'\,\tilde{\rho}\,\delta\tilde{v}_y - \\ &\quad -\tilde{B}_x'\,\delta\tilde{B}_y + \varepsilon_0\,\tilde{v}_y\,\tilde{\rho}\,\delta\tilde{v}_x'\,, \end{aligned} \tag{19.53}$$

$$\begin{aligned} \left(\delta\tilde{p} + \tilde{B}_x\,\delta\tilde{B}_x\right)' &= \varepsilon_2\,\varepsilon_3\,\tilde{\rho}\,\tilde{v}_y\left[\mathrm{i}\,\tilde{\xi}\,\tilde{\omega}_{\parallel}' - \varepsilon_1\left(\tilde{\omega}_{\parallel}\,\delta\tilde{v}_y\right)'\right] + \\ &+\varepsilon_1\,\varepsilon_3\,\tilde{\omega}_{\parallel}^2\,\tilde{\rho}\left(\tilde{\xi} + \mathrm{i}\,\varepsilon_1\,\delta\tilde{v}_y\right) - \varepsilon_1\,\tilde{k}_x\,\tilde{B}_x\left(\tilde{k}_x\,\tilde{\xi}\tilde{B}_x - \mathrm{i}\,\varepsilon_1\,\delta\tilde{B}_y\right), \end{aligned} \tag{19.54}$$

$$\begin{aligned} \mathrm{i}\,\tilde{\omega}_{\parallel}\,\delta\tilde{p} &= \mathrm{i}\,\tilde{k}_x\,\tilde{p}\,\delta\tilde{v}_x + \varepsilon_3\left[\tilde{p}\,\left(\tilde{\omega}_{\parallel}\,\delta\tilde{v}_y\right)' + \frac{1}{\gamma}\,\tilde{\omega}_{\parallel}\,\tilde{p}'\,\delta\tilde{v}_y\right] + \\ &\quad + \varepsilon_0\left[\tilde{v}_y\,\delta\tilde{p}' - 2\,(\gamma-1)\,\tilde{B}_x'\,\delta\tilde{B}_x'\right], \end{aligned} \tag{19.55}$$

$$\begin{aligned} \mathrm{i}\,\tilde{\omega}_{\parallel}\,\delta\tilde{B}_x &= \left(\tilde{B}_x\,\tilde{\omega}_{\parallel}\,\delta\tilde{v}_y\right)' + \frac{1}{\tilde{k}_x}\,\tilde{\omega}_{\parallel}'\,\delta\tilde{B}_y + \\ &\quad +\varepsilon_0\left(\tilde{v}_y\,\delta\tilde{B}_x' - \delta\tilde{B}_x''\right). \end{aligned} \tag{19.56}$$

This is the complete set of dimensionless equations valid on the reconnecting current sheet as a discontinuity surface.

19.3.4 Solution of the linearized equations

Since we are interested in the solutions of the set of Equations (19.51)–(19.56) in approximation (19.9), let us allow the small parameters ε_i (except ε_3) to tend to zero. Then the equations reduce to the following simpler ones:

$$\mathrm{i}\,\tilde{\omega}_{\parallel}\,\delta\tilde{\rho} = \mathrm{i}\,\tilde{\rho}\,\delta\tilde{v}_x\,, \tag{19.57}$$

$$\mathrm{i}\,\delta\tilde{B}_x + \delta\tilde{B}_y' = 0\,, \tag{19.58}$$

$$\mathrm{i}\,\tilde{\omega}_{\parallel}\,\tilde{\rho}\,\delta\tilde{v}_x = \mathrm{i}\,\delta\tilde{p} - \varepsilon_3\,\tilde{\omega}_{\parallel}\,\tilde{\omega}_{\parallel}'\,\tilde{\rho}\,\delta\tilde{v}_y - \tilde{B}_x'\,\delta\tilde{B}_y\,, \tag{19.59}$$

$$\left(\delta\tilde{p} + \tilde{B}_x\,\delta\tilde{B}_x\right)' = 0\,, \tag{19.60}$$

$$\mathrm{i}\,\tilde{\omega}_{\parallel}\,\delta\tilde{p} = \mathrm{i}\,\tilde{p}\,\delta\tilde{v}_x + \varepsilon_3\left[\tilde{p}\left(\tilde{\omega}_{\parallel}\,\delta\tilde{v}_y\right)' + \frac{1}{\gamma}\,\tilde{\omega}_{\parallel}\,\tilde{p}'\,\delta\tilde{v}_y\right], \tag{19.61}$$

$$\mathrm{i}\,\tilde{\omega}_{\parallel}\,\delta\tilde{B}_x = \left(\tilde{B}_x\,\tilde{\omega}_{\parallel}\,\delta\tilde{v}_y\right)' + \tilde{\omega}_{\parallel}'\,\delta\tilde{B}_y\,. \tag{19.62}$$

The terms proportional to ε_3 are retained in Equations (19.59) and (19.61), since inside the current sheet the quantities

$$\left(\tilde{\omega}_{\parallel}',\ \tilde{\omega}_{\parallel}\right) \lesssim 1/\sqrt{\varepsilon_3}$$

(see (19.7)) and $(\tilde{p},\ \tilde{p}\,') \sim 1/\varepsilon_3$ (see equality (19.38)). Besides, the expression for $\tilde{k}_x$, which follows from (19.18) and (19.19), is used

$$\tilde{k}_x = 1 + O(\varepsilon_2) + O(\varepsilon_3)\,. \tag{19.63}$$

In the set (19.57)–(19.62) the Equations (19.57) and (19.59) are not differential, but serve as the algebraic definitions of the functions $\delta\tilde{v}_x$ and $\delta\tilde{\rho}$. After the substitution of $\delta\tilde{B}_x$ from Equation (19.58) to (19.62), the latter becomes the full derivative with respect to $\tilde{y}$ and, by integrating, is brought to the form

$$\delta\tilde{B}_y + \tilde{B}_x\,\delta\tilde{v}_y = 0\,. \tag{19.64}$$

The constant of integration in this equation is put equal to zero, as the perturbation outside the sheet represents the superposition of magnetoacoustic waves, for which (19.43) holds. The integration of Equation (19.60) gives

$$\delta\tilde{p} + \tilde{B}_x\,\delta\tilde{B}_x = C_0\,. \tag{19.65}$$

The substitution of (19.59), (19.64) and (19.65) in Equation (19.61) reduces it to

$$\left[\varepsilon_3\,\tilde{p} + \tilde{B}_x^2\left(1 - \frac{\tilde{p}}{\tilde{\rho}\,\tilde{\omega}_{\parallel}^2}\right)\right]\delta\tilde{v}_y' + \left(\frac{1}{\gamma}\,\varepsilon_3\,\tilde{p}' + \tilde{B}_x\,\tilde{B}_x'\right)\delta\tilde{v}_y =$$

$$= \mathrm{i}\,C_0\left(1 - \frac{\tilde{p}}{\tilde{\rho}\,\tilde{\omega}_{\parallel}^2}\right). \tag{19.66}$$

Expressing the dimensionless values in the coefficient in front of $\delta\tilde{v}_y$ in terms of the dimensional ones, we find that they are equal to

$$\left(p + \frac{B_x^2}{8\pi}\right)'\frac{4\pi a}{(B_x^{ex})^2} = 0\,. \tag{19.67}$$

(see equality (19.38)).

Hence the solution of the set (19.58), (19.60)–(19.62) is

$$\delta\tilde{v}_y = \mathrm{i}\,C_0\int\frac{\left(1 - \tilde{p}/\tilde{\rho}\,\tilde{\omega}_{\parallel}^2\right)\,d\tilde{y}}{\varepsilon_3\,\tilde{p} + \tilde{B}_x^2\left(1 - \tilde{p}/\tilde{\rho}\,\tilde{\omega}_{\parallel}^2\right)} + C\,, \tag{19.68}$$

$$\delta\tilde{B}_y = -\,\tilde{B}_x\,\delta\tilde{v}_y\,, \tag{19.69}$$

$$\delta\tilde{B}_x = -\mathrm{i}\left(\tilde{B}_x\,\delta\tilde{v}_y\right)', \tag{19.70}$$

$$\delta\tilde{p} = C_0 - \tilde{B}_x\,\delta\tilde{B}_x\,. \tag{19.71}$$

The solution (19.68)–(19.71) has a singularity at the point $\tilde{y}_0$, in which

$$\tilde{A} \equiv \varepsilon_3\,\tilde{p} + \tilde{B}_x^2\left(1 - \frac{\tilde{p}}{\tilde{\rho}\,\tilde{\omega}_{\parallel}^2}\right) = 0\,, \tag{19.72}$$

and the function in the integral in (19.68) turns to infinity. However, it may be shown by expressing $\delta Q'$ in terms of δQ in the set (19.32)–(19.37) that it has a singularity only for $y = 0$, where $v_y = 0$. This means that in some neighborhood of $\tilde{y}_0$ we cannot neglect the small parameters in the set (19.51)–(19.56) and turn to (19.57)–(19.62). The vicinity of the point $\tilde{y}_0$ will be considered below.

Let us now find the remaining solutions of the set of Equations (19.51)–(19.56) in the domain where the formulae (19.68)–(19.71) are valid. Suppose, for the sake of definiteness, that $v_x^{in} \sim V_{\mathrm{A}}^{ex}$ (see (19.7)), i.e. $\tilde{\omega}_{\parallel}^2 \sim 1/\varepsilon_3$. Such a relation holds if x is not close to 0 and $\pm b$. The solution (19.68)–(19.71) is valid when the expression in the integral in (19.68) is of order of unity. Since, inside the current sheet $\tilde{B}_x \lesssim 1$ and $\tilde{p} \sim 1/\varepsilon_3$, it follows from (19.68) and (19.72), that in this case

$$\tilde{A} \sim 1 . \tag{19.73}$$

Then the remaining solutions of the set (19.51)–(19.56) satisfy the WKB approximation inside the reconnecting current sheet and may be found from the dispersion Equation (19.20). Let us express the dimensionless quantities in $\tilde{A}$ in terms of the dimensional ones and take into account that $k_x = \omega_{\|}^{ex} / V_s^{ex}$. Then we find that the quantity $\tilde{A}$ is related with the coefficient in front of k_y^2 in dispersion Equation (19.20) in the following way:

$$A \sim \omega_{\|}^2 \left(V_{\rm A}^{ex} \right)^2 \tilde{A} . \tag{19.74}$$

Under condition (19.73) in the zeroth order in terms of the small parameters ε_i (see definition (19.50)) the solutions of Equation (19.20) take on the form (19.21) and

$$k_y^d = \frac{\omega_{\|}}{v_y} , \tag{19.75}$$

$$k_y^- = \pm \sqrt{\frac{{\rm i} A}{V_s^2 \, \nu_m \, \omega_{\|}}} , \tag{19.76}$$

$$k_y^* = \frac{1}{A} \left[\omega_{\|} \, v_y \, F \pm \sqrt{\omega_{\|}^2 v_y^2 \, F^2 - A \left(k_x^2 \, A - \omega_{\|}^4 \right)} \right] , \tag{19.77}$$

where $F = V_{\perp}^2 \, k_x^2 - 2 \, \omega_{\|}^2$.

From the basic inequality (19.9) it follows that the wave vectors (19.21), (19.75), and (19.76) satisfy the WKB approximation inside the current sheet. The dispersion equation is valid for them, as in the limit $k_y \gg 1/a$ the terms with the derivatives of unperturbed quantities in Equations (19.32)–(19.37) are negligible.

The expressions (19.42), (19.75), and (19.76) give us four solutions of the set of Equations (19.32)–(19.37). By contrast, the perturbations (19.77) do not satisfy the WKB approximation, since they have $1/k_y \, a \to 0$. In this case we cannot neglect the derivatives of unperturbed quantities in the set of Equations (19.32)–(19.37), so we cannot use Equation (19.20). These perturbations are described by formulae (19.68)–(19.71).

Thus we have shown that

> there are four perturbations, which satisfy the WKB approximation inside the current sheet, regardless of the value of k_x.

Recall that outside the current sheet there are also four of such perturbations, in the case of normal propagation, but in the case of oblique propagation there are three. Therefore in the latter case the perturbations with $k_y \ll 1/a$ and $k_y \gg 1/a$ transform to each other.

19.4 Solution on the boundary of the RCS

To obtain the boundary conditions it is necessary to determine the value of the perturbation on the boundary of the current sheet, i.e. for $Q = Q^{ex}$. In this case $a \ll y \ll 1/k_y^{ex}$. If $Q = Q^{ex}$, then the solution (19.68)–(19.71) is not valid, since the coefficients in Equation (19.66) are much smaller than unity (see definitions (19.46)) and the small parameters cannot be neglected in deducing of this equation.

Let us find the solutions of Equations (19.51)–(19.56) in the neighborhood of the boundary of the RCS in the domain

$$\tilde{Q} \sim 1 . \tag{19.78}$$

Note that as $p^{in} \gg p^{ex}$ and $\omega_{\parallel}^{in} \gg \omega_{\parallel}^{ex}$, the value of $\tilde{y}$ exists, for which $\tilde{p} \gg 1$ and $\tilde{\omega}_{\parallel} \gg 1$, although for $\tilde{y} \gg 1$ always $\tilde{Q}'/\tilde{Q} \ll 1$.

Substitute Equation (19.52) in (19.56) and then substitute (19.56) and (19.53) in Equation (19.54), in the same way as for deduction of (19.66), but hold the terms proportional to the small parameter ε_0

$$\mathrm{i}\tilde{\omega}_{\parallel}\left(1 - \frac{\tilde{p}}{\tilde{\rho}\tilde{\omega}_{\parallel}^2}\right)\delta\tilde{p} = \tilde{\omega}_{\parallel}\varepsilon_3\left(\tilde{p}\,\delta\tilde{v}_y' + \frac{1}{\gamma}\tilde{p}'\,\delta\tilde{v}_y\right) -$$

$$- \frac{\tilde{p}}{\tilde{\rho}\tilde{\omega}_{\parallel}}\tilde{B}_x'\,\delta\tilde{B}_y + \varepsilon_0\,\tilde{v}_y\left(\frac{\tilde{p}}{\tilde{\omega}_{\parallel}}\,\delta\tilde{v}_x' + \delta\tilde{p}'\right). \tag{19.79}$$

Here we use (19.63) and the inequality $\varepsilon_0 \ll (\varepsilon_2, \varepsilon_3)$, which follows from condition (19.9).

As the derivatives $\delta\tilde{v}_x'$ and $\delta\tilde{p}'$ appear in (19.78) with small parameters, in the first order they may be expressed from Equations (19.59) and (19.60), which do not contain small parameters. Let us integrate Equation (19.59) and use (19.64) and (19.65). Then, taking into account that $\tilde{Q}' \ll 1$ and considering (19.67), we find the equation describing the function $\delta\tilde{v}_y$,

$$\mathrm{i}\,\varepsilon_0\,\tilde{B}_x^2\,\tilde{v}_y\left(1 + \frac{\tilde{p}}{\tilde{\rho}\tilde{\omega}_{\parallel}^2}\right)\delta\tilde{v}_y'' + \tilde{\omega}_{\parallel}\,\tilde{A}\,\delta\tilde{v}_y' = \mathrm{i}\,C_0\,\tilde{\omega}_{\parallel}\left(1 - \frac{\tilde{p}}{\tilde{\rho}\tilde{\omega}_{\parallel}^2}\right) \tag{19.80}$$

(cf. Equation (19.66)). Three cases differ.

(a) Let $1 - \tilde{p}/\tilde{\rho}\tilde{\omega}_{\|}^{2} \gg \varepsilon_0$, then $\tilde{A} \gg \varepsilon_0$ (see definition (19.72)), as in the domain (19.78) $\varepsilon_3 \tilde{p} \ll \varepsilon_0$, and Equation (19.66) is valid.

(b) Let $1 - \tilde{p}/\tilde{\rho}\tilde{\omega}_{\|}^{2} \lesssim \varepsilon_0$, then $\tilde{A} \lesssim \varepsilon_0$ and all the terms in Equation (19.79) are essential. In this case, in the first order, it is sufficient to substitute $\delta\tilde{p}$ in Equation (19.79) from (19.65), but not from (19.54). So the small parameter ε_1 does not enter in Equation (19.80).

(c) On the boundary of the reconnecting current sheet ($|\,\tilde{Q}\,| = 1$),

$$1 - \frac{\tilde{p}}{\tilde{\rho}\tilde{\omega}_{\|}^{2}} = 0, \qquad \tilde{A} = 0,$$

and Equation (19.80) transforms to $\delta\tilde{v}_y'' = 0$. After integrating, this equality turns to the following one:

$$\delta\tilde{v}_y = C_* \tilde{y} + C\,. \tag{19.81}$$

Expression (19.81) together with (19.69)–(19.71) defines three solutions of the set of Equations (19.51)–(19.56). The remaining three solutions for $|\,\tilde{Q}\,| = 1$ satisfy the WKB approximation with the wave vectors (19.12), (19.21), and (19.22).

* * *

Let us now return to the vicinity of the point $\tilde{y}_0$, in which $\tilde{A} = 0$. From Equation (19.38) and condition (19.7) it follows that the point $\tilde{y}_0$ may generally be situated either in the domain $\tilde{y} \lesssim 1$ or $\tilde{y} \gg 1$. If

$$\tilde{y}_0 \lesssim 1, \tag{19.82}$$

then the terms containing $\tilde{v}_y'$ appear in the equation for $\delta\tilde{v}_y$ with $\tilde{A} = 0$. As $\tilde{v}_y' \sim 1$, they are found to be comparable with the terms proportional to $\partial v_x/\partial x$, which we have neglected when deducing the set of Equations (19.32)–(19.37). Because of this, to determine $\delta\tilde{v}_y$ in the vicinity of $\tilde{y}_0$, in the present case, it is necessary to solve a partial differential equation.

Let

$$\tilde{y}_0 \gg 1, \tag{19.83}$$

then $\tilde{v}_y' \ll 1$ and for $\tilde{y} = \tilde{y}_0$, in the first order, $\delta\tilde{v}_y$ is described by an ordinary differential equation. In particular, in the domain (19.78), it is the Equation (19.80). It does not have a singularity for $\tilde{A} = 0$ and the solutions of the set of Equations (19.51)–(19.56) in the vicinity of $\tilde{y}_0$ are given by the formulae (19.81), (19.69)–(19.71), (19.12), (19.21), and (19.22).

Finally let us establish the correspondence between the perturbations outside and inside the current sheet. Assume that (19.83) holds and, for $\tilde{y} \lesssim 1$ (19.73) is true.

Solving the set of Equations (19.51)–(19.56) in the domain

$$1 \ll \left(\tilde{p}\,,\, \tilde{\omega}_{\parallel}^{2}\right) \ll 1/\varepsilon_3\,,$$

it may be shown that the following correspondence takes place. The perturbations, which are described by the wave vectors k_y^d from (19.12) and k_y^0 from (19.21) outside the reconnecting current sheet, *transform* into (19.76) and (19.21) inside it, i.e. represent the same roots of Equation (19.20) for the different values of $\tilde{y}$.

> The wave (19.22) transforms into one of the perturbations (19.76), with the sign '$-$' or '$+$' depending on the sign of v_y.

Hence the superposition of (19.23) and (19.24) corresponds to the superposition of (19.68)–(19.71) and the other perturbation (19.76).

Besides, the frequency $\omega_{\parallel}$ from the interval (19.9) may be chosen in such a way, that the solution proportional to C_0 exists inside the reconnecting current sheet for all $\tilde{y}$. In this case the solution proportional to C_*, in the domain (19.78), transforms, for $\tilde{y} \lesssim 1$ into the perturbation with the wave vector (19.76). Thus

> the three waves with $k_y^{ex} \ll 1/a$ (i.e., $\lambda_y^{ex} \gg a$) outside the reconnecting current sheet cause the perturbation inside it, for which $k_y^{in} \gg 1/a$ (i.e., $\lambda_y^{ex} \ll a$).

So now we can formulate the conditions of evolutionarity for the reconnecting current sheet.

19.5 The criterion of evolutionarity

19.5.1 One-dimensional boundary conditions

Let us now turn to the criterion of evolutionarity. With this end in view, deduce the boundary conditions on the reconnecting current sheet as a surface of a discontinuity. There are two possibilities.

(a) If the amplitudes of the perturbations (19.21), (19.75), and (19.76) with $k_y \gg 1/a$ inside the current sheet differ from zero, then the boundary conditions, similar to those which hold on one-dimensional discontinuities, do not exist on its surface. If this were not so, then the quantity δv_y would remain constant after a transition across the sheet, by virtue of condition (19.45). However, the magnitude of the perturbations (19.21), (19.75), and (19.76) changes substantially within the distance a and (19.45) is not valid in a general case.

(b) Consider below only such perturbations that the amplitudes of the modes (19.21), (19.75), and (19.76) equal zero. This requirement is obeyed by the solution of Equations (19.32)–(19.37), if the constant C_0 differs from zero, but the other constants equal zero (see the end of Section 19.4).

Let us obtain the boundary conditions which the solution proportional to C_0 satisfies. Due to (19.81), formulae (19.48) and (19.49) give the boundary values (19.44) and (19.45) for δv_y and δB_y. From (19.45) it follows that

$$\{\, \delta v_y \,\} = 0 \,. \tag{19.84}$$

As for condition (19.44), it is equivalent to (19.45) and does not result in an additional boundary condition. Expression (19.71) determines the second boundary condition

$$\left\{ \delta p + \frac{B_x \, \delta B_x}{4\pi} \right\} = 0 \,. \tag{19.85}$$

Finally formula (19.70) means that

$$\delta B_x = 0 \tag{19.86}$$

on both sides of the discontinuity, since $\delta \tilde{v}_y' = 0$ and $\tilde{B}_x' = 0$.

The appearance of the equality (19.86) is caused by the fact that we consider the perturbation, for which only the constant C_0 differs from zero, but not an arbitrary one. Given another perturbation is present inside the current sheet, the condition (19.86) is generally not satisfied. As δB_x in magnetoacoustic waves do not equal zero, condition (19.86) together with (19.84) and (19.85) represents four boundary conditions, relating the amplitudes of the waves outside the sheet. Note that equalities (19.57) and (19.58) do not give additional boundary conditions, since they are valid for the perturbations in magnetoacoustic waves.

19.5.2 Solutions of the boundary equations

Now we write Equations (19.84)–(19.86) in an explicit form, i.e. expressing all small quantities in terms of the perturbation of density. As was pointed out at the end of Section 19.4, the superposition of the waves (19.23) and (19.24) outside the reconnecting current sheet corresponds to the superposition of the solutions (19.68)–(19.71) and (19.76) inside it.

This being so, the waves (19.23) and (19.24) are present outside the current sheet, but the amplitudes of the waves (19.12), (19.21), and (19.22) equal zero, if inside it only the constant C_0 differs from zero. Using the relationship between the perturbations of MHD quantities in magnetoacoustic waves in approximation (19.9) we obtain from the boundary conditions (19.84)–(19.86), respectively

$$\sum_{i=1}^{3} \frac{k_{y+}^{(i)}}{(k^{(i)})^2} \left(\delta\rho_{+}^{(i)} + \delta\rho_{-}^{(i)} \right) = 0\,, \tag{19.87}$$

$$\sum_{i=1}^{3} \frac{1}{(k^{(i)})^2} \left(\delta\rho_{+}^{(i)} - \delta\rho_{-}^{(i)} \right) = 0\,, \tag{19.88}$$

$$\sum_{i=1}^{3} \left(\frac{k_y^{(i)}}{k^{(i)}} \right)^2 \delta\rho_{\pm}^{(i)} = 0\,. \tag{19.89}$$

Here the indexes + and − denote the quantities outside the sheet for $y = +\infty$ and $y = -\infty$, the index i specifies three waves (19.23) and (19.24); and it is taken into account that $k_{y+}^{(i)} = -k_{y-}^{(i)}$ due to the plasma flow symmetry.

Let us find the solutions of these equations for the cases of the inflowing and the outflowing of a plasma, i.e. determine the amplitudes of outgoing waves versus the amplitudes of incident ones.

If the plasma flows into the current sheet, then there are two outgoing waves (one on each side). As there are four equations, system (19.87)–(19.89) has solutions only for a definite relationship between the amplitudes of incident waves. If these amplitudes are arbitrary, then the set of Equations (19.87)–(19.89) does not have a solution. It means that for such perturbations condition (19.86) cannot be satisfied. Since equality (19.86) is valid always, when C_0 is the only constant which differs from zero, a violation of this equality results in the fact that the other constants, i.e. the amplitudes of the perturbations with $k_y^{in} \gg 1/a$, differ from zero. Hence, in this case, the boundary conditions do not exist on the surface of the current sheet, i.e. it is not a discontinuity, and the conclusion of its evolutionarity cannot be obeyed.

Let the plasma flow out from the current sheet. In this case there are four outgoing waves (two on each side). Denote them by the indexes $i = 1, 2$.

Then their amplitudes $\delta\rho_{\pm}^{(1,2)}$ are expressed in terms of the amplitudes $\delta\rho_{\pm}^{(3)}$ of incident waves in the following way

$$\delta\rho_{\pm}^{(1)} = -\frac{1}{2}\left(\frac{k^{(1)}}{k^{(3)}}\right)^2 \frac{k_y^{(2)} - k_y^{(3)}}{k_y^{(2)} - k_y^{(1)}} \times$$

$$\times\left[\frac{k_y^{(3)}}{k_y^{(1)}}\left(\delta\rho_{+}^{(3)} + \delta\rho_{-}^{(3)}\right) \pm \frac{k_y^{(2)} + k_y^{(3)}}{k_y^{(2)} + k_y^{(1)}}\left(\delta\rho_{+}^{(3)} - \delta\rho_{-}^{(3)}\right)\right], \tag{19.90}$$

$$\delta\rho_{\pm}^{(2)} = -\left(\frac{k^{(2)}}{k_y^{(2)}}\right)^2\left[\left(\frac{k_y^{(3)}}{k^{(3)}}\right)^2 \delta\rho_{\pm}^{(3)} + \left(\frac{k_y^{(1)}}{k^{(1)}}\right)^2 \delta\rho_{\pm}^{(1)}\right]. \tag{19.91}$$

In formula (19.90) all the quantities $k_y^{(i)}$ are taken for one side of the discontinuity. From (19.90) it follows that if $k_y^{(1)} = k_y^{(2)}$ and $k_y^{(2)} \neq k_y^{(3)}$, then $\delta\rho_{\pm}^{(1)}$ turns to infinity, i.e. the coefficients of refraction and reflection are not limited.

Let us find the conditions under which the wave vectors of two outgoing waves coincide. In Section 19.2 it was shown that if

$$|\, v_y^{ex} \,| < \frac{3\sqrt{3}}{16}\,\frac{V_s^{\,3}}{V_{\rm A}^{\,2}}, \tag{19.92}$$

then the resonant angle θ_0^* exists, for which the expression in the round brackets in formula (19.24) equals zero and two roots (19.24) coincide. This angle is determined by Equation (19.25).

Provided $\theta_0 = \theta_0^*$, both waves (19.24) are outgoing, since if the plasma flows out from the current sheet, then there is only one incoming wave. In the present case its wave vector is given by formula (19.23) and $k_y^{(2)} \neq k_y^{(3)}$. If condition (19.92) is not valid, then the expression in the round brackets in (19.24) is negative and the corresponding waves are surface ones for all θ_0 (see Section 19.2). In this case all wave vectors are different and $k_y^{(i)} \neq \pm\, k_y^{(j)}$ for $i \neq j$. So the coefficients of refraction and reflection are limited.

For the definite, but rather general, distribution of the unperturbed MHD properties inside the reconnecting current sheet the expressions describing the perturbation (and thus the transition between the perturbations with $k_y \ll 1/a$ and $k_y \gg 1/a$) can be found in an analytical form (Markovskii and Somov, 1996a). These solutions are represented schematically in Figure 19.4.

Horizontal solid and dotted lines represent the solutions with $k_y \ll 1/a$ and $k_y \gg 1/a$ respectively. Inclined lines represent the solutions that do not

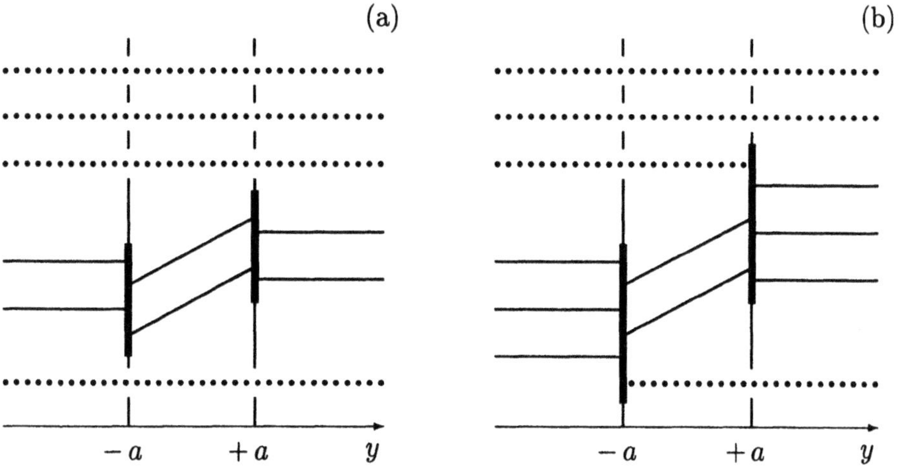

Figure 19.4: Schematic representation of solutions of the linear MHD equations in the case of normal (a) and oblique (b) propagation.

satisfy the WKB approximation. Superposition of perturbations on one side of the bold line $y = \pm a$ transforms to superposition of perturbations on the other side.

In the case of normal propagation the long waves, $k_y \ll 1/a$, do not transform to the short ones, $k_y \gg 1/a$, (see Fig. 19.4a). In this case the long waves interact with the current sheet as with a tangential discontinuity, i.e. as if v_y equals zero. The amplitudes of the waves satisfy the linearized boundary conditions for magnetoacoustic waves at a tangential discontinuity with $v_{x1} = v_{x2}$:

$$\left\{ \delta p + \frac{B_x \, \delta B_x}{4\pi} \right\} = 0 \, , \quad \{ \delta v_y \} = 0 \, . \tag{19.93}$$

There are thus two boundary equations and two outgoing waves (see Section 19.2.2) regardless of the sign of v_y. Moreover, these equations always have a unique solution, therefore the current sheet is evolutionary with respect to normally propagating waves.

Another situation arises in the case of oblique propagation. In this case long waves outside the sheet transform inside it to short waves. This imposes two additional boundary conditions on the perturbations that interact with the sheet as with a discontinuity, because for such perturbations the amplitudes of short waves must be equal zero. Therefore the reconnecting current sheet behaves like a discontinuity only with respect to a specially selected perturbation.

We emphasize that the conditions (19.93) appear as a result of the properties of the solutions of the linearized MHD equations, while the additional conditions occur due to the fact that we consider the perturbation which is not arbitrary. An otherwise additional condition generally does not hold.

With respect to these perturbations the problem of evolutionarity can be posed. However, the conclusions on non-evolutionarity are different for the domain of forward (direct) current, where the plasma flows into the current sheet, and for the domains of reverse current, where the plasma flows out.

19.5.3 Evolutionarity and splitting of current sheets

Therefore we have obtained the criterion of evolutionarity for reconnecting current sheets as a discontinuity with respect to linear magnetoacoustic waves.

If the plasma flows into the sheet (in the region DC of the forward current in Figures 19.1b and 19.3) or if inequality (19.92) does not hold, then **the conclusion of non-evolutionarity cannot hold**. In this case the current sheet either does not behave like a discontinuity or else the problem of its infinitesimal perturbation has a single solution. The last is the case when we can consider an ordinary problem of linear stability. For example, the question on the linear tearing instability always exists concerning the central part (the region of the direct forward current) of the reconnecting current sheet (see Chapter 20).

Let the relation (19.92) be valid, provided the plasma flows out from the sheet (in the regions RC of the reverse current in Figure 19.1b and 19.3), and the outflow velocity is less than the projection of the group velocity of a slow magnetoacoustic wave on the normal to the current sheet (see (19.92)). Then the perturbation exists, for which, firstly, the boundary conditions on the surface of the current sheet are true, and, secondly, the amplitudes of the outgoing waves are as large as is wished, compared with the amplitudes of the incident ones in the limit $\varepsilon_i \to 0$, i.e. when the conductivity is large enough.

Such a perturbation inside the reconnecting current sheet is the solution of the set of Equations (19.32)–(19.37) proportional to C_0, and is characterized by the resonant angle θ_0^* from (19.25) outside it. Thus the perturbation is not described by linear equations and the problem of its time evolution does not have not a single solution. Hence the current sheet is non-evolutionary, as **the initial perturbation of the MHD flow is not small**. This perturbation may be the splitting of the reconnecting current sheet into shock waves that are observed in the numerical experiments carried out by Brushlinskii, Zaborov, and Syrovatskii (1980), Podgornii and Syrovarskii (1981), Biskamp (1986, 1997).

Therefore we have found a possible cause of splitting of the reconnecting current sheet into a set of the one-dimensional MHD discontinuities observed in numerical experiments. Moreover we have obtained the condition under which the splitting takes place. This allows us to unify the two regimes of magnetic reconnection in current sheets: with attached shocks and without them. Such a unified model can be used to describe unsteady phenomena in cosmic plasma, which occur as a result of magnetic reconnection.

Chapter 20

The Tearing Instability of a Reconnecting Current Sheet

The tearing instability can play a significant role in reconnecting current sheets, but it is well stabilized in many cases of interest.

20.1 The origin of the tearing instability

20.1.1 Two necessary conditions

Among the host of instabilities appearing in a plasma with magnetic field, the tearing mode is of fundamental value for processes which transform 'free magnetic energy' into other kinds of energy. In a sense, the tearing instability is an integral part of magnetic reconnection. It is conceivable that the instability can play the role of a triggering mechanism for many of its essentially nonstationary manifestations in cosmic plasma – flares on the Sun and in magnetospheres of the Earth and other cosmic bodies.

The tearing instability has a universal character and arises in reconnecting current sheets over quite a wide range of their parameter values. In fact, it is seen from the 2D picture of the magnetic field lines shown in Figure 20.1a, that this state with the neutral current sheet at $y = 0$ is energetically high and hence it must tend to a lower one, depicted in Figure 20.1b.

Such a transition may be interpreted as a process of coalescence of parallel currents constituting the current sheet. However, for ideally conducting plasma, the process is impossible since it implies the displacement of field lines, leading to their tearing and the formation of closed loops – magnetic islands. This transition, i.e. the reconnection of field lines, is known to

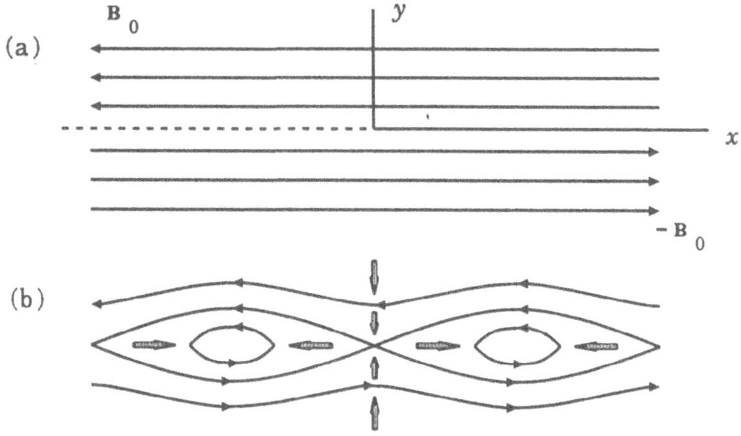

Figure 20.1: (a) Magnetic field 'reversal', a peculiarity of the configuration of field lines in a neutral current sheet. (b) Magnetic-field lines in the course of the tearing instability; the arrows show the plasma velocity directions.

be forbidden by the condition of magnetic lines freezing into plasma (Section 9.3). Such a restriction is removed given a finite (even if very high) electric conductivity. Thus

> for the tearing instability to develop, two conditions are necessary: (1) magnetic field reversal and (2) the availability of a finite electric conductivity.

The instability is called *tearing* because, as we have seen, its growth, once unbounded, causes the current sheet to tear into separate filaments.

20.1.2 Historical comments

Before giving an account of the theory of the tearing instability, let us briefly describe the history of the question. Dungey (1958) supposed that the availability of an X-type neutral line in a plasma with finite conductivity leads to the instability giving rise to the current concentration. This hypothesis was based on the consideration of a *non-equilibrium* configuration of the magnetic field with an X-line whose separatrix (forming the letter X) lines intersect at an angle not equal to $\pi/2$ (see also discussion of the paper by Zwingmann, Schindler, and Birn (1985) in Chapter 22).

The presence of the instability was experimentally found in configurations of a pinch type (Colgate and Furth, 1960), for which stability had been predicted by the ideal MHD theory. Using Dungey's mechanism, Furth (1961) qualitatively explained the current sheet tearing instability. Murty (1961) investigated the same process theoretically and found the presence of the tearing mode in a resistive current sheet for the low conductivity case. Finally, the theory of resistive MHD instabilities was thoroughly developed for the case of the neutral current sheet without plasma flows, in the famous work of Furth, Killen, and Rosenbluth (1963).

In the framework of the kinetic approach the first fundamental results on the tearing instability were obtained by Coppi, Laval, and Pellat (1966). They showed that the tearing instability arises from coupling between a negative energy wave and a dissipative process. Landau resonance of electrons inside and near the zero magnetic field plane was proposed to provide the appropriate dissipation mechanism (Section 20.6).

In parallel with the investigation of the tearing instability, mechanisms resulting in its stabilization were searched for. Why? – The point is that laboratory and numerical experiments, as well as astrophysical observations, contrary to theoretical predictions, allowed one to conclude that **reconnecting current sheets can be stable for a long time**. The appearance of such stable states is of paramount importance, in particular, for the physics of current sheets in the cosmic plasma.

Furth (1967) proposed the hypothesis that the tearing mode is suppressed by a small transversal magnetic field (i.e., perpendicular to the current sheet). Such a non-neutral current sheet, as pointed out by Pneuman (1974), cannot be topologically affected by an infinitesimal displacement, as opposed to a sheet that does not contain a transversal field. This suggests that a disturbance of *finite* amplitude is necessary to disturb the current sheet, i.e. the configuration could be *metastable*. The stabilizing effect of the transversal field was demonstrated in the frame of the kinetic approach by Schindler (1974), Galeev and Zelenyi (1975, 1976).

Janicke (1980, 1982) considered the same hypothesis in the context of MHD and drew the conclusion that the stabilizing influence was absent. This is the reason why a fundamental indecision as to the role of the transversal field remained for a long time. Somov and Verneta (1989) demonstrated a considerable stabilizing effect within the limits of the MHD approach. They also explained the reasons for negative results due to Janicke. Incidentally, Otto (1991), Birk and Otto (1991) once again confirmed the conclusion that, in the context of Janicke's model, the transversal component of the magnetic field does not change the tearing increment. A comparative review of alternative approaches is given, for example, in Somov and Verneta (1993). As we shall also see in Section 20.4, the transversal component of the magnetic field modifies the collisional tearing mode in such a way that it results in its stabilization.

Having finished this brief introduction, we come now to an account of the basic theory of the tearing instability.

20.2 The simplest problem and its analytic solution

20.2.1 The model and equations for small disturbances

We begin by obtaining a expression for the growth rate of a *pure* tearing instability without additional stabilizing or destabilizing effects. For this purpose, consider the case when the instability increment is much larger than the inverse time of magnetic diffusion τ_r (as will be shown in Section 20.5, once these quantities are of the same order, the effect of plasma compressibility becomes decisive). Provided diffusion may be ignored, plasma drift into the reconnecting current sheet becomes unimportant since its characteristic time is also τ_r. For the case $\omega \gg V/b$ (ω is the instability increment, V is the speed of plasma outflow from the current sheet, b is its half-width, see Figure 4.19), the plasma flow along the current sheet is negligible as well.

Let us consider the instability in a *linear* approximation:

$$f\,(\mathbf{r},t) = f_0\,(\mathbf{r}) + f_1\,(\mathbf{r},t)\,.$$

Unperturbed quantities in the frame of the simplest model depend only upon the y coordinate which is perpendicular to the current sheet (Fig. 20.1a):

$$f_0 = f_0\,(y)\,.$$

Hence small perturbations are of the form (20.1)

$$f_1\,(\mathbf{r},t) = f_1\,(y)\,\exp\left[\,\mathrm{i}\,(k_x x + k_z z) + \omega t\,\right], \tag{20.1}$$

provided $1/k_x \ll b$.

The set of MHD equations for an *incompressible* plasma with a finite conductivity σ is reduced to the following one:

$$\mathrm{curl}\left(\rho\,\frac{d\,\mathbf{v}}{dt}\right) = \mathrm{curl}\left(\frac{1}{4\pi}\,\mathrm{curl}\,\mathbf{B}\times\mathbf{B}\right),$$

$$\frac{\partial\mathbf{B}}{\partial t} = \mathrm{curl}\,(\mathbf{v}\times\mathbf{B}) - \mathrm{curl}\left(\frac{\eta}{4\pi}\,\mathrm{curl}\,\mathbf{B}\right),$$

$$\frac{\partial \rho}{\partial t} + \mathbf{v} \cdot \nabla \rho = 0 \,, \qquad \frac{\partial \eta}{\partial t} + \mathbf{v} \cdot \nabla \eta = 0 \,,$$

$$\operatorname{div} \mathbf{v} = 0 \,, \qquad \operatorname{div} \mathbf{B} = 0 \,.$$

Here $\eta = c^2/\sigma$ is the value proportional to magnetic diffusivity (8.24); the other symbols are conventional. This set gives the following equations for the perturbations:

$$\omega \operatorname{curl} (\rho_0 \mathbf{v}_1) = \operatorname{curl} \left\{ \frac{1}{4\pi} \left[(\mathbf{B}_0 \cdot \nabla) \mathbf{B}_1 + (\mathbf{B}_1 \cdot \nabla) \mathbf{B}_0 \right] \right\} ,$$

$$\omega \mathbf{B}_1 = (\mathbf{B}_0 \cdot \nabla) \mathbf{v}_1 - (\mathbf{v}_1 \cdot \nabla) \mathbf{B}_0 - \frac{1}{4\pi} (\nabla \eta_0 \times \operatorname{curl} \mathbf{B}_1 -$$

$$- \eta_0 \Delta \mathbf{B}_1 + \nabla \eta_1 \times \operatorname{curl} \mathbf{B}_0 - \eta_1 \Delta \mathbf{B}_0) \,,$$

$$\omega \rho_1 + \mathbf{v}_1 \cdot \nabla \rho_0 = 0 \,, \qquad \omega \eta_1 + \mathbf{v}_1 \cdot \nabla \eta_0 = 0 \,,$$

$$\operatorname{div} \mathbf{v}_1 = 0 \,, \qquad \operatorname{div} \mathbf{B}_1 = 0 \,.$$

These dimensional equations are reduced to two dimensionless equations containing y components of the velocity and magnetic field perturbations as unknown variables:

$$(\tilde{\rho} W')' = \alpha^2 \tilde{\rho}^2 W - \frac{S^2 \alpha^2}{p} (\alpha^2 F \Psi + F'' \Psi - F \Psi'') \,, \tag{20.2}$$

$$\Psi'' = \left(\alpha^2 + \frac{p}{\tilde{\eta}} \right) \Psi + \left(\frac{F}{\tilde{\eta}} + \frac{\tilde{\eta}' F'}{p \tilde{\eta}} \right) W. \tag{20.3}$$

Here

$$\Psi = \frac{B_{1y}}{B(a)} \,, \qquad W = -\mathrm{i} \, v_{1y} \, k \, \tau_r \,, \qquad \mu = \frac{y}{a} \,,$$

$$F = \frac{\mathbf{k} \cdot \mathbf{B}_0}{k \, B(a)} \,, \qquad k = \left(\mathbf{k}^2 \right)^{1/2} \,, \qquad \alpha = k a \,, \qquad \tau_r = \frac{4\pi a^2}{\langle \eta \rangle} \,,$$

$$\tau_{\mathrm{A}} = \frac{a \, (4\pi \langle \rho \rangle)^{1/2}}{B(a)} \,, \qquad S = \frac{\tau_r}{\tau_{\mathrm{A}}} \,, \qquad p = \omega \tau_r \,, \qquad \tilde{\eta} = \frac{\eta_0}{\langle \eta \rangle} \,, \qquad \tilde{\rho} = \frac{\rho_0}{\langle \rho \rangle} \,.$$

Thus we intend to solve Equations (20.2) and (20.3). As will be seen from the final results, the tearing instability is a *long-wave* mode:

$$\alpha^2 \ll 1 \,. \tag{20.4}$$

Hence this case is considered from the beginning. For definiteness, the following distribution of the unperturbed field is chosen:

$$\mathbf{B}_0 = F(\mu)\ \mathbf{e}_x\,,$$

where

$$F(\mu) = \begin{cases} -1\,, & \mu < -1\,, \\ \mu\,, & -1 < \mu < 1\,, \\ 1\,, & \mu > 1\,. \end{cases}$$

Examine the instability mode with the *fastest* growth, for which the condition

$$\mathbf{k} \parallel \mathbf{B}_0$$

holds. Assume that

$$S \gg 1\,, \tag{20.5}$$

i.e., the plasma is highly-conductive (compare definition of S with definition of the magnetic Reynolds number (8.36) where $v = V_{\rm A}$, $L = a$). What this means is that

> dissipative processes in such a regime are not large in magnitude, while they play a principle role in the tearing instability,

as was mentioned in the previous section.

20.2.2 The external non-dissipative region

Starting from some distance y from the neutral plane $y = 0$ of the current sheet, the dissipative processes may be ignored. We shall call this region the *external non-dissipative* one. In the limiting case

$$\boxed{S = \frac{\tau_r}{\tau_{\rm A}} = \frac{V_{\rm A}\,a}{\nu_m} \rightarrow \infty\,,}$$

Equation (20.2) is simplified to

$$\Psi'' - \left(\alpha^2 + \frac{F''}{F} \right) \Psi = 0\,. \tag{20.6}$$

The function Ψ should be even for reasons of symmetry:

$$\Psi\,(-\mu) = \Psi\,(\mu)\,. \tag{20.7}$$

The boundary condition for the sought-after function must be formulated for $\mu \to \infty$:

$$\Psi \to 0 . \tag{20.8}$$

Equation (20.6), under conditions (20.7)–(20.8), has the following solution, once $\mu \neq 0$:

$$\Psi = \begin{cases} A \exp\left[\alpha(\mu+1)\right], & \mu < -1, \\ A\{\left[\cosh\alpha + (1-\alpha^{-1})\sinh\alpha\right]\cosh\alpha\,\mu + \\ + \left[\sinh\alpha + (1-\alpha^{-1})\cosh\alpha\right]\sinh\alpha\,\mu\}, & -1 < \mu < 0, \\ \Psi(-\mu), & \mu > 0 . \end{cases} \tag{20.9}$$

Here A is an arbitrary constant.

The derivative Ψ' suffers a rupture at the point $\mu = 0$, with

$$\Delta' = \frac{\Psi'}{\Psi}\Big|_{-0}^{+0} \approx \frac{2}{\alpha} \tag{20.10}$$

for $\alpha^2 \ll 1$. This fact signifies that the solution applicable in the external non-dissipative region corresponds to a singular current at the $\mu = 0$ plane.

The approximation $S \to \infty$ is not applicable in a neighbourhood of the point $\mu = 0$. This will be called the *internal dissipative* region. Outside this region the solution is described by the function (20.9) which, for $\mu \to 0$ (once $\alpha^2 \ll 1$), gives the asymptotic expression

$$\Psi \sim \text{const}\left(1 + \frac{1}{\alpha}\,|\,\mu\,|\right) . \tag{20.11}$$

20.2.3 The internal dissipative region

Consider now the neighbourhood of the point $\mu = 0$ where the condition $S \to \infty$ does not hold. Since this region is sufficiently small, the quantities $\tilde{\rho}$ and $\tilde{\eta}$ may be assumed to vary weakly inside it. On using this assumption and making the change of variables

$$\theta = \left(\frac{\alpha^2 S^2}{p}\right)^{1/4} \mu , \tag{20.12}$$

$$Z = \Psi'' , \tag{20.13}$$

the set of Equations (20.2)–(20.3) results in the equation for the function $Z = Z(\theta)$

$$Z''' = \left(\nu + \theta^2\right) Z' + 4\theta Z\,. \tag{20.14}$$

This equation must be supplemented by the conditions

$$\begin{aligned} &Z(-\theta) = Z(\theta)\,,\\ &Z \to 0 \quad \text{for} \quad \theta \to \infty\,. \end{aligned} \tag{20.15}$$

We find from (20.14)–(20.15) that the sought-after function $Z(\theta)$ has the following asymptotic behaviour for $\theta \gg 1$ ($\theta \to \infty$):

$$Z \sim A_1 \exp\left(-\theta^2/2\right) + B\,\theta^{-4}\,. \tag{20.16}$$

For $\theta < 1$ the function $Z(\theta)$ has no singularities and can be expanded in a Taylor series.

In order to obtain the dispersion relation the integrals

$$I_0 = \int_0^{+\infty} \Psi''\,d\mu\,, \qquad I_1 = \int_0^{+\infty} \Psi''\mu\,d\mu \tag{20.17}$$

have to be evaluated. On normalizing the function $Z(\theta)$ by the condition

$$Z(0) = 1\,,$$

we find from (20.16) that

$$\tilde{I}_0 = \int_0^{+\infty} Z(\theta)\,d\theta \approx 1\,, \qquad \tilde{I}_1 = \int_0^{+\infty} Z(\theta)\,\theta\,d\theta \approx 1\,. \tag{20.18}$$

The integrals (20.17) are expressed through (20.18).

For the function $\Psi(\theta)$, we have

$$\Psi(\theta) = \int_0^{\theta} d\theta_1 \int_0^{\theta_1} Z(\theta_2)\,d\theta_2\,,$$

whence

$$\Psi(\mu) \sim \text{const}\left(1 + \frac{I_0}{(1/p) - I_1}\,|\,\mu\,|\right) \tag{20.19}$$

for $\theta \to \infty$. Here it is taken into account that

$$\Psi''_{\mu\mu}(0) = p\,\Psi(0)\,.$$

20.2.4 Matching of the solutions and the dispersion relation

As is seen from the asymptotic solution (20.16), the approximation $S \to \infty$ is valid once $\mu \gg \varepsilon_0$, where

$$\varepsilon_0 = \left(\frac{p}{\alpha^2 S^2}\right)^{1/4} . \tag{20.20}$$

Hence the function (20.19) must coincide with (20.12). Equating them results in the *dispersion* equation

$$\left(1 - \frac{p^{3/2}}{\alpha S}\right) - p\alpha\left(\frac{p}{\alpha^2 S^2}\right)^{1/4} = 0 . \tag{20.21}$$

There is no difficulty in understanding that, given the ratio

$$\frac{p^{3/2}}{\alpha S} \ll 1 , \tag{20.22}$$

the equation is reduced to

$$p \approx \left(\frac{S}{\alpha}\right)^{2/5} , \tag{20.23}$$

while given

$$p\alpha\left(\frac{p}{\alpha^2 S^2}\right)^{1/4} \ll 1 , \tag{20.24}$$

it reduces to

$$p = (\alpha S)^{2/3} . \tag{20.25}$$

Conditions (20.22) and (20.24) are equivalent to

$$p\alpha^2 \gg 1 \tag{20.26}$$

and

$$p\alpha^2 \ll 1 , \tag{20.27}$$

respectively. Region (20.26) may be termed that of 'short' waves, whereas region (20.27) is that of 'long' waves. In the former the growth rate increases with the increase of the wavelength, while decreasing in the latter.

At $p\alpha^2 \sim 1$, i.e., when $\alpha \sim S^{1/4}$, the growth rate reaches the maximum

$$p_{\max} \sim S^{1/2} . \tag{20.28}$$

Recall that the dimensionless parameters

$$\alpha = ka = \frac{2\pi a}{\lambda}, \qquad p = \omega\,\tau_r\,.$$

Without using the condition $\alpha^2 \ll 1$, Equation (20.6) shows that $\Delta' \approx 0$ for $\alpha \approx 1$. So, the tearing instability completely disappears for $\alpha \approx 1$ and exists in the region of the wave length

$$\boxed{\lambda > 2\pi a\,.} \tag{20.29}$$

That is why it is called a *long-wave* instability.

As $\alpha \to S^{-1}$, the increment tends to ${\tau_r}^{-1}$. As was mentioned earlier, in this case, i.e. in the region $\alpha < S^{-1}$, the effect of compressibility becomes dominant. It will be discussed in Section 20.5.

Expression (20.23) was obtained analitically by Furth, Killen, and Rosenbluth (1963); they also obtained the dependence (20.25) numerically.

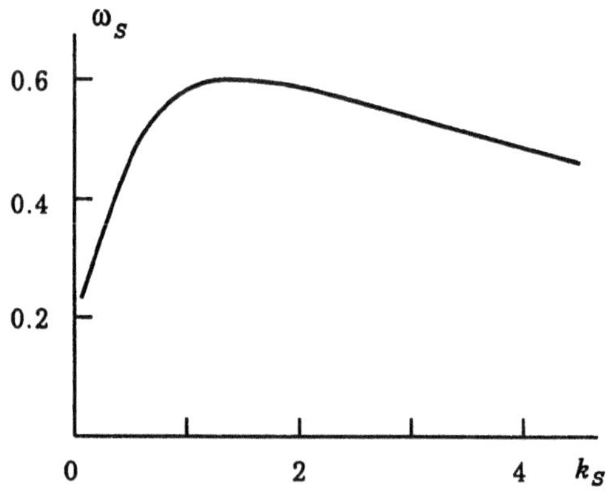

Figure 20.2: The dependence of the tearing instability increment ω_s on the wave vector k_s.

The results of the numerical solution of the general Equation (20.21) are given in Figure 20.2, using the notation

$$\omega_s = \omega\,\tau_r\,S^{-1/2}\,, \qquad k_s = ka\,S^{1/4}\,. \tag{20.30}$$

Recall that the dimensionless parameter S is the Lundquist number (17.22) but determined with respect to the current-sheet thickness a.

20.3 The physical interpretation of the instability

We now present another derivation of the dispersion relations, based on the consideration of the physical mechanism of the tearing instability (Furth, Killen, and Rosenbluth, 1963). Let us make use of the absolute system of units where the speed of light $c = 1$. Besides, every coefficient of order unity will be set equal to unity.

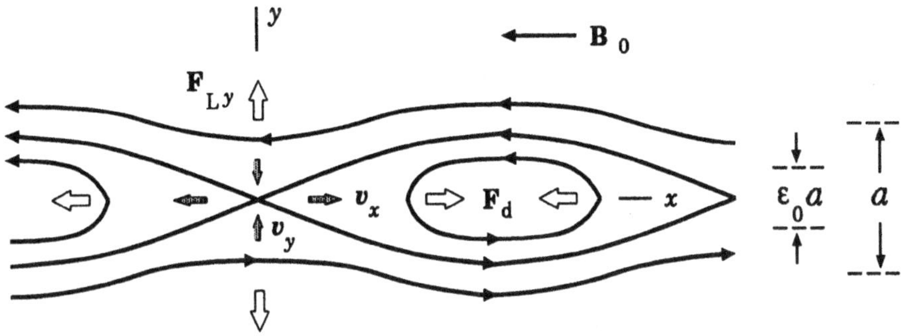

Figure 20.3: The magnetic field lines and the velocity in the course of the development of a tearing instability. The small arrows show velocity directions. Forces are shown by thick empty arrows. $\varepsilon_0 a$ is the internal region thickness. The case $\varepsilon_0 < \alpha$ is shown.

Let a small perturbation appear in the current sheet. As a consequence of the magnetic field structure (namely, antiparallel directions of reconnecting components on either side of the neutral plane), a *driving* force $\mathbf{F}_d$ of the instability arises, accelerating the plasma along the x axis, i.e. along the width of the sheet (see Figure 20.3). This force corresponds to a simple fact:

> parallel electric currents flowing inside the neutral sheet attract each other and tend to coalesce into separate current filaments.

Thus the driving force of the instability generates plasma motions inside the sheet, directed along the x axis, with a velocity v_{1x}. As this takes place, the surrounding plasma must, by virtue of the flow continuity, flow into the internal region with a velocity v_{1y}. As a consequence, the electric current j_s arises, giving rise to the corresponding Lorentz force F_{Ly}, hindering the plasma from flowing into the internal region:

$$j_s = \sigma \, v_{1y} \, \varepsilon_0 B \,, \qquad F_{Ly} = j_s \, \varepsilon_0 B = \sigma \, v_{1y} \, (\varepsilon_0 B)^2 .$$

Here we have taken into account that the reconnecting component of the field at the boundary of the internal region is equal to $B_x(y) = \varepsilon_0 B$, where $\varepsilon_0 a$ is the thickness of the internal region. The force $F_{{\rm L}y}$ is directed against the plasma motion and is comparable in magnitude with the driving force $F_{\rm d}$ of the instability. Hence the power with which the driving force performs work on a unit volume of the plasma is

$$P = v_{1y} F_{{\rm L}y} = \sigma \, v_{1y}^2 \, (\varepsilon_0 B)^2 . \tag{20.31}$$

This power goes to acceleration of the plasma; that is why

$$P = K, \tag{20.32}$$

where K is the kinetic energy acquired by the unit plasma volume in unit time:

$$K = \omega \rho \, v_{1x}^2 = \omega \rho \, \frac{v_{1y}^2}{(k \, \varepsilon_0 a)^2} \, . \tag{20.33}$$

Here use is made of the incompressibility condition div $\mathbf{v} = 0$:

$$v_{1x} = \frac{v_{1y}}{k \, \varepsilon_0 a} \, .$$

On comparing (20.31) and (20.33), an expression for the thickness of the internal dissipative region is found,

$$\varepsilon_0 = \left(\frac{\omega \rho}{k^2 a^2 B^2 \sigma} \right)^{1/4} , \tag{20.34}$$

which coincides with expression (20.9), obtained earlier from the analytical solution.

Let us now find the dispersion relations. In the dissipative region, where the motions of plasma and field lines are relatively independent, the first addendum on the right-hand side of Ohm's law

$$\eta \mathbf{j} = \mathbf{E} + \mathbf{v} \times \mathbf{B}$$

dominates the second one (though these two are of the same order of magnitude). What this means is that $\varepsilon_0 a$ must be taken in such a way that

$$\eta \, j_1 \sim E_1 \, . \tag{20.35}$$

However, the plasma and magnetic field line motions are not completely independent, even in the internal dissipative region. The electric field perturbation E_1 is related with that of the magnetic field perturbation B_1 through

$$E_1 \sim \frac{\omega B_{1y}}{k} \, .$$

Using the Maxwell's equations

$$\operatorname{curl} \mathbf{B} = \frac{4\pi}{c} \mathbf{j} \quad \text{and} \quad \operatorname{div} \mathbf{B} = 0 \, ,$$

we obtain

$$j_1 \sim \frac{B_1''}{4\pi k} \tag{20.36}$$

once $ka < 1$. Relations (20.35) and (20.36) give rise to

$$\frac{\omega B_{1y}}{\eta} \sim \frac{B_{1y}''}{4\pi} \, . \tag{20.37}$$

Now the quantity B_{1y}'' has to be evaluated. As a consequence of a partial freezing-in, magnetic field deviations during the plasma motion along the sheet in a region with a thickness

$$a\tilde{\varepsilon} \sim a^2 k \, ,$$

since $a\tilde{\varepsilon}\lambda \sim a^2$. For

$$a\tilde{\varepsilon} > a\varepsilon_0 \tag{20.38}$$

this gives the estimate

$$B_{1y}'' \sim \frac{B_{1y}'}{\varepsilon_0 a} \sim \frac{B_{1y}}{\varepsilon_0 a \, \tilde{\varepsilon} a} \sim \frac{B_{1y}}{\varepsilon_0 k a^3} \, , \tag{20.39}$$

whereas for

$$a\tilde{\varepsilon} < a\varepsilon_0 \tag{20.40}$$

one has

$$B_{1y}'' \sim \frac{B_{1y}'}{\varepsilon_0 a} \sim \frac{B_{1y}}{(\varepsilon_0 a)^2} \, . \tag{20.41}$$

It is a simple matter to see that the inequality (20.38) is equivalent to the inequality (20.26) determining the region of short-wave perturbations, while the inequality (20.40) is equivalent to (20.27) which corresponds to the long-wave region. Substituting the relations (20.39) and (20.41) in (20.37), with care taken of (20.34), leads to the dispersion relations:

$$\omega^5 = \frac{\eta^3 B^3}{a^{10} \rho} \frac{1}{k^2} \tag{20.42}$$

for the case (20.38), and

$$\omega^3 = \frac{\eta B^2}{a^2 \rho} k^2 \tag{20.43}$$

for the case (20.40). Equations (20.42) and (20.43) are easily shown to be equivalent, respectively, to Equations (20.23) and (20.25), obtained analytically in Section 20.2.

20.4 The stabilizing effect of the transversal field

While describing the effect of a transversal magnetic field, attention will be centred on the physical picture of the phenomenon. In this way we are able to understand the stabilization mechanism and easily obtain the dispersion relations for the tearing instability with a transversal field.

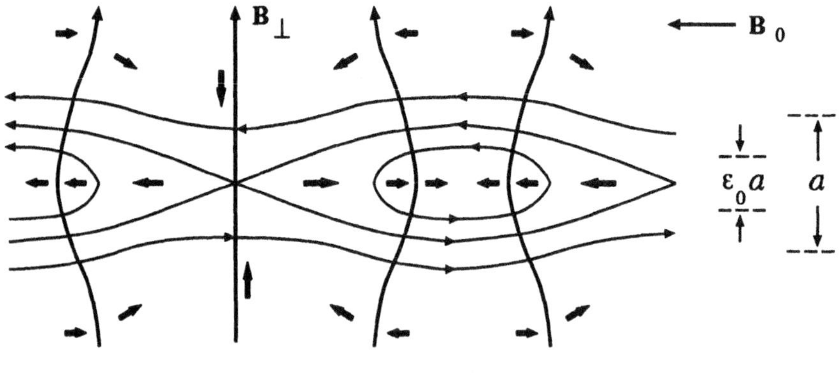

Figure 20.4: The magnetic field lines and velocities for the tearing instability in a reconnecting current sheet with a transversal magnetic field.

Given the transversal field, the plasma moves along the width of the reconnecting current sheet, overcoming the braking influence of the transversal field as shown in Figure 20.4. Taking this fact into account, we have instead of (20.32) to write down

$$P = K + \Pi \, . \tag{20.44}$$

The second term on the right is the work done in a unit of time against the force $F_{B\perp}$ related to the transversal field $B_\perp$, and it is given by

$$\Pi = v_{1x} F_{B\perp} \,. \tag{20.45}$$

Here

$$F_{B\perp} = j_{B\perp} B_\perp \quad \text{and} \quad j_{B\perp} = \sigma \, v_{1x} \, B_\perp \,. \tag{20.46}$$

Using Equations (20.45)–(20.46) and div **v** = 0, the power Π is evaluated to be

$$\Pi = \sigma B_\perp^2 \, \frac{v_{1y}^2}{(k\, \varepsilon_0 a)^2} \,. \tag{20.47}$$

Substituting the relations (20.31), (20.33), and (20.46) in the relation (20.44) gives

$$\sigma \, v_{1y}^2 \, (\varepsilon_0 B)^2 = \frac{\omega \rho \, v_{1y}^2}{(k\, \varepsilon_0 a)^2} + \sigma \, B_\perp^2 \, \frac{v_{1y}^2}{(k\, \varepsilon_0 a)^2} \,.$$

From this there immediately follows an estimate for the thickness of the internal dissipative region with the transversal field at hand:

$$\varepsilon_0 = \left(\frac{\omega \rho}{k^2 a^2 B^2 \sigma} \right)^{1/4} \left(1 + \frac{\sigma B_\perp^2}{\omega \rho} \right)^{1/4} \tag{20.48}$$

or

$$\varepsilon_0(\xi_\perp) = \varepsilon_0(0) \left(1 + \frac{\xi_\perp^2 S^2}{p} \right)^{1/4} .$$

Here $\xi_\perp = B_\perp / B$ and the internal region thickness for $B_\perp = 0$ is designated as). Now $\varepsilon_0(\xi_\perp)$ is implied in the expressions (20.36) to (20.41) by ε_0. Substituting (20.48) in (20.36)–(20.41) gives the dispersion relations:

$$\omega^5 = \frac{\eta^3 B^3}{a^{10} \rho} \, \frac{1}{k^2} - \frac{B_\perp^2}{\rho \eta} \, \omega^4$$

in the short-wave region

$$\varepsilon_0 < \alpha \,, \tag{20.49}$$

and

$$\omega^3 = \frac{\eta B^2}{a^2 \rho} \, k^2 - \frac{B_\perp^2}{\rho \, \eta} \, \omega^2$$

in the long-wave region

$$\varepsilon_0 > \alpha \,. \tag{20.50}$$

Rewrite the same dispersion relations in the dimensionless form

$$p^5 = \left(\frac{S}{\alpha}\right)^2 - \xi_\perp^2 S^2 p^4 \qquad (20.51)$$

and

$$p^3 = \alpha^2 S^2 - \xi_\perp^2 S^2 p^2 \qquad (20.52)$$

for the cases (20.49) and (20.50), respectively. It is easy to comprehend that

> the transversal component of magnetic field decreases the tearing mode increment over the whole wave range and also decreases the wavelength at which the increment peaks.

The rigorous analytic solution (Somov and Verneta, 1989) gives the dispersion relation

$$\Delta^{1/4} \left(\frac{\alpha^2 S^2}{p}\right)^{1/4} \left(1 - \frac{p^{3/2}}{\alpha S} \Delta^{-1/2}\right) - p\,\alpha \left(\frac{\pi}{2}\right)^{1/2} = 0\,, \qquad (20.53)$$

where

$$\Delta = \left(1 + \frac{\xi_\perp^2 S^2}{p}\right)^{-1}. \qquad (20.54)$$

From this the relations (20.51) and (20.52) follow, given the conditions (20.49) and (20.50), respectively.

The stabilizing influence of the transversal field is demonstrated by Figure 20.5 on which the graphs of the instability increment $\omega\,\tau_r$ dependence on the wave length λ/a are presented for $S = 10^8$ and three values of the transversal field: $\xi_{\perp 0} = 0$, $\xi_{\perp 1} = 10^{-4}$, and $\xi_{\perp 2} = 10^{-3}$. The solutions of the asymptotical Equations (20.51) and (20.52) are shown by the straight dotted lines, the solutions of the exact Equation (20.53) are shown by solid curves. The figure shows that,

> as the transversal field increases, the increment of the tearing instability decreases and its maximum moves to the short-wave region.

Nishikawa and Sakai (1982) have numerically solved a set of eigenmode equations in a reconnecting current sheet with the transversal magnetic field. The mode associated with magnetic island formation was investigated. It was found that the transversal component strongly modifies this mode and has a significant stabilizing effect on the collisional tearing mode.

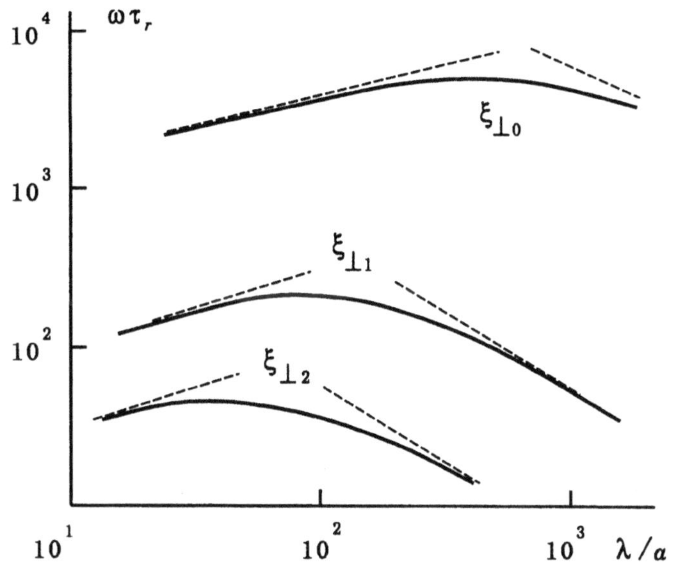

Figure 20.5: The dependence of the collisional tearing instability increment on the wavelength and the transversal component of magnetic field.

20.5 Compressibility and a longitudinal field

20.5.1 Neutral current sheets

Let us find the conditions under which compressibility should be taken care of and show the effect of compressibility on the tearing instability of the reconnecting current sheet. For simplicity's sake, we first restrict our attention to the case $B_y = B_\perp = 0$ and $B_z = B_\parallel = 0$.

During development of the tearing instability, the plasma starts moving along the width of the current sheet as shown in Figure 20.3. Given the finite value of the sound velocity, V_s, the plasma in the neighbourhood $|\,\delta x\,| < V_s \,/\, \omega$ of the reconnection point is drawn into the motion in a characteristic time of the instability growth ω^{-1}. Provided $V_s \,/\, \omega > \lambda$, the plasma may be considered incompressible. In the opposite case

$$\frac{V_s}{\omega} < \lambda \tag{20.55}$$

the compressibility of the plasma must be accounted for: div $\mathbf{v} \neq 0$. In this case the estimate

$$\frac{v_{1x}}{(V_s \,/\, \omega)} \sim \frac{v_{1y}}{\varepsilon_0 a} \tag{20.56}$$

holds, where $\varepsilon_0 a$ is the internal region dimension.

Compare the work done by the driving instability force (Section 20.3) in unit time on unit volume,

$$P \sim \sigma v_{1y}^2 \, (\varepsilon_0 B)^2 \, ,$$

with the kinetic energy acquired in unit time by the unit plasma volume drawn into the motion along the current sheet within the neighbourhood $|\,\delta x\,| < V_s/\omega$ of the reconnection point,

$$K \sim \omega \rho_0 \, v_{1x}^2 \sim \omega \rho_0 \left(\frac{V_s}{\omega} \frac{1}{\varepsilon_0 a} \right)^2 v_{1y}^2 \, .$$

Here relation (20.56) is used. Equating P and K gives an estimate for ε_0:

$$\varepsilon_0 \sim \left(\frac{\rho_0 V_s^{\,2}}{\omega a^2 \sigma B^2} \right)^{1/4} \sim \left(\frac{1}{\omega \tau_r} \frac{V_s^2}{V_{Ax}^2} \right)^{1/4} , \tag{20.57}$$

where $V_{Ax} = B_x / \sqrt{4\pi\rho}$ is the Alfvén speed.

Now substituting the quantity (20.57) for ε_0 in formulae (20.37)–(20.41) immediately results in the dispersion relation

$$\omega \approx \frac{1}{\tau_r} \frac{V_{Ax}^2}{V_s^2} \, .$$

Thus it is seen that

> because of compressibility of the plasma, a new branch of the tearing instability arises

in the region

$$\lambda > \lambda_0 \approx \frac{V_s}{\omega} \sim 2\pi a \, S \left(\frac{V_{Ax}}{V_s} \right)^{-3} , \tag{20.58}$$

which was absent for an incompressible plasma ($\omega \to 0$ for $\lambda > \lambda_0$). Recall that so far we have treated the case $B_\perp = 0$, $B_\parallel = 0$, i.e. the magnetically neutral current sheet.

20.5.2 Non-neutral current sheets

In the context of the above treatment, the role of a longitudinal field $B_z = B_\parallel \neq 0$ (along the electric current in the reconnecting sheet) becomes clear. While compressing a plasma with a longitudinal magnetic field which is in fact frozen into the plasma, **the work is to be done to compress the longitudinal field** (Somov and Titov, 1985b). Thus, given the longitudinal field, the plasma pressure is supressed by the sum of the plasma pressure and the magnetic one (connected with the longitudinal field). This leads to the change

$$V_s \to \left(V_s^2 + V_{A\|}^2 \right)^{1/2} , \tag{20.59}$$

where $V_{A\|} = B_{\|}/\sqrt{4\pi\rho}$, which describes the stabilizing influence of the longitudinal field. Once

$$B_{\|} > B_x(a) , \tag{20.60}$$

the instability caused by the compressibility becomes suppressed.

Note that the values obtained for the growth rate of the instability are comparable with the inverse time of magnetic diffusion τ_r^{-1}. Magnetic diffusion, however, is neutralized by the plasma drift into the current sheet (see Section 3.5 in Somov, 1992) and the stationary zero configuration persists for a time $t_s \gg \tau_r$. If the condition

$$\rho_{\text{out}} \ll \rho_{\text{in}} \tag{20.61}$$

is satisfied, where ρ_{out} and ρ_{in} are the plasma densities inside and outside the sheet, respectively, the plasma drift into the current sheet cannot usually suppress the tearing instability (see, however, Pollard and Taylor, 1979). Hence the tearing instability of the reconnecting current sheet plays an essential role as a universal dynamic instability (e.g., Somov and Verneta, 1993).

The rigorous analytic solution of the problem concerning the compressibility effect on the tearing mode development was given by Verneta and Somov (1993).

In actual reconnecting current sheets, the plasma continuously flows into the sheet through its wide surfaces and flows out through the narrow side boundaries (see Figure 17.3).

> The fast outflow of plasmas from the current sheet can be of principal importance for its tearing stability

(see reviews by Syrovatskii, 1981). The accelerating outflow along the main (B_x) magnetic field, with is present in the configuration with the velocity stagnation point, causes a substantial decrease in the magnitude of the linear growth rate and, for some parameter ranges, stabilization (e.g., Ip and Sonnerup, 1996).

20.6 The kinetic approach

20.6.1 The tearing instability of neutral sheet

We now desribe the tearing instability in the framework of the collisionless plasma model, starting from the Vlasov equation (Section 2.4)

$$\frac{\partial f_k}{\partial t} + \mathbf{v}\,\frac{\partial f_k}{\partial \mathbf{r}} + \frac{\mathbf{F}_k}{m_k}\,\frac{\partial f_k}{\partial \mathbf{v}} = 0\,. \tag{20.62}$$

Here

$$\mathbf{F}_k = q_k \left(\mathbf{E} + \frac{1}{c}\,\mathbf{v} \times \mathbf{B} \right)$$

and symbols $k =$ e, i denote electrons and ions, respectively.

As equilibrium distribution functions describing the reconnecting current sheet, it is appropriate to choose (see Harris, 1962)

$$f_k^{(0)}(y) = n_0 \, \exp \left\{ -\frac{1}{k_{\rm B} T_k} \left[\frac{1}{2}\, m_k v^2 - \vartheta_k \left(m_k v_z + \frac{1}{c}\, q_k A^{(0)} \right) \right] \right\}. \tag{20.63}$$

The notation is conventional. Here the vector potential $\mathbf{A} = \mathbf{e}_z A$ for a two-dimensional magnetic field $\mathbf{B} = \operatorname{curl} \mathbf{A}$ is introduced. The scalar potential is excluded by choosing $\vartheta_{\rm i}/T_{\rm i} = -\vartheta_{\rm e}/T_{\rm e}$. $\vartheta_{\rm e}$ and $\vartheta_{\rm i}$ are the flow velocities of electrons and ions.

Such distribution functions (as can be shown using Maxwell's equations) specify a current sheet with the following characteristics:
(a) the equilibrium magnetic field

$$\mathbf{B} = B_0\,(y)\, \mathbf{e}_x\,,$$

where

$$\boxed{B_0\,(y) = B_0 \, \tanh \frac{y}{a}} \tag{20.64}$$

on choosing

$$A^{(0)}\,(y) = \text{const} \times \ln \cosh \frac{y}{a}\,;$$

(b) the plasma density in the current sheet

$$n^{(0)}\,(y) = n_0 \, \cosh^{-2} \frac{y}{a}\,, \tag{20.65}$$

where

$$n_0 = \frac{1}{k_{\mathrm{B}}(T_{\mathrm{e}} + T_{\mathrm{i}})} \, \frac{B_0^2}{8\pi} \, ;$$

(c) the current-sheet half-thickness

$$a = \frac{2ck_{\mathrm{B}}(T_{\mathrm{e}} + T_{\mathrm{i}})}{eB_0 \left(\vartheta_{\mathrm{i}} - \vartheta_{\mathrm{e}}\right)} \, . \tag{20.66}$$

Therefore a magnetically-neutral one-dimemsional current sheet of the Harris type is considered.

Near the plane $y = 0$ where $B_0 = 0$, particle motion is almost free inside a non-adiabatic region of thickness $2d_k$ (cf. definition (18.28)). Outside this region the particles are magnetized. The quantity d_k can be evaluated as follows (see also Section 18.1). The local Larmor radius of a particle at the boundary of the region is

$$r_{\mathrm{L}}^{(\mathrm{k})}\left(d_k\right) = \frac{V_{Tk}\, m_k\, c}{q_k\, B_0\left(d_k / a\right)} \, .$$

Equating it to the internal dissipative region thickness

$$r_{\mathrm{L}}^{(\mathrm{k})}\left(d_k\right) \approx d_k \, ,$$

we find

$$\boxed{d_k \approx \sqrt{a r_{\mathrm{L}}^{(\mathrm{k})}} \, ,} \tag{20.67}$$

where $r_{\mathrm{L}}^{(\mathrm{k})}$ is the Larmor radius in the B_0 field. Thus the motion of particles of kind k is assumed to be free inside the region $|\, y \,| < d_k$, whereas they are magnetized once $|\, y \,| > d_k$.

⋆ ⋆ ⋆

Equations (20.62) will be solved in a *linear* approximation. The Fourier components of the perturbations are of the form

$$f^{(1)}(\mathbf{r}, t) = f^{(1)}(y) \, \exp\left(\omega t + \mathrm{i} k x\right) . \tag{20.68}$$

Recall that the case $\mathbf{k} \parallel \mathbf{B}_0$ is considered. The initial Equations (20.62) give, for perturbations,

$$\left(\omega + \mathrm{i} k v_x\right) f_k^{(1)} = -\frac{1}{m_k} \, \mathbf{F}_k^{(1)} \cdot \frac{\partial f_k^{(0)}}{\partial \mathbf{v}} \, .$$

These equations determine the approximate form of the perturbed distribution function, the connection between $f_k^{(1)}$, $E^{(1)}$, and $A^{(1)}$:

$$f_k^{(1)} = \frac{q_k f_k^{(0)}}{k_{\mathrm{B}} T_k} \left\{ \vartheta_k A^{(1)} + E^{(1)} \frac{v_z}{\omega + ikv_x} \right\} . \tag{20.69}$$

The first term on the right-hand side represents the influence of the magnetic field perturbation and the second one represents the interaction between the electric field of a wave and particles. The latter contribution is negligible outside the current sheet as the particle motion becomes adiabatic and there is no electric field along the magnetic field lines.

From Maxwell's equations, the perturbation electric field

$$\boxed{E^{(1)} = -\frac{1}{c}\, \omega A^{(1)} .} \tag{20.70}$$

Final results show that the instability growth rate complies with the condition

$$\omega < k\, V_{Tk} \, , \tag{20.71}$$

where (different from the mean thermal velocity introduced in Section 6.1)

$$V_{Tk} = \sqrt{\frac{2k\, T_k}{m_k}} \, . \tag{20.72}$$

Therefore we consider a low-frequency mode of the instability. This is the reason for assuming that

$$\frac{1}{v_x - \mathrm{i}\,(\omega/k)} \approx \mathrm{i}\pi\, \delta\,(v_x) + \mathrm{Vp}\left(\frac{1}{v_x}\right) \tag{20.73}$$

(the Sokhotsky formula). Here Vp is the principal value of an integral (see Vladimirov, 1967, Ch. 2, § 7).

* * *

If W is the total kinetic energy of the particles in the perturbation, then

$$\frac{dW}{dt} = \sum_k q_k \int E^{(1)}\, v_z\, f_k^{(1)}\, d^3\mathrm{v}\, dy \, . \tag{20.74}$$

On the other hand, the energy conservation law gives

$$\frac{dW}{dt} = -\frac{1}{8\pi}\frac{d}{dt}\int \left(B^{(1)}\right)^2 dy\,. \tag{20.75}$$

Substituting (20.69) and (20.73) in formula (20.74), we get

$$\frac{dW}{dt} = \frac{\pi}{k}\sum_k \frac{q_k}{k_{\rm B}T_k}\int\limits_{-d_k}^{+d_k}\left[\int f_k^{(0)}\,\delta(v_x)\left(E^{(1)}v_z\right)^2 d^3\mathrm{v}\right]dy\,-$$

$$-\frac{1}{4\pi}\frac{d}{dt}\int\limits_{-\infty}^{+\infty}\frac{n(y)}{n(0)}\left(\frac{A^{(1)}}{a}\right)^2 dy \stackrel{\rm def}{=} \sum_k \frac{d}{dt}W_k^{\rm r} - \frac{d}{dt}W^{\rm m}\,. \tag{20.76}$$

Here $dW_k^{\rm r}/dt$ is the growth rate of the kinetic energy of the resonant particles of kind k in the region $|\,y\,| < d_k$, whereas $dW^{\rm m}/dt$ is the rate of energy decrease of the remaining particles.

The electron resonance term is $(r_{\rm L}^{(i)}/r_{\rm L}^{(e)})^{1/2}$ times greater than the ion one. Taking this fact into account, we find from formulae (20.75) and (20.76) for electrons ($k = \mathrm{e}$)

$$W^{\rm r} = \omega\int\limits_{-d_{\rm e}}^{+d_{\rm e}}\left[\int f_{\rm e}^{(0)}\,\delta(v_x)\left|\left(A^{(1)}\right)^2 v_z\right|^2 d^3\mathrm{v}\right]dy =$$

$$= \frac{k_{\rm B}T_{\rm e}}{8\pi e^2}\int\limits_{-\infty}^{+\infty}\left\{\left|\frac{\partial A^{(1)}}{\partial y}\right|^2 + \left|A^{(1)}\right|^2\left(k^2 - \frac{2}{a^2\cosh^2(y/a)}\right)\right\}dy =$$

$$= W^{\rm m} - \frac{1}{8\pi}\int\left(B^{(1)}\right)^2 dy\,. \tag{20.77}$$

From this it follows that the *energy transfer to electrons* exists in the region

$$ka < 1 \quad \text{or} \quad \lambda > 2\pi a \tag{20.78}$$

(cf. condition (20.29)). This process constitutes the development of the *electron* mode of the tearing instability.

It arises from the coupling of a negative energy perturbation (associated with filamentation of the original magnetically-neutral current sheet) to the electron energization due to Landau resonance (see Section 5.1.2).

Formula (20.77) gives us the following estimate for the growth rate of the electron tearing instability:

$$\omega \approx \left(\frac{a}{r_{\rm L}^{(\rm e)}} \right)^2 \frac{d_{\rm e}}{V_{\rm Te}} \, . \tag{20.79}$$

The presented derivation was originally suggested by Coppi, Laval, and Pellat (1966). They first proposed the *electron tearing* instability as a mechanism of explosive reconnection in the Earth magnetotail during substorm break-up (Section 20.6.3).

20.6.2 Stabilization by the transversal field

As we saw above, Landau resonance of electrons inside the neutral current sheet was proposed to provide the appropriate collisionless dissipation necessary for the spontaneous reconnection in the geomagnetic tail during a substorm (Coppi *et al.*, 1966). However, Schindler (1974) showed that nonzero magnetic field component $B_\perp$ normal to the current sheet *magnetizes* the electrons and restricts them from being resonant. As a result, the required dissipation relies upon the ions that are still unmagnetized. So, Schindler proposed the so called *ion tearing* instability, in which the dissipation is due to ion Landau resonance. In this model the electrons act only as a charge neutralizing background.

Galeev and Zelenyi (1975, 1976) found, however, that the magnetized electrons can change the basic character of the tearing perturbation, thus making the ion energization invalid as a driver for the instability. Therefore the kinetic tearing instability can be suppressed by the transversal (i.e. perpendicular to the current sheet plane) magnetic field. Let us consider this effect in some detail.

(a) We begin by considering sufficiently small values of the transversal field $B_\perp$, for which the inequality

$$\omega_{\rm L}^{(\rm e)} = \frac{eB_\perp}{m_{\rm e}c} < \omega \tag{20.80}$$

holds. Here $\omega_{\rm L}^{(\rm e)}$ is the electron gyro-frequency in the transversal magnetic field $\mathbf{B}_\perp$; recall that ω is the instability increment.

In this case **electrons** in the region $|\,y\,| < d_{\rm e}$, where the reconnecting magnetic field components tend to zero, **are in Landau resonance with**

the electric field perturbation (20.70). As a consequence, the electron tearing mode develops in the reconnecting current sheet (see above).

(b) As the transversal magnetic field increases, the Larmor frequency $\omega_{\rm L}^{(\rm e)}$ increases as well. When $\omega_{\rm L}^{(\rm e)} > \omega$ the electron resonance with the electric field perturbation breaks down and the electron mode of the instability becomes stabilized (Schindler, 1974). This takes place for

$$\frac{B_{\perp}}{B_0} = \xi_{\perp} > \left(\frac{r_{\rm L}^{(\rm e)}}{a} \right)^{5/2} \left(1 + \frac{T_{\rm i}}{T_{\rm e}} \right) . \tag{20.81}$$

If the electron mode of the tearing is stabilized, there remains the possibility for ions to become the resonant particles, gaining energy. However, electron gyration also stabilizes the ion mode up to the values (Galeev and Zelenyi, 1976):

$$\frac{B_{\perp}}{B_0} < \left(\frac{r_{\rm L}^{(\rm e)}}{a} \right)^{1/4} \left(1 + \frac{T_{\rm i}}{T_{\rm e}} \right)^{-1/2} . \tag{20.82}$$

Thus there exists a 'split' – a range of values of the magnetic field transversal component

$$\left(\frac{r_{\rm L}^{(\rm e)}}{a} \right)^{5/2} \left(1 + \frac{T_{\rm i}}{T_{\rm e}} \right) < \frac{B_{\perp}}{B_0} = \xi_{\perp} < \left(\frac{r_{\rm L}^{(\rm e)}}{a} \right)^{1/4} \left(1 + \frac{T_{\rm i}}{T_{\rm e}} \right)^{-1/2} . \tag{20.83}$$

Here the **linear kinetic tearing instability becomes suppressed** (Galeev and Zelenyi, 1976). Somov and Verneta (1988) have shown that

> the transversal magnetic field effect ensures the tearing stability of high-temperature reconnecting turbulent-current sheets

during the 'main' or 'hot' phase of solar flares (for a review see Somov and Verneta, 1993; see also Section 3.5 in Somov, 1992).

20.6.3 The tearing instability of the geomagnetic tail

Although the tearing instability was first proposed as à clue mechanism of magnetospheric substorms more than three decades ago (Coppi *et al.*, 1966), during the last years its prime role among other substorm processes was persistently challenged. The main theoretical reason was the proof by Lembege and Pellat (1982) that the sign of the energy of the tearing mode perturbations can be changed from negative to positive one due to the drift motion of magnetized electrons inside the reconnecting current sheet.

In fact, this conclusion is similar to that one of Galeev and Zelenyi (1976) but Lembege and Pellat showed in particularly that this effect stabilizes the tearing instability under the condition

$$\xi_{\perp} = \frac{B_{\perp}}{B_0} < \frac{\pi}{4}\, ka \tag{20.84}$$

regardless the temperature ratio T_e/T_i. Here a corresponds to the current-sheet half-thickness according to the Harris formula (20.64).

Condition (20.84) shows that in the case of adiabatic electrons the tearing instability can be stabilized only for very short wavelengths

$$\lambda < \lambda_{\min} = \frac{\pi^2}{2}\, \frac{a}{\xi_{\perp}} \, . \tag{20.85}$$

They are too short to be relevant to the underlying spontaneous reconnection process in the geomagnetic tail current sheet. In fact, condition (20.85) coincides with that of the WKB approximation in the stability analysis and as a result has made the *linear* tearing instability as the substorm mechanism suspect.

There were many attempts to restore necessarily the linear ion instability as a clue substorm process. All of them look however pretty inconsistent with a general representation of the substorm as a relatively fast unloading process in the tail of the magnetosphere. The substorm is usually preceded and prepared by the quasi-static changes in the tail during the growth phase (e.g., Nagai *et al.*, 1998; Kokubun and Kamide, 1998).

From a consideration of observational constraints on the onset mechanism Sitnov, Malova, and Lui (1997), Sitnov and Sharma (1998) conclude that

the tearing instability must have a considerable initial stage when the equilibrium magnetic field topology is still conserved.

Moreover, the instability is shown to have no linear stage. Instead, either the explicitly nonlinear or pseudolinear instability of negative energy eigenmode can develope. So, the unavoidable *nonlinearity* is a key element of the substorm.

Sitnov *et al.* use the theory of catastrophes (Haken, 1978; Guckenheimer and Holmes, 1983) to consider a substorm as backward bifurcation in an open nonlinear system. In general, the theory of catastrophes is widely accepted as an appropriate mathematical tool to describe abrupt changes in a low-dimensional system driven by quasi-stationary evolution of a set of control parameters. The theory can be applied if we treat the tearing instability as a process for the growth of a large-scale one-mode perturbation.

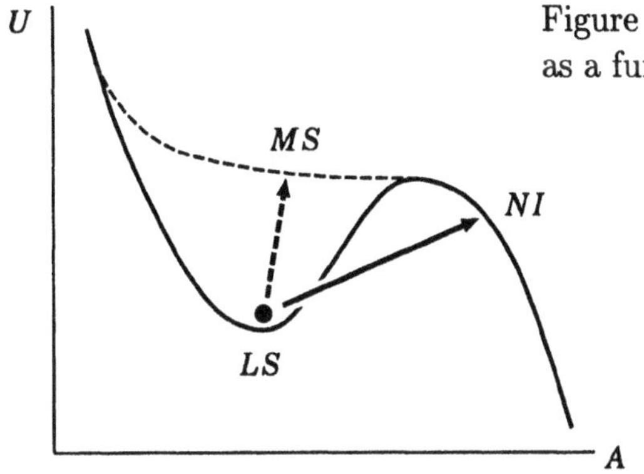

Figure 20.6: The effective potential U as a function of the state parameter A.

In Figure 20.6 the effective potential U of the geomagnetic tail current sheet near the marginal state of a tearing instability is shown as a function of the state parameter A. A process of quasi-stationary transformation of the potential minimum (LS) into the point of inflection (MS) is shown by the dashed arrow.

Being located near the bottom of the potential U well before the catastrophe, the **system is linearly stable** (LS) because of positive energy of small perturbations from the minimum. The transition to instability is possible only at the moment of the catastrophe or before the catastrophe under the influence of a *finite* amplitude perturbation (the large solid arrow) necessary to surmount the potential barrier. In both cases the destabilization of the system proves to be nonlinear.

Many difficulties of the substorm theory have arisen presumably not from the incorrect physics involved but rather from irrelevant mathematical treatment of the instability problem. Suitable treatment of the tearing instability as a backward bifurcation can resolve some long-standing problems in the theory including the consistent description of both triggered and spontaneous onsets. Much more can be done due to further elaboration of this promising approach to the magnetospheric substorm mechanism.

Chapter 21

Selected Trends in Cosmic Plasma Physics

The open issues focused on in this Chapter presumably will determine the nearest future and most interesting perspectives of cosmic plasma physics.

21.1 Reconnection and magnetic helicity

21.1.1 General properties of complex MHD systems

In this section we are going to consider some properties of the reconnection process in complex magnetic field configurations containing many places (points or lines) where reconnection occurs. Such a situation frequently appears in space plasmas, for example in a set of *closely packed* flux tubes suggested by Parker (1972). The tubes tend to form many reconnecting current sheets at their interfaces. This may be the case of active regions on the Sun when the field-line footpoint motions are slow enough to consider the evolution of the coronal magnetic field as a series of equilibria, but fast enough to explain coronal heating (see Sections 21.1.3 and 21.4.2).

Another example of a similar complex structure is the 'spaghetti' model of solar flares suggested by De Jager (1986) or the 'avalanche' model of them (Parker, 1988; Lu and Hamilton, 1991; see also Zirker and Cleveland, 1993). The last assumes that the energy release process in flares can be understood as **avalanches of many small reconnection events**. LaRosa and Moore (1993) propose that the large production rate of energetic electrons in solar flares (Section 18.1) is achieved through MHD *turbulent cascade* (see

Section 5.2.2) of the bulk kinetic energy of the outflows from many separate reconnecting current sheets (see also Antonucci *et al.*, 1993, 1996).

How can we estimate the rate of magnetic energy release due to reconnection in such a very complex system of flux tubes? The inherent complexity of the field configuration which can be used as a model does not allow any optimism in an attempt to solve the dissipative MHD problem numerically.

> An alternative approach to that of solving the MHD equations as they stand is to reformulate them in terms of invariant quantities.

As we have seen in Chapter 8, the mass, momentum and energy are conserving quantities and can be used to construct invariants. For example, the total energy of a system before reconnection is equal to the total energy after reconnection plus dissipation. A less familiar invariant in ideal MHD is the *magnetic helicity* or, more exactly, the *global* magnetic helicity (see Problem 14.2):

$$\mathcal{H} = \int_V \mathbf{A} \cdot \mathbf{B} \, d^3\mathbf{r} \,. \tag{21.1}$$

Here $\mathbf{A}$ is a vector potential for field $\mathbf{B}$, and V is the plasma volume bounded by a magnetic surface S, i.e.

$$\mathbf{B} \cdot \mathbf{n} \,|_S = 0 \,. \tag{21.2}$$

Woltjer (1958) showed that

> in ideal magnetohydrodynamic motions the global magnetic helicity $\mathcal{H}$ is conserved in any closed magnetic flux tube.

Woltjer's theorem may be extended to open-end flux tubes as well, provided the ends do not suffer any motion. In order to explain the observed toroidal field reversal in reversed-field pinches, Taylor (1974) generalized the ideal MHD result derived by Woltjer to a class of dissipative motions. Woltjer's theorem can also be used to show that the fields which minimize the magnetic energy subject to given initial and boundary conditions are in general force-free fields (Problem 14.2).

The magnetic helicity, defined by definition (21.1), provides a measure of the *linkage* or knottedness of field lines (e.g., Berger, 1988). **The helicity is a topological property of a magnetic field** (see, for example, Problem 21.1). In ideal MHD there is no reconnection. For this reason, the magnetic helicity is conserved.

If we do not have ideal MHD there is some reconnection, and helicity is not conserved. However,

> reconnection at a large magnetic Reynolds number generally conserves the global magnetic helicity to a great extent.

In laboratory (Taylor, 1974, 1986), solar (Berger, 1984) and magnetospheric (Wright and Berger, 1989) plasmas the fraction of helicity dissipated is normally very small.

The approximate conservation of magnetic helicity has been successful in calculating heating rates in the solar corona (Section 21.1.3). The main idea here is that the magnetic field tends to minimize its energy, subject to the constraint that its topological characteristic – helicity – is fixed. Reconnection gives the fastest way for this relaxation. The magnetic configuration in the region which is subject to reconnection should relax towards a constant-α force-free field. Such a field is also called the *linear* force-free field (see Problem 14.1 and 14.2). Taylor (1974) used this conjecture – Taylor's hypothesis – to predict the formation of a Lundquist field in actively reconnecting fusion devices.

Interestingly, however, it is observed in some laboratory experiments that the relaxation can take place without the conservation of global magnetic helicity. Presumably such unexpected loss of helicity may be related to a *self-organization* effect in a reversed field plasma (e.g., Hirano *et al.*, 1997). Even if the value of $\mathcal{H}$ is null at the initial stage, the plasma relaxes to a certain field configuration by producing the toroidal magnetic field and $\mathcal{H}$.

21.1.2 Helical scaling in turbulence

The possibility of the self-similar cascade transfer of the hydrodynamic helicity flux over the spectrum was first introduced by Brissaud *et al.* (1973). The following two scenarios were analyzed from the standpoint of the dimensionality method: (a) the simultaneous transfer of energy and helicity with constant fluxes over the spectra of both parameters, (b) a constant helicity flux determining the energy distribution.

The influence of the hydrodynamic helicity is obvious from a physical standpoint:

> two helical vortices with strong axial motion in one direction have a tendency to merge because of the Bernoulli effect.

In other words, helicity should result in redistribution of the chaotic energy. Moreover, a helicity flux that characterizes the variation of the mean helicity should also appear. Above all, helicity has an effect on the spectral features of turbulence. As for the spectra, variations occur in incompressible, compressible, and stratified media, as shown by Moiseev and Chkhetiani (1996). One of the tendencies inherent in helical media is the **energy transfer to the long-wavelength region** due to the tendency of helical vortices to merge.

According to Moiseev and Chkhetiani (1996), the mechanism that generates the mean hydrodynamic helicity leads to a second cascade range in addition to the Kolmogorov range (see Section 5.2.2). The constant that does not depend on the scale of the helicity here is its flux. Nevertheless, this requirement, like the requirement that the energy flux F be constant in the Kolmogorov range, is not inflexible. The spectral characteristics undergo significant changes. They are associated, as we understand, with at least a partial **reverse cascade** into the large-scale region.

There is a broad class of effects that generate both hydrodynamic helicity itself and large helicity fluctuations under terrestrial and astrophysical conditions. In particular, the simultaneous presence of such factors as temperature and density gradients, shearing flows, and nonuniform rotation is sufficient.

21.1.3 Coronal heating in solar active regions

Heyvaerts and Priest (1984), Browning (1988) developed the model of current dissipation by reconnection, adapting Taylor's hypothesis to the conditions in a solar active region. They assumed that at any time the most relaxed accessible magnetic configuration is **a linear force-free field** which can be determined from the evolution of magnetic helicity. By so doing, Heyvaerts and Priest illustrated the role of the velocity v of photospheric motions in coronal heating. No heating is produced if these motions are very slow, and negligible heating is also produced when they are very fast. So

> coronal heating presumably results from photospheric motions which build up magnetic stresses in the corona at a rate comparable to that at which reconnection relaxes them.

The corresponding heating rate can be estimated in order of magnitude by:

$$F \approx \frac{B^2}{4\pi} \, v \left(\frac{l_b}{l_b + l_v} \right) \left(\frac{\tau_d \, v}{l_b} \right) , \tag{21.3}$$

where τ_d is the effective dissipative time, l_b and l_v are scale lengths for the magnetic field and velocity at the boundary. (Terms in brackets are limiting factors smaller than 1.) The results showed that a substantial contribution to coronal heating can come from current dissipation by reconnection.

Reconnection with a small magnetic Reynolds number can produce significant dissipation of helicity, of course.

Wright and Berger (1991) proved that helicity dissipation in two-dimentional configurations is associated with the retention of some of the inflowing magnetic flux by the reconnection region R_r. When the reconnection site is a simple Ohmic conductor, all the field parallel to the reconnection line (the longitudinal component of magnetic field) that is swept into the region R_r is retained (Somov and Titov, 1985). In contrast, the inflowing magnetic field perpendicular to the line is annihilated. Wright and Berger (1991) relate the amount of helicity dissipation to the retained magnetic flux.

21.1.4 Reconnection and helicity in solar flares

Flares in a solar or stellar atmosphere predominantly arise from the release of coronal magnetic energy. Since magnetic field lines may have fixed endpoints in the photosphere, observations of photospheric quantities such as shear and twist become important diagnostics for energy storage in the corona. The magnetic energy of an equilibrium field in the corona can be related to measures of its net shear and twist. For example, the energy of a linear force-free field is proportional to its magnetic helicity (see Problem 21.2). Berger (1988b) presented a formula for the energy of a non-linear force-free field in terms of linking field lines and electric currents. This allows us to partition the magnetic energy among different current sources in a well-defined way. For example, energy due to current sheets may be compared to energy due to field-aligned currents (see Chapter 22).

In principle, there may be an application in observational models of the field structure of an active region with vector magnetogram data supplying information on the force-free field parameter α. This would provide a check on the model's insight as to the true topology of the field.

Using vector magnetograms and X-ray morfology, Pevtsov *et al.* (1996) determine the helicity density of the magnetic field in active region NOAA 7154 during 1992 May 5–12. The observations show that a long, twisted X-ray structure retained the same helicity density as the two shorter structures, but its greater length implies a higher coronal twist. The measured length and α value combine to imply a twist that exceeds the threshold for the MHD kink instability. It appears that such simple models, which have found that the kink instability does not lead to global dissipation, do not adequately address the physical processes that govern coronal fields.

It is believed that the **excess energy**, which is the energy difference between the contained energy and the minimum energy predicted by the Taylor hypothesis, **is more rapidly dissipated than the magnetic helicity**. It is also believed that reconnection may lead to the fast MHD relaxation process to the minimum energy state, creating flares. However, this theoretical preposition should be subject to careful observational examination.

21.2 Reconnection in weakly-ionized plasma

21.2.1 Some observations and classical models

Magnetic reconnection, while being firmly established as a means of energy release during solar flares (Section 16.5), is frequently invoked for explanation of various phenomena in the solar atmosphere. A particular example of these is the *prominence* phenomenon. Prominences are defined as dense ($\approx 10^{11}$ cm^{-3}) and cool ($\approx$ 6000 K) plasma 'clouds' visible in Hα above the solar surface. Pneuman (1983) suggested that both the material necessary for their formation and the magnetic field topology supporting them are the result of reconnection.

According to Pneuman (see also Syrovatskii, 1982) a neutral line of the magnetic field is produced in the corona owing to some kind of plasma motion in the photosphere. Reconnection at this line gives rise to a helical magnetic field configuration. As this takes place, chromospheric material flows into the reconnection region and is then carried up by the reconnected field lines which are concave upward. The material is thereupon radiatively cooled to form a prominence that nests in the helical field topology.

An interesting modification of this model is due to van Ballegooijen and Martens (1989, 1990) who conjectured that the reconnection place is in fact located at the photospheric boundary. The point is that

> if reconnection takes place deep enough in the solar atmosphere, a sufficient quantity of material can easily be supplied to the corona,

thus facilitating the process of prominence formation. On the observational side this conjecture is substantiated by the fact (Martin, 1986) that for several hours before the formation of a filament, small-scale fragments of opposite polarity flux were seen to cancel in the region below the eventual filament.

So the model accounts for the cancelling magnetic features that are usually observed to be present in the photosphere below prominences. The scenario of the phenomenon has three phases: (a) a pre-interaction phase in which two opposite polarity photospheric magnetic fragments are unconnected magnetically, (b) an interaction phase when the fragments reconnect in the corona and create a filament, (c) a cancellation phase when reconnection in the photosphere produces the cancelling magnetic features.

Roumeliotis and Moore (1993) have developed a linear, analytical model for reconnection at an X-type neutral line (cf. Chapter 16). The reconnection process is assumed to be driven by converging or diverging motions applied at the photosphere. The gas pressure has been ignored (without much justifications) in the vicinity of the neutral line, and only small perturbations have been considered. The model relates the flows around the diffusion region, where dissipative effects are important, to the photospheric driving motions. The calculations based on this linear theory support the possibility of the laminar, slow reconnection occuring low in the solar atmosphere.

None of the above-mentioned authors considered the details of the reconnection process. Therefore it is still unclear whether the process can occur effectively enough in low-temperature plasma to ensure the upward flux of matter that is sufficient for prominence formation in the corona. In this section we treat the reconnection process in the chromosphere and the photosphere in greater detail.

The reconnecting current sheet is envisaged to be formed in consequence of centre-to-boundary flows of weakly ionized plasma in convective cells. It is in such a current sheet that field lines reconnect to change the field topology in the way suggested by Syrovatskii (1982) and Pneuman (1983). As distinct from the coronal case (e.g., Somov, 1992), we treat the current sheet in the chromosphere and photosphere. We find that the reconnection efficiency is highest in the temperature minimum region, where the classical electric conductivity of weakly ionized plasma reaches its minimum.

21.2.2 Balance equations and their solution

Consider the stationary current sheet in the chromosphere and photosphere (Litvinenko and Somov, 1994b; Litvinenko, 1999). To find its characteristics, let us write down the order-of-magnitude relations stemming from the one-fluid equations of continuity, momentum conservation (both across and along the sheet) and magnetic field diffusion into the sheet:

$$n_0 v_0 \, b = n v_1 a \, , \tag{21.4}$$

$$(1 + x\,(T_0))\, n_0 \, k_{\rm B} T_0 + \frac{{B_0}^2}{8\pi} = (1 + x\,(T))\, n \, k_{\rm B} T \, , \tag{21.5}$$

$$(1 + x(T))\, n\, k_{\rm B} T = m_p\, n\, \frac{v_1^2}{2} + (1 + x(T_0))\, n_0\, k_{\rm B} T_0 \,, \tag{21.6}$$

$$\frac{c^2}{4\pi\, \sigma(T)\, a} = v_0 \,. \tag{21.7}$$

Here a and b are the sheet half-thickness and half-width. n_0 and n are the plasma concentrations outside and inside the sheet, x is the ionisation degree, v_0 and v_1 are the plasma inflow and outflow velocities, m_p is the proton mass (hydrogen being assumed to be the main component of the medium), T_0 and T are the temperatures outside and inside the current sheet. σ is the collisional conductivity in the sheet where the magnetic field perpendicular to the electric current is zero. B_0 is the field in the vicinity of the current sheet.

The set of Equations (21.4)–(21.7) should be supplemented by the energy balance equation. However, it is not an easy matter to do this. On the one hand, thermal conductivity is unlikely to play a significant role in the energy balance of the low-temperature current sheet. On the other hand, there are no reliable calculations for the radiative loss function $L(T)$ in the temperature domain $< 10^4$ K. An attempt to solve the radiative transfer equation for such a current sheet in the dense plasma of the low solar atmosphere would be an unjustified procedure given the order-of-magnitude character of the model at hand.

Therefore let us adhere to the simplest assumption, namely that the cooling processes are effective enough to ensure the approximate equality of plasma temperatures inside and outside the current sheet. Hence we postulate that

$$T = T_0 \,. \tag{21.8}$$

Now the sought-after quantities can be expressed with the aid of Equations (21.4)–(21.8) via the external parameters n_0, T, x, σ, v_0, and B_0:

$$a = \frac{c^2}{4\pi\, \sigma(T)\, v_0} \,, \tag{21.9}$$

$$b = (1 + \beta^{-1})\, a\, \frac{v_1}{v_0} \,, \tag{21.10}$$

$$n = n_0\, (1 + \beta^{-1}) \,, \tag{21.11}$$

$$v_1 = V_{\rm A,s} \equiv B_0 \left[4\pi\, m_p\, n_0\, (1 + \beta^{-1}) \right]^{-1/2} . \tag{21.12}$$

Here

$$\beta = (1 + x\,(T))\, n_0\, k_{\rm B} T\, \frac{8\pi}{{B_0}^2} \qquad (21.13)$$

and $V_{\rm A,s}$ is the Alfvén speed defined by formula (17.7).

Returning to the question posed in the introduction of this section, it is now straightforward to calculate the mass flux into the corona through the reconnecting current sheet (assuming the latter to be vertically orientated):

$$F = 2m_p\, n v_1\, a l = 2m_p\, n_0\, (1 + \beta^{-1})\, \frac{c^2\, l\, V_{\rm A,s}}{4\pi\, \sigma v_0}\,, \qquad (21.14)$$

$l \sim 10^9$ cm being a typical value of the current sheet length.

To find numerical values of the current sheet parameters, we make use of the chromosphere model due to Vernazza, Avrett, and Loeser (1981). This model gives us the input parameters n_0, x and T as functions of the height h above the lower photospheric boundary, i.e. the level where the optical column depth in continuum $\tau_{5000} = 1$. The collisional conductivity, σ, for this model was calculated by Kubát and Karlický (1986). A typical value of the field is assumed to be $B_0 = 100$ G. As for the inflow velocity, it is a free parameter. Its magnitude is of the order of the photospheric convective flow velocity $\approx$ 100 m/s. Table 21.1 presents the current sheet characteristics predicted by our model using these data and the sheet length $l = 10^9$ cm.

21.2.3 Characteristics of the reconnecting current sheet

Apart from variation of the inflow velocity, we have considered three levels in the solar atmosphere, in an attempt to clarify the physical picture of the reconnection process. These are the lower photosphere ($h = 0$ km), the temperature minimum ($h = 350$ km), and the upper chromosphere ($h = 2113$ km). The properties of the reconnection process were found to be drastically different at these different levels. Different regimes of *linear* reconnection (see Craig and McClymont, 1993; Priest *et al.*, 1994) seem to be possible, including very slow (very small magnetic Reynolds number) reconnection.

The remarkable thing about our solution is that reconnection is predicted to effectively occur only in a thin layer (not thicker than several hundred km), coinciding with the temperature minimum region. Here

> a relatively thick current sheet can be formed, where reconnection proceeds at a rate imposed by the converging plasma flows.

Table 21.1: Parameters of the reconnecting current sheet in the chromosphere and photosphere

Height	h, km	0	0	350	350	2110	2110
Temperature	T, 10^3 K	6.4	6.4	4.5	4.5	18.5	18.5
Conductivity	σ, 10^{11} s^{-1}	6	6	1.5	1.5	140	140
Inflow velocity	v_0, $10\,\mathrm{m\,s^{-1}}$	1	100	1	100	1	100
Half-thickness	a, 10^4 cm	10	0.1	50	0.5	0.5	0.005
Half-width	b, 10^7 cm	0.8	10^{-4}	10	10^{-3}	3000	0.3
Concentration	n, 10^{16} $\mathrm{cm^{-3}}$	10	10	1	1	0.02	0.02
Outflow velocity	v_1, $\mathrm{km\,s^{-1}}$	0.6	0.6	2	2	20	20
Mass flux	F, $10^{10}\,\mathrm{g\,s^{-1}}$	300	3	300	3	0.4	0.004

Since the magnetic field is relatively weak, the flow is practically incompressible. Magnetic energy is transformed into the thermal and kinetic energy of the resulting plasma motion. The upward flux of matter through the current sheet into the corona is capable of supplying 10^{16} g of cold chromospheric material in a time of 10^4 s. This is amply sufficient for the formation of a huge prominence.

An interesting peculiarity of the solution obtained is the inverse proportionality of the mass flux to the inflow velocity. The physical reason for this is that decreasing v_0 leads to a decrease of the electric current in the current sheet and hence the magnetic field gradient. Since B_0 is kept fixed, the sheet thickness $2a$ has to increase, thus augmenting the matter flux.

Below the temperature minimum, in spite of the small conductivity σ, the current sheet does not form; $a \approx b$ because the plasma density is very high there. That diminishes the Alfvén speed and prevents the magnetic field from playing a significant role in the plasma dynamics. The overall geometry of the field is that of an X-point, so that the inflow magnetic field is highly nonuniform. This regime corresponds to the 'nonuniform' reconnection class according to classification given by Priest *et al.* (1994).

As for reconnection in the upper chromosphere, it is not efficient either, at least at the velocities under consideration. The reason for this is the relatively high temperature, resulting in the high conductivity (Table 21.1), which makes magnetic diffusion into the current sheet too slow for any observable consequences related to the mass flux into the corona.

* * *

Several remarks are in order here, concerning our initial assumptions. First, we have assumed the reconnecting current sheet to be purely neutral, that is no magnetic field perpendicular to the sheet has been taken into account. Allowing for a non-zero transversal field $\xi_{\perp} B_0$, Equation (21.4) might be rewritten as follows:

$$n_0 v_0 \, b = n v_1 \, (a + \xi_{\perp} \, b) \, . \tag{21.15}$$

Since our model predicts the sheet to be rather thick ($a \, / \, b \, > \, 10^{-2}$) this correction is of no importance: a small transversal field does not considerably increase the effective cross-section of the matter outflow from the current sheet.

Second, formula (21.8) needs some justification. By way of example, suppose that the influx of magnetic energy is balanced by radiative losses:

$$\frac{{B_0}^2}{4\pi} \, v_0 \, b = L \, (T) \, x n^2 \, ab \, . \tag{21.16}$$

A crude estimate for the loss function $L \, (T) \; = \; \chi T^{\alpha}$ has been given by, for example, Peres *et al.* (1982). Using this estimate together with the above current sheet characteristics, one could find $T \approx 10^4$ K (for $h \; = \; 350$ km). Given the order-of-magnitude character of our model, it seems reasonable to presume that radiative losses can balance the Joule heating, so that (21.8) is valid as a first approximation. Anyway, although we expect the plasma heating to have some impact on our results, it is not likely to considerably alter the conclusions concerning reconnection efficiency. This is well supported by numerical results obtained in the more accurate model by Oreshina and Somov (1999a).

Finally, we have implicitly assumed the plasma flow in the reconnection region to be well coupled. What this means is that both neutral and charged plasma components participate in the plasma flow (see, however, Section 21.2.5). As a consequence, the total density appears in the expression for the Alfvén speed determining the outflow velocity. If the coupling were weak, the ion Alfvén speed would have to be used in Equation (21.12), giving a faster outflow of ions.

Zweibel (1989) investigated reconnection in partially ionized plasmas and introduced the parameter Q defining the degree of coupling:

$$Q = \frac{v_0}{a\,\nu_{ni}}\,, \tag{21.17}$$

ν_{ni} being the frequency of neutral-ion collisions. The smaller Q is, the stronger is the coupling. It is easy to check that for the reconnecting current sheet in the temperature minimum region we have $Q \approx 10^{-5} - 10^{-1}$ for $v_0 = 10^3 - 10^5$ cm s^{-1}. This value of Q substantiates the assumption of **strong coupling for reasonably slow inflows**. In fact, a more self-consistent consideration of the magnetic reconnection region is necessary to take account of the generalized Ohm's law in a weakly-ionized plasma with a magnetic field near the temperature minimum.

21.2.4 Reconnection under solar prominences

The idea that reconnection in the dense cool plasma of the solar atmosphere is a mechanism of the so-called quiescent prominence (filament) formation was put forward years ago. The model of prominence formation by dint of the reconnection process was shown to predict realistic field topologies near filaments. However, no investigation has been performed on the value of the upward flux of matter into the corona. As has been proved in this section, the flux can be high enough to explain the filament formation in a reasonable time: $F \approx 10^{11} - 10^{12}$ g s^{-1}. This seems to be a strong argument in favour of the Pneuman–van Ballegooijen–Martens model. However there are only circumstantial evidences in its favour.

Direct indications of reconnection in the temperature minimum can be found on the basis of the study of photospheric and chromospheric magnetograms together with dopplergrams in the same spectral lines. Liu *et al.* (1995) have obtained magnetograms in the Hβ (λ4861.34 A) and FeI (λ5324.19 A) lines. A comparative study of such magnetograms has revealed the existence of reverse polarity features. The appearance and behaviour of these features can be explained by the twisting of the magnetic flux tubes and reconnection of them in the layer between the photosphere and the chromosphere, i.e. in the temperature minimum region.

We have seen that current sheets can be formed in the temperature minimum region in response to photospheric flows. Reconnection efficiency is determined by the low classical (collisinal) conductivity rather than by the turbulent one, as opposed to the coronal case. As a final speculation, high-speed flows which are predicted by our model in regions of strong magnetic fields ($B_0 > 300$ G) might be identified with spicules.

* * *

Observations reviewed by Martin (1998) confirm the **necessary conditions** for the formation and maintenance of the filaments previously cited: (a) location of filaments at a boundary between opposite-polarity magnetic fields, (b) a system of overlying coronal loops, (c) a magnetically-defined channel beneath, (d) the convergence of the opposite-polarity network of magnetic fields towards their common boundary within the channel, and (e) cancelation of magnetic flux at the common polarity boundary.

Evidence is put forth for **three additional conditions** associated with fully developed filaments: (A) field-aligned mass flows parallel with their fine structure, (B) a multi-polar background source of a small-scale magnetic field necessary for the formation of the filament barbs, and (C) a handedness property known as *chirality* which requires them to be either of two types, dextral or sinistral.

> In the northern hemisphere most quiescent filaments are dextral, and in the southern hemisphere most are sinistral.

This refers to the direction of the magnetic field when standing on the positive polarity and gives the two possible orientations for the axial field: namely to the right for a dextral structure and to the left for a sinistral one.

One-to-one relationships have been established between the chirality of filaments and the chirality of their filament channels and overlying coronal arcades. These findings reinforce either evidence that every filament magnetic field is separate from the magnetic field of the overlying arcade but both are parts of a larger magnetic field system. The larger system has **at least quadrupolar footprints in the photosphere** (cf. Fig. 22.1) and includes the filament channel and subphotospheric magnetic fields (Martin, 1998).

To explain the hemispheric pattern, Mackay *et al.* (1998) consider the emergence of a sheared activity complex. The complex interacts with a remnant flux and, after convergence and flux cancellation, the filament forms in the channel. A key feature of the model is the net magnetic helicity of the complex. With the correct sign a fillament channel can form, but with the opposite sign no filament channel forms after convergence because a transversal structure of the field is obtained across the polarity inversion line. This situation is quite similar to that one shown in Figure 22.3.

Three-dimensional quasi-dissipative MHD simulations (Galsgaard and Longbottom, 1999) show that a thin current sheet is created above the polarity inversion line. When the current becomes strong enough, magnetic reconnection starts. In the right parameter regime,

> with the correct sign of helicity, the reconnected field lines are able to lift plasma several pressure scale heights against gravity.

The lifted plasma forms a region with an enhanced density above the current sheet along the polarity inversion line.

21.2.5 Element fractionation by reconnection

It is observationally established that element abundances of the solar corona and solar wind obey a systematic fractionation pattern with respect to their original photospheric abundances. This pattern is organized in such a way that elements with a low first ionization potential (FIP), the so-called low-FIP elements, are enriched by a factor of about four. Apparently the elements are enriched or depleted by a process that dependes on the FIP or perhaps even more clearly on the characteristic first ionization time and the relative diffusion length for the neutrals of the minor species colliding with the dominant hydrogen atoms.

When two regions of opposite magnetic polarity come into contact with each other in a partially ionized plasma, ions drifting in response to the Lorentz force fall into the minimum of the magnetic field, and then the drifting ions force the neutrals to take part in the flow. This is the case considered by Arge and Mullan (1998). An essential aspect of reconnection in weakly-ionized plasma is that

> the atoms have no trouble flowing accross the magnetic field lines; the ions are not entirely constrained to follow the field lines as this should be in ideal MHD.

Instead, they have a significant component across the field lines. The reason is **dissipation in the form of ion-atom collisions**. In view of the fact that the atoms move across field lines freely, and in the view of the fact that collisional coupling connects the atom fluid and the ion fluid, it is not surprising that ions are *not* tied strictly to the field lines. As a result, departures from ideal MHD behaviour are an inevitable feature of the process we discuss here.

Because of the finite time required for ion-atom collisions to occur, the plasma which emerges from the reconnecting current sheet has an ion/atom ratio which may be altered relative to that in the ambient medium. Arge and Mullan show that in chromospheric conditions, outflowing plasma exhibits enhancements in ion/atom ratios which may be as large as a factor of ten or more. The results are relevant in the context of the Sun, where the coronal abundances of elements with low FIP are systematically enhanced in certain magnetic structures.

The first ionization potential gives the energy scale of an atomic species, hence many atomic parameters and the chemical behaviour of elements are closely related to it. Thus, in principle,

> very different physical mechanisms could be imagined which would produce an FIP dependence of elemental abundance

(see, for example, Section 21.3.3). It is important that the observed FIP enhancement varies from one type of solar magnetic features to another, ranging from unity (i.e., no enhancement) in impulsive flares to as much as 10 in diverging field structures. The last suggests that **magnetic field topology plays a role in creating the FIP effect in the Sun**.

If the magnetic field can trap the solar material and confine it (such as in a loop), the FIP effect apparently does not occur. On the other hand, if the field is such that a free outflow of material is allowed (e.g., in divergent field), then the FIP effect develops to a large amplitude. For this reason, when we model magnetic interactions in the chromosphere, for example the fine magnetic-flux tube formation (Section 21.3.3) we have to choose a topology which allows material to flow out freely.

In stars other than the Sun, EUV data have allowed to search for the FIP effect. Some stars with magnetic activity levels significantly higher than the Sun show evedence for FIP enhancement. This is consistent with a magnetic origin of FIP enhancement. Moreover the same FIP-based compositional fractionation mechanism at work in the solar atmosphere is presumably operational in the coronae of significantly more active stars (e.g., Laming and Drake, 1999).

21.3 The photospheric dynamo

21.3.1 Current generation mechanisms

In the deep photosphere, under the temperature minimum, particles are well coupled by collisions. That is why the physics of the deep photosphere, including the physics of magnetic flux tubes, is often described by the resistive one-fluid MHD approach. The same is valid even more for under-photospheric layers.

In the temperature minimum region, there are many neutral atoms which collide with ions and bring them into macroscopic motion. However the electrons remain frozen in the magnetic field. Therefore a treatment of this region as

> an ensemble of three fluids (electrons, ions and neutrals) is necessary to give a clear physical insight on the mechanisms of current generation near the temperature minimum

in the photoshere – the *photospheric dynamo* effect. Moreover, higher in the solar chromosphere, significant effects arise due to the density decrease that leads to a decoupling of the motions of ions and neutrals, that cannot be described by the one-fluid approximation.

For an axially symmetrical magnetic field, the horizontal velocities of electrons, ions and neutrals can be found analytically by solving the equations which describe the balance of the horizontal forces acting on each particle fluid (Hénoux and Somov, 1991). The horizontal velocities of ions and neutrals derived from these equations are relative to the horizontal velocities in the convective zone – the primary source of motion. It has been shown that, in an initially weak magnetic field,

> a radial inflow of neutrals can generate azimuthal DC currents, and an azimuthal velocity field can create radial DC currents leading to the circulation of vertical currents.

The effects of such velocity fields on the intensity and topology of electric currents flowing in thin magnetic flux tubes will be discussed below.

21.3.2 Physics of thin magnetic flux tubes

A simplified schematic representation of an open flux tube S is given in Figure 21.1, which shows the location of the solar chromosphere Ch and photosphere Ph with the temperature minimum region T. Such a semi-empirical model follows, for example, from the He I ($\lambda 10830$ A) triplet observations (Somov and Kozlova, 1998).

Consider the electric currents generated by azimuthal motions with the velocity v_φ in a partially ionized plasma in the region T. Since it is the relative azimuthal velocity between the magnetic field lines and the plasma, these currents can result either from azimuthal motions of the photosheric plasma around a fixed magnetic field or from the rotation around the flux tube axis of the magnetic field inserted in a static partially ionized atmosphere. Anyway, the azimuthal motions generate the radial currents j_r.

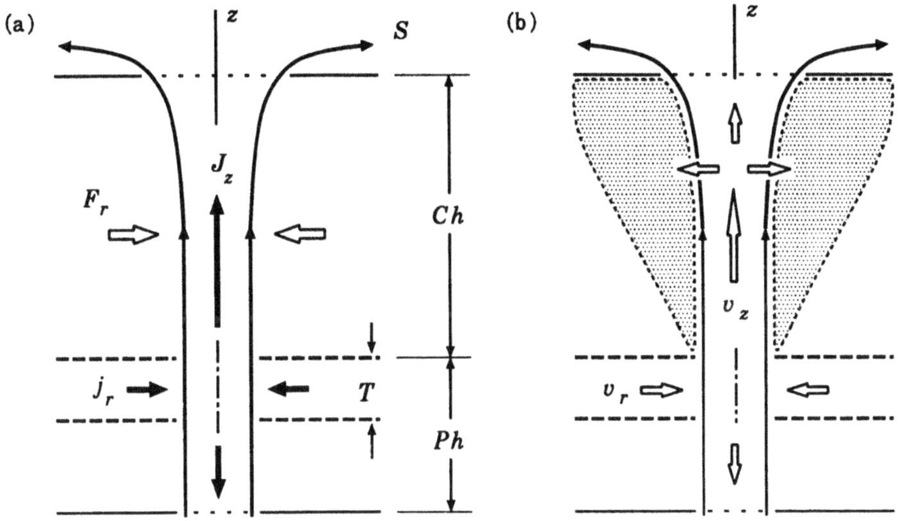

Figure 21.1: An oversimplified semi-empirical model of an open flux tube in the solar atmosphere. (a) The generation of electric currents and the pinch effect. (b) The motion of neutrals and their diffusion across the magnetic field lines in the chromosphere.

An inflow of the radial current density j_r is related to the vertical current density j_z by continuity equation

$$\frac{\partial j_z}{\partial z} = -\frac{1}{r}\,\frac{\partial\,(r j_r)}{\partial r}\,. \tag{21.18}$$

The vertical electric current

$$J_z = \int 2\pi\, r j_z(r)\, dr \tag{21.19}$$

cannot be derived locally, i.e. independently of the contribution of the other neighbouring (in height z) layers in the solar atmosphere. Every layer in the temperature minimum region T acts as a current generator in a circuit that extends above and below this layer. So, a circuit model is necessary to relate the total current J_z to the current densities. However, in all cases the contributions of every layer to the circuit regions placed above and below it are proportional to the inverse ratio of the resistances of these parts of the circuit.

The magnetic forces produced by these currents play a significant role in the structure and dynamics of flux tubes. Even for moderate values of the azimuthal photospheric velocities v_{φ}, the current J_z created is strong enough to prevent by the *pinch effect* (an action of the Lorentz force component F_r) an opening of the flux tube with height (Hénoux and Somov, 1997).

> Despite the decrease of the ambient gas pressure with height, the thin magnetic flux tube extends into the solar atmosphere high above the temperature minimum.

In the internal part of the tube, the rise from the photosphere of a partially ionized plasma is found to have four effects.

First, the upflow of this plasma is associated to a leak of neutrals across the field lines as shown in Figure 21.1b and leads to an increase of the ionization degree with altitude typical for the chromosphere. Moreover, the upflow brings above the temperature minimum an energy flux comparable to the flux required for chromospheric heating.

Second, the outflow of neutrals takes place at the chromospheric level across the field lines. Here the neutrals occupy an extensive area shown by the shadow in Figure 21.1b outside the tube. This outflow of neutrals leads to ion-neutral separation and may explain the observed abundance anomalies in the corona by enhancing in the upper part of the tube the abundances of elements of a low ionization potential (Section 21.3.3).

Third, the upward motion velocities are high enough to lift the matter to an altitude characteristic of spicules or even macrospicules.

Fourth, it seems plausible that if the footpoints of the flux tubes are twisted by the photosphere, then when they emerge into the transition region and release their magnetic energy some rotational component is retained. Strong evidence has been found from SOHO's CDS (the Coronal Diagnostic Spectrometer) observations to support the hypothesis that rotation plays a role in the dynamics of transition region features. These observations are interpreted as indicating the presence of a rotating plasma, a sort of *solar tornado* (Pike and Mason, 1998).

21.3.3 FIP fractionation theory

The flux-tube model predicts qualitatively the formation of closed or open structures with higher-temperature ionization state and higher low FIP to high FIP elements abundance ratios than the surrounding. A strong pressure gradient across the field lines can be present in the flux tubes where electric currents are circulating (Hénoux and Somov, 1991, 1997). Since they produce **two of the ingredients that are required for ion-neutral fractionation by magnetic fields**, i.e. small scales and strong pressure gradients perpendicular to the field lines (Hénoux and Somov, 1992), these currents can lead to the efficient ion-neutral fractionation.

Azimutal motions of the partially ionized photospheric plasma, with velocity v_φ at the boundary of the tube, $r = r_0$, generate a system of two current shells: S_{in} and S_{out} in Figure 21.2. The vertical currents j_z in these shells flow in opposite directions, such that the azimutal component of the field, B_φ, vanishes at infinity. This result can be easily understood in the case of a fully ionized atmosphere where the field lines are frozen in the plasma. However, the study of a partially ionized atmosphere gives insight into questions that cannot be tackled in the hypothesis of a fully ionized plasma, i.e. the possible difference in velocities perpendicular to the field lines of ions and neutrals.

The internal current system and the azimutal component of the magnetic field, B_φ, create an inward radial force $B_\varphi j_z$ that enhances, by the pinch effect discussed in Section 21.3.2, the pressure inside the internal part of the tube.

> The pinch effect is present from the photosphere to the chromosphere but its effects are different in these two regions.

In the photosphere, collisions couple ions and neutrals; so they do not cross the field lines. Above the photosphere, due to the exponential decrease of the density and, as a result, of the ion-neutral friction force with height, the difference in radial velocities of neutrals and ions increases with height.

The current densities and magnetic fields in the flux tube are such that, at hydrogen densities lower than 10^{13} cm^{-3}, the collisional coupling is low enough to allow the neutrals to cross the field lines and to escape from the internal current shell with high velocities. In usual plane-parallel-atmosphere

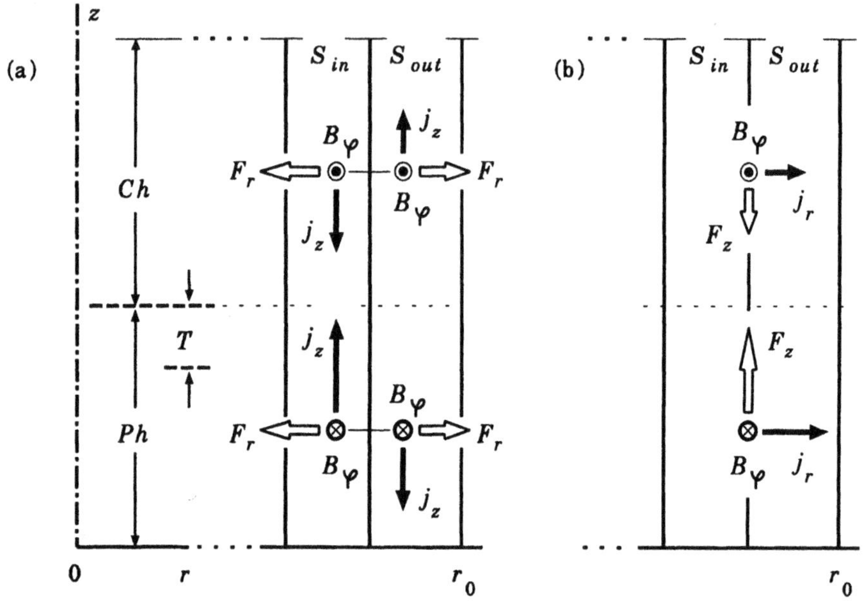

Figure 21.2: A simplified model of a thin magnetic flux tube in the solar atmosphere. (a) The vertical current density j_z and azimutal component of field B_φ create the pinch effect in the internal part of the tube. (b) The radial current density j_r and azimutal magnetic field B_φ produce the upward force in the photosphere.

models, the fractionation starts in the temperature minimum region T in Figure 21.1a at a temperature of about 4000 K. So the population of ionized low FIP species begin to be enhanced inside the internal current shell just at heights where the usual models place the chromospheric temperature rise and where the separation between the hot and cool components of the Ayres (1996) bifurcation model starts to take place.

Between the two opposite currents flowing vertically, the upwards Lorentz force component $B_\varphi j_r$ is present. Since the change of the direction of the vertical currents goes with the change of direction, from the photosphere to the chromosphere, of the transversal current j_r carried by ions, the $B_\varphi j_r$ force always produces a net ascending action. The intensity of this force is compatible with an ejection of matter up to heights of about 10 000 km, and therefore with the formation of spicules. This force acts in a shell, between the two neutralizing currents, where the gas pressure and collisional friction forces are reduced; it acts on ions and may then lead to a FIP effect in spicules by rising up preferentially the ionized low FIP species. A quantitative study of all these effects remains to be done.

21.4 Mechanisms of coronal heating

21.4.1 Heating of the quiet solar corona

The high temperature of the solar corona was originally interpreted as due to the steady dissipation of various kinds of waves coming from the lower layers (e.g., Ulmschneider *et al.*, 1991). Later on, heating by a myriad of very small flares releasing magnetic energy by reconnection has also been proposed (Gold, 1964; Priest, 1982; Parker, 1988). However, these microflares or *nanoflares* have not yet been identified.

It is difficult to detect the smallest flares in active regions, but in the quiet corona the background flux and stray light are smaller, and sensitive observations, for example, by the EIT (the Extreme ultraviolet Imaging Telescope) on the SOHO satellite can be used (Benz and Krucker, 1998). The thermal radiation of the quiet corona in high-temperature iron lines is found to fluctuate significantly, even on the shortest time scale as short as 2 min and in the faintest pixels. These observations give us an evidence that

> a significant fraction of the 'steady heating' in the quiet coronal regions is, in fact, impulsive.

The most prominent enhancements are identified with the X-ray flares above the network of the quiet chromosphere. Presumably, these X-ray flares above network elements are caused by additional plasma injected from below and heated to slightly higher temperatures than the preexisting corona.

Magnetic flux tubes in the photosphere are subject to constant buffeting by convective motions, and as a result, flux tubes experience random walk through the photosphere. From time to time, these motions will have the effect that a flux tube will come into contact with another tube of opposite polarity. We refer to this process as reconnection in weakly-ionized plasma (Section 21.2). Another possibility is the photospheric dynamo effect (Section 21.3) which, in an initially weak field, generates thin flux tubes of strong magnetic fields. Such tubes extend high into the chromosphere and can contribute to the mass and energy balance of the quiet corona.

SOHO's MDI (the Michelson Doppler Imager) observations show that the magnetic field in the quiet network of the solar photosphere is organized into relatively small 'concentrations' (magnetic elements, small loops etc.) with fluxes in the range of 10^{18} Mx up to a few times 10^{19} Mx, and an intrinsic field strength of the order of a kilogauss. These concentrations are embedded in a superposion of flows, including the granulation and supergranulation. They *fragment* in response to sheared flows, *merge* when they collide with others of the same polarity, or *cancel* against concentrations of opposite polarity. Newly emerging fluxes replace the canceled ones.

Schrijver *et al.* (1997) present a quantitative statistical model that is consistent with the histogram of fluxes contained in concentrations of magnetic flux in the quiet network as well as with estimated collision frequencies and fragmentation rates. Based on the model, Schrijver *et al.* estimate that as much flux is cancelled as is present in quiet-network elements in 1.5 to 3 days. This time scale is close to the timescale for flux replacement by emergence in ephemeral regions. So that this appears to be the most important source of flux for the quiet network. Schrijver *et al.* (1997) point out that the reconnection process appears to be an important source of outer-atmosphere heating.

Direct evidence that the 'magnetic carpet' (Day, 1998), an ensemble of magnetic concentrations in the photosphere, really can heat the corona comes from the two other SOHO instruments, the Coronal Diagnostic Spectrometer (CDS) and the Extreme ultraviolet Imaging Telescope (EIT). Both instruments have recorded local brightenings of hot plasma that coincide with disappearaces of the carpet's elements. This indicates that just about all the elements reconnect and cancel, thereby releasing magnetic energy, rather than simply sick back beneath the photosphere.

The coronal transition region and chromospheric lines observed by SOHO together with centimeter radio emission of the quiet Sun simultaneously observed by the VLA show that the corona above the magnetic network has a higher pressure and is more variable than that above the interior of supergranular cells. Comparison of multiwavelength observations of quiet Sun emission shows good spatial correlations between enhanced radiations originating from the chromosphere to the corona. Furthermore, **the coronal heating events follow the properties of regular solar flares** and thus may be interpreted as microflares and nanoflares (Benz and Krucker, 1999). The differences seem to be mainly quantitative (Krucker and Benz, 2000).

21.4.2 Coronal heating in active regions

The soft X-ray observations of the Sun from *Yohkoh* have revealed that roughly half of the solar X-ray luminosity comes from a tiny fraction (~ 2 %) of the solar disk (Acton, 1996). Virtually all of the X-ray luminosity is concentrated within active regions, where the magnetic field is the strongest. While the corona is evidently heated everywhere, there is no question that it is heated most intensively within active regions. So this Section will focus entirely on active regions.

The energy that heats the corona almost certainly propagates upward across the photosphere. Since the magnetic field plays a dominant role, the required energy flux can be expessed in terms of the electromagnetic Poynting vector (8.27) in an ideal MHD medium:

$$\mathbf{G}_{\mathrm{P}} = \frac{1}{4\pi}\,\mathbf{B} \times (\,\mathbf{v} \times \mathbf{B}\,)\,. \tag{21.20}$$

Assuming that the plasma vertical velocity v_z vanishes, we have the following expression for the vertical component of the energy flux:

$$G_z = -\frac{1}{4\pi}\,(\,\mathbf{v} \cdot \mathbf{B}\,)\,B_z\,. \tag{21.21}$$

A value of $G_z \sim 10^7$ erg cm^{-2} s^{-1} is frequently used to account for the X-ray flux from active regions.

Detailed models of heating typically invoke mechanisms belonging to one of the **two broadly defined categories**: wave (AC) or stress (DC) heating. In wave heating, the large-scale magnetic field surves essentially as a conductor for small-scale Alfvén waves propagating into the corona. So, the average flux of wave energy can be written as

$$< G_z > = -\sqrt{\frac{\rho}{4\pi}}\, < v^2 > \,B_z\,. \tag{21.22}$$

Here B_z is the large-scale, stationary field, and $< v^2 >$ is the mean square velocity amplitude of the Alfvén waves.

In stress heating, the coronal magnetic field stores energy in the form of DC electric currents until it can be dissipated through, for example, nanoflares (e.g., Parker, 1988). Estimating the rate of energy storage results in a Poynting flux of the form

$$G_z = c_d\,|\,v\,|\,B_z^2\,. \tag{21.23}$$

Here the constant c_d describes the efficiency of magnetic dissipation, which might involve the random velocity v or the magnetic field geometry. Anyway, the Poynting flux in Equations (21.22) and (21.23) **scales differently** with the magnetic field B_z. While the constants of proportionality in each case may vary due to numerous other factors,

> we might expect a large enough sample to be capable of distinguishing between the two mechanisms of coronal heating.

To analyze whether active region heating is dominated by slow (DC) or rapid (AC) photospheric motions of magnetic footpoints, the so-called reduced magnetohydrodynamic (RMHD) equations are used. They describe the dynamic evolution of the macroscopic structures of coronal loops assuming a fully turbulent state in the coronal plasma (e.g., Milano *et al.*, 1997). The boundary condition for these equations is the subphotospheric velocity field which stresses the magnetic field lines, thus replenishing the magnetic energy that is continuously being dissipated inside the corona. In a turbulent scenario, energy is efficiently transferred by a direct cascade to the 'microscale', where viscous and Joule dissipation take place (see, however, Section 21.1.2).

Therefore, for the macroscopic dynamics of the fields, the net effect of turbulence is to produce a dramatic enhancement of the dissipation rate. Milano *et al.* (1997) integrated the large-scale evolution of a coronal loop and computed the effective dissipation coefficients by applying the eddy-damped closure model. They conclude that

> for broadband power-law photospheric power spectra, the heating of coronal loops is DC dominated.

Nonetheless a better knowledge of the photospheric power spectrum as a function of both frequency and wavenumber will allow for more accurate predictions of the heating rate from the theory.

21.5 Practice: Problems and Answers

Problem 21.1. Consider two interconnected tubes C_1 and C_1 with magnetic fluxes Φ_1 and Φ_2 inside of them but without a magnetic field outside (Figure 21.3). Show that the global magnetic helicity of the system is given

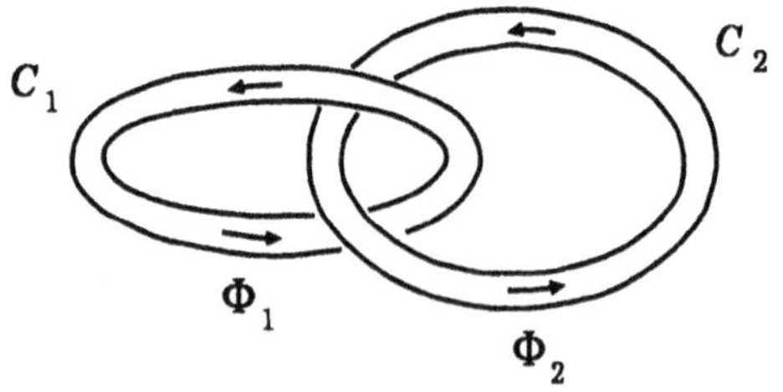

Figure 21.3: Two interconnected magnetic flux tubes.

by the formula

$$\mathcal{H} = 2\Phi_1\Phi_2 \tag{21.24}$$

(e.g., Moffatt, 1978).

Answer. First, calculate the helicity of the tube C_1 by integrating formula (21.1) over the volume V_1 of the tube C_1 and replacing $\mathbf{B}\,d^3\mathbf{r}$ by $\Phi_1\,d\mathbf{r}$ where $d\mathbf{r}$ is the length along the circuit C_1:

$$\mathcal{H}_1 = \int_{V_1} \mathbf{A} \cdot \mathbf{B}\, d^3\mathbf{r} = \Phi_1 \oint_{C_1} \mathbf{A} \cdot d\mathbf{r}\,. \tag{21.25}$$

By virtue of the Stokes theorem

$$\mathcal{H}_1 = \Phi_1 \int_{S_1} \text{curl}\mathbf{A} \cdot d\mathbf{S} = \Phi_1 \int_{S_1} \mathbf{B} \cdot d\mathbf{S} = \Phi_1\Phi_2\,. \tag{21.26}$$

Since the other tube C_2 makes the same contribution to the helicity, we obtain the Moffatt formula (21.24).

Therefore the global helicity depends only on the fact that the two magnetic fluxes are interlinked. The value of the helicity does not change if we deform the flux tubes as long as the linkage remains the same. If, however, by reconnection the tubes would be cut and removed so that the linkage between them were broken, then we see that the global helicity would go to zero. So we conclude that

> as long as the topology of magnetic fluxes does not change, the magnetic helicity is an invariant.

Problem 21.2. Show that for the force-free fields with constant α, the magnetic energy is proportional to the helicity (Woltjer, 1959):

$$\mathcal{M} = \alpha\,\mathcal{H}\,\frac{1}{8\pi}\,. \tag{21.27}$$

Here

$$\mathcal{M} = \int_V \frac{B^2}{8\pi}\, dV\,, \tag{21.28}$$

V is the volume of a simply connected region bounded by a magnetic surface S where $\mathbf{B} \cdot \mathbf{n} = 0$ (see Section 21.1.1).

Discuss a kind of a surface integral which must be added to expression (21.27) in the case of a multiply connected volume such as a torus (e.g., Reiman, 1980).

Chapter 22

Magnetic Reconnection of Electric Currents

Magnetic reconnection reconnects field lines together with field-aligned electric currents. This process may play a significant role in the dynamics of cosmic plasma.

22.1 Introductory comments

In this chapter we consider the general idea of interruption and redistribution of electric currents which are aligned with magnetic-field lines (the field-aligned currents in what follows), for example in the solar atmosphere. The currents are created under the photosphere and/or inside it, as well as they are generated in the corona. However, independently of their origin, electric currents distributed in the solar atmosphere reconnect together with magnetic field lines. So the currents are interrupted and redistributed in a topological way.

This phenomenon is discussed in the classical example of a 2D configuration with four magnetic sources of interchanging polarity and with the 3D topological model described in Section 16.5.2. Converging or diverging motions in the photosphere create a thin reconnecting current sheet at the separator – the line where separatrix surfaces are crossing. Shearing motions generate highly concentrated currents at the separatrices. We discuss their properties and point out that

> the interruption of field-aligned electric currents by the magnetic reconnection process at the separator can be responsible for fast energy release

in solar flares, in active regions with observed large shear as well as in quiet regions above the 'magnetic carpet' responsible for heating of the quiet corona.

22.2 Flare energy storage and release

22.2.1 From early models to future investigations

It has for a long time been clear that the energy released in flares is stored originally as magnetic energy of electric currents in the solar atmosphere. At least, there do not appear to be any other sources of energy which are adequate. Simple estimates of the *free* magnetic energy content in typical active regions (e.g., Den and Somov, 1989) show that it generally exceeds the observed energy of flares as well as the energy which is necessary for coronal heating in active regions. Free magnetic energy can, in principle, be converted into kinetic and thermal energy of the solar plasma with particle acceleration to high energies and other things that can be observed in the solar atmosphere and interplanetary space. This is the flare or, more exactly, the solar flare problem.

Jacobsen and Carlqvist (1964), Alfvén and Carlqvist (1967) were the first to suggest that

> the interruption of electric currents in the solar corona creates strong electric fields that accelerate particles during flares.

This mechanism of magnetic energy release and its conversion into thermal and supra-thermal energies of particles has been considered and well developed by many authors (e.g., Baum, Bratenahl, and Kamin, 1978). The interruption of current was described as the formation of an electrostatic *double layer* within a current system – an electric circuit – storing the flare energy.

The formation of the double layer *locally* leads to a direct acceleration of particles. However, because the potential, which gives this acceleration, must be maintained by the external system, the *global* effects of the double layers are not small. In general, they lead to an MHD relaxation of the surrouding magnetic field-plasma configuration providing the influx of energy which is dissipated by the double layers (Raadu, 1984).

* * *

An alternative approach to the solar flare problem was introduced by Giovanelli (1946, 1947, 1948), Dungey (1958) and Sweet (1958). After them, it was believed that the solar flare energy can be accumulated as magnetic energy of reconnecting current sheets

in the place of magnetic flux interaction and redistribution, more exactly, at the *separators* (Sweet, 1958). This idea was well supported by many analytical investigations, by laboratory and numerical experiments (for a review see Syrovatskii, 1981; Priest, 1985; Somov, 1992), by observations of the reconnection process in space plasmas (e.g., Hones, 1984; Berger, 1988) and especially on the Sun (e.g., Tsuneta, 1993; Demoulin *et al.*, 1993; Bagalá *et al.*, 1995).

In fact, the laboratory experiment by Stenzel and Gekelman (1984) clearly indicated the appearance of double layers in the reconnecting current sheet. This means that local interruptions of the electric current, induced by reconnection, can exist in the place of magnetic-field line reconnection. In what follows, we will consider another effect – magnetic reconnection of electric currents – the physical phenomenon which is different from the creation of an ordinary double layer in the reconnecting current sheet or in the field-aligned current.

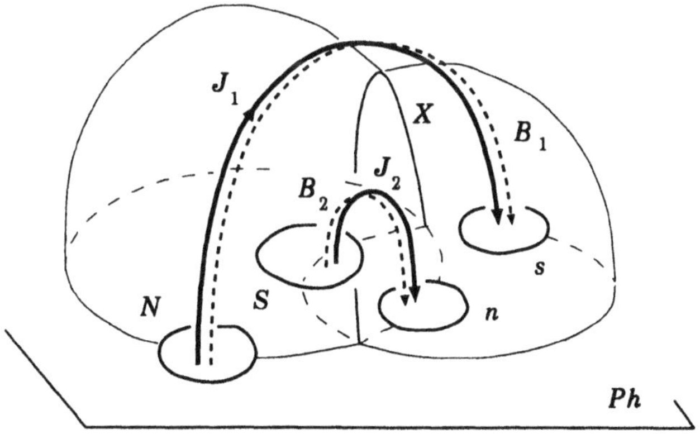

Figure 22.1: A 3D topological model of the coronal magnetic field in an active complex with four magnetic sources of interchanging polarities in the photosphere and with the two systems of currents in the corona.

Hénoux and Somov (1987) considered two systems of coronal currents J_1 and J_2 distributed inside two different magnetic cells interacting along the separator X as shown in Figure 22.1. Such a model for an active region complex is, in fact, the case of the magnetic topology described in Section 16.5.2. The two field lines B_1 and B_2 connect the 'old' (N, S) and 'new' (n, s) centres of activity (active regions). The coronal currents that flow from one magnetic flux region to the other (from the old region to the new one) are distributed inside the two different cells and shown schematically as the total currents J_1 and J_2 along the field lines B_1 and B_2.

For simplicity, in Figure 22.2 the geometry of the same magnetic field lines and currents is illustrated in the case where the old and new bipolar regions are aligned. The field lines B_1 and B_2 near the reconnecting current

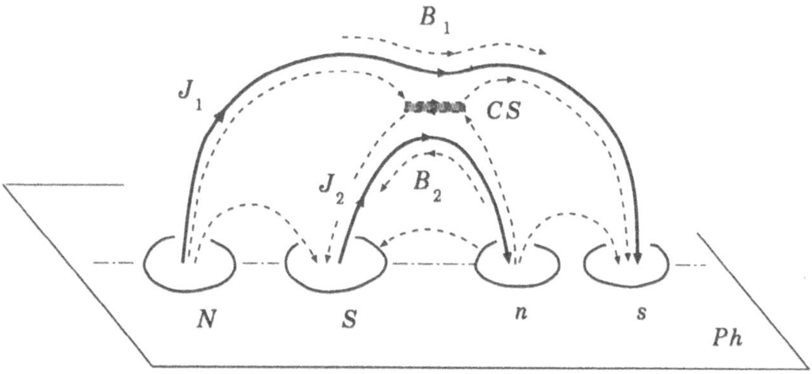

Figure 22.2: Coronal currents for the aligned old and new bipolar regions.

sheet CS along the separator (cf. Figure 16.8) have an opposite direction and can be reconnected. The two current systems J_1 and J_2 can be close to each other near the separator. Moreover, in the case under consideration, the currents flow in the same direction. Therefore, as in Gold and Hoyle (1960), Sakai and de Jager (1996), they attract each other. So the field-aligned electric currents have to modify the equilibrium conditions for the reconnecting current sheet CS along the separator (Hénoux and Somov, 1987).

> The components of the magnetic field transversal to the separator reconnect together with electric currents flowing along them.

In this way, with a perpendicular (see Figure 7.1) magnetic field inside the place of interruption, magnetic reconnection creates local interruptions of the electric currents in the solar atmosphere. If these currents are highly concentrated, their interruption can give rise to strong electric fields accelerating particles and can contribute significantly to the flare energetics.

Consider the magnetic fields created by the currents. These additional or secondary fields are perpendicular to the currents; hence they are parallel to the separator. So they play the role of the longitudinal magnetic field near the reconnecting current sheet (Section 17.2.2). Being superimposed on the potential magnetic field, the additional field components B_φ create two field line spirals: left-handed and right-handed (Figure 22.3a). When looking

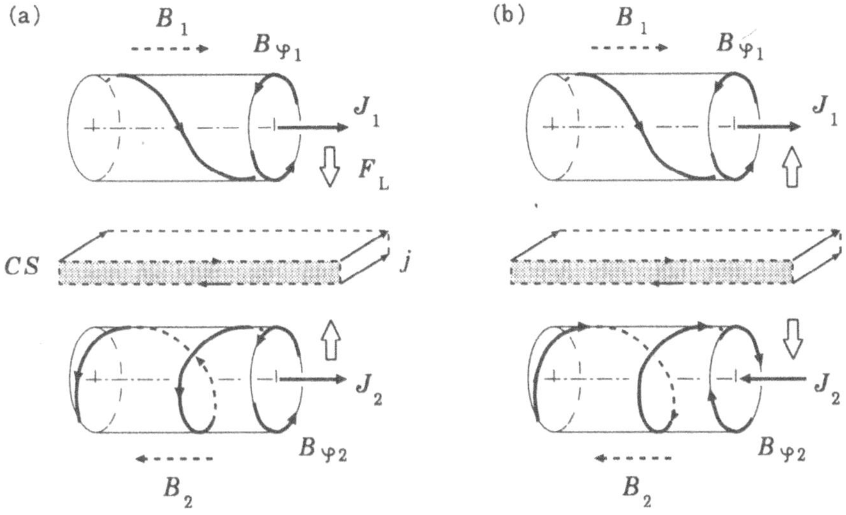

Figure 22.3: Two possible orientations of twist in two interacting magnetic flux-tubes with field-aligned electric currents.

along the positive direction of the main field lines B_1 and B_2, we see the two opposite orientations for the spirals: namely to the right for the dextral structure and to the left for the senistral one.

When the currents flow in the same direction, as was shown in Figure 22.2, the azimuthal components $B_{\varphi 1}$ and $B_{\varphi 2}$ have the same direction of rotation. Being opposite inside the reconnecting current sheet CS, they reconnect fully or partially. At the same time, the Lorentz force F_L pushes the parallel currents one to another. Therefore the case shown in Figure 22.3a is the most favourable for reconnection of magnetic fields and field-aligned electric currents.

On the contrary, if the currents are antiparallel, as shown in Figure 22.3b, the azimuthal components $B_{\varphi 1}$ and $B_{\varphi 2}$ cannot be reconnected. They are compressed and they decrease the reconnection rate for the main components of the magnetic fields B_1 and B_2, as it was discussed in Section 17.2.2. Hence a handedness property known as chirality does influence upon the magnetic reconnection of electric currents.

This is a qualitative picture of reconnection of the field-aligned electric current according to Hénoux and Somov (1987). Physical properties of the electric current reconnection in a highly-magnitized plasma have not been investigated yet. Many of them remain to be understood, in particular, the role of Hall's and perpendicular conductivities (Section 7.3) at the place of the electric current rupture and the role of plasma motions generated there. However, it is clear that magnetic reconnection changes the path of an electric current circuit. Because of large dimensions, the current circuit in the corona has a huge inductance. So a large inductive voltage can be generated locally, leading to a complex electrodynamic phenomenon with particle acceleration to high energies.

The review of the present situation in the solar flare theory will help us to understand the basic features of the electric current reconnection phenomenon in Section 22.4, see also Somov and Hénoux (1999).

22.2.2 Some alternative trends in the flare theory

A potential field in an active region contains a minimal energy which cannot be extracted from the plasma-magnetic field system. It was a question whether or not it is possible to explain the pre-flare energy storage in the force-free approximation, i.e only with electric currents aligned with the magnetic field (Section 8.3). This idea never looked too promissing, except in some investigations (see Sturrock, 1991) that suggested that the energy of a force-free field (FFF) generated by footpoint shearing motions can exceed the energy of the 'completely open' field having the same boundary condition (the same vertical component) in the photospheric plane. If this were true, we could expect an explosive opening of such an FFF configuration with a fast release of excess energy. Then spontaneous eruptive opening could be a good model for coronal transients or coronal mass ejections (CMEs).

Aly (1984), by using the virial theorem (Section 14.1), as well as without it (Aly, 1991), has shown that the energy of any FFF occupying a 'coronal half-space' is either infinite or smaller than the energy of the open field. So **the opening costs energy** and cannot occur spontaneously. The initial field must have free energy in excess of the threshold set by the open field limit. Only that excess is available to lift and drive the expelled plasma in CMEs or other similar phenomena (Sturrock, 1991).

This conclusion seems to be natural and could actually have important consequences for our understanding of non-steady phenomena with the opening of the coronal magnetic fields. Let us mention some of these consequences, bearing in mind, however, that coronal fields are never completely open or completely closed (see Low and Smith, 1993).

Generally, the electric currents flowing *across* the field allow the corona to have a magnetic energy in excess of the Aly's limit. These currents can be related to any non-magnetic forces; for example, the gravity force, the gradient of gas pressure or inertia forces. The problem arises because such forces are normally relatively weak in comparison with the magnetic force in the corona. Therefore the related effects can be considered as small corrections to the FFF (Section 8.3).

Another possibility is that the real currents in the corona comprise two different types: (i) **distributed currents** that are necessarily parallel to the magnetic field, so that the field is locally force-free; (ii) **thin current sheets** of different origin, in which the gas pressure gradient or other forces are significant. If, following Aly (1984, 1991), we could recognize the low efficiency of the smooth FFF (i) in energetics and dynamics of global eruptive events in the corona, we could well replace them by potential fields in evolution (e.g., Syrovatskii and Somov, 1980). This means that, to some extent, it is possible to neglect the field-aligned current in (i); we may call this approximation the **minimum current corona**. However, at least one exception can be important. It is discussed in the next Section.

If we do not consider flares or other flare-like events that open coronal fields, and if we do not investigate how to extract the accumulated energy from the FFF, then it is easy to conclude that the free magnetic energy can well be accumulated in FFFs, even if they are smoothly distributed. The basic idea here, used by many authors, is that photospheric footpoint motions stress the coronal field lines, inflate them, thereby producing free magnetic energy. For example, Porter *et al.* (1992) have studied the energy build-up in the stressed coronal fields possessing cylindrical symmetry. In the non-linear FFF approximation ($\alpha \neq const$), they have shown that

> a reasonable amount of the photospheric twist can produce enough free magnetic energy to power of a typical solar flare.

The rate of the energy build-up is enhanced if the greatest twist and/or the magnetic flux is concentrated closer to the photospheric neutral line.

22.2.3 Current sheets at separatrices

Analytically, by using the Grad-Shafranov equation, and numerically, by quasi-static MHD computations, Zwingmann, Schindler, and Birn (1985) have shown the occurence of current sheets near the separatrix in sheared field structures containing an X-type neutral point – the place where the separatrices cross. They interpret the break-down of the quasi-static theory near the separatrix as evidence for the appearance of a *boundary* layer with the current flowing parallel to the *poloidal* (Section 22.3) magnetic field.

Low (1991), Vekstein and Priest (1992) demonstrated analytically, in the force-free approximation, that shearing motions can produce current sheets along separatrices with or without neutral points. Numerical solutions of the time-dependent MHD equations by Karpen, Antiochos, and De Vore (1991), generally, confirmed the formation of currents in the frame of the line tying approximation. However, they showed that *true* (reconnecting) current sheets do not form in the solar corona when a more realistic atmospheric model is considered without a null point present in the initial potential field. These authors found more distributed currents, related to plasma inertia and the absence of a *true* static equilibrium, that cannot be considered as thin current sheets.

Therefore shearing motions in the photosphere can generate highly-concentrated electric currents flowing along and near separatrices. In this context, we can suggest a new mechanism of flare energy release – the *topological interruption* of electric currents in the solar atmosphere and their redistribution (Section 22.4). We shall consider two stages of its development. In the first, the electric currents are produced by photospheric shearing motions and the magnetic energy is stored in the system of concentrated field-aligned currents. In the second stage, the flare energy release takes place because a strong electric current system is approaching the separator and disrupted by the magnetic field line reconnection process in the separator region.

22.3 Current sheet formation mechanisms

22.3.1 Magnetic footpoints and their displacements

Let us discuss the possibility of topological interruption of coronal electric currents by using the classical example of a potential field in the plane (x, y) shown in Figure 22.4. Here e_i are the 'magnetic charges' placed on the x axis at the points with coordinates $(x_i\,,0)$, $i = 1$, 2, 3, and 4 at the underphotospheric plane $y = 0$. For simplicity we assume that they have interchanging balanced polarities: $\mathrm{e}_1 = -\,\mathrm{e}_4 = Q$ and $-\,\mathrm{e}_2 = \mathrm{e}_3 = q$. So these are the same magnetic charges as in Figure 16.6 but placed along a straight line – the x axis. This relative position of magnetic sources corresponds to the idealized one shown in Figure 16.5.

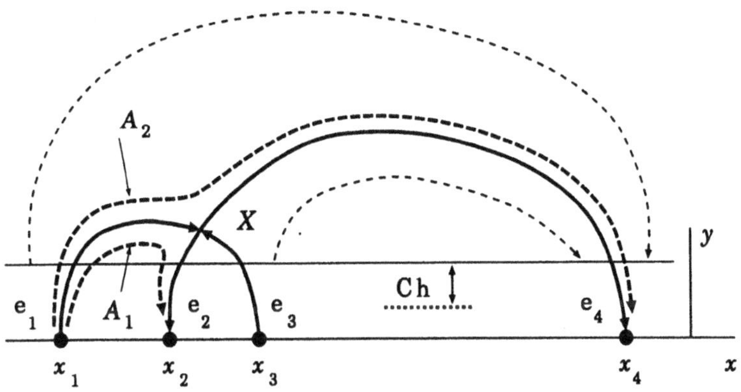

Figure 22.4: A 2D model of the magnetic field of four sources of interchanging polarities.

The solid curves show two separatrices (cf. Figure 4.17) crossing at the neutral point X which is the special topological line in the z direction – the separator (see also Figure 9.1). Two field lines are shown by the dashed curves A_1 and A_2. They start from the magnetic charge q_1, go near the neutral point, but arrive at different charges: q_2 and q_4 respectively. So they have different magnetic connectivity.

This is the *initial* configuration of a magnetic field. Just to keep the same notation as in the early works related with the controlled nuclear fusion (Morozov and Solov'ev, 1966a; Shafranov, 1966), we refer to a magnetic field in the plane (x, y) as the *poloidal* one. This part of the magnetic field $\mathbf{B}_p^{(0)}(x, y)$ can be described, according to (9.18), by the z component of the vector potential $\mathbf{A}$:

$$\mathbf{B}_p^{(0)}(x,y) = \left(\frac{\partial A^{(0)}}{\partial y}, -\frac{\partial A^{(0)}}{\partial x}, 0 \right), \quad \mathbf{A}^{(0)}(x,y) = \left(0, 0, A^{(0)}(x,y) \right).$$

In the case under consideration

$$A^{(0)}(x,y) = \sum_{i=1}^{4} \ln r_i \,, \quad \text{where} \quad r_i = \left[(x - x_i)^2 + y^2 \right]^{1/2} \tag{22.1}$$

(see Lavrent'ev and Shabat, 1973, Ch. 3, § 2).

Near the X-type point, where the field equals zero, the vector-potential can be written as (cf. formula (16.2)):

$$A^{(0)}(x,y) = \frac{1}{2} h_0 \left[-(x - x_0)^2 + (y - y_0)^2 \right], \tag{22.2}$$

with x_0 and y_0 being the coordinates of the neutral point. The constant which can be added to the vector-potential is selected in such a way that $A = 0$ on the sepatrices – the lines that separate the magnetic fluxes of different linkage (or connectivity).

The main aim of our treatment is to understand the relative efficiency in generation and dissipation of electric currents of different origin. Bearing this aim in mind we will consider different motions in the photosperic plane, i.d. different displacements of field line footpoints.

Following Low (1991), we will consider **three classes of displacements.** The displacements of the first class are strictly on the line of the magnetic charges – the x axis in Figure 22.4. These displacements model the converging, diverging or emerging motions of the magnetic sources in the photosphere. They keep the magnetic field lines in the plane of the initial field – the plane (x, y).

Shearing motions in the z direction belong to the second and third classes. The displacements of the second class are only 'antisymmetric in x', i.e. the photospheric velocity in the z direction is an odd function of x. No symmetry is prescribed for the third class of displacements.

22.3.2 Classical 2D reconnection

The displacements of the *first class* defined above do not create reconnecting current sheets in the absence of a neutral point X shown in Figure 22.4. The appearance of such a point on the boundary (for example, in the photosheric plane) is a necessary condition for the creation of a current sheet. A sufficient condition is the existence of a non-zero electric field in this point (Section 9.3). The magnetic field remains potential above the photospheric plane if the boundary conditions prohibit the appearance of a neutral point. In general, however, 'a neutral point begins to appear' on the boundary surface (Somov and Syrovatskii, 1972b; Low, 1991) and the reconnecting current sheet is generated in it by the electric field.

Let us consider, as the simplest example, a symmetrical initial distribution of magnetic charges shown in Figure 22.5a and the small symmetrical displacements of footpoints x_2 and x_3 as follows $\delta x_2 = -\delta x_3 = \delta x(t)$; they are shown in Figure 22.5b. In the presence of the neutral line X, in its vicinity, the electromagnetic field can be expressed through the vector-potential (Syrovatskii, 1966a, 1971)

$$A\,(x, y, t) = A^{(0)}\,(x, y) + \delta A(t)\,. \tag{22.3}$$

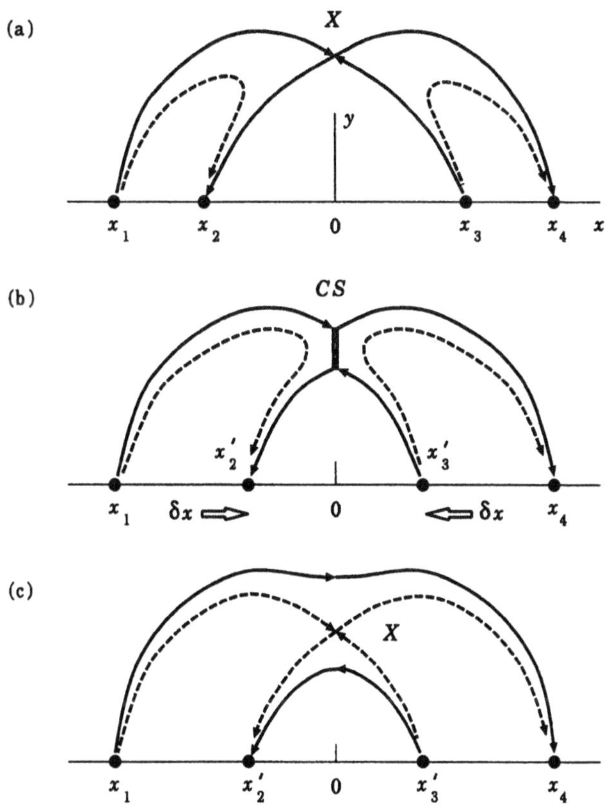

Figure 22.5: (a) The initial field configuration; (b) the formation of the current sheet CS under the converging motion of footpoints x_2 and x_3; (c) the disappearance of the current sheet when the field relaxes to the new potential state.

Here $\delta A(t)$ is the value of the magnetic flux which has to be reconnected in the current sheet at the neutral point. Then, after the reconnection time τ_r, the magnetic field will be potential one again, but with new positions of the footpoints $x_2 + \delta x$, $x_3 - \delta x$. The value $\delta A(t)$ is proportional to the displacement δx.

It is clear from formula (22.3) that in the vicinity of the neutral line there is a uniform electric field directed along the line:

$$\mathbf{E} = -\frac{1}{c}\frac{\partial}{\partial t}\mathbf{A} = (0, 0, E_z), \quad \text{where} \quad E_z = -\frac{1}{c}\frac{\partial\, \delta A(t)}{\partial t}. \tag{22.4}$$

It is just this field which produces an electric current $\mathbf{J}$ along the neutral line (Figure 4.18b) as well as a drift motion of plasma outside the line (Figure 4.18a). In a time of the order of the Alfvén time $\tau_{\rm A}$, the current sheet is formed along the neutral line.

Figure 22.5b schematically illustrates the process of the current sheet formation induced by the photospheric displacements δx of the first class. The relaxation of the magnetic field which contains the current sheet to the potential field corresponding to the new boundary conditions is shown in Figure 22.5c.

22.3.3 The creation of current sheets by shearing motions

Consider some general properties of the field component B_z from the initial field (Figure 22.4) generated by a shearing displacement $\delta z\,(x)$ in the FFF approximation. To study plasma equilibrium and stability, it is convenient to use the *specific* volume of the magnetic flux tube (Section 14.3). This is the ratio of the geometrical volume of the flux tube $d\,V$ to the enclosed magnetic flux $d\,\Phi$, i.e.

$$U = \frac{d\,V}{d\,\Phi}\,.$$

For a field line specified by a given value of vector-potential A, by invoking the conservation of magnetic flux inside the tube, the specific volume is

$$U\,(A) = \int \frac{d\,l}{B}\,. \tag{22.5}$$

The integral in (22.5) is taken along the field line between two certain appropriate points corresponding to the beginning and the end of the tube. For the example considered in Figure 22.4, the beginning and the end of a tube are defined by the photospheric points x_1 and x_2 for all field lines connecting these points above the photospheric plane:

$$U\,(A) = \int\limits_{x_1}^{x_2} \frac{d\,l}{B_p^{(0)}\,(x,y)}\,. \tag{22.6}$$

By integrating the differential equation for a magnetic field line

$$\frac{dz}{B_z} = \frac{d\,l}{B_p^{(0)}\,(x,y)}\,,$$

taking account of (22.6), we see that the *toroidal* component B_z is given by the displacement of field line footpoints at the boundary plane $y = 0$:

$$B_z(A) = \frac{\delta z(A)}{U(A)}. \tag{22.7}$$

We see from (22.7) that, even if the displacement δz is a continuous function of x, a problem may arise for the following reason. In the presence of topological features like X-type points, the different field lines, by having different footpoints x_i in the photosphere and different footpoint displacements δx_i, may have the same values of A. Therefore discontinuities of B_z may appear above the photospheric plane.

Zwingmann, Schindler, and Birn (1985) have illustrated this important feature of sheared magnetic fields analytically by considering the FFF locally near a hyperbolic X-point of the form (cf. formula (22.2)):

$$A^{(0)}(x,y) = -\frac{ax^2}{2} + \frac{by^2}{2} \quad \text{with} \quad a \neq b. \tag{22.8}$$

They showed that the specific volume has a logarithmic divergence for A corresponding to the separatrices that cross at the X-point, i.e. for $A = 0$. This means, first of all, that one of the diverging physical quantities is the poloidal current density

$$\mathbf{j}_p = \operatorname{curl} \mathbf{B}_z = \frac{d\mathbf{B}_z(A)}{dA} \cdot \mathbf{B}_p^{(0)} \propto \frac{1}{A \ln^2 A}. \tag{22.9}$$

The total current integrated in the direction perpendicular to the initial poloidal field $\mathbf{B}_p^{(0)}$ is finite:

$$J_t = \int_{A_1}^{A_2} \frac{d\mathbf{B}_z(A)}{dA} \, dA = \mathbf{B}_z(A_2) - \mathbf{B}_z(A_1). \tag{22.10}$$

We are therefore led to the conclusion that

> shearing motions induce current sheets extending along the separatrices, with the current flowing parallel to the poloidal field.

This fact was also tested by numerical computations (Zwingmann, Schindler, and Birn, 1985) which take into account the physical effects that in real plasmas keep the current density from becoming infinitely large (see also Section 22.4).

22.3.4 Antisymmetrical shearing motions

The conclusion made above is valid even in the cases of very high symmetry, e.g. if the displacements are *antisymmetric*, and the initial potential field is symmetric (Figure 22.5) with respect to the y axis. This is clear from the following example. Let $x_1 = -x_4$, $x_2 = -x_3$, and $\delta z_1 = -\delta z_4 = \delta Z$, $\delta z_2 = -\delta z_3 = \delta z$ as shown in Figure 22.6.

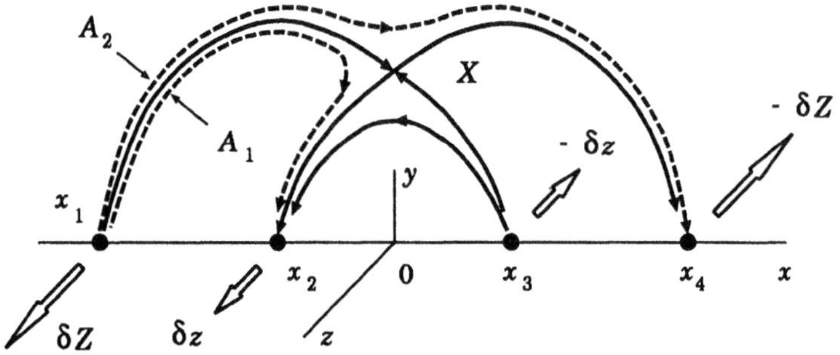

Figure 22.6: A 2D initial magnetic field configuration and the antisymmetric shearing motions of footpoints δZ and δz.

The specific volume of the magnetic flux tube which goes along the field line A_1 from the point x_1 very near the neutral X-point to the point x_2 consists of two terms

$$U(A_1) = \int_{x_1}^{X} \frac{dl}{B_p^{(0)}(x,y)} + \int_{X}^{x_2} \frac{dl}{B_p^{(0)}(x,y)} \equiv U_{1,X} + U_{X,2}\,.$$

According to (22.7) the toroidal (or longitudinal) component of the magnetic field is equal to

$$B_z(A_1) = \frac{\delta z_2 - \delta z_1}{U_{1,X} + U_{X,2}}\,. \tag{22.11}$$

For the field line A_2 which goes from x_1 to x_4 very near the X-point, with account of the symmetry described above, we find the specific volume

$$U(A_2) = U_{1,X} + U_{X,4} = 2\,U_{1,X}$$

and the relative displacement $\delta z = \delta z_4 - \delta z_1 = -2\,\delta z_1$. So

$$B_z(A_2) = -\frac{\delta z_1}{U_{1X}} \neq B_z(A_1)\,. \tag{22.12}$$

Hence an antisymmetric shear creates the discontinuity of the toroidal field, i.e. the current sheet with total current (22.10) along the separatrices, in the presence of X-type point even if the initial potential field is symmetric.

Consider another example. Let the shearing motions be antisymmetric and the initial magnetic field be symmetric, but with the neutral point placed below the level of the photosheric plane (Low, 1991). In this case the separatrix surface separates two 'magnetic islands' from each other at the point $x = 0$ and $y = 0$ as well as separating them from the surrounding field at the total separatrix surface in Figure 22.7. In this way the connectivity of

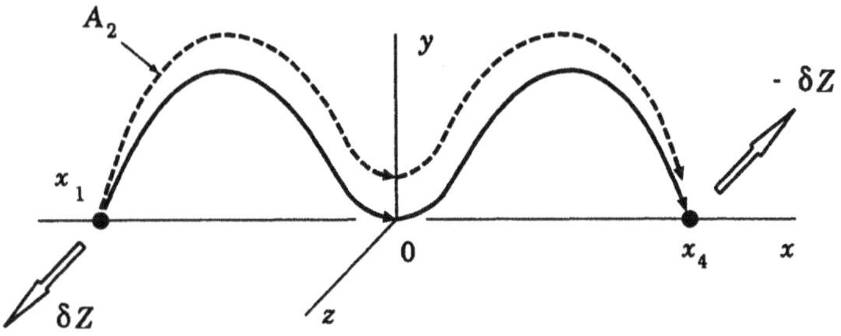

Figure 22.7: A 2D potential magnetic field of the quadrupole type without a neutral point above the photospheric plane.

the magnetic field is discontinuous, and one may in principal expect the creation of magnetic field discontinuities. However, because of the symmetry, the specific volume is

$$U\,(A_2) = U_{1,O} + U_{O,4} = 2\,U_{1,O}$$

with a relative displacement $\delta z = \delta z_4 - \delta z_1 = -2\,\delta z_1$, therefore

$$B_z\,(A_2) = B_z\,(A_1)\,. \tag{22.13}$$

We see that the second class of boundary motions cannot create current sheets in the absence of neutral points (Figure 22.7). However, an antisymmetric shear creates current sheets with the currents flowing along separatrices in the plane (x, y) in the presence of a neutral point, even if the initial potential field is symmetrical one (Figure 22.6).

All the other shearing boundary displacements directed in the z direction are called the third class, according to the classification by Low (1991), and are discussed in the next Section.

22.3.5 The third class of displacements

Several examples of the third class of displacements, including those which are symmetrical in x, were studied by Low (1991); Vekstein and Priest (1992). It was shown that these shearing displacements can create discontinuities of the B_z component which are related with electric currents along separatrices. The displacements can generate such current sheets even in the absence of a neutral point, but the separatrices are necessary of course.

The general boundary displacement is a superposition of displacements from all these three classes. Titov, Priest, and Demoulin (1993) have demonstrated the existence of sections of the photospheric polarity inversion line where the overlying field lines are parallel to the photosphere (like in Figure 22.7). Such sections, called 'bald patches', may exist for a wide range of fields created by four concentrated sources of magnetic flux (Gorbachev and Somov, 1989, 1990; Lau, 1993). Bald patches appear, for example, when the photospheric neutral line is bent too much in an S-like manner, because this is the case of the separator appearance (Somov, 1985; Somov and Merenkova, 1999). The field lines touching a patch belong to a separatrix surface along which a current sheet may be formed by shearing motions of magnetic footpoints at the photosphere.

In the next Section we will discuss the mechanisms which determine the real thickness and other properties of the current sheets.

22.4 The shear and reconnection of currents

22.4.1 Physical processes related to shear and reconnection

Let us start by discussing the second and third classes of displacements. Since the current density $\mathbf{j}_p$ is parallel to the poloidal field $\mathbf{B}_p^{(0)}$ (see formula (22.9)), the plasma velocity $\mathbf{v}_z$ and the total magnetic field $\mathbf{B}_t = \mathbf{B}_p^{(0)} + \mathbf{B}_z$ are parallel to the discontinuity surface which coincides locally with the plane tangential to the separatrix. In this case, all the MHD boundary conditions are satisfied identically except one (Section 11.2.2):

$$p_1 + \frac{\mathbf{B}_1^2}{8\pi} = p_2 + \frac{\mathbf{B}_2^2}{8\pi} \, . \tag{22.14}$$

This means that the velocity and the magnetic field may experience arbitrary jumps in magnitude and direction, being parallel to the discontinuity surface. The only requirement is that the total pressure, i.e. the sum of the gas pressure and the magnetic one, remains continuous at the discontinuity surface.

According to the general classification of MHD discontinuities given in Section 11.2, these discontinuities, generated by shearing motions, are usual tangetial discontinuities, except that the plasma velocities in the z direction are small in comparison with the Alfvén speed in the solar corona because the magnetic field is strong there. Therefore, until we take into account the effect discussed at the end of Section 22.4.3,

> we can consider MHD tangential discontinuities as a good model for highly concentrated currents at separatrices, generated by shearing motions in the photosphere.

As treated in MHD, tangential discontinuities have several remarkable properties. One of them is important for what follows. Even in a plasma of very high conductivity, such as the coronal plasma, a tangential discontinuity is a *non-evolutionary* discontinuity (Section 12.1). This means that, in contrast to the behaviour of the reconnecting current sheet, there is not a steady solution, the stability of which can be considered in the linear approximation.

The origin of this effect lies in the fact that the thickness of a tangential discontinuity is a continuously growing value if the electrical conductivity is finite. After its creation the $\mathbf{B}_z$ component starts to evolve in accordance with the diffusion equation

$$\frac{\partial B_z}{\partial t} = \frac{\partial}{\partial s}\left(\nu_m \frac{\partial B_z}{\partial s}\right). \tag{22.15}$$

Here ν_m is the magnetic diffusivity, s is the coordinate ortogonal to the discontinuity surface. By virtue of Equation (22.15), the total magnetic flux of $\mathbf{B}_z$ does not change:

$$\frac{\partial}{\partial t}\int_{-\infty}^{+\infty} B_z\, ds = \nu_m \left.\frac{\partial B_z}{\partial s}\right|_{-\infty}^{+\infty} = 0. \tag{22.16}$$

The thickness of a tangential discontinuity is increasing, but a part of the excess magnetic energy related with a tangential discontinuity is released in the continuous process in the form of Joule heating at a rate

$$\frac{\partial}{\partial t}\int_{-\infty}^{+\infty} \frac{{B_z}^2}{8\pi}\, ds = -\frac{1}{4\pi}\int_{-\infty}^{+\infty} \nu_m \left(\frac{\partial B_z}{\partial s}\right)^2 ds \neq 0. \tag{22.17}$$

Magnetic diffusion always acts to smooth out gradients in both the magnetic field and the electric current density, not to concentrate them. This property has been well demonstrated by many numerical computations.

In the reconnecting current sheet, however, the process of magnetic diffusion away from the discontinuity is compensated by the plasma drift motions into the sheet. That is why the steady state for the reconnecting current sheet can exist with the sheet width

$$a = \nu_m \, v_d^{-1} \, , \tag{22.18}$$

where v_d is the drift velocity, and the reconnecting current sheet at separator can be considered as an evolutionary discontinuity (see Chapter 19). So

> there is a principal difference between the reconnecting current sheet at the separator and the current sheets at separatrices.

It is important that it is not possible to consider the reconnecting current sheet as a one-dimensional discontinuity because the plasma coming into the sheet has to be compensated by plasma outflow from it. These two conditions are necessary for the existence of steady states for the reconnecting current sheet.

As for tangential discontinuities generated by shearing motions in the photoshere, their electric currents are always spreading out in both directions from sepatrix surfaces into the surrounding coronal plasma. By doing so, a part of the electric current flowing along the separatrices appears on the field lines which have already been reconnected (see Fig. 22.4), but the remaining

> part of the electric current will be reconnected later on together with the field lines which have not been reconnected yet.

Hence we have to consider how electric currents flowing along the magnetic field lines reconnect with them.

We will not discuss here all other mechanisms (except presumably the most important one in Section 22.4.3) which make the tangential discontinuity currents more distributed rather than concentrated. Neither will we discuss the generation of the electric currents of different origin in the solar corona, for example, currents due to variations in plasma response time (because of plasma inertia) at different heights in the solar atmosphere, nor currents related to the absence of a *true* static equilibrium (Karpen, Antiochos, and De Vore, 1991). We only would like to point out that electric currents of different origin, being field-aligned after their generation (see Spicer, 1982), may participate in the process of magnetic field line reconnection.

22.4.2 The topological interruption of electric currents

The magnetic reconnection process does the same with electric currents as with magnetic field lines, i.e. it disrupts them and connects them in a different way. Physical consequences of the phenomenon have not yet been well investigated, but some of them look clear and unavoidable.

The first of these, an interruption of the electric current, produces an electric field. It is neccesary to note here that if reconnection of magnetic field lines would create symmetrical reconnection of currents, then one electric current, J_1, should replace another one, J_2, which is equal to the first current, and no electric field could be induced in such a way. Such coincidence has zero probalility.

In general, the reconnected currents are not equal among themselves; hence the current $(J_1 - J_2)$ is actually interrupted at the X point of reconnection. This process creates an electric field at the separator.

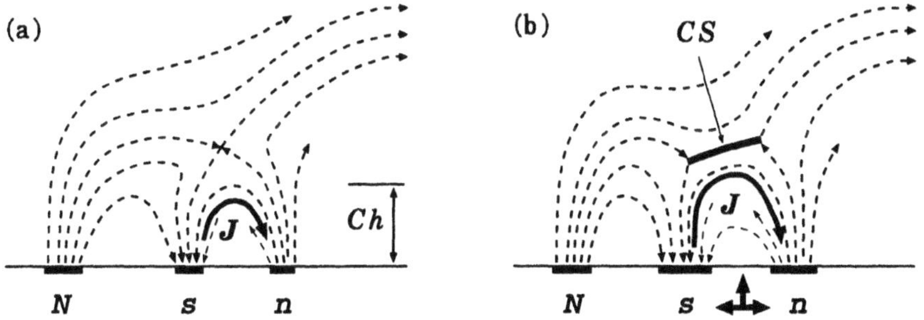

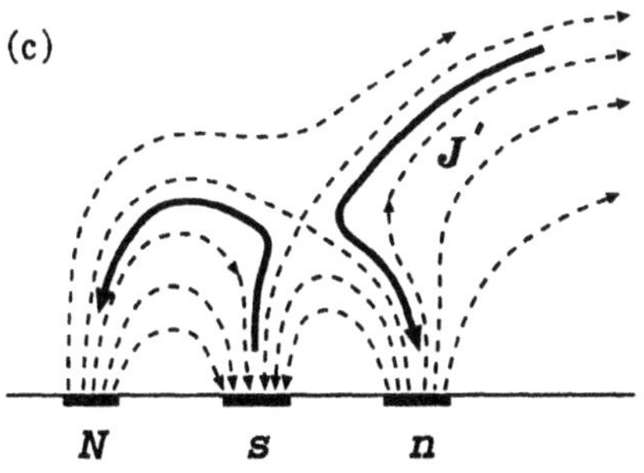

Figure 22.8: A reconnecting field with electric current: (a) the initial state is mainly potential but contains a loop of new emerging flux which carries a current J, (b) the *pre-reconnection* state, (c) the final state after recoonection of the magnetic field lines and field-aligned electric currents.

The simplest but realistic example is the case where we can neglect one of the currents; e.g., $J_2 = 0$. Figure 22.8 shows such example. A new emerging magnetic flux moves upward together with electric current J. This current is disrupted by the reconnection process in the current sheet CS and appears to be connected into new electric circuits.

22.4.3 The inductive change of energy

The second consequence of non-symmetrical reconnection of electric currents is related to the fact that the current $(J_1 - J_2)$ is connected in another electric circuit which, in general, has another self-inductance L. Hence the reconnection of the current $(J_1 - J_2)$ changes the energy of the current system $W = LJ^2/2$ and its inductive time scale $\tau_L = L/R$. A larger circuit implies a larger energy, but a longer inductive time scale.

Zuccarello *et al.* (1987) pointed out that the energy release in a flare should not be attributed to current dissipation but rather to a change in the current pattern that reduces the stored magnetic energy. They introduced an example of how self-inductance and energy storage can be changed in a sheared FFF arcade. In fact, the inductive change of energy can be reversed, with the stored energy being resupplied on the inductive time scale.

There is an essential advantage in our model of reconnecting electric currents. The topological interruption of large-scale electric currents flowing along and near separatrix surfaces does not require an increase of the total resistivity R everywhere the currents flow but only in the place where these surfaces cross, i.e. along the separator line. More exactly, the plasma resistivity must be increased, for example by excitation of plasma turbulence, only inside the very thin reconnecting current sheet at the separator. Otherwise the reconnection process will be too slow and the rate of energy release insufficient for a typical flare.

Another important property of the model under consideration is that magnetic reconnection, when it is fast enough, restricts the current density j_p of electric currents flowing along the separatrix surfaces and near them. The mechanism of this restriction is the same topological one.

If the characteristic time τ_x of δx displacements which drive reconnection is comparable with the reconnection time scale τ_r, then the field lines connecting the footpoints x_i with the X-type point (see Figure 22.5a) will not play the role of separatrices anylonger after the time τ_r. New magnetic field lines, shown by the dashed curves in Figure 22.5c, with footpoints $x_i' = x_i + \delta x_i$ will be the place where a new portion of shearing motions will produce a new portion of highly concentrated currents along these field lines, but not the previous ones. Therefore the real velocities of the footpoint displacements and the real reconnection rate determine the real distribution of concentrated electric currents generated by shearing motions in the photosphere.

22.4.4 To the future observations by *Solar-B*

Magnetic reconnection of electric currents generated by shearing motions in the photoshere may play significant role in the energetics of solar flares related to *observed* photospheric shear. Other flares may not have any significant shear. So they have a different kind of electric currents related, for example, to diverging and converging motions in the photosphere near the region of a newly emerging flux, which we called the first class displacements.

To understand the relative role of different electric currents in the energetics and dynamics of an active region,

> it is necessary to study the evolution of its magnetic structure in and above the photoshere.

This will allow us to determine not only the magnetic fluxes of certain magnetic links, but also their changes – redistribution and reconnection. Such a study will also give us information, at least qualitative, about the structure and evolution of the electric field in an active region.

As well known, NASA has announced the selection of three experiments to be flown on the Japan Institute of Space and Astronautical Science (ISAS) *Solar-B* mission planned for launch in 2004. The objective of *Solar-B* is to study the origin of the corona and the coupling between the fine magnetic structure in the photosphere and the dynamic processes occuring in the corona.

As currently envisaged by the ISAS, the *Solar-B* mission will include the equipment for the three investigations: a 50-cm optical telescope with sophisticated focal plane instrumentation, an X-ray telescope for imaging the high-temperature coronal plasma, and an extreme-ultraviolet imaging spectrometer for diagnosing events observed. The main telescope will give quantitative measurements of the magnetic fields in features as small as 100 km in size thereby providing 10 times better resolution than other space- and ground-based magnetic field measurements.

Epilogue

Most of the known matter in the Universe is in an ionized state, and many naturally occuring plasmas, such as the atmosphere of the Sun and magnetic stars, the magnetospheres of the Earth and other planets, the magnetospheres of pulsars and other relativistic objects, galactic and extragalactic jets, exibit distinctively plasma-dynamical phenomena arising from the effects of magnetic and electric forces. The science of cosmic plasma physics was born and developed to provide an understanding of these naturally occurring plasmas and those which will be discovered and investigated in future space observations. With this aim, from the very beginning, **many of the conceptual tools** and many different approaches were introduced and developed in the course of general fundamental research on the plasma state or independently. How can we understand the interconnection between different descriptions of cosmic plasma behavior?

I was frequently asked by my students to give them a quick introduction to the theory of cosmic plasma. It turned out that it is not easy to do for many reasons. The most important of them is that the usual way of such an introduction is generalization. This means that we go from simple well-known things to more complicated ones, for example, we generalize the ordinary hydrodynamics to magnetohydrodynamics. Though this certainly makes a textbook easier to read, it does not give the reader complete knowledge of the subject, the tools esspecially. For a long time, my goal was to write a book which I would myself had liked when I first took up the subject, physics of cosmic plasma, and which I could recommend to my students to provide them an **accessible introduction** to physics of cosmic plasma at least at an intuitive level of the basic concepts.

We began a long journey together, when we first started this book, and we are now almost at that journey's end. A unifying theme of the book was the attempt at a deeper understanding of the underlying physics.

Starting from the most general physical principles, we have seen the consecutive simplifications of them and of simplifying assumptions which allowed us to obtain a simpler description of plasma under cosmic conditions. In so doing, the boundaries of the domain of applicability for the approximation at hand were well outlined from the viewpoint of physics and possible applications.

On the basis of this approach we find the answers to the key questions: (1) what approximation is the simplest but a sufficient one for a description of a phenomenon in cosmic plasma; (2) how to build an adequate model for the phenomenon, for example, a solar flare.

Practice is really important in the theory of cosmic plasma; related exercises (problems and answers supplemented to each chapter) surved to better understanding of cosmic plasma physics. Most of the problems for students have been used as homework in the lecture course. A particular feature of the problems is that they widely range in difficulty from fairly straightforward (useful for an exam) to quite challenging. This property is not an advantage or disadvantage of the book but rather **a current state of modern astrophysics**.

As for applications, evidently preference was given to physical processes in the solar plasma. The Sun is unique in the astrophysical realm for the great diversity of the diagnostic data that are available. Much attention to solar plasma physics was and will be conditioned by the possibility of the all-round observational test of theoretical models.

Some thirty-fourty years ago it was still possible, as Alfvén and Fälthammar (1963) so ably demonstrated, to write a single book on cosmic plasma theory concerning practically everything worth knowing of the subject. The subsequent development has been explosive, and today a corresponding comprehensive coverage would require a hole library. The present book is an earnest attempt to a general overview of the whole area but big gaps unavoidably appear. Important and interesting effects and problems have been skipped because I either felt to go too far beyond an introductory text or, worse, I have not been aware of them.

> There would be infinitely more to say about observations, numerical simulations, and analytical investigations of cosmic plasma.

Any reader who, after having read this book, would like to become acquainted with profound results of cosmic plasma physics should keep this fact in mind. I hope, however, that he/she, having learned sufficiently many topics of this textbook, will willingly and easily fill the gaps. Good luck!

Appendix 1. Notation

Latin alphabet

Symbol	*Description*	*Introduced in Section (Formula)*
a	current sheet half-thickness	6.3
$\mathbf{A}$	vector potential of a magnetic field	4.4
b	current sheet half-width	6.3
$\mathbf{b}$	perturbation of a magnetic field	16.1
$\mathbf{B}$	magnetic field	1.1
$\mathbf{B}_\tau$	tangential magnetic field	11.2
c	speed of light	4.1
d	thickness of non-adiabatic region	18.1
e, e_a	electric charge	1.1
$\mathbf{e}_c$	unit vector from the curvature centre	4.2
$\mathcal{E}$	energy of a particle	4.1
$\mathbf{E}$	electric field	1.1
$\mathbf{E}_u$	electric field in the plasma rest-frame	7.3
f, f_k	averaged distribution function for particles of kind k	1.2
f_{kl}	binary correlation function	2.2
f_{kln}	triple correlation function	2.3
$\hat{f}_k$	exact distribution function for particles of kind k	2.2
F	complex potential	9.2
$\mathbf{F}, \mathbf{F}_k$	force	1.2
$\langle \mathbf{F}_k \rangle_v$	mean force per unit volume	7.1
$\mathbf{F}_{\mathbf{kl}}$	force density in the phase space	2.2

$\mathbf{F}'$	fluctuating force	2.1
g	velocity-integrated correlation function	2.5
$\mathbf{g}$	gravitational acceleration	4.1
G	gravitational constant	1.1
$\mathbf{G}$	energy flux density	(1.41)
h	magnetic field gradient	4.4
$\mathbf{h}$	magnetic field at a wave front	15.1
H	Hamiltonian	18.2
$\mathcal{H}$	magnetic helicity	21.1
Ha	Hartmann number	15.2
$\mathbf{j}$	electric current density	1.1
$\mathbf{j}'$	current density in plasma the rest-frame	7.3
$\mathbf{j}_k^{\mathrm{q}}$	current density due to particles of kind k	7.1
$\mathbf{j}_k$	particle flux density in the phase space	2.4
J	electric current	14.3
k	friction coefficient	1.2
$\mathbf{k}$	wave vector	10.1
K	curvature of a magnetic field line	18.2
$\mathcal{K}$	kinetic energy of a particle	(4.92)
l	current sheet length	21.2
L	magnetic trap length	4.3
$L(T)$	radiative loss function	21.2
m	magnetic dipole moment	9.4
m, m_a	particle mass	1.1
M	mass of star	14.1
$\mathcal{M}$	magnetic moment of a particle	4.2
	magnetic energy of a system	14.1
n, n_k	number density	6.1
$\mathbf{n}$	unit vector along a magnetic field	4.1
N_k	number of particles of kind k	1.2
p	plasma pressure	7.2
p_k	gas pressure of particles of kind k	7.1
p_{m}	magnetic pressure	10.1
$p_{\alpha\beta}$	pressure tensor	7.1
$\mathbf{p}$	particle momentum	4.1
$\mathbf{P}$	generalized momentum	4.3
q	generalized coordinate	4.3
	electric charge	18.1

$\mathbf{q}$	heat flux density	7.2
$\mathbf{q}_k$	heat flux density due to particles of kind k	7.1
Q_k	rate of energy release in a gas of particles of kind k	7.1
r_{D}	Debye radius	6.2
r_{L}	Larmor radius	4.1
$\mathbf{r}$	radius vector	1.1
$\mathbf{r}_a$	coordinates of ath particle	1.1
R	radius of star	9.4
$R_{\perp}$	guiding centre spiral radius	4.2
$\mathcal{R}$	rigidity of a particle	4.1
$\mathbf{R}$	guiding centre vector	4.2
Re	Reynolds number	8.2
Re_{m}	magnetic Reynolds number	8.2
s	entropy per unit mass	8.1
T	temperature	8.1
	kinetic energy of a macroscopic motion	14.1
T_{B}	period of the Larmor rotation	4.2
$T_{\alpha\beta}$	Maxwellian stress tensor	7.2
$\mathbf{u}$	relative velocity	4.1
	velocity of the centre-of-mass system	7.3
	local Alfvén speed	10.1
	electric current velocity	16.3
$\mathbf{u}_e$	mean electron velocity	7.3
$\mathbf{u}_i$	mean ion velocity	7.3
$\mathbf{u}_k$	mean velocity of particles of kind k	7.1
U	interaction potential	6.1
	volume of a fluid particle	9.2
	specific volume of a magnetic tube	14.3
U_{th}	thermal energy	14.1
$\mathbf{U}$	velocity of the moving reference frame	11.2
	shock speed	12.1
$\mathbf{v}$	macroscopic velocity of a plasma	8.1
$\mathbf{v}, \mathbf{v}_a$	particle velocity	1.1
$\mathbf{v}_{\mathrm{d}}$	drift velocity	4.1
v_{n}	normal component of the velocity	11.2
v_x	velocity orthogonal to a discontinuity surface	11.1
$\mathbf{v}'$	deviation of particle velocity from its mean value	7.1
$\mathbf{v}_{\tau}$	tangential velocity	11.1

$v_{\parallel}$	velocity component along the magnetic field lines	4.1
$v_{\perp}$	transversal velocity	4.1
V	velocity of the plasma flow	21.2
V_{a}	gradient of the Alfvén speed	16.1
V_{A}	Alfvén speed	8.3
$\mathbf{V}_{\mathrm{gr}}$	group velocity of a wave	10.1
$\mathbf{V}_{\mathrm{ph}}$	phase velocity of a wave	10.1
V_{s}	sound speed velocity	10.1
V_{Te}	mean thermal velocity of electons	(4.88)
V_{Ti}	mean thermal velocity of ions	(4.87)
V_{Tp}	mean thermal velocity of protons	(4.89)
$V_{\pm}$	speed of a fast (slow) magnetoacoustic wave	10.1
w	probability density	2.1
w, w_k	heat function per unit mass	7.1
W	energy density of an electromagnetic field	(1.40)
x	ionisation degree	21.2
X	phase space	1.2
Z	ion charge number	6.2

Greek alphabet

Symbol	*Description*	*Introduced in Section (Formula)*
α_{B}	parameter of the magnetic field inhomogeneity	4.2
α_{E}	parameter of the electric field inhomogeneity	4.2
β	coefficient in an expulsion force	15.3
γ	dimensionless parameter of ideal MHD	8.3
γ_g	ratio of specific heats	11.1
Γ	$6N$-dimensional phase space	2.1
δ	dimensionless parameter of ideal MHD	8.2
ε	mean kinetic energy of a chaotic motion	7.2
	dimensionless parameter of ideal MHD	8.3
	dimensionless electric field	18.1
ε_α	small parameter of expansion	19.3
ζ	second viscosity coefficient	8.1
ζ_{i}	interaction parameter	2.4
ζ_{p}	plasma parameter	2.4
η	first viscosity coefficient (dynamic viscosity)	8.1
θ	pitch-angle	4.1
	angle between a wave vector and the magnetic field	10.1
κ_{e}	classical electron conductivity	6.3
λ	mean free path	6.1
$\ln\Lambda$	Coulomb logarithm	6.1
ν	collisional frequency	6.1
ν	kinematic viscosity	8.1
ν_{ei}	electron-ion mean collisional frequency	7.3
ν_{kl}	mean collisional frequency	7.1
ν_{ni}	neutral-ion mean collisional frequency	21.2
ν_{m}	magnetic diffusivity	8.1
ξ	column depth	6.3
	displacement of a current sheet	19.3
$\xi_{\parallel}$	dimensionless longitudinal magnetic field	18.1
$\xi_{\perp}$	dimensionless transverse magnetic field	18.1
$\boldsymbol{\xi}$	displacement of the medium	16.1

$\pi_{\alpha\beta}^{(k)}$	viscous stress tensor	7.1
Π	work against the Lorentz force	20.4
$\Pi_{\alpha\beta}^{*}$	total momentum flux density tensor	8.1
ρ	plasma mass density	7.1
ρ_k	mass density for particles of kind k	7.1
ρ^{q}	electric charge density	1.1
ρ_k^{q}	charge density due to particles of kind k	7.1
$\boldsymbol{\rho}$	rotational motion vector	4.2
σ	isotropic electric conductivity	7.3
σ_{H}	Hall conductivity	7.3
$\sigma_{\parallel}$	conductivity parallel to the magnetic field	7.3
$\sigma_{\perp}$	conductivity perpendicular to the magnetic field	7.3
$\sigma_{\alpha\beta}^{\mathrm{v}}$	viscous stress tensor	8.1
τ	characteristic time scale	4.2
τ_{ee}	electron collisional time	6.3
τ_{ei}	electron-ion collisional time	6.3
τ_{ii}	ion collisional time	6.3
τ_r	reconnection time scale	22.4
ϕ	gravitational potential	1.1
φ	electrostatic potential	6.2
φ	angle in the spherical frame	9.4
ϕ, φ	angle in the cylindrical frame	14.2
$\hat{\varphi}_k$	deviation of the exact distribution function from an averaged distribution function	2.2
Φ	magnetic flux	9.2
	stream function	9.4
χ	deflection angle	6.1
ψ	angle to the x axis	9.4
	potential of an electric current	15.3
Ψ	potential of a current-free magnetic field	8.3
ω	wave frequency	10.1
ω_0	wave frequency in a moving frame of reference	10.1
ω_{B}	cyclotron or Larmor frequency	4.1
ω_{pl}	electron plasma frequency	6.2
Ω	gravitational energy	14.1
$\boldsymbol{\omega}$	vector of angular velocity	15.1

Appendix 2
Useful Expressions

Some important characteristics of cosmic plasma

Source formulae

Larmor frequency of a non-relativistic electron (4.11), (4.85)

$$\omega_{_B}^{(e)} = \frac{eB}{m_e c} \approx 1.76 \times 10^7 \, B\,(\mathrm{G})\,, \quad \mathrm{rad\ s^{-1}}\,.$$

Larmor frequency of a non-relativistic proton (4.86)

$$\omega_{_B}^{(p)} \approx 9.58 \times 10^3 \; B\,(\mathrm{G})\,, \quad \mathrm{rad\ s^{-1}}\,.$$

Larmor radius of a non-relativistic electron (4.14), (4.93)

$$r_{_L}^{(e)} = \frac{c p_{\perp}}{eB} \approx 5.69 \times 10^{-8} \, \frac{v\,(\mathrm{cm\,s^{-1}})}{B\,(\mathrm{G})}\,, \quad \mathrm{cm}\,.$$

Larmor radius of a non-relativistic proton (4.14), (4.95)

$$r_{_L}^{(p)} \approx 1.04 \times 10^{-4} \, \frac{v\,(\mathrm{cm\,s^{-1}})}{B\,(\mathrm{G})}\,, \quad \mathrm{cm}\,.$$

Mean thermal velocity of electrons (4.88)

$$V_{\mathrm{Te}} = \left(\frac{3 k_{_B} T_e}{m_e} \right)^{1/2} \approx 6.74 \times 10^5 \sqrt{T_e\,(\mathrm{K})}\,, \quad \mathrm{cm\ s^{-1}}\,.$$

Mean thermal velocity of protons (4.89)

$$V_{\mathrm{Tp}} \approx 1.57 \times 10^4 \sqrt{T_p\,(\mathrm{K})}\,, \quad \mathrm{cm\ s^{-1}}\,.$$

Larmor radius of non-relativistic *thermal* electrons (4.90)

$$r_{\mathrm{L}}^{(\mathrm{e})} = \frac{V_{\mathrm{Te}}}{\omega_{\mathrm{B}}^{(\mathrm{e})}} \approx 3.83 \times 10^{-2} \, \frac{\sqrt{T_{\mathrm{e}}\,(\mathrm{K})}}{B\,(\mathrm{G})}\,, \quad \mathrm{cm}\,.$$

Larmor radius of non-relativistic *thermal* protons (4.91)

$$r_{\mathrm{L}}^{(\mathrm{p})} = \frac{V_{\mathrm{Tp}}}{\omega_{\mathrm{B}}^{(\mathrm{p})}} \approx 1.64 \, \frac{\sqrt{T_{\mathrm{p}}\,(\mathrm{K})}}{B\,(\mathrm{G})}\,, \quad \mathrm{cm}\,.$$

Drift velocity (4.20)

$$\mathbf{v}_{\mathrm{d}} = \frac{c}{e}\,\frac{\mathbf{F} \times \mathbf{B}}{B^2}\,.$$

Magnetic moment of a particle on the Larmor orbit (4.49)

$$\mathcal{M} = \frac{1}{c}\,JS = \frac{e\,\omega_{\mathrm{B}}\,r_{\mathrm{L}}^2}{2c} = \frac{p_\perp^2}{2mB} = \frac{\mathcal{E}_\perp}{B}\,.$$

Debye radius ($T_{\mathrm{e}} = T\,,\; T_{\mathrm{i}} = 0$ or $T_{\mathrm{e}} \gg T_{\mathrm{i}}$) (6.28)

$$r_{\mathrm{D}} = \left(\frac{k_{\mathrm{B}} T}{4\pi\, n e^2} \right)^{1/2}.$$

Debye radius in electron-proton thermal plasma ($T_{\mathrm{e}} = T_{\mathrm{p}} = T$) (6.75)

$$r_{\mathrm{D}} = \left(\frac{k_{\mathrm{B}} T}{8\pi\, e^2\, n} \right)^{1/2} \approx 4.9 \left(\frac{T}{n} \right)^{1/2}, \quad \mathrm{cm}\,.$$

Coulomb logarithm (6.70)

$$\ln \Lambda = \ln \left[\left(\frac{3 k_{\mathrm{B}}^{3/2}}{2\pi^{1/2}\, e^3} \right) \left(\frac{T_{\mathrm{e}}^3}{n_{\mathrm{e}}} \right)^{1/2} \right] \approx \ln \left[1.25 \times 10^4 \left(\frac{T_{\mathrm{e}}^3}{n_{\mathrm{e}}} \right)^{1/2} \right].$$

Electron plasma frequency (6.76)

$$\omega_{pl}^{(\mathrm{e})} = \left(\frac{4\pi\, e^2\, n_{\mathrm{e}}}{m_{\mathrm{e}}} \right)^{1/2} \approx 5.64 \times 10^4 \,\sqrt{n_{\mathrm{e}}}\,, \quad \mathrm{rad\ s^{-1}}\,.$$

Thermal electron collisional time (6.72)

$$\tau_{\mathrm{ee}} = \frac{m_{\mathrm{e}}^2}{0.714\, e^4\, 8\pi \,\ln \Lambda}\,\frac{V_{\mathrm{Te}}^3}{n_{\mathrm{e}}} \approx 4.04 \times 10^{-20}\, \frac{V_{\mathrm{Te}}^3}{n_{\mathrm{e}}}\,, \quad \mathrm{s}\,.$$

Thermal proton collisional time (6.73)

$$\tau_{\rm pp} = \frac{m_{\rm p}^2}{0.714\, e^4\, 8\pi\, \ln\Lambda}\, \frac{V_{\rm Tp}^3}{n_{\rm p}} \approx 1.36 \times 10^{-13}\, \frac{V_{\rm Tp}^3}{n_{\rm p}}\,, \quad {\rm s}\,.$$

Electron-ion collision (energy exchange) time Section 6.3

$$\tau_{\rm ei}\,(\mathcal{E}) = \frac{m_{\rm e} m_{\rm i}\,[\,3k_{\rm B}\,(T_{\rm e}/m_{\rm e} + T_{\rm i}/m_{\rm i})\,]^{3/2}}{e_{\rm e}^2\, e_{\rm i}^2\,(6\pi)^{1/2}\, 8\,\ln\Lambda}\,.$$

Time of energy exchange between electrons and protons (6.39)

$$\tau_{\rm ep}\,(\mathcal{E}) \approx 22\,\tau_{\rm pp} \approx 950\,\tau_{\rm ee}\,.$$

Dreicer field (6.78)

$$E_{\rm Dr} = \frac{4\pi e^3 \ln\Lambda}{k_{\rm B}}\,\frac{n_{\rm e}}{T_{\rm e}} \approx 6.54 \times 10^{-8}\,\frac{n_{\rm e}}{T_{\rm e}}\,, \quad {\rm V\,cm^{-1}}\,.$$

Conductivity of magnetized plasma Section 7.3.2

$$\sigma_{\parallel} = \sigma = \frac{e^2 n}{m_{\rm e}}\,\tau_{\rm ei} \approx 2.53 \times 10^8\, n\,({\rm cm^{-3}})\,\tau_{\rm ei}\,({\rm s})\,, \quad {\rm s^{-1}}\,,$$

$$\sigma_{\perp} = \sigma\,\frac{1}{1 + \left(\omega_{\rm B}^{\rm (e)}\,\tau_{\rm ei}\right)^2}\,, \qquad \sigma_{\rm H} = \sigma\,\frac{\omega_{\rm B}^{\rm (e)}\,\tau_{\rm ei}}{1 + \left(\omega_{\rm B}^{\rm (e)}\,\tau_{\rm ei}\right)^2}\,.$$

Magnetic diffusivity (or viscosity) (8.24)

$$\nu_{\rm m} = \frac{c^2}{4\pi\sigma} \approx 7.2 \times 10^{19}\,\frac{1}{\sigma}\,, \quad {\rm cm^2\,s^{-1}}\,.$$

Magnetic Reynolds number (8.36)

$${\rm Re_m} = \frac{L^2}{\nu_{\rm m}\,\tau} = \frac{vL}{\nu_{\rm m}}$$

Alfvén speed (8.63), (8.84)

$$V_{\rm A} = \frac{B}{\sqrt{4\pi\rho}} \approx 2.18 \times 10^{11}\,\frac{B}{\sqrt{n}}\,, \quad {\rm cm\ s^{-1}}\,.$$

Lundquist number Section 17.1.3

$$N_{\rm L} = {\rm Re_m}(V_{\rm A}, L) = \frac{V_{\rm A} L}{\nu_{\rm m}}\,.$$

Sound speed in electron-proton plasma (11.95)

$$V_s = \left(\gamma_g \frac{p}{\rho}\right)^{1/2} \approx 1.66 \times 10^4 \sqrt{T(\mathrm{K})}\,, \ \ \mathrm{cm\ s^{-1}}\,.$$

Electric field in magnetized plasma (14.73)

$$E \approx \frac{1}{c} vB \approx 10^{-8}\, v\,(\mathrm{cm\ s^{-1}})\, B\,(\mathrm{G})\,, \ \ \mathrm{V\ cm^{-1}}\,.$$

Appendix 3. Constants

Fundamental physical constants

Speed of light	c	$2.998 \times 10^{10}\,\mathrm{cm\ s^{-1}}$
Electron charge	e	4.802×10^{-10} CGSE
Electron mass	m_e	$9.109 \times 10^{-28}\,\mathrm{g}$
Proton mass	m_p	$1.673 \times 10^{-24}\,\mathrm{g}$
Boltzmann constant	k_B	$1.381 \times 10^{-16}\,\mathrm{erg\ K^{-1}}$
Gravitational constant	G	$6.673 \times 10^{-8}\,\mathrm{dyne\ cm^2\ g^{-2}}$
Planck's constant	h	$6.625 \times 10^{-27}\,\mathrm{erg\ s}$

Some useful constants and units

Ampere (current)	A	3×10^9 CGSE
Angström (length)	A	$10^{-8}\,\mathrm{cm}$
Electron Volt (energy)	eV	$1.602 \times 10^{-12}\,\mathrm{erg}$
	eV	11605 K
Gauss (magnetic induction)	G	3×10^{10} CGSE
Henry (inductance)	H	$1.111 \times 10^{-12}\,\mathrm{s^2\ cm^{-1}}$
Ionization potential of hydrogen		13.60 eV
Joule (energy)	J	$10^7\,\mathrm{erg}$
Maxwell (magnetic flux)	M	3×10^{10} CGSE
Ohm (resistance)	Ω	$1.111 \times 10^{-12}\,\mathrm{s\ cm^{-1}}$
Tesla (magnetic induction)		10^4 Gauss
Volt (potential)	V	3.333×10^{-3} CGSE
Watt (power)	W	$10^7\,\mathrm{erg\ s^{-1}}$
Weber (magnetic flux)	Wb	10^8 Maxwell

Some astrophysical constants

Astronomical unit	AU	1.496×10^{13} cm
Mass of the Sun	$M_\odot$	1.989×10^{33} g
Mass of the Earth	M_E	5.98×10^{27} g
Solar radius	$R_\odot$	6.960×10^{10} cm
Solar surface gravity	$g_\odot$	2.740×10^{4} cm s^{-2}
Solar luminosity	$L_\odot$	3.827×10^{33} erg s^{-1}
Mass loss rate	$\dot{M}_\odot$	10^{12} g s^{-1}
Rotation period of the Sun	$T_\odot$	26 days (at equator)

Bibliography

Acton, L.: 1996, Coronal structures, local and global, in *Magnetohydrodynamic Phenomena in the Solar Atmosphere: Prototypes of Stellar Magnetic Activity*, eds Y. Uchida, T. Kosugi, and H. Hudson, Dordrecht, Kluwer Academic Publ., p. 3–11.

Akhiezer, A.I., Lyubarskii, G.Ya., and Polovin, R.V.: 1959, On the stability of shock waves in MHD, *Soviet Physics-JETP*, v. 8, No. 3, 507–512.

Akimov, V.V., Ambroz, P., Belov, A.V., Berlicki, A., Chertok, I.M., Karlický, M., Leikov, N.G., Litvinenko, Yu.E., Maggun, A., Rompolt, B., and Somov, B.V.: 1996, Evidence for prolongated acceleration in the solar flare of June 15, 1991, *Solar Phys.*, v. 166, No. 1, 107–134.

Alekseyev, I.I. and Kropotkin, A.P.: 1970, Passage of energetic particles through a MHD discontinuity, *Geomagn. Aeron.*, v. 10, No. 6, 755–758.

Alfvén, H.: 1950, *Cosmic Electrodynamics*, Oxford, Clarendon Press, p. 228.

Alfvén, H.: 1981, *Cosmic Plasma*, Dordrecht, D. Reidel Publ., p. 164.

Alfvén, H. and Carlqvist, P.: 1967, Currents in solar atmosphere and a theory of flares, *Solar Phys.*, v. 1, No. 1, 220–228.

Alfvén, H. and Fälthammar, C.-G.: 1963, *Cosmic Electrodynamics*, Oxford, Clarendon Press, p. 228.

Altyntsev, A.T., Krasov, V.I., and Tomozov V.M.: 1977, Magnetic field dissipation in neutral current sheets, *Solar Phys.*, v. 55, No. 1, 69–81.

Aly, J.J.: 1984, On some properties of force-free fields in infinite regions of space, *Astrophys. J.*, v. 283, No. 1, 349–362.

Aly, J.J.: 1991, How much energy can be stored in a force-free field? *Astrophys. J.*, v. 375, No. 1, L61–L64.

Anderson, J.E.: 1963, *Magnetohydrodynamic Shock Waves*, Cambridge, Massachusetts; M.I.T. Press, p. 226.

Andres, U.T., Polak, L.S., and Syrovatskii, S.I.: 1963, Electromagnetic expulsion of spherical bodies from a conductive fluid, *Soviet Phys.-Technical Physics*, v. 8, No. 3, 193–196.

Anile, A.M.: 1989, *Relativistic Fluids and Magneto-Fluids*, Cambridge Univ. Press, p. 336.

Anosov, D.V.: 1969, *Geodesic Flows on Closed Riemannian Manifolds with Negative Curvature*, Providence, Amer. Math. Soc., p. 235.

Antiochos, S.K.: 1998, The magnetic topology of solar eruptions, *Astrophys. J.*, v. 502, L181–L184.

Antiochos, S.K., Karpen, J.T., and DeVore, C.R.: 1996, The nature of magnetic reconnection in the corona, in *Magnetic Reconnection in the Solar Atmosphere*, eds R.D. Bentley and J.T. Mariska, Astron. Soc. of Pacific, Conf. Series, v. 111, p. 79–81.

Antonucci, E. and Somov, B.V.: 1992, A diagnostic method for reconnecting magnetic fields in the solar corona, in *Coronal Streamers, Coronal Loops, and Coronal and Solar Wind Composition*, Proc. First SOHO Workshop, ESA SP-348, p. 293–294.

Antonucci, E., Benna, C., and Somov, B.V.: 1996, Interpretation of the observed plasma 'turbulent' velocities as a result of reconnection in solar flares, *Astrophys. J.*, v. 456, No. 2, 833–839.

Arge, C.N. and Mullan, D.J.: 1998, Modelling of magnetic interactions in partially-ionized gas, *Solar Phys.*, v. 182, No. 2, 293–332.

Aschwanden, M.J., Kliem B., Schwarz, U., Kurths, U., and Schwartz, R.A.: 1998, Wavelet analysis of solar flare hard X-rays, *Astrophys. J.*, v. 505, No. 2, 941–956.

Atoyan, A.M. and Aharonian, F.A.: 1999, Modelling of the non-thermal flares in the Galactic microquasar GRS 1915+105, *Mon. Not. Royal Astron. Soc.*, v. 302, No. 1, 253–276.

Axford, W.I., Leer, E., and Skadron, G.: 1977, The acceleration of cosmic rays by shock waves, *Proc. 15th Int. Cosmic Ray Conf.*, Plovdiv, v. 11, p. 132–137.

Ayres, T.R.: 1996, Thermal bifurcation of the solar chromosphere, in *Stellar Surface Structure*, eds K.G. Strassmeier and J.L. Linsky, IAU Symp. **176**, Dordrecht, Kluwer Academic Publ., p. 371–384.

Bachiller, R.: 1996, Bipolar molecular outflows from young stars and protostars, *Ann. Rev. Astron. Astrophys.*, v. 34, 111–154.

Bagalá, L.G., Mandrini, C.H., Rovira, M.G., Démoulin, P., and Hénoux, J.C.: 1995, A topological approach to understand a multi-loop flare, *Solar Phys.*, v. 161, No. 1, 103–121.

Bai, T. and Sturrock, P.A.: 1989, Classification of solar flares, *Ann. Rev. Astron. Astrophys.*, v. 27, 421–467.

Bai, T., Hudson, H.S., Pelling, R.M., Lin, R.P., Schwartz, R.A., and Von Rosenvinge, T.T.: 1983, First-order Fermi acceleration in solar flares as a mechanism for the second-step acceleration of protons and electrons, *Astrophys. J.*, v. 267, No. 1, 433-441.

Balesku, R.: 1963, *Statistical Mechanics of Charged Particles*, London, New York, Sydney; Interscience Publ., p. 477.

Balikhin, M., Gedalin, M., and Petrukovich, A.: 1993, New mechanism for electron heating in shocks, *Phys. Rev. Lett.*, v. 70, 1259–1262.

Balogh, A. and Erdös, G.: 1991, Fast acceleration of ions at quasi-perpendicular shocks, *J. Geophys. Res.*, v. 96, No. A9, 15853–15862.

Barenblatt, G.I.: 1979, *Similarity, Self-Similarity, and Intermediate Asymptotics*, New York, Consultants Bureau, Plenum.

Baum, P.J., Bratenahl, A., and Kamin, G.: 1978, Current interruption and flux transfer solar flare models, *Astrophys. J.*, v. 226, No. 1, 286–300.

Bednarek, W. and Protheroe, R.J.: 1999, Gamma-ray and neutrino flares produced by protons accelerated on an accretion disc in active galactic nuclei, *Mon. Not. Royal Astron. Soc.*, v. 302, 373–380.

Begelman, M.C., Blandford, R.D., and Rees, M.J.: 1984, Theory of extragalactic radio sorces, *Rev. Mod. Phys.*, v. 56, No. 2, 255–351.

Beloborodov, A.M.: 1999, Plasma ejection from magnetic flares and the X-ray spectrum of Cygnus X-1, *Astrophys. J.*, v. 510, L123–L126.

Benz, A.: 1993, *Plasma Astrophysics: Kinetic Processes in Solar and Stellar Coronae*, Dordrecht, Kluwer Academic Publ., p. 299.

Benz, A. and Krucker, S.: 1998, Heating events in the quiet solar corona, *Solar Phys.*, v. 182, No. 2, 349–363.

Benz, A. and Krucker, S.: 1999, Heating events in the quiet solar corona: Multiwavelength correlations, *Astron. Astrophys.*, v. 341, No. 1, 286–295.

Bertin, G.: 1999, *The Dynamics of Galaxies*, Cambridge Univ. Press, p. 448.

Bethe, H.A.: 1942, Office of Scientific Research and Development, Rep. No. 445.

Berger, M.A.: 1984, Rigorous limits on magnetic helicity dissipation in the solar corona, *Geophys. Astrophys. Fluid Dyn.*, v. 30, No. 1, 79–104.

Berger, M.A.: 1988a, 3D reconnection from a global viewpoint, in *Reconnection in Space Plasma*, eds T.D. Guyenne and J.J. Hunt, ESA SP-285, v. 2, p. 83–86.

Berger, M.A.: 1988b, An energy formula for nonlinear force-free fields, *Astron. Astrophys.*, v. 201, No. 1, 355–361.

Binney, J. and Tremaine, S.: 1987, *Galactic Dynamics*, Princeton, New Jersey; Princeton Univ. Press.

Birk, G.T. and Otto, A.: 1991, The resistive tearing instability for generalized resisitivity models, *Phys. Fluids*, v. 3, No. B7, 1746–1754.

Birkinshaw, M.: 1997, Instabilities in astrophysical jets, in *Advanced Topics on Astrophysical and Space Plasmas*, eds E.M. de Gouveia Dal Pino *et al.*, Dordrecht, Kluwer Academic Publ., p. 17–91.

Biskamp, D.: 1986, Magnetic reconnection via current sheets, *Phys. Fluids*, v. 29, No. 5, 1520–1531.

Biskamp, D.: 1997, *Nonlinear Magnetohydrodynamics*, Cambridge Univ. Press, p. 378.

Biskamp, D. and Welter, H.: 1989, Magnetic arcade evolution and instability, *Solar Phys.*, v. 120, No. 1, 49–77.

Blackman, E.G.: 1999, On particle energization in accretion flow, *Mon. Not. Royal Astron. Soc.*, v. 302, No. 4, 723–730.

Blandford, R.D.: 1994, Particle acceleration mechanisms, *Astrophys. J., Suppl.*, v. 90, No. 2, 515–520.

Bliokh, P., Sinitsin, V., and Yaroshenko, V.: 1995, *Dusty and Self-Gravitational Plasmas in Space*, Dordrecht, Kluwer Academic Publ., p. 250.

Blokhintsev, D.I.: 1945, Moving receiver of sound, *Doklady Akademii Nauk SSSR (Soviet Physics Doklady)*, v. 47, No. 1, 22–25 (in Russian).

Bobrova, N.A. and Sasorov, P.V.: 1993, MHD equations for a fully ionized plasma of complex composition, *Plasma Phys. Rep.*, v. 19, No. 6, 409–412.

Bobrova, N.A. and Syrovatskii, S.I.: 1979, Singular lines of 1D force-free field, *Solar Phys.*, v. 61, No. 2, 379–387.

Bogachev, S.A., Somov, B.V., and Masuda, S.: 1998, On the velocity of a hard X-ray source in the solar corona. *Astronomy Letters*, v. 24, No. 4, 543–548.

Bogdanov, S.Yu., Frank, A.G., Kyrei, N.P., and Markov, V.S.: 1986, Magnetic reconnection, generation of plasma fluxes and accelerated particles in laboratory experiments, *Plasma Astrophys.*, ESA SP-251, 177-183.

Bolcato, R., Etay, J., Fautrelle, Y., and Moffatt, H.K.: 1993, Electromagnetic billiards, *Phys. Fluids*, v. 5, No. A7, 1852–1853.

Bontemps, S., André, P., Terebey, S., Carbit, S.: 1996, Evolution of outflow activity around low-mass embedded young stellar objects, *Astron. Astrophys.*, v. 311, 858–875.

Boyd, T.J.M. and Sanderson, J.J.: 1969, *Plasma Dynamics*, New York, Barnes and Noble, Inc. p. 348.

Braginskii, S.I.: 1965, Transport processes in plasma, in *Reviews of Plasma Physics*, ed. M. Leontovich, New York, Consultants Bureau, v. 1, 205–311.

Bridgman, P.W.: 1931, *Dimensional Analysis*, New Haven, Yale Univ. Press, p. 113.

Brissaud, A., Frisch, U., Leorat, J., Lesieur, M., and Masure, A.: 1973, Helicity cascades in fully developed isotropic turbulence, *Phys. Fluids*, v. 16, 1366-1367.

Brown, J.C.: 1971, The deduction of energy spectra of non-thermal electrons in flares from the observed spectra of hard X-ray bursts, *Solar Phys.*, v. 18, No. 2, 489–502.

Brown, J.C.: 1972, The directivity and polarization of thick target X-ray bremsstrahlung from flares, *Solar Phys.*, v. 26, No. 2, 441–459.

Brown, J.C., McArthur, G.K., Barrett, R.K., McIntosh, S.W., and Emslie, A.G.: 1988a, Inversion of the thick-target bremsstrahlung spectra from non-uniformly ionised plasmas, *Solar Phys.*, v. 179, No. 2, 379–404.

Brown, J.C., Conway, A.J., and Aschwanden, M.J.: 1988b, The electron injection function and energy-dependent delays in thick-target hard X-rays, *Astrophys. J.*, v. 509, No. 2, 911–917.

Browning, P.K.: 1988, Helicity injection and relaxation in a coronal magnetic loop with a free surface, *J. Plasma Phys.*, v. 40, No. 2, 263–280.

Brushlinskii, K.V., Zaborov, A.M., and Syrovatskii, S.I.: 1980, Numerical analysis of the current sheet near a magnetic null line, *Soviet J. Plasma Physics*, v. 6, No. 2, 165–173.

Büchner, J. and Zelenyi, L.: 1989, Regular and chaotic particle motion in magnetotail field reversal, *J. Geophys. Res.*, v. 94, No. A9, 11821–11842.

Cadjan, M.G. and Ivanov, M.F.: 1999, Langevin approach to plasma kinetics with collisions, *J. Plasma Phys.*, v. 61, No. 1, 89–106.

Cai, H.J. and Lee, L.C.: 1997, The generalized Ohm's law in collisionless reconnection, *Phys. Plasmas*, v. 4, No. 3, 509–520.

Camenzind, M.: 1995, Magnetic fields and the physics of active galactic nuclei, *Rev. Mod. Astron.*, v. 8, 201–233.

Campbell, C.G.: 1997, *Magnetohydrodynamics of Binary Stars*, Dordrecht, Kluwer Academic Publ., p. 306.

Chakrabarti, S.K. (ed.): 1999, *Observational Evidence for Black Holes in the Universe*, Dordrecht, Kluwer Academic Publ., p. 399.

Chandrasekhar, S.: 1943a, Stochastic problems in physics and astronomy, *Rev. Mod. Phys.*, v. 15, No. 1, 1–89.

Chandrasekhar, S.: 1943b, *Astrophys. J.*, v. 97, No. 1, 255–267.

Chandrasekhar, S.: 1981, *Hydrodynamic and Hydromagnetic Stability*, New York, Dover Publ., p. 654.

Chandrasekhar, S. and Fermi, E.: 1953, Problems of gravitational stability in the presence of a magnetic field, *Astrophys. J.*, v. 118, No. 1, 116–141.

Chapman, S. and Kendall, P.C.: 1963, Liquid instability and energy transformation near magnetic neutral line, *Proc. Roy. Soc. London*, v. A271, 435–448.

Chen, J. and Palmadesso, P.J.: 1986, Chaos and nonlinear dynamics of single particle orbits in a magnetotail field, *J. Geophys. Res.*, v. 91, No. A2, 1499–1508.

Chernov, A.A. and Yan'kov, V.V.: 1982, Electron flow in low-density pinches, *Soviet J. Plasma Phys.*, v. 8, No. 5, 522–528.

Chew, G.F., Goldberger, M.L., and Low, F.E.: 1956, The Boltzmann equation and the one-fluid hydromagnetic equations in the absence of particle collisions, *Proc. Royal Soc. London*, v. A236, No. 1, 112–118.

Choudhuri, A.R.: 1998, *The Physics of Fluids and Plasmas: An Introduction for Astrophysicists*, Cambridge Univ. Press, p. 427.

Chupp, E.L.: 1996, in *High Energy Solar Physics*, eds R. Ramaty, N. Mandzhavidze, and X.-M. Hua, AIP Conf. Proc. 374, AIP, Woodbury, New York, 3–9, 1996.

Clausius, R.: 1870, On a mechanical theorem applicable to heat, *Philosophical Magazine* (Series 4), v. 40, No. 1, 122–127.

Cole, J.D. and Huth, J.H.: 1959, Some interior problems of hydromagnetics, *Phys. Fluids*, v. 2, No. 6, 624–626.

Colgate, S.A.: 1988, Relationship between high-energy phenomena on the Sun and in astrophysics, *Solar Phys.*, v. 118, No. 1, 1–15.

Colgate, S.A. and Furth, H.P.: 1960, Stabilization of pinch discharges, *Phys. Fluids*, v. 3, No. 6, 982–1000.

Collins, G.W.: 1978, *The Virial Theorem in Stellar Astrophysics*, Tucson, Pachart.

Coppi, B., Laval, G., and Pellat, R.: 1966, Dynamics of the geomagnetic tail, *Phys. Rev. Lett.*, v. 6, No. 26, 1207–1210.

Courant, R. and Friedrichs, K.O.: 1985, *Supersonic Flow and Shock Waves*, New York, Berlin, Heidelberg, Tokyo; Springer-Verlag, p. 464.

Cowling, T.G.: 1976, *Magnetohydrodynamics*, Bristol, Adam Hilger, p. 135.

Cox, D.P. and Tucker, W.H.: 1969, Ionization equilibrium and radiative cooling of a low-density plasma, *Astrophys. J.*, v. 157, No. 3, 1157–1167.

Craig, I.J.D. and McClymont, A.N.: 1993, Linear theory of reconnection at an X-type neutral point, *Astrophys. J.*, v. 405, No. 1, 207–215.

Crooker, N., Joselyn, J.A., and Feynman, J. (eds): 1997, *Coronal Mass Ejections*, Washington, Amer. Geophys. Un., p. 299.

Darwin, C.: 1949, Source of the cosmic rays, *Nature*, v. 164, 1112–1114.

Davis, L.Jr.: 1956, Modified Fermi mechanism for the acceleration of cosmic rays, *Phys. Rev.*, v. 101, 351–358.

Day, C.: 1998, SOHO observations implicate 'magnetic carpet' as source of coronal heating in quiet Sun, *Physics Today*, March issue, 19–21.

de Hoffmann, F. and Teller, E.: 1950, MHD shocks, *Phys. Rev.*, v. 80, No. 4, 692–703.

de Jager, C.: 1986, Solar flares and particle acceleration, *Space Sci. Rev.*, v. 44, No. 1, 43–90.

Debye, P. and Hückel, E.: 1923, *Phys. Zs.*, v. 24, 185.

Demoulin, P., Raadu, M.A., Malherbe, J.M., and Schmieder, B.: 1987, Fine structures in solar filaments, *Astron. Astrophys.*, v. 183, No. 1, 142–150.

Demoulin, P., van Driel-Gestelyi, L., Schmieder, B., Hénoux, J.-C., Csepura, G., and Hagyard, M.J.: 1993, Evidence for reconnection in solar flares. *Astron. Astrophys.*, v. 271, No. 1, 292–307.

Den, O.G. and Somov, B.V.: 1989, Magnetic field dissipation in a high-temperature plasma as a mechanism of energy release in a solar flare, *Soviet Astronomy-AJ*, v. 33, No. 2, 149–155.

Dennis, B.R.: 1985, Solar hard X-ray bursts. *Solar Phys.*, v. 100, No. 2, 465–490.

Dennis, B.R.: 1988, Solar flare hard X-ray observations, *Solar Phys.*, v. 118, No. 1, 49–94.

Diakonov, S.V. and Somov, B.V.: 1988, Thermal electrons runaway from a hot plasma during a flare in the reverse-current model and their X-ray bremsstrahlung, *Solar Phys.*, v. 116, No. 1, 119–139.

Di Matteo, T., Celotti, A., and Fabian, A.C.: 1999, Magnetic flares in accretion disc coronae and the spectral states of black hole candidates: the case of GX339-4, *Mon. Not. Royal Astron. Soc.*, v. 304, 809–820.

Dobrowolny, M.: 1968, Instability of a neutral sheet, *Nuoco Cimento*, v. B55, 427–438.

Dokuchaev, V.P.: 1964, Emission of magnetoacoustic waves in the motion of stars in cosmic space, *Soviet Astronomy-AJ*, v. 8, No. 1, 23–31.

Drake, J.F. and Kleva R.G.: 1991, Collisionless reconnection and the saw-tooth crash, *Phys. Rev. Lett.*, v. 66, No. 11, 1458–1461.

Dreicer, H.: 1959, Electron and ion runaway in a fully ionized gas, *Phys. Rev.*, v. 115, No. 2, 238–249.

Duijveman, A., Hoyng P., and Ionson, J.A.: 1981, Fast plasma heating by anomalous and inertial resistivity effects in the solar atmosphere, *Astrophys. J.*, v. 245, No. 2, 721–735.

Duijveman, A., Somov, B.V., and Spektor, A.R.: 1983, Evolution of a flaring loop after injection of fast electrons, *Solar Phys.*, v. 88, No. 1, 257–273.

Dungey, J.W.: 1958, *Cosmic Electrodynamics*, England, Cambridge Univ. Press, p. 183.

D'yakov, S.P.: 1954, *Zhurnal Exper. Teor. Fiz.*, v. 27, 288–297 (in Russian).

Eichler, D.: 1979, Particle acceleration in solar flares by cyclotron damping of cascading turbulence, *Astrophys. J.*, v. 229, No. 1, 413–418.

Elsasser, W.M.: 1956, Hydromagnetic dynamo theory, *Rev. Mod. Phys.*, v. 28, No. 2, 135–163.

Erdös, G. and Balogh, A.: 1994, Drift acceleration at interplanetary shocks, *Astrophys. J., Suppl.*, v. 90, No. 2, 553–559.

Fedoryuk, V.M.: 1985, *Ordinary Differential Equations*, Moscow, Nauka (in Russian).

Feldman, W.C., Bame, S.J., Gary, S.P., Gosling, et al.: 1982, Electron heating within the Earth's bow shock, *Phys. Rev. Lett.*, v. 49, 199–202.

Fermi, E.: 1949, On the origin of cosmic radiation, *Phys. Rev.*, v. 75, 1169–1174.

Fermi, E.: 1954, Galactic magnetic fields and the origin of cosmic radiation, *Astrophys. J.*, v. 119, No. 1, 1–6.

Field, G.B.: 1965, Thermal instability, *Astrophys. J.*, v. 142, No. 2, 531–567.

Fokker, A.D.: 1914, Die mittlere Energie rotieren der elektrischer Dipole im Strahlungsfeld, *Ann. der Physik*, v. 43, No. 5, 810–820.

Fox, D.C. and Loeb, A.: 1997, Do the electrons and ions in X-ray clusters share the same temperature? *Astrophys. J.*, v. 491, No. 2, 459–466.

Froyland, J.: 1992, *Introduction to Chaos and Coherence*, Bristol, Inst. of Phys. Publ., p. 156.

Furth, H.P.: 1961, Sheet pinch instabilities caused by finite conductivity, *Bull. Amer. Phys. Soc.*, v. 6, No. 2, p. 193.

Furth, H.P.: 1967, *Proc. ESRW Conf. of the Stability of Plane Plasmas*, Frascaty, Eur. Space Res. Inst., p. 22–25.

Furth, H.P., Killen, J., and Rosenbluth, M.N.: 1963, Finite-resistivity instabilities of a sheet pinch, *Phys. Fluids*, v. 6, No. 4, 459–484.

Galeev, A.A. and Zelenyi, L.M.: 1975, Metastable states of neutral sheet and the substorms, *JETP-Lett.*, v. 22, No. 7, 170–172.

Galeev, A.A. and Zelenyi, L.M.: 1976, Tearing instability in plasma configurations, *Soviet Physics-JETP*, v. 43, No. 6, 1113–1123.

Galeev, A.A., Rosner, R., and Vaiana, G.S.: 1979, Structured coronae of accretion discs, *Astrophys. J.*, v. 229, No. 1, 318–326.

Gal'per, A.M., Zemskov, V.M., Luchkov, B.I., Ozerov, Yu.V., and Khodarovich, A.M.: 1994, *Pis'ma v ZhETF* v. 59, 145–147 (in Russian).

Galsgaard, K. and Longbottom, A.W.: 1999, Formation of solar prominences by flux convergence, *Astrophys. J.*, v. 510, No. 1, 444–459.

Gedalin, M. and Griv, E.: 1999, Collisionless electrons in a thin high Much number shocks, *Ann. Geophysicae*, v. 17, No. 10, p. 1251–1259.

Gel'fand, I.M.: 1959, Some problems of the theory of quasilinear equations, *Usp. Mat. Nauk*, v. 14, No. 2, 87–158 (in Russian).

Gerbeth, G., Thess, A., and Marty, P.: 1990, The MHD flow around a cylinder in crossed electric and magnetic fields, *Eur. J. Mech., B/Fluids*, v. 9, No. 3, 239–257.

Germain, P.: 1960, Shock waves and shock-wave structure in MHD, *Rev. Mod. Phys.*, v. 32, No. 4, 951–958.

Ginzburg, V.L. and Zheleznyakov, V.V.: 1958, *Soviet Astonomy-AJ*, v. 2, 653–662.

Giovanelli, R.G.: 1946, A theory of chromospheric flares, *Nature*, v. 158, No. 4003, 81–82.

Giovanelli, R.G.: 1947, Magnetic and electric phenomena in the sun's atmosphere associated with sunspots, *Mon. Not. Royal Astron. Soc.*, v. 107, No. 4, 338–355.

Giovanelli, R.G.: 1948, Chromospheric flares, *Mon. Not. Royal Astron. Soc.*, v. 108, No. 2, 163–176.

Giovanelli, R.G.: 1949, Electron energies resulting from an electric field in a highly ionized gas, *Phil. Mag.*, Seventh Series, v. 40, No. 301, 206–214.

Gieseler, U.D.J., Kirk, J.G., Gallant, Y.A., and Achtenberg, A.: 1999, Particle acceleration at oblique shocks and discontinuities of the density profile, *Astron. Astrophys.*, v. 435, No. 1, 298–306.

Gisler, G. and Lemons, D.: 1990, Electron Fermi acceleration in collapsing magnetic traps, *J. Geophys. Res.*, v. 95, No. A9, 14925–14938.

Glasstone, S. and Loveberg, R.H.: 1960, *Controlled Thermonuclear Reactions*, Princeton, Van Nostrand, p. 523.

Gold, T.: 1964, in *AAS-NASA Symp. in the Physics of Solar Flares*, ed. W.N. Hess, NASA-SP 50, Washington DC, p. 389–396.

Gold, T. and Hoyle, F.: 1960, On the origin of solar flares, *Monthly Not. Royal Astron. Soc.*, v. 120, No. 2, 89–105.

Goldreich, P. and Sridhar, S.: 1997, MHD turbulence revisited, *Astrophys. J.*, v, 485, No. 2, 680–688.

Goldston, R.J. and Rutherford, P.H.: 1995, *Introduction to Plasma Physics*, Bristol, Inst. of Phys. Publ., p. 492.

Golub, L. and 11 co-authors: 1999, A new view of the solar corona from TRACE, *Phys. Plasmas*, v. 6, No. 5, 2205–2212.

Gombosi, T.I.: 1999, *Physics of the Space Environment*, Cambridge Univ. Press, p. 339.

Gopasyuk, S.I.: 1990, Solar magnetic fields and large-scale electric currents in the active regions, *Adv. Space Res.*, v. 10, No. 9, 151–160.

Gorbachev, V.S. and Kel'ner, S.R.: 1988, Formation of plasma condensations in fluctuating strong magnetic field, *Soviet Physics–JETP*, v. 67, No. 9, 1785–1790.

Gorbachev, V.S. and Somov, B.V.: 1989, Solar flares of November 5, 1980, as the result of magnetic reconnection at a separator, *Soviet Astronomy–AJ*, v. 33, No. 1, 57–61.

Gorbachev, V.S. and Somov, B.V.: 1990, Magnetic reconnection on the separator as a cause of a two-ribbon flare, *Adv. Space Res.*, v. 10, No. 9, 105–108.

Guckenheimer, J. and Holmes, P.: 1983, *Nonlinear Oscillations, Dynamical Systems and Bifurcations of Vector Fields*, New York, Springer-Verlag.

Gurevich, A.V.: 1961, On the theory of runaway electrons, *Soviet Physics–JETP*, v. 12, No. 5, 904–912.

Gurevich, A.V. and Zhivlyuk, Y.N.: 1966, Runaway electrons in a non-equilibrium plasma, *Soviet Physics–JETP*, v. 22, No. 1, 153–159.

Gurevich, A.V. and Istomin, Y.N.: 1979, Thermal runaway and convective heat transport by fast electrons in a plasma, *Soviet Physics–JETP*, v. 50, No. 3, 470–475.

Haisch, B.M., Strong, K.T., and Rodonò M.: 1991, Flares on the Sun and other stars, *Ann. Rev. Astron. Astrophys.*, v. 29, 275–324.

Haken, H.: 1978, *Synergetics*, New York, Springer-Verlag.

Hargreaves, J.K.: 1992, *The Solar-Terrestrial Environment*, Cambridge Univ. Press, p. 420.

Harris, E.G.: 1962, On a plasma sheath separating regions of oppositely directed magnetic field, *Nuovo Cimento*, v. 23, No. 1, 115–121.

Hawley, J.F. and Balbus, S.A.: 1999, Instability and turbulence in accretion discs, in *Numerical Astrophysics*, eds S.M. Miyama et al., Dordrecht, Kluwer Academic Publ., p. 187–194.

Hénoux, J.-C. and Somov, B.V.: 1987, Generation and structure of the electric currents in a flaring activity complex, *Astron. Astrophys.*, v. 185, No. 1, 306–314.

Hénoux, J.-C. and Somov, B.V.: 1991, The photospheric dynamo. 1. Magnetic flux-tube generation, *Astron. Astrophys.*, v. 241, No. 2, 613–617.

Hénoux, J.-C. and Somov, B.V.: 1992, First ionization potential fractionation, in *Coronal Streamers, Coronal Loops, and Coronal and Solar Wind Composition*, Proc. First SOHO Workshop, ESA SP-348, p. 325–330.

Hénoux, J.-C. and Somov, B.V.: 1997, The photospheric dynamo. 2. Physics of thin magnetic flux tubes, *Astron. Astrophys.*, v. 318, No. 3, 947–956.

Hesse, M., Birn, J., Baker, D.N., and Slavin, J.A.: 1996, MHD simulation of the transition of reconnection from closed to open field lines, *J. Geophys. Res.*, v. 101, No. A5, 10805–10816.

Heyvaerts, J. and Priest, E.R.: 1984, Coronal heating by reconnection in DC current systems, *Astron. Astrophys.*, v. 137, No. 1, 63–78.

Hirano, Y., Yagi, Y., Maejima, Y., Shimada T., and Hirota, I.: 1997, Self-organization and its effect on confinement in a reversed field pinch plasma, *Plasma Phys. Control. Fusion*, v. 39, No. 5A, A393–A400.

Hirotani, K. and Okamoto, I.: 1998, Pair plasma production in a force-free magnetosphere around a supermassive black hole, *Astrophys. J.*, v. 497, No. 2, 563–572.

Hoh, F.C.: 1966, Stability of sheet pinch, *Phys. Fluids*, v. 9, 277–284.

Hollweg, J.V.: 1986, Viscosity and the CGL equations in the solar corona, *Astrophys. J.*, v. 306, No. 2, 730–739.

Holman, G.D.: 1995, DC electric field acceleration of ions in solar flares, *Astrophys. J.*, v. 452, No. 2, 451–456.

Hones, E.W.Jr.(ed.): 1984, *Magnetic Reconnection in Space and Laboratory Plasmas*, Washington, Amer. Geophys. Un., p. 386.

Horiuchi, R. and Sato, T.: 1994, Particle simulation study of driven reconnection in a collisionless plasma, *Phys. Plasmas*, v. 1, No. 11, 3587–3597.

Horiuchi, R. and Sato, T.: 1997, Particle simulation study of collisionless reconnection in a sheared field, *Phys. Plasmas*, v. 4, No. 2, 277–289.

Horwitz, J.L., Gallagher, D.L., and Peterson, W.K. (eds): 1998, *Geospace Mass and Energy Flow*, Washington, Amer. Geophys. Un., p. 393.

Hoyng, P., Brown, J.C., and van Beek, H.F.: 1976, High time resolution analysis of solar hard X-ray flares observed on board the ESRO TD-1A satellite, *Solar Phys.*, v. 48, No. 2, 197–254.

Hudson, H. and Ryan, J.: 1995, High-energy particles in solar flares, *Ann. Rev. Astron. Astrophys.*, v. 33, 239–282.

Hudson, P.D.: 1965, Reflection of charged particles by plasma shocks, *Mon. Not. Royal Astron. Soc.*, v. 131, No. 1, 23–50.

Ichimoto, K., Hirayama, T., Yamaguchi, A., Kumagai, K., et al.: 1992, Effective geometrical thickness and electron density of a flare of 1991 December 2, *Publ. Astron. Soc. Japan*, v. 44, No. 5, L117–L122.

Imshennik, V.S. and Syrovatskii, S.I.: 1967, 2D flow of an ideally conducting gas in the vicinity of the zero line of a magnetic field, *Soviet Physics-JETP*, v. 25, No. 4, 656–664.

Innes, D.E., Inhester, B., Axford, W.I., and Wilhelm, K.: 1997, Bi-directional jets produced by reconnection on the Sun, *Nature*, v. 386, 811–813.

Iordanskii, S.V.: 1958, On compression waves in MHD, *Soviet Physics-Doklady*, v. 3, No. 4, 736–738.

Ip, J.T.C. and Sonnerup, B.U.: 1996, Resistive tearing instability in a current sheet with coplanar viscous stagnation-point flow, *J. Plasma Phys.*, v. 56, No. 2, 265–284.

Iroshnikov, R.S.: 1963, *Astronom. Zhur.*, v. 40, 742 (in Russian).

Isliker, H.: 1992, Structural properties of the dynamics in flares, *Solar Phys.*, v. 141, No. 2, 325–334.

Jacobsen, C. and Carlqvist, P.: 1964, Solar flares caused by circuit interruptions, *Icarus*, v. 3, No. 3, 270–272.

Janicke, L.: 1980, The resistive tearing mode in weakly two-dimentional neutral sheets, *Phys. Fluids*, v. 23, No. 9, 1843–1849.

Janicke, L.: 1982, Resistive tearing mode in coronal neutral sheets, *Solar Phys.*, v. 76, No. 1, 29–43.

Jeans, J.: 1929, *Astronomy and Cosmogony*, Cambridge Univ. Press.

Jones, F.C. and Ellison D.C.: 1991, The plasma physics of shock acceleration, *Space Sci. Rev.*, v. 58, No. 3, 259–346.

Jones, M.E., Lemons, D.S., Mason, R.J., Thomas, V.A., and Winske, D.: 1996, A grid-based Coulomb collision model for PIC codes, *J. Comput. Phys.*, v. 123, No. 1, 169–181.

Kadomtsev, B.B.: 1960, Convective instability of a plasma, in *Plasma Physics and the Problem of Controlled Thermonuclear Reactions*, ed. M.A. Leontovich, London, Oxford; Pergamon Press, v. 4, p. 450–453.

Kadomtsev, B.B.: 1966, Hydrodynamic stability of a plasma, in *Reviews of Plasma Physics*, ed. M.A. Leontovich, New York, Consultants Bureau, v. 2, p. 153–198.

Kan, J.R., Akasofu, S.I., and Lee, L.C.: 1983, A dynamo theory of solar flares, *Solar Phys.*, v. 84, No. 1, 153–167.

Kandrup, H.E.: 1998, Collisionless relaxation in galactic dynamics and the evolution of long-range order, in *Annals of the New York Acad. of Sci.*, v. 848, 28–47.

Karpen, J.T., Antiochos, S.K., and De Vore, C.R.: 1991, Coronal current sheet formation: The effect of shears, *Astrophys. J.*, v. 382, No. 1, 327–337.

Karpen, J.T., Antiochos, S.K., De Vore, C.R., and Golub, L.: 1998, Dynamic responses to reconnection in solar arcades, *Astrophys. J.*, v. 495, No. 1, 491–501.

Kivelson, M.G. and Russell, C.T. (eds): 1995, *Introduction to Space Physics*, Cambridge Univ. Press, p. 568.

Klein, K.-L., Chupp, E.L., Trottet, G., Magun, A., Dunphy, P.P., Rieger, E., and Urpo, S.: 1999, Flare-associated energetic particles in the corona and at 1 AU, *Astron. Astrophys.*, v. 348, No. 1, 271–285.

Klimontovich, Yu.L.: 1986, *Statistical Physics*, New York, Harwood Academic.

Klimontovich, Yu.L.: 1998, Two alternative approaches in the kinetic theory of a fully ionized plasma, *J. Plasma Phys.*, v. 59, No. 4, 647–656.

Koide, S., Shibata, K., and Kudoh, T.: 1999, Relativistic jet formation from black hole magnetized accreation discs, *Astrophys. J.*, v. 522, 727–752.

Kokubun, S. and Kamide, Y. (eds): 1998, *Substorms-4*, Dordrecht, Kluwer Academic Publ.; Tokyo, Terra Scientific Publ., p. 823.

Kolmogorov, A.N.: 1941, The local structure of turbulence in incompressble viscous fluid for very large Reynolds numbers, *C.R. Acad. Sci. USSR*, v. 30, 201–206.

Kontorovich, V.M.: 1959, On the interaction between small perturbations and the discontinuities in MHD and the stability of shock waves, *Soviet Physics-JETP*, v. 8, No. 5, 851–858.

Korchak, A.A.: 1971, On the origin of solar flare X-rays, *Solar Phys.*, v. 18, No. 2, 284–304.

Korchak, A.A.: 1980, Coulomb losses and the nuclear composion of the solar flare accelerated particles, *Solar Phys.*, v. 66, No. 1, 149–158.

Kosugi, T. and Somov, B.: 1998, Magnetic reconnection and particle acceleration in solar flares, in *Observational Plasma astrophysics: Five Years of Yohkoh and Beyond*, eds. T. Watanabe, T. Kosugi, and A.C. Sterling, Dordrecht, Kluwer Academic Publ., p. 297–306.

Kotchine, N.E.: 1926, Rendiconti del Circolo Matematico di Palermo, v. 50, 305–314.

Kraichnan, R.H.: 1965, Inertial-range spectrum of hydromagnetic turbulence, *Phys. Fluids*, v. 8, No. 7, 1385–1389.

Krucker, S. and Benz, A.O.: 2000, Are heating events in the quiet solar corona small flares? *Solar Phys.*, in press.

Krymskii, G.F.: 1977, A regular mechanism for the acceleration of charged particles on the front of a shock wave, *Sov. Phys. Dokl.*, v. 22, No. 6, 327–328.

Kubát, J. and Karlický, M.: 1986, Electric conductivity in the solar photosphere and chromosphere, *Bull. Astron. Inst. Czechosl.*, v. 37, No. 3, 155–163.

Kudriavtsev, V.S.: 1958, Energetic diffusion of fast ions in plasma, *Soviet Physics-JETP*, v. 7, No. 6, 1075–1079.

Kulikovskii, A.G. and Liubimov, G.A.: 1961, On the structure of an inclined MHD shock wave, *Appl. Math. Mech.*, v. 25, No. 1, 171–179.

Kunkel, W.B.: 1984, Generalized Ohm's law for plasma including neutral particles, *Phys. Fluids*, v. 27, No. 9, 2369–2371.

Kurths, J. and Herzel, H.: 1986, Can a solar pulsation event be characterized by a chaotic attractor? *Solar Phys.*, v. 10, No. 1, 39–45.

Kurths, J., Benz, A., and Aschwanden, M.J.: 1991, The attractor dimension of solar decimetric radio pulsation, *Astron. Astrophys.*, v. 248, No. 1, 270–276.

Kusano, K. and Nishikawa, K.: 1996, Bifurcation and stability of coronal arcades in a linear force-free field, *Astrophys. J.*, v. 461, No. 1, 415–423.

Lahav, O., Terlevich, E., and Terlevich, R.J. (eds): 1996, *Gravitational Dynamics*, Cambridge Univ. Press, p. 270.

Laming, J.M. and Drake, J.J.: 1999, Stellar coronal abundances. VI. The FIP effect and ξ Bootis A—Solar-like anomalies, *Astrophys. J.*, v. 516, No. 1, 324–334.

Landau, L.D.: 1937, Kinetic equation in the case of Coulomb interaction, *Zh. Exp. Teor. Fiz.*, v. 7, No. 1, 203–212 (in Russian).

Landau, L.D.: 1946, On the vibrations of the electron plasma, *J. Phys. USSR*, v. 10, No. 1, 25–30.

Landau, L.D. and Lifshitz, E.M.: 1959, *Fluid Mechanics*, Oxford, London; Pergamon Press, p. 536.

Landau, L.D. and Lifshitz, E.M.: 1959, *Statistical Physics*, London, Paris; Pergamon Press, p. 478.

Landau, L.D. and Lifshitz, E.M.: 1960, *Mechanics*, Oxford, London, Paris; Pergamon Press, p. 165.

Landau, L.D. and Lifshitz, E.M.: 1971, *Classical Theory of Field*, Oxford, New York; Pergamon Press, p. 374.

Landau, L.D., Lifshitz, E.M., and Pitaevskii, L.P.: 1984, *Electrodynamics of Continuous Media*, Oxford, New York; Pergamon Press, p. 460.

Langmuir, I.: 1928, *Proc. Nat. Acad. Sci. U.S.A.*, v. 14, 627.

LaRosa, T.N. and Moore, R.L.: 1993, A mechanism for bulk energization in solar flares: MHD turbulent cascade, *Astrophys. J.*, v. 418, No. 2, 912–918.

LaRosa, T.N., Moore, R.L., Miller, J.A., and Shore, S.N.: 1996, New promise for electron bulk energization in solar flares: Preferential Fermi acceleration of electrons, *Astrophys. J.*, v. 467, No. 1, 454–464.

Lau, Y.-T.: 1993, Magnetic nulls and topology in solar flare models, *Solar Phys.*, v. 148, No. 2, 301–324.

Lavrent'ev, M.A. and Shabat, B.V.: 1973, *Methods of the Theory of Complex Variable Functions*, Moscow, Nauka, p. 736 (in Russian).

Lax, P.: 1957, Hyperbolic systems of conservation laws, *Comm. Pure Appl. Math.*, v. 10, No. 4, 537–566.

Lax, P.: 1973, Hyperbolic Systems of Conservation Laws and the Mathematical Theory of Shock Waves, *SIAM*.

Leenov, D. and Kolin, A.: 1954, Theory of electromagnetophoresis. 1. MHD forces experienced by spheric and cylindrical particles, *J. Chemical Phys.*, v. 22, No. 4, 683–688.

Leith, C.E.: 1967, Diffusion approximation to inertial energy transfer in isotropic turbulence, *Phys. Fluids*, v. 10, No. 7, 1409–1416.

Lembege, B. and Pellat R.: 1982, Stability of a thick 2D quasi-neutral sheet, *Phys. Fluids*, v. 25, No. 11, 1995–2004.

Leontovich, M.A. (ed.): 1960, *Plasma Physics and the Problem of Controlled Thermonuclear Reactions*, London, Oxford, New York, Paris; Pergamon Press, v. 1–4.

Lesch, H. and Pohl, M.: 1992, A possible explanation for intraday variability in active galactic nuclei, *Astron. Astrophys.*, v. 254, No. 1, 29–38.

Liberman, M.A.: 1978, On actuating shock waves in a completely ionized plasma, *Soviet Physics–JETP*, v. 48, No. 5, 832–840.

Liboff, R.L.: 1969, *Introduction to the Theory of Kinetic Equations*, New York, London; J. Willey and Sons, Inc., p. 376.

Lichnerowicz, A.: 1967, *Relativistic Hydrodynamics and Magnetohydrodynamics*, New York, Amsterdam, Benjamin, p. 196.

Lichtenberg, A.J. and Lieberman, M.A.: 1983, *Regular and Stochastic Motion*, New York, Springer-Verlag, p. 314.

Lifshitz, E.M. and Pitaevskii, L.P.: 1981, *Physical Kinetics*, Oxford, New York; Pergamon Press, p. 452.

Lin, R.P., Schwartz, R.A., Pelling, R.M., and Hurley, K.C.: 1981, A new component of hard X-rays in solar flares, *Astrophys. J.*, v. 251, No. 2, L109–L114.

Lin, Y., Wei, X., and Zhang, H.: 1993, Variations of magnetic fields and electric currents associated with a solar flare, *Solar Phys.*, v. 148, No. 1, 133–138.

Litvinenko, Yu.E.: 1993, Regular versus chaotic motion of particles in non-neutral current sheets, *Solar Phys.*, v. 147, No. 2, 337–342.

Litvinenko, Yu.E.: 1997, Interpretation of particle acceleration in a simulation study of reconnection, *Phys. Plasmas*, v. 4, No. 9, 3439–3441.

Litvinenko, Yu.E.: 1999, Photospheric reconnection and canceling magnetic features on the Sun, *Astrophys. J.*, v. 515, No. 1, 435–440.

Litvinenko, Yu.E. and Somov, B.V.: 1991a, Solar flares and virial theorem, *Soviet Astronomy-AJ*, v. 35, No. 2, 183–188.

Litvinenko, Yu.E. and Somov, B.V.: 1991b, Electron acceleration in current sheets in solar flares, *Soviet Astronomy Lett.*, v. 17, No. 5, 353–356.

Litvinenko, Yu.E. and Somov, B.V.: 1991c, Nonthermal electrons in the thick-target reverse-current model for hard X-ray bremsstrahlung, *Solar Phys.*, v. 131, No. 2, 319–336.

Litvinenko, Yu.E. and Somov, B.V.: 1993, Particle acceleration in reconnecting current sheets, *Solar Phys.*, v. 146, No. 1, 127–133.

Litvinenko, Yu.E. and Somov, B.V.: 1994a, Electromagnetic expulsion force in cosmic plasma, *Astron. Astrophys.*, v. 287, No. 1, L37–L40.

Litvinenko, Yu.E. and Somov, B.V.: 1994b, Magnetic reconnection in the temparature mimimum and prominence formation, *Solar Phys.*, v. 151, No. 2, 265–270.

Litvinenko, Yu.E. and Somov B.V.: 1995, Relativistic acceleration of protons in current sheets of solar flares, *Solar Phis.*, v. 158, No. 1, 317–330.

Liu, Y., Srivastava, N., Prasad, D., Li, W., and Ai, G.: 1995, A possible explanation of reversed magnetic field features observed in NOAA AR 7321, *Solar Phys.*, v. 158, No. 1, 249–258.

Liubarskii, G.Ya. and Polovin, R.V.: 1958, Simple magnetoacoustic waves, *Soviet Physics-JETP*, v. 8, No. 2, 351.

Longcope, D.W.: 1996, Topology and current ribbons: A model for current, reconnection and flaring, *Solar Phys.*, v. 169, No. 1, 91–121.

Longcope, D.W. and Cowley, S.C.: 1996, Current sheet formation along 3D magnetic separators, *Phys. Plasmas*, v. 3, No. 8, 2885–2897.

Longcope, D.W. and Silva, A.V.R.: 1998, A current ribbon model for energy storage and release with application to the flare of 7 January 1992, *Solar Phys.*, v. 179, No. 2, 349–377.

Longmire, C.L.: 1963, *Elementary Plasma Physics*, New York, London; Interscience Publ., p. 296.

Lovelace, R.V.E.: 1976, *Nature*, v. 262, 649–652.

Low, B.C.: 1987, Electric current sheet formation in a magnetic field induced by footpoint displacements, *Astrophys. J.*, v. 323, No. 1, 358–367.

Low, B.C.: 1991, On the spontaneous formation of current sheets above a flexible solar photosphere, *Astrophys. J.*, v. 381, No. 1, 295–305.

Low, B.C. and Smith, D.F.: 1993, The free energies of partially open coronal magnetic fields, *Astrophys. J.*, v. 410, No. 1, 412–425.

Low, B.C. and Wolfson, R.: 1988, Spontaneous formation of current sheets and the origin of solar flares, *Astrophys. J.*, v. 324, No. 1, 574–581.

Lu, E.T. and Hamilton, R.J.: 1991, Avalanches and distribution of solar flares, *Astrophys. J.*, v. 380, No. 2, L89–L92.

Lundquist, S.: 1951, Magneto-hydrostatic fields, *Ark. Fys.*, v. 2, No. 35, 361–365.

Macdonald, D.A., Thorne, K.S., Price, R.H., and Zhang, X.H.: 1986, Astrophysical applications of black-hole electrodynamics, in *Black Holes: The Membrane Paradigm*, eds. K.S. Thorne, R.H. Price, and D.A. Macdonald, New Haven, London; Yale Univ. Press, p. 121–137.

MacDonald, W.M., Rothenbluth, M.N., and Chuck, W.: 1957, Relaxation of a system of particles with Coulomb interactions, *Phys. Rev.*, v. 107, No. 2, 350–353.

Mackay, D.H., Priest, E.R., Gaizauskas, V., and van Ballegooijn A.A.: 1998, Role of helicity in the formation of intermediate filaments, *Solar Phys.*, v. 180, No. 1, 299–312.

MacNeice, P., McWhirter, R.W.P., Spicer, D.S., Burgess, A.: 1984, A numerical model of a solar flare based on electron beam heating of the chromosphere, *Solar Phys.*, v. 90, No. 2, 357–353.

Mandrini, C.H. and Machado, M.E.: 1993, Large-scale brightenings associated with flares, *Solar Phys.*, v. 141, No. 1, 147–164.

Mandrini, C.H., Demoulin, P., Hénoux, J.C., and Machado, M.E.: 1991, Evidence for the interaction of large scale magnetic structures in solar flares, *Astron. Astrophys.*, v. 250, No. 2, 541–547.

Mandrini, C.H., Rovira, M.G., Demoulin, P., Hénoux, J.C., Machado, M.E., and Wilkinson, L.K.: 1993, Evidence for reconnection in large-scale structures in solar flares, *Astron. Astrophys.*, v. 272, No. 2, 609–620.

Markovskii, S.A. and Somov, B.V.: 1996a, A criterion for splitting of a reconnecting current sheet into MHD discontinuities, *J. Plasma Phys.*, v. 55, No. 3, 303–325.

Markovskii, S.A. and Somov, B.V.: 1996b, MHD discontinuities in space plasmas: Interrelation between stability and structure, *Space Sci. Rev.*, v. 78, No. 3–4, 443–506.

Marsh, G.E.: 1996, *Force-Free Magnetic Fields: Solutions, Topology and Applications*, River Edge, London; World Scientific Publ., p. 159.

Martens, P.C.H.: 1988, The generation of proton beams in two-ribbon flares, *Astrophys. J.*, v. 330, No. 2, L131–L133.

Martin, S.F.: 1986, Recent observations of the formation of filaments, in *Coronal and Prominence Plasmas*, NASA CP-2442, p. 73–80.

Martin, S.F.: 1998, Conditions for the formation and maintenance of filaments, *Solar Phys.*, v. 182, No. 1, 107–137.

Marty, P. et Alemany, A.: 1983, Écoulement dû à des champs magnétique et électrique croisés autour d'un cylindre de conductivité quelconque, *Journal de Mécanique Théorique et Appliquée*, v. 2, No. 2, 227–243.

Masuda, S., Kosugi, T., Hara, H., Tsuneta, S., and Ogawara, Y.: 1994, A loop-top hard X-ray source in a compact solar flare as evidence for reconnection, *Nature*, v. 371, 495–487.

Masuda, S., Kosugi, T., Hara, H., Sakao T., and Tsuneta, S.: 1995, Hard X-ray sources and the energy-release site in solar flares, *Publ. Astron. Soc. Japan*, v. 47, 677–689.

McDonald, L., Harra-Murnion, L.K., and Culhane, J.L.: 1999, Nonthermal electron energy deposition in the chromosphere and the accompanying soft X-ray flare emission, *Solar Phys.*, v. 185, No. 2, 323–350.

McKenzie, D.E. and Hudson, H.S.: 1999, X-ray observations of motions and structure above a solar flare arcade, *Astrophys. J.*, v. 519, L93-L96.

Michel, F.C.: 1991, *Theory of Neutron Star Magnetospheres*, Chicago, London; Chicago Univ. Press, p. 456.

Mikhailovskii, A.B.: 1979, Nonlinear excitation of electromagnetic waves in a relativistic electron-positron plasma, *Soviet J. Plasma Phys.*, v. 6, No. 3, 336–340.

Mikhailovskii, A.B., Onishchenko, O.G., and Tatarinov, E.G.: 1985, Alfvén solitons in a relativistic electron-positron plasma, *Plasma Physics and Controlled Fusion*, v. 27, No. 5, 539–556.

Milano, L.J., Gómez, D.O., and Martens, P.C.H.: 1997, Solar coronal heating: AC versus DC, *Astrophys. J.*, v. 490, No. 1, 442-451.

Miller, J.A. and Reames, D.V.: 1996, Heavy ion acceleration by cascading Alfvén waves in impulsive flares, in *High Energy Solar Physics*, eds. R. Ramaty, N. Mandzhavidze, and X.-M. Hua, New York, AIP, 450–460.

Miller, J.A., LaRosa, T.N., and Moore, R.L.: 1996, Stochastic electron acceleration by cascading fast mode waves in impulsive solar flares, *Astrophys. J.*, v. 461, No. 1, 445–464.

Mirabel, I.F. and Rodriguez, L.F.: 1998, Microquasars in our Galaxy, *Nature*, v. 392, 673–676.

Moffatt, H.K.: 1978, *Magnetic Field Generation in Electrically Conducting Fluids*, London, New York; Cambridge Univ. Press, p. 343.

Moiseev, S.S. and Chkhetiani, O.G.: 1996, Helical scaling in turbulence, *JETP*, v. 83, No. 1, 192–198.

Moreau, R.: 1990, *Magnetohydrodynamics*, Dordrecht, Kluwer Academic Publ., p. 328.

Moreton, G.E. and Severny, A.B.: 1968, Magnetic fields and flares, *Solar Phys.*, v. 3, No. 2, 282–297.

Morozov, A.I. and Solov'ev, L.S.: 1966a, The structure of magnetic fields, in: Leontovich M.A. (ed.), *Reviews of Plasma Physics*, New York, Consultans Bureau, v. 2, 1–101.

Morozov, A.I. and Solov'ev, L.S.: 1966b, Motion of particles in electromagnetic fields, in: Leontovich M.A. (ed.), *Reviews of Plasma Physics*, New York, Consultants Bureau, v. 2, 201–297.

Murata, H.: 1991, Magnetic field intensification and formation of field-aligned current in a non-uniform field, *J. Plasma Physics*, v. 46, No. 1, 29–48.

Murty, G.S.: 1961, Instabilities of a conducting fluid slab carrying uniform current in the presence of a magnetic field, *Ark. Fysik*, v. 19, No. 6, 499–510.

Nagai, T., Fuijimoto, M., Saito, Y., Machida, S. et al.: 1998, Structure and dynamics of magnetic reconnection for substorm onsets with Geotail observations, *J. Geophys. Res.*, v. 103, 4419–4428.

Nakano, T.: 1998, Star formation in magnetic clouds, *Astrophys. J.*, v. 494, No. 2, 587–604.

Narayan, R., Garcia, M.R., and McClintock, J.E.: 1997, Advection-dominated accretion and black hole horizons, *Astrophys. J.*, v. 478, No. 2, L79-L82.

Negoro, H., Kitamoto, S., Takeuchi, M., and Mineshige, S.: 1995, Statistics of X-ray fluctuations from Cygnus X-1: Reservoirs in the disk? *Astrophys. J.*, v. 452, No. 1, L49–L52.

Nishida, A. and Nagayama, N.: 1973, Synoptic survey for the neutral line in the magnetotail during the substorm expansion phase, *J. Geophys. Res.*, v. 78, No. 19, 3782–3798.

Nishida, A., Baker, D.N., and Cowley, S.W.H. (eds): 1998, *New Perspectives on the Earth's Magnetotail*, Washington, Amer. Geophys. Un., p. 339.

Nishikawa, K.I. and Sakai, J.: 1982, Stabilizing effect of a normal magnetic field on the collisional tearing mode, *Phys. Fluids*, v. 25, No. 8, 1384–1387.

Nishikawa, K.I., Frank, J., Christodoulou, D.M., et al.: 1999, 3D relativistic MHD simulations of extragalactic jets, in *Numerical Astrophysics*, eds S.M. Miyama et al., Dordrecht, Kluwer Academic Publ., p. 217–218.

Northrop, T.G.: 1963, *The Adiabatic Motion of Charged Particles*, New York, John Wiley, Interscience.

Novikov, I.D. and Frolov, V.P.: 1989, *Physics of Black Holes*, Dordrecht, Kluwer Academic Publ., p. 341.

Novikov, I.D. and Thorne, K.S.: 1973, in *Black Holes*, eds C.D. Dewitt and B. Dewitt, New York, Gordon and Breach, p. 345–354.

Obertz, P.: 1973, 2D problem of the shape of the magnetosphere, *Geomagn. Aeron.*, v. 13, No. 5, 758–766.

Ono, Y., Yamada,M., Akao, T., Tajima, T., and Matsumoto, R.: 1996, Ion acceleration and direct ion heating in three-component reconnection, *Phys. Rev. Lett.*, v. 76, No. 18, 3328–3331.

Oreshina, A.V. and Somov, B.V.: 1999a, Slow and fast magnetic reconnection. I. Role of radiative cooling, *Astron. Astrophys.*, v. 331, 1078–1086.

Oreshina, I.V. and Somov, B.V.: 1999b, On the conform transformation method for solution of cosmic electrodynamics problems, *Bull. Russ. Acad. Sci., Phys. Series*, No. 8, 1543–1549.

Ostriker, E.C.: 1999, Dynamical friction in a gaseous medium, *Astrophys. J.*, v. 513, No. 1, 252–258.

Otto, A.: 1991, The resistive tearing instability for generalized resistive models, *Phys. Fluids*, v. 3B, No. 7, 1739–1745.

Ozernoy, L.M. and Somov, B.V.: 1971, The magnetic field of a rotating cloud and magneto-rotational explosions, *Astrophys. Space Sci.*, v. 11, No. 2, 264–283.

Paesold, G. and Benz, A.O.: 1999, Electron firehose instability and acceleration of electrons in solar flares, *Astron. Astrophys.*, v. 351, 741–746.

Palmer, I. D. and Smerd, S. F.: 1972, Evidence for a two-component injection of cosmic rays from the solar flare of 1969, March 30, *Solar Phys.*, v. 26, No. 2, 460–467.

Palmer, P.L.: 1994, *Stability of Collisionless Stellar Systems*, Dordrecht, Kluwer Academic Publ., p. 349.

Park, B.T., Petrosian, V., and Schwartz, R.A.: 1997, Stochastic acceleration and photon emission in electron-dominated solar flares, *Astrophys. J.*, v. 489, No. 1, 358–366.

Parker, E.N.: 1972, Topological dissipation and the small-scale fields in turbulent gases, *Astrophys. J.*, v. 174, No. 1, 499–510.

Parker, E.N.: 1979, *Cosmic Magnetic Fields. Their Origin and Their Activity*, Oxford, Clarendon Press, p. 841.

Parker, E.N., 1988, Nanoflares and the solar X-ray corona, *Astrophys. J.*, v. 330, No. 1, 474–479.

Parks, G.K.: 1991, *Physics of Space Plasma*, Redwood City, Addison-Wesley Publ., p. 538.

Peacock, J.A.: 1999, *Cosmological Physics*, Cambridge Univ. Press, p. 682.

Peratt, A.L.: 1992, *Physics of the Plasma Universe*, New York, Berlin, Heidelberg; Springer-Verlag, p. 342.

Peres, G., Rosner, R., Serio, S., and Vaiana, G.S.: 1982, Coronal closed structures. 4. Hydrodynamical stability and response to heating perturbations, *Astrophys. J.*, v. 252, No. 2, 791–799.

Petschek, H.E.: 1964, Magnetic field annihilation, in *AAS-NASA Symp. on the Physics of Solar Flares*, ed. W.N. Hess, Washington, NASA SP-50, p. 425–439.

Pevtsov, A.A. and Longcope, D.W.: 1998, NOAA 7926: A kinked Ω-loop? *Astrophys. J.*, v. 508, No. 2, 908–915.

Pevtsov, A.A., Canfield, R.C., and Zirin, H.: 1996, Reconnection and helicity in a solar flare, *Astrophys. J.*, v. 473, No. 1, 533–538.

Pfaffelmoser, K.: 1992, Global classic solutions of the Vlasov-Poisson system in 3D for general initial data, *J. Diff. Equations*, v. 95, 281–303.

Pike, C.D. and Mason, H.E.: 1998, Rotating transion region features observed with the SOHO CDS, *Solar Phys.*, v. 182, No. 2, 333–348.

Plank, M.: 1917, *Sitz. der Preuss. Akad.*, 324.

Pneuman, G.W.: 1974, Magnetic structure responsible for disturbances, in *Coronal Disturbances*, ed. G. Newkirk, Dordrecht, Boston; D. Reidel Publ., p. 35–68.

Pneuman, G.W.: 1983, The formation of solar prominences by magnetic reconnection and condensation, *Solar Phys.*, v. 88, No. 2, 219–239.

Podgornii, A.I. and Syrovatskii, S.I.: 1981, Formation and development of a current sheet for various magnetic viscosities and gas pressures, *Soviet J. Plasma Phys.*, v. 7, No. 5, 580–584.

Pollard, R.K. and Taylor, Y.B.: 1979, Influence of equilibrium flows on tearing modes, *Phys. Fluids*, v. 22, No. 1, 126–131.

Polovin, R.V.: 1961, Shock waves in MHD, *Soviet Phys. Usp.*, v. 3, No. 5, 677–688.

Polovin, R.V. and Demutskii, V.P.: 1990, *Fundamentals of MHD*, New York, Consultants Bureau.

Polovin, R.V. and Liubarskii, G.Ya.: 1958, Impossibility of rarefaction shock waves in MHD, *Soviet Physics-JETP*, v. 8, No. 2, 351–352.

Porter, L.J., Klimchuk, J.A., and Sturrock, P.A.: 1992, Cylindrically symmetric force-free fields, *Astrophys. J.*, v. 385, No. 2, 738–745.

Priest, E.R.: 1982, *Solar Magnetohydrodynamics*, Dordrecht, D. Reidel Publ. Co., p. 472.

Priest, E.R., Titov, V.S., Vekstein, G.E., and Rickard, G.J.: 1994, Steady linear X-point magnetic reconnection, *J. Geophys. Res.*, v. 99, No. A11, 21467–21479.

Raadu, M.A.: 1984, Global effects of double layers, in *Second Symp. on Plasma Double Layers and Related Topics*, eds R. Schrittwieser and G. Eder; Innsbruck, p. 3–27.

Ramos, J.I. and Winowich, N.S.: 1986, MHD channel flow study, *Phys. Fluids*, v. 29, No. 4, 992–997.

Reiman, A.: 1980, *Phys. Fluids*, v. 23, 230–239.

Rodrigues-Pacheco, J., Sequeiros, J., del Peral, L., Bronchalo, E., and Cid, C.: 1998, Diffusive-shock-accelerated interplanetary ions during the solar cycle 21 maximum, *Solar Phys.*, v. 181, No. 1, 185–200.

Roikhvarger, Z.B. and Syrovatskii, S.I.: 1974, Evolutionarity of MHD discontinuities with allowance for dissipative waves, *Soviet Physics–JETP*, v. 39, No. 4, 654–656.

Rose, W.K.: 1998, *Advanced Stellar Astrophysics*, Cambridge Univ. Press, p. 494.

Rosenbluth, M. and Longmire, C.: 1957, Stability of plasmas confined by magnetic fields, *Ann. Phys.*, v. 1, No. 1, 120–140.

Roumeliotis, G. and Moore, R.L.: 1993, A linear solution for reconnection driven by converging or diverging footpoint motions, *Astrophys. J.*, v. 416, No. 1, 386–391.

Ruderman, M.: 1971, Matter in superstrong magnetic fields: The surface of a neutron star, *Phys. Rev. Lett.*, v. 27, No. 19, 1306–1308.

Ruderman, M.A. and Sutherland, P.G.: 1975, Theory of pulsars: Polar gaps, sparks, and coherent radiation, *Astrophys. J.*, v. 196, No. 1, 51–72.

Rüdiger, G. and von Rekowski, B.: 1998, Differential rotation and meridional flow for fast-rotating solar-type stars, *Astrophys. J.*, v. 494, No. 2, 691–699.

Ruffolo, D.: 1999, Transport and acceleration of energetic particles near an oblique shock, *Astrophys. J.*, v. 515, No. 2, 787–800.

Sakai, J.I. and de Jager, C.: 1996, Solar flares and collisions between current-carring loops, *Space Sci. Rev.*, v. 77, No. 1, 1–192.

Sakao, T., Kosugi, T., and Masuda, S.: 1998, Energy release in solar flares with respect to magnetic loops, in *Observational Plasma Astrophysics: Five Years of Yohkoh and Beyond*, eds T. Watanabe, T. Kosugi, and A.C. Sterling, Dordrecht, Kluwer Academic Publ., p. 273-284.

Sarris, E.T. and Van Allen, J.A.: 1974, Effects of interplanetary shocks on energetic particles, *J. Geophys. Res.*, v. 79, No. 28, 4157–4173.

Schabansky, V.P.: 1971, Some processes in the magnetosphere, *Space Sci. Rev.*, v. 12, No. 3, 299–418.

Schindler, K.: 1974, A theory of the substorm mechanism, *J. Geophys. Res.*, v. 79, No. 19, 2803–2810.

Schlüter, A.: 1951, Dynamic des Plasmas, *Zeitschrift für Naturforschung*, v. 6A, No. 2, 73–78.

Schmidt, G.: 1979, *Physics of High Temperature Plasmas*, New York, London; Academic Press, p. 408.

Schou, J., Antia, H.M., Basu, S., Bogart, R.S., Bush, R.I., Chitre, S.M., et al.: 1998, *Astrophys. J.*, v. 505, No. 1, 390–399.

Schram, P.P.J.: 1991, *Kinetic Theory of Gases and Plasmas*, Dordrecht, Kluwer Academic Publ., p. 426.

Schrijver, C.J. and Zwaan, C.: 1999, *Solar and Stellar Magnetic Activity*, Cambridge Univ. Press, p. 400.

Schrijver, C.J., Title, A.M., van Ballegooijen, A.A., Hagenaar, H.J., and Shine, R.A.: 1997, Sustaining the quiet photospheric network: The balance of flux emergence, fragmentation, merging, and cancellation, *Astrophys. J.*, v. 487, No. 1, 424–436.

Schuster, H.G.: 1984, *Deterministic Chaos. An Introduction*, Weinheim, Physik-Verlag, p. 220.

Sedov, L.I.: 1973, *Mechanics of Continuous Medium*, Moscow, Nauka, v. 1, p. 536; v. 2, p. 584 (in Russian).

Severnyi, A.B.: 1963, The stability of plasma layer with a neutral-point magnetic field, *Soviet Astronomy-AJ*, v. 6, No. 6, 770–773.

Shafranov, V.D.: 1966, Plasma equilibrium in a magnetic field, in *Reviews of Plasma Physics*, ed. M.A. Leontovich, New York, Consultants Bureau, v. 2, 103–151.

Shakura, N.I. and Sunyaev, R.A.: 1973, Black holes in binary systems, *Astron. Astrophys.*, v. 24, No. 2, 337–355.

Shercliff, A.J.: 1965, *A Textbook of MHD*, Oxford, Pergamon Press, p. 265.

Shkarofsky, I.P., Johnston, T.W., and Bachynski, M.P.: 1966, *The Particle Kinetics of Plasma*, Reading, Mass.; Addison-Wesley Publ., p. 518.

Shoub, E.C.: 1983, Invalidity of local thermodynamic equilibrium for electrons in solar transition region, *Astrophys. J.*, v. 266, No. 1, 339–369.

Shu, F.H.: 1992, *The Physics of Astrophysics, v. 2, Gas Dynamics*, Mill Valley, California; Univ. Science Books, p. 476.

Silin, V.P.: 1971, *Intoduction to the Kinetic Theory of Gases*, Moscow, Nauka, p. 332 (in Russian).

Simon, A.L.: 1959, *An Introduction to Thermonuclear Research*, London, Pergamon Press, p. 182.

Sirotina, E.P. and Syrovatskii, S.I.: 1960, Structure of low intensity shock waves in MHD, *Soviet Physics–JETP*, v. 12, No. 3, 521–526.

Sitnov, M.I. and Sharma, A.S.: 1998, Role of transient electrons and microinstabilities in the tearing instability of the geomagnetotail current sheet, in *Substorms-4*, eds S. Kokubun and Y. Kamide, Dordrecht, Kluwer Academic Publ.; Tokyo, Terra Sci. Publ., p. 539–542.

Sitnov, M.I., Malova, H.V., and Lui, A.T.Y.: 1997, Quasi-neutral sheet tearing instability induced by electron preferential acceleration from stochasticity, *J. Geophys. Res.*, v. 102, No. A1, 163–173.

Sivukhin, D.V.: 1965, Motion of particles in electromagnetic fields in the drift approximation, in *Reviews of Plasma Physics*, ed. M.A. Leontovich, New York, Consultants Bureau, v. 1, 1–104.

Sivukhin, D.V.: 1966, Coulomb collisions in a fully ionized plasma, in *Reviews of Plasma Physics*, ed. M.A. Leontovich, New York, Consultants Bureau, v. 4, 93–341.

Sivukhin, D.V.: 1977, *A Course of General Physics. 3. Electricity*, Moscow, Nauka, p. 688 (in Russian).

Smirnov, B.M.: 1981, *Physics of Weakly Ionized Gases: Problems and Solutions*, Moscow, Mir Publ., p. 432.

Smirnov, V.I.: 1965, *A Course of Higher Mathematics*, v. 2, Oxford, New York; Pergamon Press.

Somov, B.V.: 1981, Fast reconnection and transient phenomena with particle acceleration in the solar corona, *Bull. Acad. Sci. USSR, Phys. Ser.*, v. 45, No. 4, 114–116.

Somov, B.V.: 1982, Accumulation and release of flare energy, in *Proc. 12th Leningrad Seminar on Space Physics*, 'Complex Study of the Sun', Leningrad, LIYaF, p. 6–49 (in Russian).

Somov, B.V.: 1985, New theoretical models of solar flares, *Soviet Phys. Usp.*, v. 28, No. 3, 271–272.

Somov, B.V.: 1986, Non-neutral current sheets and solar flare energetics, *Astron. Astrophys.*, v. 163, No. 1, 210–218.

Somov, B.V.: 1992, *Physical Processes in Solar Flares*, Dordrecht, Boston, London; Kluwer Academic Publ., p. 248.

Somov, B.V.: 1994a, *Fundamentals of Cosmic Electrodynamics*, Dordrecht, Boston, London; Kluwer Academic Publ., p. 364.

Somov, B.V.: 1994b, Features of mass supply and flows related with reconnection in the solar corona, *Space Sci. Rev.*, v. 70, No. 1, 161–166.

Somov, B.V. and Hénoux J.C.: 1999, Generation and interaction of electric currents in the quiet photospheric network, in *Magnetic Fields and Solar Processes*, Proc. 9th European Meeting on Solar Physics, ESA SP-448, 659–663, 1999.

Somov, B.V. and Kosugi, T.: 1997, Collisionless reconnection and high-energy particle acceleration in solar flares, *Astrophys. J.*, v. 485, No. 2, 859–868.

Somov, B.V. and Kozlova, L.M.: 1998, Fine structure of the solar chromosphere from infrared He I line observations, *Astronomy Reports*, v. 42, No. 6, 819–826.

Somov, B.V. and Litvinenko, Yu.E.: 1993, Magnetic reconnection and particle acceleration in the solar corona, in *Physics of Solar and Stellar Coranae*, eds. J. Linsky and S. Serio, Dordrecht, Kluwer Academic Publ., p. 603–606.

Somov, B.V. and Merenkova, E.Yu.: 1999, Model computations of magnetic fields in solar flares, *Bull. Russ. Acad. Sci., Phys. Ser.*, v. 63, No. 8, 1512–1515 (in Russian).

Somov, B.V. and Oreshina, A.V.: 2000, Slow and fast magnetic reconnection. II. High-temperature turbulent-current sheet, *Astron. Astrophys.*, v. 354, 703–713.

Somov, B.V. and Syrovatskii, S.I.: 1972a, Plasma motion in an increasing strong dipolar field, *Soviet Phys.-JETP*, v. 34, No. 2, 332–335.

Somov, B.V. and Syrovatskii, S.I.: 1972b, Appearance of a current sheet in a plasma moving in the field of a two-dimensional magnetic dipole, *Soviet Phys.-JETP*, v. 34, No. 5, 992–997.

Somov, B.V. and Syrovatskii, S.I.: 1975, Electric and magnetic fields arising from the rupture of a neutral current sheet. *Bull. Acad. Sci. USSR, Phys. Series*, v. 39, No. 2, 109–111.

Somov, B.V. and Syrovatskii, S.I.: 1976a, Hydrodynamic plasma flows in a strong magnetic field, in *Neutral Current Sheets in Plasma*, Proc. P.N. Lebedev Phys. Inst., v. 74, ed. N.G. Basov, New York and London, Consultants Bureau, p. 13–71.

Somov, B.V. and Syrovatskii, S.I.: 1976b, Physical processes in the solar atmosphere associated with flares, *Soviet Physics Usp.*, v. 19, No. 10, 813–835.

Somov, B.V. and Syrovatskii, S.I.: 1977, Current sheets as the source of heating for solar active regions, *Solar Phys.*, v. 55, No. 2, 393–399.

Somov, B.V. and Syrovatskii, S.I.: 1982, Thermal trigger for solar flares and coronal loops formation, *Solar Phys.*, v. 75, No. 1, 237–244.

Somov, B.V. and Tindo, I.P.: 1978, Polarization of hard X-rays from solar flares, *Cosmic Research*, v. 16, No. 5, 555-564.

Somov, B.V. and Titov, V.S.: 1983, Magnetic reconnection as a mechanism for heating the coronal loops, *Soviet Astronomy Letters*, v. 9, No. 1, 26–28.

Somov, B.V. and Titov, V.S.: 1985a, Effect of longitudinal magnetic field in current sheets on the Sun, *Soviet Astronomy-AJ*, v. 29, No. 5, 559–563.

Somov, B.V. and Titov, V.S.: 1985b, Magnetic reconnection in a high-temperature plasma of solar flares. 2. Effects caused by transverse and longitudinal magnetic fields. *Solar Phys.*, v. 102, No. 1, 79–96.

Somov, B.V. and Verneta, A.I.: 1988, Magnetic reconnection in a high-temperature plasma of solar flares. 3. Stabilization effect of a transverse field in non-neutral current sheets, *Solar Phys.*, v. 117, No. 1, 89–95.

Somov, B.V. and Verneta, A.I.: 1989, Magnetic reconnection in a high-temperature plasma of solar flares. 4. Resistive tearing mode in non-neutral current sheets, *Solar Phys.*, v. 120, No. 1, 93–115.

Somov, B.V. and Verneta, A.I.: 1993, Tearing instability of reconnecting current sheets in space plasmas, *Space Sci. Rev.*, v. 65, No. 3, 253–288.

Somov, B.V., Kosugi, T., and Sakao, T.: 1998, Collisionless 3D reconnection in impulsive solar flares. *Astrophys. J.*, v. 497, No. 2, 943–956.

Somov, B.V., Syrovatskii, S.I., and Spektor, A.R.: 1981, Hydrodynamic response of the solar chromosphere to elementary flare burst. 1. Heating by accelerated electrons, *Solar Phys.*, v. 73, No. 1, 145–155.

Somov, B.V., Litvinenko, Y.E., Kosugi, T., Sakao, T., Masuda, S., Bogachev, S.A.: 1999, Coronal hard X-rays in solar flares: Yohkoh observations and interpretation, in *Magnetic Fields and Solar Processes*, Proc. 9th European Meeting on Solar Physics, ESA SP-448, 701-708, 1999.

Speiser, T.W.: 1965, Particle trajectories in model current sheets. 1. Analytical solutions, *J. Geophys. Res.*, v. 70, No. 17, 4219–4226.

Speiser, T.W.: 1968, On the uncoupling of parallel and perpendicular particle motion in a neutral sheet, *J. Geophys. Res.*, v. 73, No. 3, 1112–1113.

Speiser, T.W. and Lyons, L.R.: 1984, Comparison of an analytical approximation for particle motion in a current sheet with precise numerical calculations, *J. Geophys. Res.*, v. 89, No. A1, 147–158.

Spicer, D.S.: 1982, Magnetic energy storage and conversion in the solar atmosphere, *Space Sci. Rev.*, v. 31, No. 1, 351–435.

Spicer, D.S. and Emslie, A.G.: 1988, A new quasi-thermal trap model for solar hard X-ray bursts, *Astrophys. J.*, v. 330, No. 2, 997–1007.

Spitzer, L.: 1940, The stability of isolated clusters, *Mon. Not. Royal Astron. Soc.*, v. 100, No. 5, 396–413.

Spitzer, L.: 1962, *Physics of Fully Ionized Gases*, New York, Wiley Interscience, p. 170.

Stenzel, R.L. and Gekelman, W.: 1984, Particle acceleration during reconnection in laboratory plasmas. *Adv. Space Res.*, v. 4, No. 2, 459–470.

Stewart, R.T. and Labrum, N.R.: 1972, Meter-wavelength observations of the solar radio storm of August 17–22, 1968. *Solar Phys.*, v. 27, No. 1, 192–202.

Störmer, C.: 1955, *The Polar Aurora*, Oxford, Clarendon Press.

Strittmatter, P.A.: 1966, Gravitational collapse in the presence of a magnetic field, *Monthly Not. Royal Astron. Soc.*, v. 132, No. 3, 359–378.

Sturrock, P.A.: 1991, Maximum energy of semi-infinite magnetic field configurations, *Astrophys. J.*, v. 380, No. 2, 655–659.

Sturrock, P.A.: 1994, *Plasma Physics: An Introduction to the Theory of Astrophysical, Geophysical and Laboratory Plasmas*, Cambridge, Cambridge Univ. Press, p. 335.

Subramanian, P., Becker, P.A., and Kazanas, D.: 1999, Formation of relativistic outflows in shearing black hole accretion coronae, *Astrophys. J.*, v. 523, No. 1, 203–222.

Sutton, G.W. and Sherman, A.: 1965, *Engineering Magnetohydrodynamics*, New York, McGraw-Hill Book Co., p. 548.

Sweet, P.A.: 1958, The production of high energy particles in solar flares, *Nuovo Cimento Suppl.*, v. 8, *Serie* 10, 188–196.

Sweet, P.A.: 1969, Mechanisms of solar flares, *Ann. Rev. Astron. Astrophys.*, v. 7, 149–176.

Syrovatskii, S.I.: 1953, On the stability of tangential discontinuities in MHD medium, *Zhur. Exper. Teor. Fiz.*, v. 24, No. 6, 622–630 (in Russian).

Syrovatskii, S.I.: 1954, Instability of tangential discontinuities in a compressive medium, *Zhur. Exper. Teor. Fiz.*, v. 27, No. 1, 121–123 (in Russian).

Syrovatskii, S.I.: 1956, Some properties of discontinuity surfaces in MHD, *Proc. P.N. Lebedev Phys. Inst.*, v. 8, 13–64 (in Russian).

Syrovatskii, S.I.: 1957, Magnetohydrodynamics, *Uspehi Fiz. Nauk*, v. 62, No. 3, 247–303 (in Russian).

Syrovatskii, S.I.: 1959, The stability of shock waves in MHD, *Soviet Physics–JETP*, v. 8, No. 6, 1024–1028.

Syrovatskii, S.I.: 1962, The stability of plasma in a nonuniform magnetic field and the mechanism of solar flares, *Soviet Astronomy–AJ*, v. 6, No. 6, 768–769.

Syrovatskii, S.I.: 1966a, Dynamic dissipation of a magnetic field and particle acceleration, *Soviet Astronomy-AJ*, v. 10, No. 2, 270–276.

Syrovatskii, S.I.: 1966b, Dynamical dissipation of magnetic energy in the vicinity of a neutral line, *Soviet Physics-JETP*, v. 23, No. 4, 754–762.

Syrovatskii, S.I.: 1968, MHD cumulation near a zero field line, *Soviet Physics-JETP*, v. 27, No. 5, 763–766.

Syrovatskii, S.I.: 1971, Formation of current sheets in a plasma with a frozen-in strong field, *Soviet Physics-JETP*, v. 33, No. 5, 933–940.

Syrovatskii, S.I.: 1976a, Neutral current sheets in laboratory and space plasmas, in *Neutral Current Sheets in Plasmas*, Proc. P.N. Lebedev Phys. Inst., v. 74, ed. N.G. Basov, New York, London; Consultants Bureau, p. 2–10.

Syrovatskii, S.I.: 1976b, Current-sheet parameters and a thermal trigger for solar flares, *Soviet Astron. Lett.*, v. 2, No. 1, 13–14.

Syrovatskii, S.I.: 1981, Pinch sheets and reconnection in astrophysics, *Ann. Rev. Astron. Astrophys.*, v. 19, 163–229.

Syrovatskii, S.I.: 1982, Model for flare loops, fast motions, and opening of magnetic field in the corona, *Solar Phys.*, v. 76, No. 1, 3–20.

Syrovatskii, S.I. and Chesalin, L.S.: 1963, Electromagnetic generation of conductive fluid flows near bodies and expulsive force, in *Questions of Magnetohydrodynamics*, Riga, Zinatne, p. 17–22 (in Russian).

Syrovatskii, S.I. and Shmeleva, O.P.: 1972, Heating of plasma by high-energy electrons, and the non-thermal X-ray emission in solar flares, *Soviet Astronomy-AJ*, v. 16, No. 2, 273–283.

Syrovatskii, S.I. and Somov, B.V.: 1980, Physical driving forces and models of coronal responses, in *Solar and Interplanetary dynamics*, eds M. Dryer and E. Tandberg-Hanssen, IAU Symp. **91**, p. 425–441.

Takahara, F. and Kusunose, M.: 1985, Electron-positron pair production in a hot accretion plasma around a massive black hole, *Progr. Theor. Phys.*, v. 73, No. 6, 1390–1400.

Takizawa, M.: 1998, A two-temperature model of the intracluster medium, Astrophys. J., v. 509, No. 2, 579–584.

Tamm, I.E.: 1989, *Basic Theory of Electricity*, 10th edition, Moscow, Nauka, p. 504 (in Russian).

Tanaka, K.: 1987, Impact of X-ray observations from the Hinitori satellite on solar flare research, *Publ. Astron. Soc. Japan*, v. 39, No. 1, 1–45.

Tandberg-Hanssen, E.: 1995, *The Nature of Solar Prominences*, Dordrecht, Kluwer Academic Publ., p. 308.

Taylor, J.B.: 1974, Relaxation of toroidal plasma and generation of reverse magnetic fields. *Phys. Rev. Lett.*, v. 33, No. 19, 1139–1141.

Taylor, J.B.: 1986, Relaxation and magnetic reconnection in plasmas, *Rev. Mod. Phys.*, v. 58, No. 3, 741–763.

Tidman, D.A. and Krall, N.A.: 1971, *Shock Waves in Collisionless Plasma*, New York, London; Wiley-Interscience, p. 175.

Titov, V.S. and Priest, E.R.: 1993, The collapse of an X-type neutral point to form a reconnecting current sheet. *Geophys. and Astrophys. Fluid Dynamics*, v. 72, 249–276.

Titov, V.S., Priest, E.R., and Démoulin, P.: 1993, Conditions for the appearance of 'bald patches' at the solar surface, *Astron. Astrophys.*, v. 276, No. 2, 564–570.

Toptyghin, I.N.: 1980, Acceleration of particles by shocks in a cosmic plasma, *Space Sci. Rev.*, v. 26, No. 1, 157–213.

Treumann, R.A. and Baumjohann, W.: 1997, *Advanced Space Plasma Physics*, London, Imperial College Press, p. 381.

Trubnikov, B.A.: 1965, Particle interactions in a fully ionized plasma, in *Reviews of Plasma Physics*, ed. M.A. Leontovich, New York, Consultants Bureau, v. 1, 105–204.

Tsuneta, S.: 1993, Solar flares as an ongoing reconnection process, in *ASP Conf. Series*, v. 46, eds H. Zirin, G. Ai, and H. Wang, p. 239–248.

Tsuneta, S.: 1996, Structure and dynamics of reconnection in a solar flare, *Astrophys. J.*, v. 456, No. 2, 840–849.

Tsuneta, S., Nitta, N., Ohki, K., Takakura, T., Tanaka, K., Makishima, K., Murakami, T., Oda, M., and Ogawara, Y.: 1984, Hard X-ray imaging observations of solar hot thermal flares with the *Hinotori* spacecraft, *Astrophys. J.*, v. 284, No. 2, 827–832.

Tsuneta, S., Hara, H., Shimuzu, T., Acton, L., Strong, K.T., Hudson, H.S., and Ogawara, Y.: 1992, Observation of a solar flare at the limb with the Yohkoh soft X-ray telescope, *Publ. Astron. Soc. Japan*, v. 44, No. 5, L63–L69.

Tsuneta, S., Masuda, S., Kosugi, T., and Sato, J.: 1997, Hot and superhot plasmas above an impulsive-flare loop, *Astrophys. J.*, v. 478, No. 2, 787–796.

Tsurutani, B.T., Gonzalez, W.D., Kamide, Y., and Arballo, J.K. (eds): 1997, *Magnetic Storms*, Washington, Amer. Geophys. Un., p. 266.

Tverskoy B.A.: 1969, Main mechanisms in the formation of the Earth's radiation belts, *Rev. Geophys.*, v. 7, No. 1, 219–231.

UeNo, S.: 1998, Comparison between statistical features of X-ray fluctuations from the solar corona and accretion disks, in *Observational Plasma Astrophysics: Five Years of Yohkoh and Beyond*, eds T. Watanabe, T. Kosugi, and A.C. Sterling, Dordrecht, Kluwer Academic Publ., p. 45–50.

Ulmschneider, P., Rosner, R., and Priest, E.R. (eds): 1991, *Mechanisms of Chromospheric and Coronal Heating*, Berlin, Springer-Verlag.

van Ballegooijen, A.A. and Martens, P.C.H.: 1989, Formation and eruption of solar prominences, *Astrophys. J.*, v. 343, No. 3, 971–984.

van Ballegooijen, A.A. and Martens, P.C.H.: 1990, Magnetic fields in quiescent prominences, *Astrophys. J.*, v. 361, No. 1, 283–289.

van de Hulst, H.C.: 1951, Interstellar polarization and MHD waves, in *Problems of Cosmical Aerodynamics*, eds J.M. Burgers and H.C. van de Hulst, p. 45–57.

van den Oord, G.H.J.: 1990, The electrodynamics of beam/return current systems in the solar corona, *Astron. Astrophys.*, v. 234, No. 2, 496–518.

Vekstein, G.E. and Priest, E.R.: 1992, MHD equilibria and cusp formation at an X-type neutral line by footpoint shearing, *Astrophys. J.*, v. 384, No. 1, 333–340.

Vernazza, J.E., Avrett, E.H., and Loeser, R.: 1981, Structure of the solar chromosphere. 3. Models of the EUV brightness components of the quiet Sun, *Astrophys. J. Suppl.*, v. 45, 635–725.

Verneta, A.I. and Somov, B.V.: 1993, Effect of compressibility on the development of the tearing instability in a non-neutral current sheet in the solar atmosphere, *Astronomy Reports*, v. 37, No. 3, 282–285.

Vladimirov, V.S.: 1971, *Equations of Mathematical Physics*, New York, M. Dekker, p. 418.

Vlasov, A.A.: 1938, On the oscillation properties of an electron gas, *Zhur. Eksp. Teor. Fiz.*, v. 8, No. 1, 29-33 (in Russian).

Vlasov, A.A.: 1945, On the kinetic theory of an ansembly of paricles with collective interactions, *Soviet J. Phys.*, v. 9, No. 1, 25–28.

Volkov, T.F.: 1966, Hydrodynamic description of a collisionless plasma, in *Reviews of Plasma Physics*, ed. M.A. Leontovich, New York, Consultant Bureau, v. 4, 1–21.

Wagner, J.S., Kan, J.R., and Akasofu, S.J.: 1979, Particle dynamics in the plasma sheet, *J. Geophys. Res.*, v. 84, No. A3, 891–897.

Walt, M.: 1994, *Introduction to Geomagnetically Trapped Radiation*, Cambridge Univ. Press, p. 188.

Wang, J.X., Shi, Z.X., Wang, H., and Lu, Y.: 1996, Flares and the magnetic non-potentiality, *Astrophys. J.*, v. 456, No. 2, 861–878.

Webb, G.M.: 1986, Similarity considerations and conservation laws for magnetostatic atmospheres, *Solar Phys.*, v. 106, No. 2, 287–313.

Webb, G.M., Zank, G.P., Ko, C.M., and Donohue, D.J.: 1995, Multidimensional Green's functions and the statistics of diffusive shock acceleration, *Astrophys. J.*, v. 453, No. 1, 178–189.

Wiita, P.J.: 1999, Accretion disks around black holes, in *Black Holes, Gravitational Radiation and the Universe*, eds B.R. Iyer and B. Bhawal, Dordrecht, Kluwer Academic Publ., p. 249–263.

Woltjer, L.: 1958, A theorem on force-free magnetic fields, *Proc. Nat. Acad. Sci. USA*, v. 44, No. 6, 489–491.

Woltjer, L.: 1959, Hydromagnetic equilibrium: II. Stability in the variational formulation, *Proc. Nat. Acad. Sci. USA*, v. 45, No. 6, 769–771.

Wright, A.N. and Berger, M.A.: 1989, The effect of reconnection upon the linkage and interior structure of magnetic flux tubes, *J. Geophys. Res.*, v. 94, No. A2, 1295–1302.

Wright, A.N. and Berger, M.A.: 1991, A physical description of magnetic helicity evolution in the presence of reconnection lines, *J. Plasma Phys.*, v. 46, No. 1, 179–199.

Zank, G.P.: 1991, Weyl's theorem for MHD, *J. Plasma Phys.*, v. 46, No. 1, 11–14.

Zel'dovich, Ya.B. and Novikov, I.D.: 1971, *Relativistic Astrophysics. Vol. 1, Stars and Relativity*, Chicago, Univ. of Chicago Press.

Zel'dovich, Ya.B. and Raizer, Yu.P.: 1966, *Physics of Shock Waves and High-Temperature Hydrodynamic Phenomena*, New York, Academic Press, v. 1, p. 464; v. 2, p. 452.

Zhang, H.: 1995, Configuration of magnetic shear and vertical current in the active region NOAA 5395 in 1989 March, *Astron. Astrophys. Suppl.*, v. 111, No. 1, 27–40.

Zhang, H.-Q. and Chupp, E. L.: 1989, Studies on post-flare prominence of 1981 April 27, *Astrophys. and Space Sci.*, v. 153, No. 1, 95–108.

Zheleznyakov, V.V.: 1996, *Radiation in Astrophysical Plasmas*, Dordrecht, Kluwer Academic Publ., p. 462.

Zhou, Y. and Matthaeus, W.H.: 1990, Models of inertial range spectra of MHD turbulence, *J. Geophys. Res.*, v. 95, No. A9, 14881–14892.

Zirker, J.B. and Cleveland, F.M.: 1993, Avalanche models of active region heating and flaring, *Solar Phys.*, v. 145, No. 1, 119–128.

Zuccarello, F., Burm, H., Kuperus, M., Raadu, M., and Spicer, D.S.: 1987, Varying self-inductance and energy storage in a sheared force-free arcade, *Astron. Astrophys.*, v. 180, No. 1, 218–222.

Zweibel, E.G.: 1989, Magnetic reconnection in partially ionized gases, *Astrophys. J.*, v. 340, No. 2, 550–557.

Zwingmann, W., Schindler, K., and Birn, J.: 1985, On sheared magnetic field structures containing neutral points, *Solar Phys.*, v. 99, No. 1, 133–143.

Index

GPSR Compliance

The European Union's (EU) General Product Safety Regulation (GPSR) is a set of rules that requires consumer products to be safe and our obligations to ensure this.

If you have any concerns about our products, you can contact us on ProductSafety@springernature.com

In case Publisher is established outside the EU, the EU authorized representative is:

Springer Nature Customer Service Center GmbH
Europaplatz 3
69115 Heidelberg, Germany

Batch number: 10823951

Printed by Printforce, the Netherlands